S0-ASW-679

INDUSTRIAL ANALYSIS WITH VIBRATIONAL SPECTROSCOPY

RSC Analytical Spectroscopy Monographs

The series aims to provide a tutorial approach to the use of spectrometric and spectroscopic measurement techniques in analytical science, providing guidance and advice to individuals on a day-to-day basis during the course of their work with the emphasis on important practical aspects of the subject.

Flame Spectrometry in Environmental Chemical Analysis: A Practical Guide, by Malcolm S. Cresser, *Department of Plant and Soil Science, University of Aberdeen, UK*

Chemometrics in Analytical Spectroscopy, by Mike J. Adams, *School of Applied Sciences, University of Wolverhampton, UK*

Inductively Coupled and Microwave Induced Plasma Sources for Mass Spectrometry, by E. Hywel Evans, *Department of Environmental Sciences, University of Plymouth, UK*; Jeffrey J. Giglio, Theresa M. Castillano, and Joseph A. Caruso, *University of Cincinnati, Ohio, USA*

Industrial Analysis with Vibrational Spectroscopy, by John M. Chalmers, *ICI Technology, Research & Technology Centre, Wilton, UK*; Geoffrey Dent, *Zeneca Specialties, Blackley, UK*

Ionization Methods in Organic Mass Spectrometry, by Alison E. Ashcroft, *Micromass UK Ltd., Altrincham, UK*

How to obtain future titles on publication

A standing order plan is available for this series. A standing order will bring delivery of each new volume immediately upon publication. For further information, please write to:

Turpin Distribution Services Ltd., Blackhorse Road, Letchworth, Hertfordshire SG6 1HN, UK. Telephone: +44 (0) 1462 672555; Fax: +44 (0) 1462 480947

Industrial Analysis with Vibrational Spectroscopy

John M. Chalmers
ICI Technology, Research & Technology Centre, Wilton, UK

Geoffrey Dent
Zeneca Specialties, Blackley, UK

A catalogue record for this book is available from the British Library.

ISBN 0-85404-565-1

Published by The Royal Society of Chemistry,
Thomas Graham House, Science Park, Milton Road, Cambridge CB4 4WF, UK

Typeset by Paston Press Ltd., Loddon, Norfolk
Printed by Athenaeum Press Ltd., Gateshead, Tyne and Wear, UK

Preface

Infrared spectroscopy has been a mainstay of laboratory industrial analysis, materials characterisations, and research support for over forty years; while, Raman spectroscopy, following a couple of inroads, is now truly establishing its position as an invaluable, cost-effective resource, alongside its vibrational spectroscopy partner. For us, the many distinctions that once clearly existed between infrared and Raman applications and their industrial value have, in the last few years, rapidly become blurred, and the technique pair are now totally integrated and complementary tools within our work environments. It is for this reason, and a strong belief that industrial Raman spectroscopy is here to stay in a big way and will supersede infrared in some areas of application, that we have in this monograph tried, wherever possible, to integrate and contrast our discussions of the two vibrational spectroscopy techniques.

We are aware that to the purist or well-informed expert some of our descriptions and discussions may seem a little too superficial, but, in trying to present a broad relevant perspective, this has been deliberate. At the practical, implementation level, the industrial laboratory-based vibrational spectroscopist needs a sound grasp of the principles and differences underlying the techniques, together with a working knowledge of the instrumentation and equipment being used and its limitations. This we have tried to impart, using for the latter technical information that is essentially freely available through published articles and manufacturers' literature, while, through references and bibliographies, pointing the reader to sources of more theoretical treatments of vibrational spectroscopy. One of our main aims has been to illustrate our discussions with examples of practical 'real-world' applications, emphasising, where our experience allowed, as much 'what can go wrong' as 'how to get it right'. This experience, and the personal comments throughout the text, are the result of two long careers spent thus far in industrial vibrational spectroscopy laboratories, in which one of the things we have learned is that time spent experimenting with and comparing different sampling techniques and differing sample forms, even the most simple, will be hugely rewarded. Spectra produced are the consequence of sample presentation and morphology, and if the practical spectroscopist is not aware of the anomalies and artefacts that they may contain, then he/she is treading a very dangerous path. Interpreting artefacts can be extremely costly not only to one's employers, but also to one's career! In discussing interpretation we have been pragmatic, since industrial spectroscopists are primarily concerned with problem-solving and support research, rather than developing full spectral band-assignments. For this reason, we have also regularly emphasised the importance of complementing vibrational spectroscopy evidence with that from other analytical techniques. To work in a 'tube' is also extremely hazardous! Quantification, to many applications as important as qualitative studies, has been developed from the simple single analyte measurement through to multivariate analysis approaches.

We are acutely aware (and frustrated) that, of necessity in a monograph of this size, 'much is missing', in particular we have avoided other than a passing mention of Near-Infrared Analysis (NIRA) and process in-line and environmental monitoring applications, all of which are extremely important industrially, but are likely to be the subjects of future monographs in this series. It has been our intention to provide a monograph that would be an informative and useful working tool to practical spectroscopists, in both industry and universities.

Contents

Acknowledgements

We are indebted to our respective employers, ICI Technology and Zeneca Specialties, for permissions to publish this monograph.

We would also like to thank many colleagues and mentors, both present and past, with whom we have worked in our respective companies' vibrational spectroscopy laboratories. Many have stimulated our interests and discussed and argued with us about industrial vibrational spectroscopy techniques and their applications, over many years. In particular, we would like to acknowledge Neil Everall, for his patience and advice through many discussions and interruptions. We would also like to thank many associates in the instrument and accessory companies, all of whom, when asked, have been generous in their offerings of diagrams and technical information, particularly David Coombs (Graseby Specac) and Marita Sweeney (Spectra-Tech).

Clearly, in our attempts to illustrate as wide an area of industrial applications as feasible in a book of this size, we have included some application examples which we have 'borrowed' and we thank the providers. In addition, there are others for which we have very limited personal practical experience. We hope for the latter that we have selected wisely, in keeping with the quotation following:

About the most originality that any writer can hope to achieve honestly is to steal with good judgment.

Josh Billings

CHAPTER 1

Introduction, Basic Theory, and Principles

1 Introduction

The vibrational spectroscopies, infrared and Raman, are techniques that are widely used in industry. They provide information on the chemical structures and physical characteristics of materials; they are used for the identification of substances by 'fingerprinting'; and they are used to provide quantitative or semi-quantitative information on products and processes. Samples may be examined in bulk or microscopic amounts over a wide range of temperatures, from very hot to very cold, in a whole range of physical states, *e.g.* as vapours, liquids, latexes, powders, films, and fibres, or as a surface or an embedded layer. The techniques have a very broad range of applications and provide solutions to a host of interesting, commercially important, and challenging analytical problems. They are used to analyse and characterise feedstocks, catalysts, by-products, end and formulated products, processed and fabricated materials, and in deformulation (reverse engineering) studies of competitors' products.

In the Research Laboratory, vibrational spectroscopies are frequently used for reaction following, or for giving chemical group information on new compounds. They are amongst the few techniques which can assist molecular interaction studies such as hydrogen bonding, and provide molecular orientation information for surface studies. In the Process Development and Works environments, quantitative information for process monitoring and product quality assurance/control (QA/QC) can be very important; an increasing development here is in the use of multicomponent analysis of infrared and Raman data, employing regressional analysis and other chemometric treatments. Simple 'fingerprinting' techniques are used extensively in identifying raw materials, but more often in characterising formulated products. The latter is an important QA technique, and a common first step in dealing with customer problems in support of Technical Service/Marketing Departments. Industrially, Raman has been a much less used technique than infrared spectroscopy, largely due to problems associated with colour and fluorescence. However, with recent advances in instrument technology, coupled with the ability to use Raman to effectively examine aqueous solutions and samples inside glass containers, there is a rapid increase in industrial applications of the Raman technique.

Infrared (IR) spectroscopy and Raman spectroscopy are both vibrational techniques; the former is concerned essentially with the absorption of radiation, the latter with scattered radiation. The two techniques are complementary. Their spectra may be considered as being recorded essentially over the same spectral range; both give rise to bands in similar positions originating from the same chemical group. Generally, vibrations which have large changes of dipole moment, *e.g.* the stretching of a carbonyl group (νC=O), gives rise to strong infrared bands and much weaker Raman bands, whilst for vibrations from groups which cause large changes in polarisability, *e.g.* the symmetrical unsaturated group (νC=C), the reverse is true, *i.e.* the band within the Raman spectrum will be relatively much stronger, and weak or even absent in its infrared counterpart. (Here, ν is the notation used to describe a fundamental stretching vibration frequency.)

Although the theory and basic principles of each technique and interpretation of their spectra have similarities, important differences need to be highlighted. However, it is not our intention in this chapter to give an in-depth theoretical approach to the interpretation of vibrational spectra. Basic principles will be set out to enable an initial assessment of spectra and to help avoid a few pitfalls. In infrared spectroscopy, a spectrum is recorded of the absorption of energy from photons by the vibrations of molecular bonds, as the irradiating frequency/wavelength is varied. Raman spectroscopy records the spectrum of light scattered by the molecule when excited by a monochromatic beam of radiation. This latter record contains radiation at the exciting line frequency and bands shifted by amounts equal to the frequency of the molecular vibrations. The strength and shape of the bands within a spectrum are dependent on the chemical and physical state of the molecules, the sample preparation method, the accessory used to mount or contain the prepared sample, and the operating conditions of the spectrometer. Spectra may also be presented in different ways after manipulation by computer software packages. Interpretation of a spectrum requires knowledge of all these factors, which may be affecting the spectrum. This may appear to be a daunting task, but by observing and remembering a few basic principles everything else becomes essentially a variation on a basic theme. The *black art* element sometimes ascribed to vibrational spectroscopy is then much diminished. In attempting interpretation, in an industrial context, a too focused theoretical assignment of each individual band can lead to loss of sight of the overall picture. Simple practical and supplementary information may be ignored and answers are sometimes generated which common sense should tell us are impossible, *e.g.* a black powder cannot be acetone, but may be 'wet' with the solvent! (*True: we have been presented with interpretation examples like this.*) In this chapter we will deal with the simple theory, describe the factors affecting interpretation, and then attempt to set out a step-by-step approach to interpretation, which should be sufficient to establish a working knowledge and a practical approach to problem-solving. Once this is established, the interested, or *insatiable*, should delve further with the help of the bibliography at the end of the chapter.

2 Simple Theory

We will start with the presumption that the spectrum obtained is meant to solve a problem, and therefore contains some information of value. (All correctly acquired vibrational spectra contain some information, even if it is negative information.) Obtaining a spectrum for its own sake is not the fundamental purpose, in most industrial situations. As vibrational spectroscopy employs some apparently strange units, we will start by determining its position in the electromagnetic spectrum.

γ-rays	X-rays	UV	Visible	Near-IR	Mid-IR	Far-IR	Micro	Radio
10^{-11}	10^{-9}	10^{-5}	10^{-4}	10^{-3}	10^{-2}	1	10	(λ)cm

Electromagnetic Spectrum

The so-called mid-infrared range, as the wavelength (λ) range is usually referred to, is approximately 2.5–25 microns (μm), which is equivalent approximately to 4000–400 cm^{-1}. The latter units are wavenumbers ($\tilde{\nu}$) or reciprocal centimetres (cm^{-1}), not frequency (ν). (The equivalent frequency values are 120 THz to 12 THz, (1.2×10^{14} Hz to 12×10^{12} Hz); in terms of photon energy this corresponds to ~0.5 eV to ~0.05 eV.) Wavenumbers is the commonly used notation; strictly speaking Raman shifts should be referred to as delta cm^{-1} (Δcm^{-1}). Early infrared spectrophotometers recorded spectra which were linear in wavelength; (and a few aficionados may still be found who employ these units, particularly if they are entrenched in long-established, hard-copy reference libraries of spectra). However, the majority of information of interest to organic chemists occurs in the 1800–600 cm^{-1} (~5.5–15 μm) range. Linear wavenumber scales make band positions generally easier to resolve and discuss; 1645 ('sixteen forty-five') and 1650 ('sixteen fifty') are significantly easier numbers to remember and debate than 6.055 and 6.095. A linear wavelength scale compresses the broad νOH and νNH bands, in the 2.5–4 μm region, compared to a linear wavenumber (4000–2500 cm^{-1}) plot, whilst the bands below 10 μm (1000 cm^{-1}) are relatively more spread out. Notwithstanding, a key basic point to remember at this stage is that we are dealing with infrared wavelengths of the order of 1–15 microns. This becomes particularly important and relevant when discussing particle size effects, penetration depths, spectral artefacts, and microscopy in later chapters.

3 Molecular Vibrations

Molecules vibrate when struck by a photon. If the frequency (ν) and hence the energy (E) of the photon matches a fundamental vibration of the molecule and causes resonance, the molecule absorbs energy from the photon, see Figure 1,

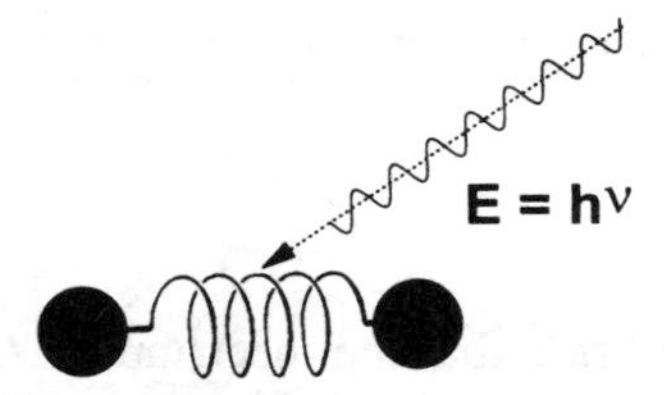

Figure 1 *Simple schematic depicting a photon–molecule interaction*

(*cf.* tuning forks and tables, sopranos and broken glass); h is Planck's constant; $E = h\nu = hc/\lambda. = hc\tilde{\nu}$, where c represents the speed of light in vacuo. This interaction results in an infrared absorption band. If the photon is not absorbed, but inelastically scattered, this scattered radiation contains some information about the molecular vibrations. This information appears generally as very weak bands compared to the intensity of the elastically scattered radiation, with wavenumber shifts from that of the irradiating wavelength. These shifts have values similar to the positions of related infrared absorption bands and the phenomenon is known as Raman scattering. Initially we will concentrate on the infrared interactions.

A linear triatomic molecule, such as CO_2, has three fundamental modes of vibration, a symmetric stretch, a bending or deformation mode, and an antisymmetric stretch. These are labelled as ν_1, ν_2 and ν_3 respectively. A bent molecule, such as H_2O, has four normal modes of vibration, two of which (orthogonal bending motions) are degenerate (*i.e.* of identical frequency), see Figure 2. (Deformation modes are more commonly denoted by the symbol, δ).

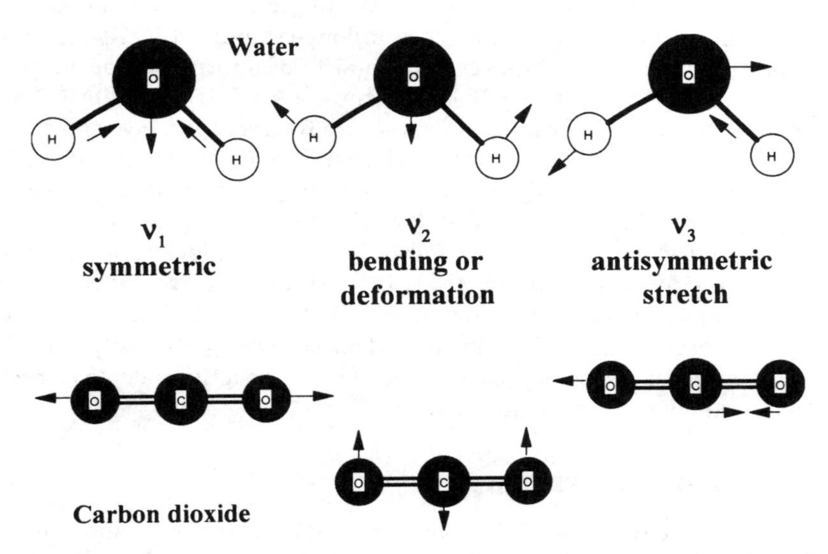

Figure 2 *Models of fundamental modes of vibration for water and carbon dioxide molecules*

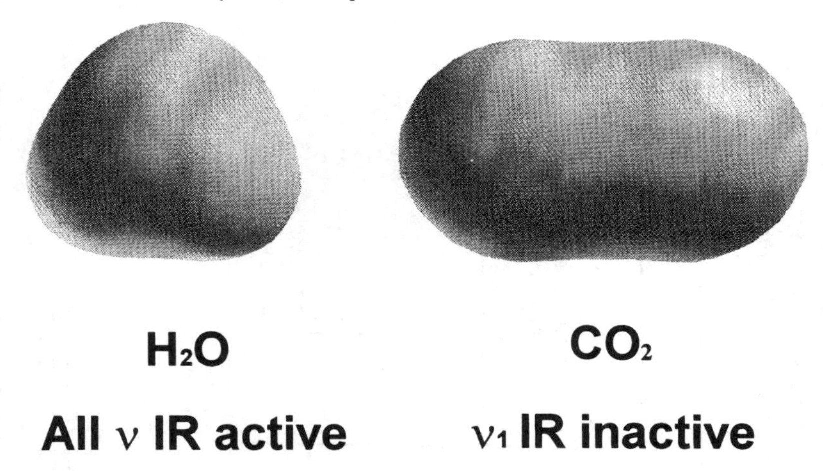

Figure 3 *Electron cloud models of water and carbon dioxide molecules. (Pictures are equivalent contours of constant electron density)*

In Figure 2 these vibrations are depicted as 'linear spring and ball-like' models, whereas, in a three-dimensional sense, the molecules can be viewed as electron clouds with sausage- and kidney-like shapes. Each molecule is now shown, in Figure 3, as a cloud of charges. As the molecule vibrates, changes in dipole moments may occur. Only those vibrations of a molecule which exhibit a strong change of dipole moment have strong infrared absorption bands. For example, carbonyl groups, >C=O, which are both asymmetric and ionic in nature have strong bands in the infrared spectrum. Symmetrical groups such as unsaturated —C=C—, and the dithio, —S—S—, are weak infrared absorbers, but have strong Raman bands. Although overall the dipole moment may not change in the stretching of symmetrical molecular moieties, the electron cloud may deform. This change of polarisability leads to the Raman band, as explained in the next section. It can be seen from Figures 2 and 3 that the symmetrical stretch of a CO_2 molecule leads to no overall change in dipole moment, and hence this vibration is infrared inactive. (Vibrations of diatomic molecules, such as N_2 and O_2 are infrared inactive, but give rise to Raman bands.)

4 Selection Rules

In energy level terms, the transition from the ground state to the next energy level by the absorption of one quantum of light is a fundamental vibration. This is the mechanism that causes the appearance of absorption bands in the IR spectrum, when a sample is exposed to polychromatic light. If a sample is exposed to monochromatic light, which is not absorbed, additional frequencies may be observed in the scattered light. The latter can be considered as three simple cases, see Table 1.

Table 1 *Comparisons of transitions in infrared and Raman spectroscopy ($\nu = 0$, vibrational ground state; $\nu = 1$, first excited state; broken lines indicate hypothetical states). In this table: ν' denotes frequency; ν denotes wavenumber; E denotes energy; υ is the vibrational quantum number, which distinguishes the vibrational levels, and ν'_{vib} is the frequency of the oscillation*

Infrared spectroscopy	Raman spectroscopy		
	Energy balance: $h\nu'_0 + \frac{1}{2}mv_0^2 + E_0 = h\nu'_1 + \frac{1}{2}mv_1^2 + E_1$ The energy change of the molecule $\Delta E_M = E_1 - E_0$ is well approximated by: $\Delta E_M = h(\nu'_0 - \nu'_1)$		
Absorption	(a) Inelastic collision $E_1 > E_0$; $\Delta E_M > 0$	(b) Elastic collision $E_1 = E_0$; $\Delta E_M = 0$	(c) Inelastic collision $E_1 < E_0$; $\Delta E_M < 0$
$\nu'_{vib} = \nu'_{LO}$	$\nu'_2 = \nu'_0 - \nu'_{vib} < \nu'_0$	$\nu'_1 = \nu'_0$	$\nu'_3 = \nu'_0 + \nu'_{vib} > \nu'_0$
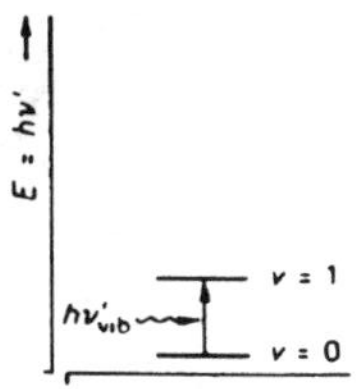 $E = h\nu'$; $h\nu'_{vib}$; $v = 1$; $v = 0$	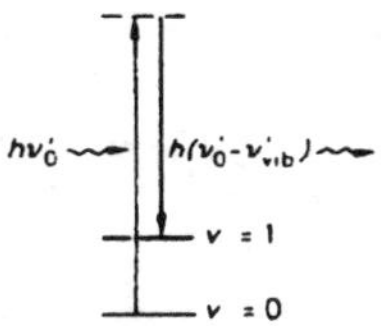 $h\nu'_0$; $h(\nu'_0 - \nu'_{vib})$; $v = 1$; $v = 0$	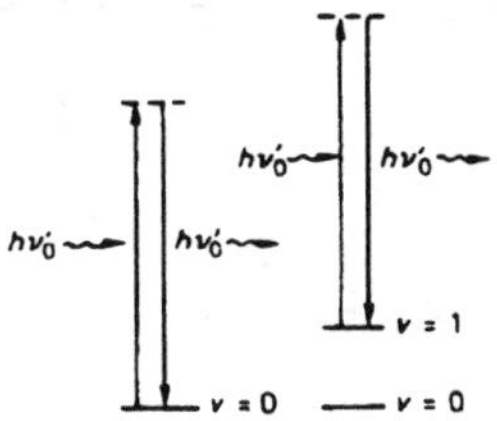 $h\nu'_0$; $h\nu'_0$; $v = 0$; $h\nu'_0$; $h\nu'_0$; $v = 1$; $v = 0$	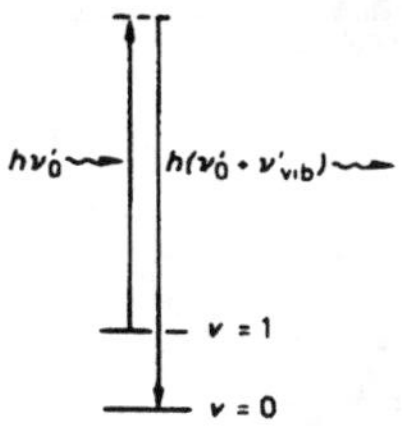 $h\nu'_0$; $h(\nu'_0 + \nu'_{vib})$; $v = 1$; $v = 0$
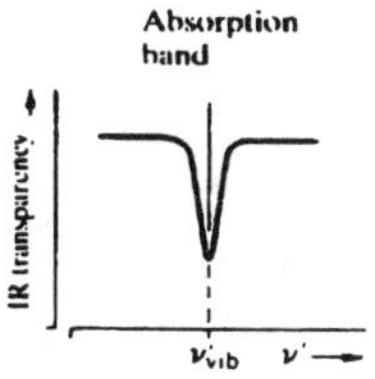 Absorption band; IR transparency; ν'_{vib}; ν'	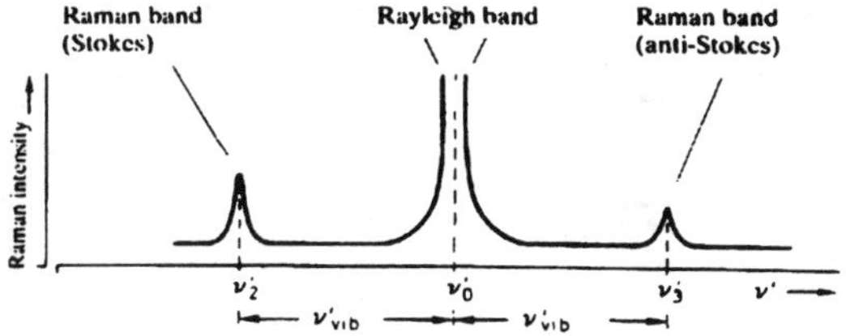 Raman band (Stokes); Raman intensity; ν'_2; ν'_{vib}	Rayleigh band; ν'_0; ν'_{vib}	Raman band (anti-Stokes); ν'_3; ν'

(Reproduced from *Vibrational Spectroscopy: Methods and Applications*, A. Fadini and F.-M. Schnepel, p. 21, see bibliography, with kind permission of Ellis Horwood Ltd)

A molecule is instantaneously excited to a *virtual* (unstable) state by a photon and immediately drops back to its original vibrational energy level. Overall no energy change occurs and the scattered photon maintains the original frequency. This elastic collision is the source of the intense 'Rayleigh' line (exciting line) in the scattered radiation spectrum. If the molecule drops back from the excited (virtual) state not to the ground state but to the first energy level, the photon energy is reduced. This appears as a 'Stokes' line. A few molecules exist already in an excited state at the first energy level. These may, by a similar process, be excited to an unstable energy level, but fall back to the ground state. The energy of the photon increases, which causes the appearance of an 'anti-Stokes' line. These are comparatively much weaker, progressively so with increasing wavenumber shifts, as the majority of molecules are in the ground state at room temperature. These effects were first observed experimentally, in 1929, by C.V. Raman and K.S. Krishnan. Since then the phenomenon has been referred to as Raman spectroscopy.

At the beginning of this section molecular vibrations were seen to be due to stretching, bending, and scissors-like motions, with strong dipole moment changes causing strong IR bands, and weak Raman bands, while symmetrical polarisability changes are the source of strong Raman bands with weak IR bands. Figure 4 illustrates the complementary information from the simple molecule, carbon disulphide.

Predicting the principal IR absorption bands for small molecules is relatively simple. The number of normal vibrations (B), which is related to the vibrational degrees of freedom, can be calculated from the formula:

$$B = 3N - 6, \text{ or, for a linear molecule, } B = 3N - 5,$$

where N is the number of atoms in the molecule. For large polyatomic molecules, the number of bands becomes very large, and will also include overtone (harmonics) and combination bands. The spectrum would thus be a plethora of peaks and impossible to interpret, except for the fact that fortunately many of these bands overlap, and what we see at room temperature are broad envelopes with recognisable positions and shapes. Bands in a spectrum arise from the absorption of energy or radiation scattering, caused by chemical groups of two or more atoms, *i.e.* not individual atoms vibrating. Band (vibration) positions and shapes will also be influenced by the overall shape of the molecule, molecular conformations and orientations, and molecular packing (crystallinity, density). Much time can be spent in assigning these 'fingerprint' bands to the bending, stretching, or scissors modes, but this does not necessarily help in the majority of first attempts to identify materials from their vibrational spectra. For the molecule and its spectrum shown in Figure 5, for example, all the νOH groups give characteristic strong bands in the 3500–2000 cm^{-1} region; a carbonyl band at approximately 1700 cm^{-1} is consistent with a carboxylic acid; and sharp bands in the 3100, 1600 and 800–700 cm^{-1} regions are attributable to the aromatic ring. A band at 950 cm^{-1} is present due to carboxylic acids forming dimers. Many of the bands below 1500 cm^{-1} appear in the spectrum arising from the total molecule vibrations *i.e.* the 'fingerprint'

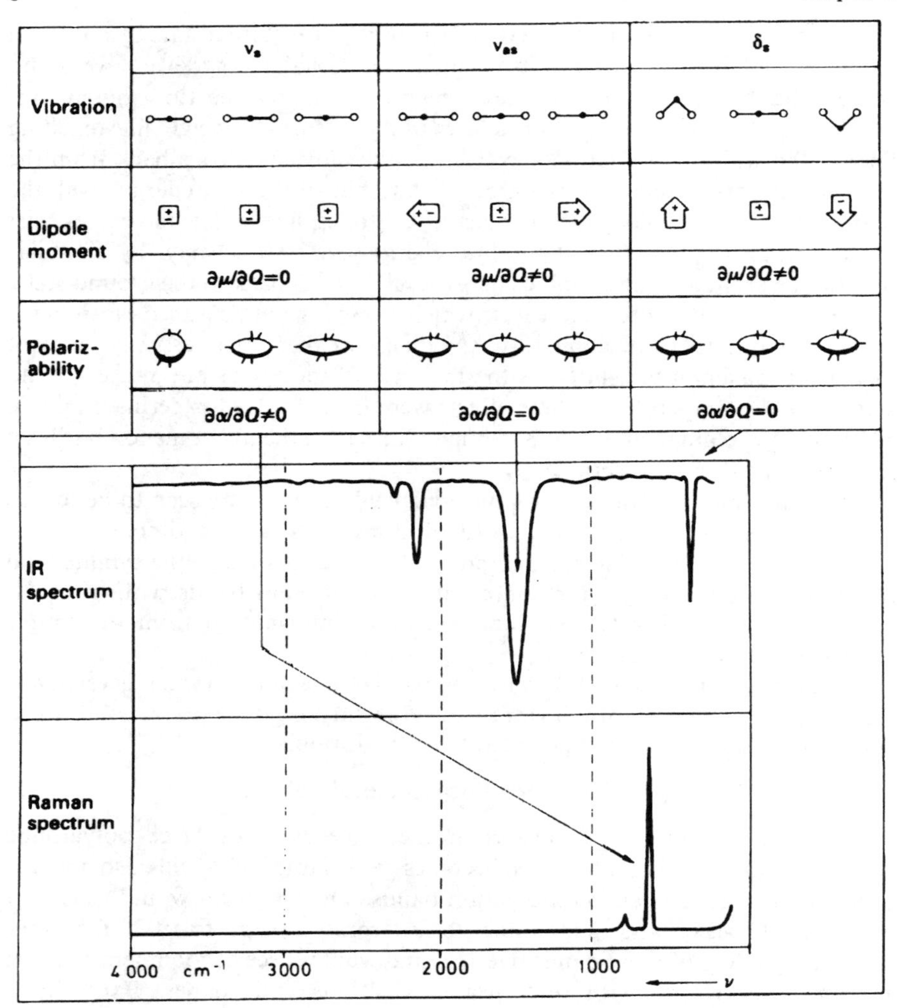

Figure 4 *The changes of dipole moment and polarisability during the normal vibrations of the carbon disulphide molecule (S=C=S) and the resulting appearance of the infrared (IR) and Raman spectra.* μ *is the molecular dipole moment;* α *is the molecular polarisability; Q represents a normal coordinate describing the motion of the atoms during a normal vibration.* ν_s, ν_{as} *and* δ_s *symbolise symmetric stretch, antisymmetric stretch, and symmetric deformation modes of vibration respectively*

(Reproduced from *Vibrational Spectroscopy: Methods and Applications*, A. Fadini and F.-M. Schnepel, p. 27, see bibliography, with kind permission of Ellis Horwood Ltd)

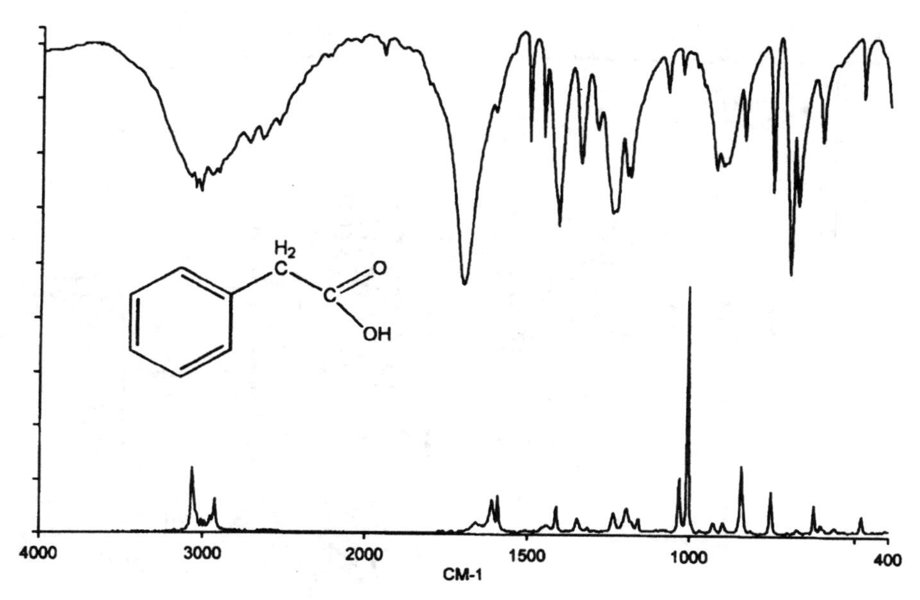

Figure 5 *Infrared and Raman spectra of phenyl acetic acid*

bands. The Raman spectrum appears somewhat simpler as the strong bands in the IR spectrum due to hydrogen bonding and asymmetry are much reduced. The strong bands in the Raman spectrum are largely due to the mono-substituted aromatic group. The band at 2900 Δcm^{-1}, due to the CH_2 group, is hidden somewhat by the strong νOH bands in the IR spectrum, but can be clearly seen in the Raman spectrum.

5 Practical Interpretation of Spectra

The phrase 'interpretation of vibrational spectra' can be employed in many different ways. The spectrum of a molecule can be the subject of a detailed mathematical interpretation which might take several years. At the other extreme, a cursory five second look at a spectrum will, based on familiarity, produce the interpretation, 'Yes, that is ethanol'. In this chapter we are concerned with the practitioner who wishes to solve a problem by using vibrational spectroscopy to identify a substance or to provide structural information. We do not have time or space for a full, rigorous, mathematical approach. Instead, we will concentrate on assisting the reader to attain a practical and pragmatic, if not theoretically polished, result to the interpretation of a vibrational spectrum. This is achieved by being aware of the features which can lead the unwary to an erroneous result.

The positions of bands from specific groups, and the overall shape of the spectrum are dependent on the chemical and physical environment of the sample being examined. Whether the molecules are in a gaseous, liquid, solid,

Table 2(a) νX—H *region (3600–2600* cm^{-1}*) functional group correlation chart*

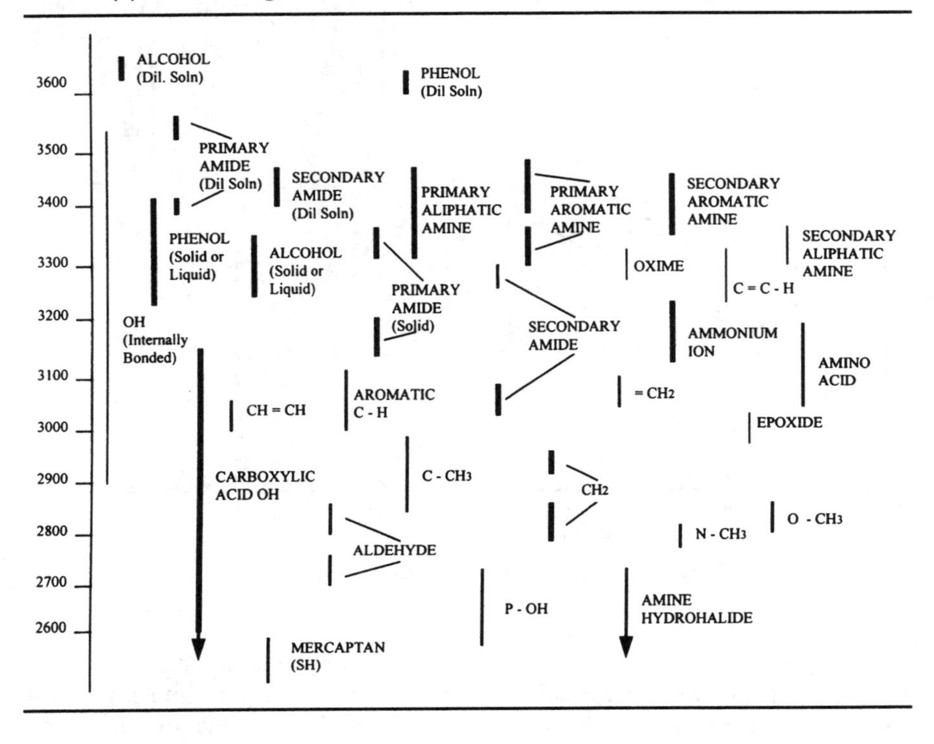

or polymeric form will affect their ability and degrees of freedom to vibrate. In general, vapours and crystalline solids have sharper spectra, whilst liquids and polymers have broader spectra. Temperature, pressure, and polymorphism will also most likely affect the physical forms, and hence the spectra. Chemical groups and total molecular shape may also respond to hydrogen bonding and pH changes. Tables 2(a) to 2(d) give an indication of where the more common groups have bands in the mid-infrared region of the spectrum. The bars indicate likely positions for guidance and are not absolute. The intensity of the bars give very approximate indications of relative band strengths. Band shape, strength, splittings, and associated group correlations must also be taken into account.

The example of an IR spectrum shown as Figure 6 indicates the likely origin of bands due to organic chemical groups in the spectrum. (Figure 6 is a *fictitious* spectrum, drawn many years ago by an expert colleague of ours, purely for teaching the *art* of interpretive spectroscopy.) The 4000–2500 cm^{-1} is the region where single bonds (νX—H) absorb. The 2500–2000 cm^{-1} is referred to as the multiple, triple bond (*e.g.* νN=C=O) region. The 2000–1500 cm^{-1} region is where double bonds (νC=O, νC=N, νC=C—) occur. Below 1500 cm^{-1}, whilst some groups do have specific fundamental bands (*e.g.* νS=O, νC—F, νSi—O, δC—CH_3), is generally referred to as the 'fingerprint' region. Signifi-

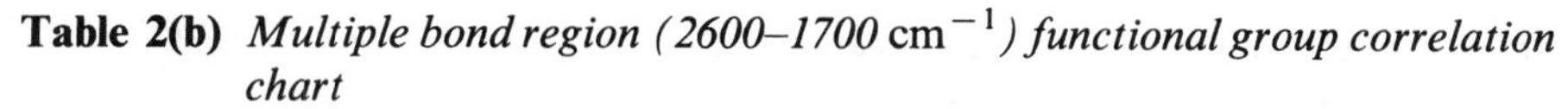

Table 2(b) *Multiple bond region (2600–1700* cm^{-1}*) functional group correlation chart*

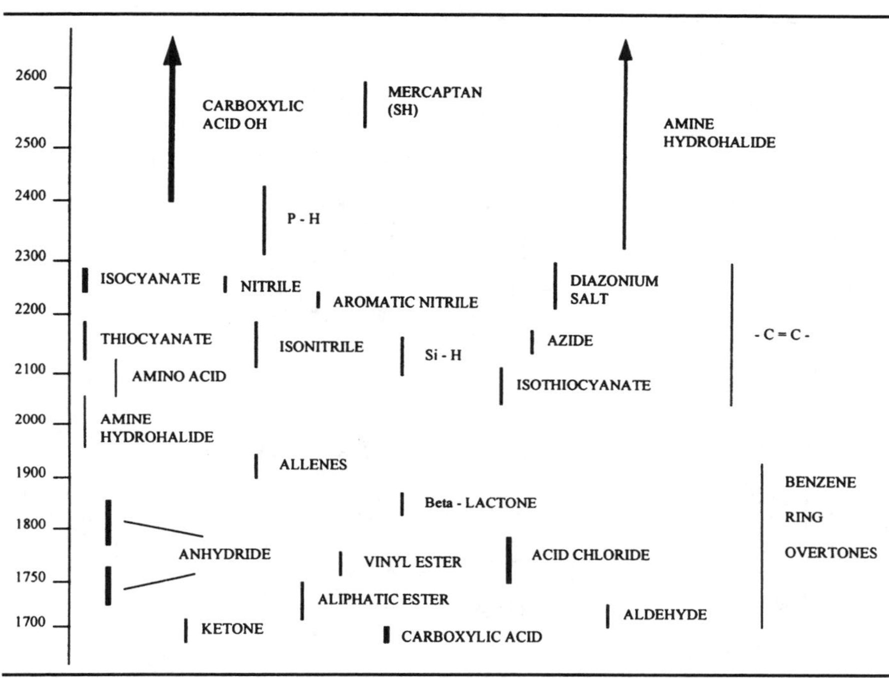

cant bands below 650 cm^{-1} usually arise from inorganic or metal-organic groups. Whilst some specific Raman bands (νS—S—) occur in this region, it is often more confusing than helpful to the organic chemist for the analysis of organic molecules. However, as emphasised above, the individual bands and the overall spectrum are very much dependent on their environment; the physical history of the sample, the sample preparation method, and the spectrometer conditions are likely to affect the resultant spectrum.

Nearly all organic materials and many inorganic materials are suitable for vibrational spectroscopic analysis. Although these can be solids, liquids, polymers, or vapours, probably the majority of bulk, industrial laboratory samples examined are in the form of powders or liquids and consequently may be examined simply by direct IR transmission or Raman measurement at room temperature. Other samples may have to be pre-prepared at different temperatures, cast from solution, or require accessories in order to obtain their IR reflection spectra, particularly in the case of organic polymers. Water and glass are both strong absorbers of infrared radiation. However, both are weak Raman scatterers, attributes which make the complementary technique particularly suitable for examining samples in aqueous environments or in glass bottles.

Table 2(c) *Double bond and fingerprint region (1700–1300* cm^{-1}*) functional group correlation chart*

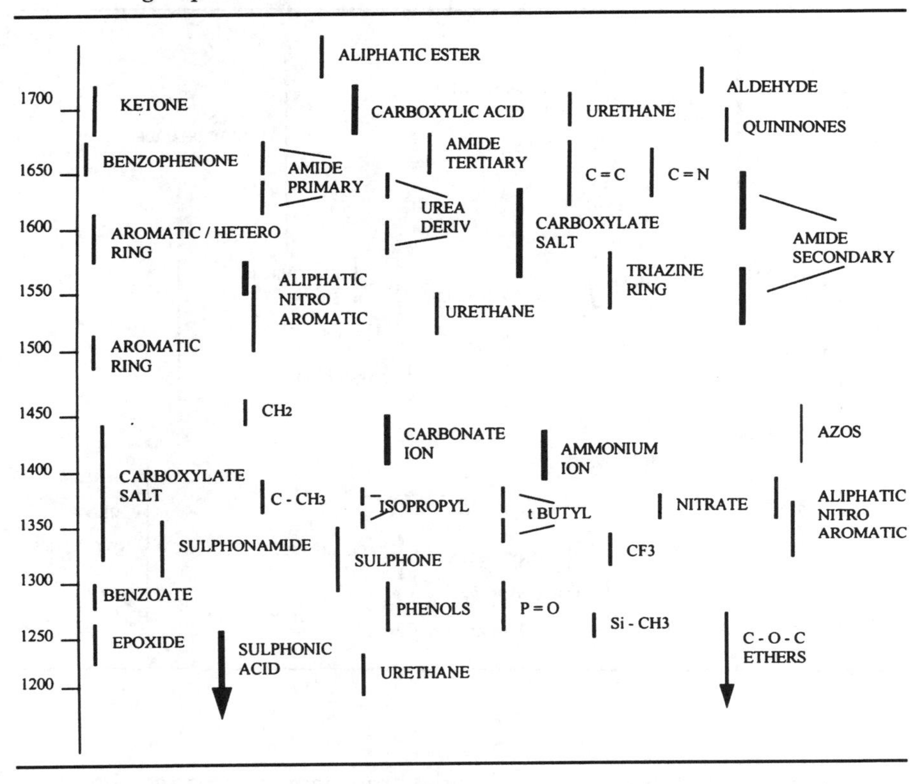

Sample Presentation Factors Affecting Interpretation

To measure a good, relevant infrared spectrum, we must apply the same principles to all samples examined under normal conditions. Preparation of a sample is determined by the physical state and the information required. For example, copper phthalocyanine (CuPc) pigments can be prepared as alkali-halide disks for simple identification (see Chapter 4 for details of sampling techniques). However the alpha/beta characterisation of the polymorphic state of a CuPc requires preparation as a mull, or by another minimal preparation technique such as diffuse reflectance FTIR spectroscopy (DRIFTS), photo-acoustic spectroscopy (PAS), or Raman spectroscopy, since grinding will convert the alpha polymorph to the beta form.

All samples must be presented into an instrument's infrared beam in such a way that the total cross-section of the beam is covered. A neat sample must have a cross-section greater than that of the beam, without holes in the cross-section,

Table 2(d) *Fingerprint region (1300–700* cm^{-1}*) functional group correlation chart*

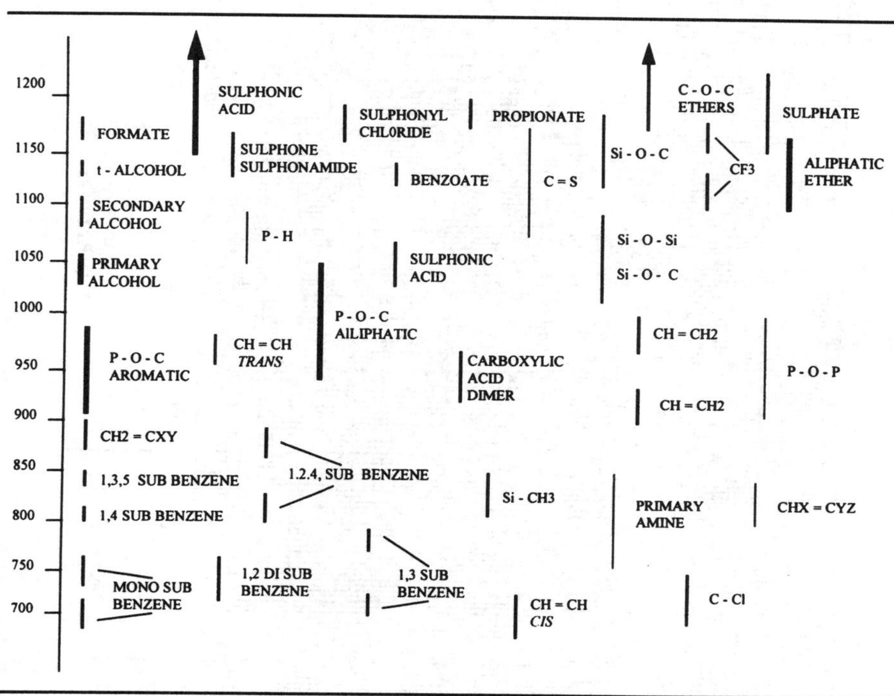

or be appropriately masked so that no stray light by-passes the specimen. Samples with a significantly smaller cross-section than the beam diameter/width may need to be examined with a beam-condenser or with a microscope. A diluted sample, (*e.g.* one dispersed in an alkali-halide powder, or in a liquid paraffin mull), must be homogeneously dispersed, with a particle size small enough not to cause excessive scatter. Ideally, the diluent will not have absorption bands in the same regions of the spectrum as the sample. The path-length (thickness) of a sample used will be dependent on the information required, but should be even across the sample. A sample with truly parallel faces may show interference fringes from which the sample thickness can be calculated. In practice, minor deviations from parallelism destroys fringes and aids interpretation, see pp. 129–130.

Raman spectroscopy is less demanding since the radiation of interest is scattered. The sample position can be optimised in the beam to maximise signal intensity. The collection solid angle and its relationship to the exciting beam are important; 90° and 180° collection geometries are common, but one needs to be aware that these may enhance different orientation effects. Probably the largest problems encountered with industrial samples for Raman spectroscopy occur

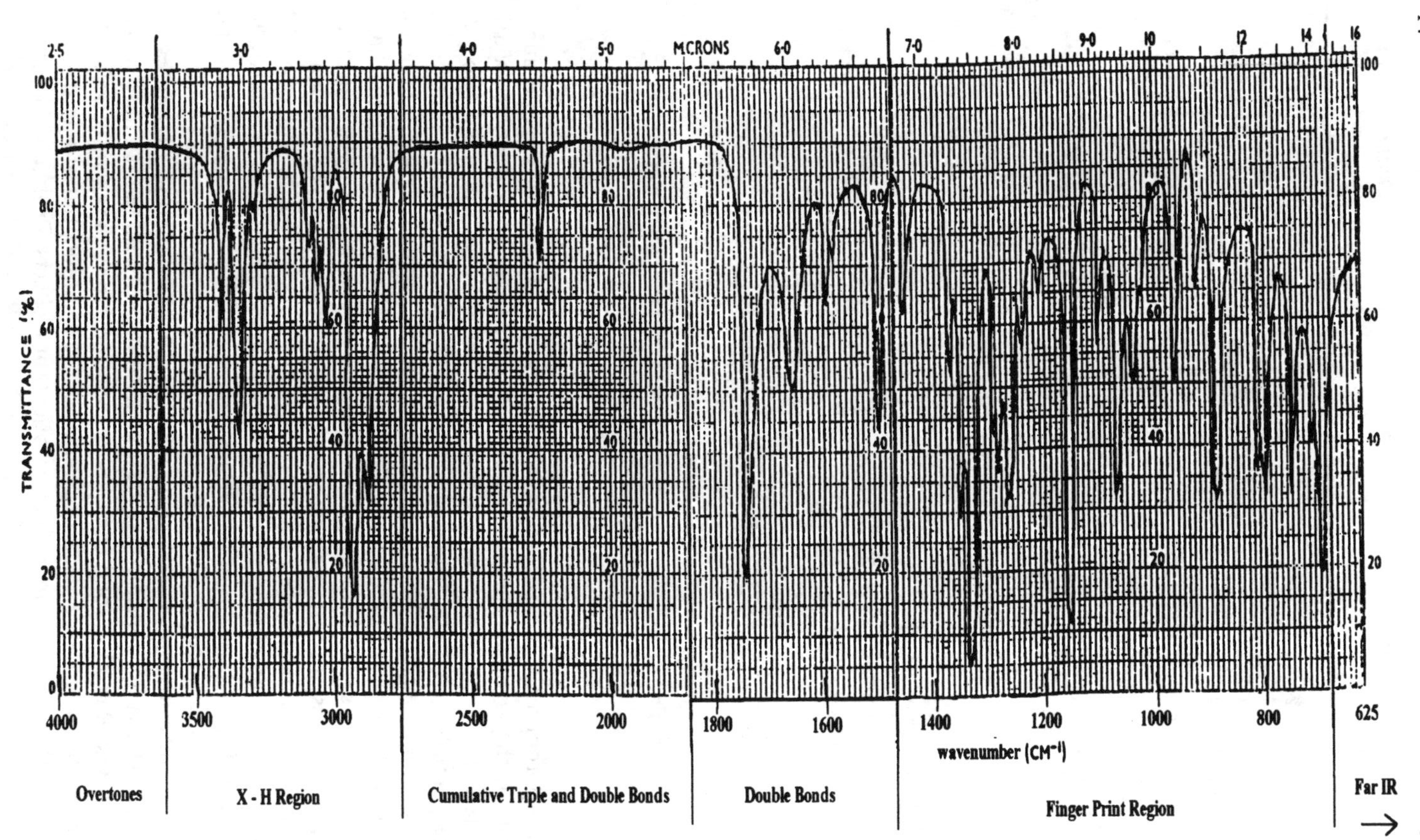

Figure 6 *Some principal functional group correlation regions overlayed onto a typical (fictitious) infrared spectrum*

from fluorescence, charring, or burning. Fluorescence can in some cases be 'burnt-out' (reduced) by leaving the sample in the beam for periods from a few minutes to overnight. With visible laser sources colour has to be considered, to avoid high energy absorption burning-up the sample, or coincidences or near-coincidences with electronic absorption bands, which may lead to selective intensity enhancements of moieties due to resonance effects. Near-IR (NIR) lasers used in FT-Raman spectrometers can cause band shape distortions or relative band intensity variations in some samples. This can occur with samples which have their own NIR absorption band close to or at the laser wavelength or Raman emission wavelengths, *e.g.* water at ~ 1.4 μm.

Instrumentation Features

Infrared spectrometers in general operation in industrial laboratories fall into two categories, dual beam dispersive instruments and single beam Fourier transform infrared (FTIR) instruments. For the reader's convenience, these will be discussed briefly here, but Chapter 3 contains a more detailed discussion.

In a dispersive instrument a single broad band source produces radiation which is split into two beams, a reference beam and a sample beam. The beams are recombined and passed through a monochromator onto a detector, which constantly compares the energy in the two beams. When a sample is placed in the sample beam, a plot of the difference in energy between the beams against wavelength/wavenumber is recorded. The standard plot is per cent Transmission (%T) or Transmittance (0–1T) against wavenumber (cm^{-1}). The latter unit is proportional to the reciprocal of wavelength; $\bar{\nu} = 10^4/\lambda$. Although the transmission/transmittance scale is plotted linearly and well suited for qualitative observations, for quantitative measurements and computer library searching/matching, absorbance (A) must be used, which is logarithmically related to transmission, $A = \log_{10} 1/T$, see pp. 177–181. Dispersive instruments attempt to maintain a linear energy profile, and an open beam plot should be essentially flat (constant transmission at 100%). This is achieved by the slits being programmed to open and close, maintaining the energy profile, as the spectrum is scanned, but this means that the resolution is variable across the spectrum.

Most industrial FTIR instruments are single beam instruments. The beam is passed through an interferometer, then through the sample, and onto the detector. An interferogram is recorded, by a single scan, over the full wavenumber range, in a few seconds or less. To improve signal to noise ratio, several scans are recorded, co-added and then averaged, which reduces the noise contribution. The interferogram is transformed by Fourier transform mathematics into an infrared absorption spectrum. The spectrum recorded is a single beam spectrum, so bands due to atmospheric water vapour and carbon dioxide (CO_2) are present, see Figure 7 and pp. 98–100. FTIR spectrometers operate with a fixed resolution and a variable energy profile. The sample spectrum is superimposed on that background shape. A

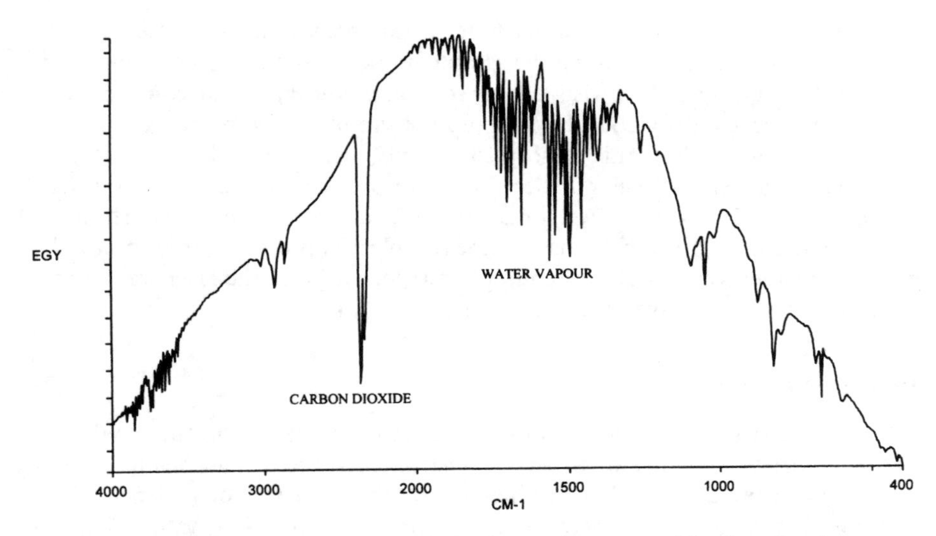

Figure 7 *FTIR spectrometer single beam background spectrum. The sharp bands between about 3900* cm^{-1} *and 3500* cm^{-1} *are also due to atmospheric water vapour; the band near 670* cm^{-1} *is also due to* CO_2*; some other sharp features may be attributed to a protective coating on an optical element. (EGY represents energy throughput)*

FTIR transmission sample spectrum is generated by ratioing it against a background spectrum, which has been recorded under the same conditions as the sample spectrum, but without the sample in place. In most FTIR experiments the background spectrum is recorded first, and in many spectrometers the default operation mode may be set such that when the sample single beam spectrum has been recorded it is automatically ratioed against the background spectrum. The resultant sample spectrum is then displayed directly as a transmission or absorbance plot (which resembles its dual beam dispersive instrument spectrum equivalent).

Raman spectrometers have recently been the subject of a number of significant technological advances (see Chapter 3). Conventional dispersive spectrometers have recently been fitted with CCD (charge coupled device) detectors, and holographic filters instead of gratings, which has vastly improved sensitivity, and led to an increase in imaging techniques. To obtain a Raman spectrum an intense, monochromatic light source is required. The advent in the 1960s of lasers provided the ideal sources for Raman spectrometers. Most spectrometers are dispersive instruments with laser sources operating in the visible region. Spectra are plotted conventionally as Raman scatter intensity *vs.* linear wavenumber shift (Δcm^{-1}), and, while they are free from water vapour and CO_2 bands, sharp spectral lines due to N_2 and O_2, fluorescent room light emissions, and, in some circumstances, cosmic rays may appear, and should be avoided wherever possible. Whilst both Stokes and many anti-Stokes bands may be observed, mostly only the Stokes bands are plotted and interrogated. Colour and fluorescence difficulties have been much reduced, in many instances, by employing near-IR lasers as sources, coupled to interferometers which enable fast scanning and use of Fourier transform techniques.

Beware the Computer – Spectrum Manipulation

Whilst interpretation of a spectrum depends on a knowledge of the sample history and other *a priori* facts such as physical state, the method of sample preparation and the spectrometer operating conditions, spectra can also be subjected to computer manipulation processes by expansion, smoothing, flattening, spectral subtraction routines, and numerous other software programs, not forgetting the effects of the apodisation functions and the like in Fourier transform. Any manipulation should be noted and taken into account before interpretation is attempted. Manipulation of spectra in this way may improve the 'aesthetics' for visual inspection, reports *etc*., but valuable spectral information can be lost. These routines should only be used with great care and an understanding of what is occurring!

Production of Spectra

As discussed above, spectra recorded as single beam spectra may contain bands due to atmospheric gases and vapours. The overall profile of the spectrum is affected by the performance and characteristics of the source, detector and other optical components and instrumental parameters. For example, Figure 7 also exhibits weak absorption bands arising from a thin protective polymeric coating(s) applied to the surface of an optical component, (which will ratio out in a properly conducted FT experiment). The general profile of a single beam infrared measurement is dominated by the polychromatic source characteristics, the output of which varies with wavenumber (see Chapter 3, Figure 1). In dispersive double beam infrared instruments the energy throughput at each wavelength is instantaneously compared between the response of the sample beam and the response of the second, reference, beam. This removes most of the single beam features and, for a sample which is essentially non-scattering and has minimal reflection losses, produces a transmission spectrum with a flat background at around 90–95% T in regions where there is no absorption by the sample. The raw data produced by an FT spectrometer is an interferogram. In a FTIR instrument, spectra have actually undergone a number of computer manipulations, such as apodisation, phase correction, and ratioing (see Chapter 3), to produce similar looking spectra to those recorded from a dispersive instrument. An interferogram would theoretically have wings which extend to infinity. In practice they are mathematically clipped by and convoluted with an apodisation function. In the early days of FTIR spectrometers the effect of these functions was the subject of much debate, particularly for quantitative studies. The 'normal or standard' apodisation functions, *e.g.* Happ–Genzel, Norton–Beer, as defaulted by an instrument manufacturer can be applied in most circumstances without noticeable effects on the spectrum, unless very high resolution spectra with sharp bands (*e.g.* gases) are being studied, (see Figure 13 of Chapter 3). The interferogram is Fourier transformed, probably by a computer chip, into a single beam spectrum, which as stated already is then referenced by computer to a background spectrum to produce the 'equivalent' to a double beam spectrum. If the background spectrum is not recorded under similar conditions to the sample a very distorted spectrum will probably result. Since the FT sample spectrum has fixed resolution across all wavelengths, unlike a dispersive spectrum which has fixed energy but variable resolution, there can be difficulties in comparing a FTIR spectrum with a spectrum previously recorded using a dispersive instrument, particularly if the fine details such as sharp bands are important.

Since the source for Raman measurements is monochromatic, backgrounding is much less of a problem with spectra recorded with Raman instruments. In the absence of fluorescence emission, and excepting when scanning close to the exciting line, the background profile from a dispersive Raman measurement is essentially flat and not in need of correction; this is particularly so for photomultiplier detection, although CCD detectors will probably need to be flat-fielded, which is often accomplished with white-light illumination. In FT-Raman measurements instrument components such as filters, beam-splitters, and detectors will affect the measurement and overlay non-linear profiles onto the spectrum, with detector type and operating temperature affecting particularly the relative strengths of the bands in the 3000 Δcm^{-1} region. Background correction spectra can be recorded with a white light source. In the ideal world this would have a known, invariable, operating temperature. In practice, sources do vary with time and can cause differences in the background correction spectrum recorded. Many practitioners dispense with using these corrections for simple qualitative purposes. The effects are most critical for quantitative measurements, (see Chapter 3, Section 7). The effects of apodisation, resolution, *etc.*, will similarly affect FT-Raman measurements as they do FTIR measurements.

Spectra can be 'interpreted' for identification of substances, by pattern-matching, either by computer or by the hard work of visually searching through hard-copy reference collections. There are now many commercial software packages and libraries available for rapid database searching. If computer searching is used, 'answers' *must be cross-checked* by visually comparing the sample and reference spectra. Do not rely on lists of nearest hits. Spectra can also be interpreted or at least generically 'fingerprinted' using the correlation table principles outlined earlier, by determining the chemical structural groups present from significant band positions, remembering that many of these have significant band shape differences, and should be associated with less intense supplementary bands. However, with this type of interpretation a knowledge of the type of chemistry involved is almost always a useful aid, as may information from an elemental analysis. Cost-effective interpretation of spectra often requires experience, familiarity with and an understanding of the problem posed, and an appreciation of the relevant answer required, particularly when mixtures, formulations, competitive products, or impure samples are examined.

Spectrum Scales

Once the initial spectrum has been produced, any further software manipulation or enhancement should be approached with great care and a healthy cynicism. In many cases the manipulation can make the spectrum look aesthetically better, but degrades spectral contrast or loses information vital for interpretation. Originally, infrared spectra were plotted as recorded analogue signals on an axis of 0–100% *T* or 0–3 *A*. If the spectrum was weak this was instantly obvious. Nowadays it is all too easy to present an apparently high intensity, high signal-to-noise spectrum by simply changing the scale. If care is not taken to read the ordinate scale the information that the spectrum is weak can be missed. A weak spectrum may be due to too little sample, poor preparation, or the fact that the 'sample' is loaded with a diluent such as salt. This information can easily be overlooked when automatic expansion routines are used, which, for example, output a spectrum plot for which the highest transmission or lowest absorbance value is set at one extreme of the ordinate visual display or chart paper ordinate axis length, adjusting the scale such that the peak maximum of the strongest band is set at the other extreme of the ordinate axis length, (for example, see the spectrum plotted as Figure 9 of Chapter 2). Use of such routines should be clearly recorded on the spectrum. Also, beware of spectra with ordinate scales plotted to >3 absorbance units; 3 *A* is equivalent to 0.1% *T* – *i.e.* not much energy! In addition, do not assume detectors are linear over more than one absorbance unit; generally they are not.

Not all scaling routines lose information. One of the most useful and commonly used is the Transmission to Absorbance, Absorbance to Transmission (*T*/*A* or *A*/*T*) transform. Most instrument manufacturers allow operators the choice of displaying directly recorded infrared spectra in transmission or absorbance. Many published collections of reference spectra display the spectra in % *T*, making direct comparison with absorbance spectra difficult. Being able to 'flick' the spectrum over is very convenient. Conversely, if the spectrum is recorded in transmission but is going to be quantitatively analysed, or arithmetically manipulated, the spectrum *must* almost always be converted to absorbance.

Spectra can be scaled for comparison with each other by overlaying. This is usually carried out by choosing a band common to both spectra. The absorbance of this band in the strongest spectrum remains fixed, while the equivalent band in the comparison spectrum is scaled to the same intensity. The remainder of the spectrum is then scaled by the same amount. This is referred to as *normalising* and is discussed in more detail on pp. 186–188.

Spectral Enhancement/Loss of Data

A number of data handling packages available to manipulate spectra are meant to enhance the appearance of the spectrum, particularly for inclusion in reports or publications. Although many claim to improve, or facilitate interpretation, if not used with great care they can achieve the opposite effect. Many samples or accessories, even when analysed and used correctly, produce spectra with a sloping base line. Software packages are available which will reduce or remove the slope from all or part of the spectrum. If this operation is not clearly noted then any attempt to interpret a spectrum may lose important information, which may relate to the presence of fillers, reinforcement materials, or particle size effects. Some data systems offer ATR (see pp. 145–153) background correction algorithms, which store a reference spectrum of a clean ATR crystal. This feature is useful in very fixed situations where repetitive, similar, samples are being examined. In many other cases the type of surface contact achieved affects the background, making this type of correction erroneous.

A spectral *smooth* function is one which can be used as a 'precision tool to enhance a spectrum, or a blunt instrument to totally destroy it'. If a spectrum has a low signal-to-noise ratio the effects of the noise can be reduced with a light smooth to make the bands appear more clearly. A heavy smooth can not only lose shoulders, but may remove some bands completely, as shown in Figure 8.

Many data systems contain routines to produce peaks from within a broad envelope of overlapping bands. These are referred to as *deconvolution* routines.

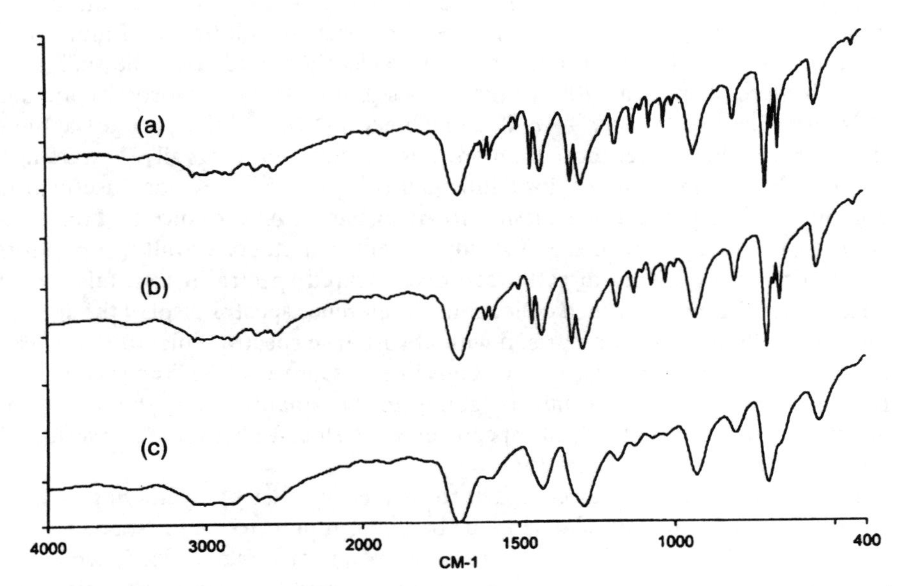

Figure 8 *Infrared spectrum (benzoic acid) with different smoothing levels.* (a) *unsmoothed spectrum,* (b) *'light' smooth (9-point), and* (c) *'heavy' smooth (32-point). Spectra have been offset for clarity*

There are many approaches, which are described in more detail in Chapter 5, Section 6. They can be useful for quantitative studies when the number of components is known. Sound judgement based on chemical common sense will indicate when to stop these processes if the number of components is unknown; otherwise, far from information being lost spurious bands will be generated, particularly if one has an eye on the pre-conceived answer, but not on the facts!

Step-by-step Approach to Interpretation

All honest, experienced, vibrational spectroscopists have their own horror stories of getting it wrong, and indeed there are examples of studies of artefacts and interpretation of contaminants being fully refereed and published as good spectra; we decided it was better here to err on the side of discretion, and not list those references we know of! The basic reasons behind these horrors are the making of incorrect assumptions, anticipating the result, having narrow vision, not having or disregarding basic information, or being given wrong or incomplete information about the sample or spectrum history. The more information available about a sample, the greater the chance of making a correct interpretation of the resultant spectrum. Always ask or consider some basic questions such as those given in the next section before attempting to interpret a spectrum for identification. Not all can be answered or are appropriate for every sample, but an awareness of gaps in the background information can also be used to advantage.

Knowledge of the Sample

Remember always that before all others, the most important questions to be considered are those relating to safety, toxicity, hazards, corrosivity, etc., *before the sample is handled and analysed.*

A lot of information can be gained by understanding the way in which a sample arrived in the state in which it was received:

How was the sample produced/collected/packaged?
What is known of the reaction scheme?
Are there possible side-reactions?
Could solvents be present?
Were there work-up procedures which could introduce impurities?
What was the type of equipment or where in the process did the sample come from? (Grease, drum linings, coupling tubes and filter aids can all appear in or as the sample spectrum.)
Is any elemental information available; does the sample contain N, S, or halogens?
How pure is the sample thought to be?
Is the sample homogeneous?
Do we have supporting NMR and/or MS data.

For solids: Is it 'dry', or a paste?
Has it been washed with a solvent or re-crystallised?
What is its thermal/processing history?
Are there likely to be any dichroic or polymorphic effects?
Is the surface representative of the bulk?
For liquids: Is it volatile?
Is it alkaline, neutral, or acidic?
For vapours: What temperatures/pressures are involved?

As stated above, the answers to these questions are not always available but should be kept in mind, especially if the spectrum doesn't appear as expected. The spectrum of a sample without any known history or source should be approached with great care and consideration. For instance, if you don't give some thought to relative absorption coefficients and sample path-length, then at first glance a spectrum of a mixture may look as though it is essentially a functionalised aromatic, when in reality it is 90% a saturated aliphatic hydrocarbon!

Sample Preparation Effects

As mentioned previously the problem-solving information required may dictate the sample preparation method used, since the way the sample is prepared and examined may affect the resultant spectrum. This is especially important if one wants to preserve the integrity of characteristics such as state of hydration, inter/intra molecular hydrogen bonding, and molecular conformation, orientation, and packing. Being aware of the likely effects on the spectrum arising from the sample preparation or mounting method is as important as knowledge about the sample itself; it helps ascertain whether the spectrum is worthy of interpretation.

Knowing the preparation method used is particularly important in infrared studies. It should provide some clues about a sample, but beware. For instance, if the sample is prepared as a KBr disk then, although it is probably a solid, bands in the spectrum may be present from residual solvent or, in the case of a polymer, residual entrained monomer. Some important preliminary considerations might be:

For powdered solids: Has it been prepared as an alkali-halide disk or liquid paraffin mull? If it is a disk, then you should probably discount broad weak OH bands near 3400 cm^{-1}. If it is a mull, mark off the bands from the mulling agent.

For cast films: Is the film of even thickness? Is there any solvent trapped or encapsulated? Can polymer films exhibit orientation/crystallisation effects?

For liquids: Is the sample a pure liquid or a solution? If the latter, mark the solvent bands.

For ATR spectra: Are the relative band strengths and positions affected by refractive index dispersion effects? Do the sample/ATR element properties satisfy the requirements for internal reflection throughout the wavenumber region being studied – the refractive index of a band with a high absorption coefficient may rise near its maximum to be above that of the ATR element, such that apparent shoulders may appear on the high wavenumber side of the band, or, in extreme cases, absorption minima may be observed where a band maximum was expected, (see Chapter 4, Section 6).

Diffuse reflectance: Are all bands real or are some arising from specular reflection? Are weak bands appearing stronger due to normal bands saturating? Is spectral contrast poor because of particle size effects? Are relative band intensities unusual because of sample packing density effects?

Photoacoustic: Is spectral contrast poor because of signal saturation effects?

Microscopy examination: Are the bands observed characteristics of the sample, interference fringes, or due to the window material, *e.g.* diamond?

A number of artefacts may also be introduced by misaligned accessories or contaminated mirrors on accessories, or spectral contrast may be degraded and relative band intensities distorted if a sample is improperly masked so that the detector sees stray-light, that is, infrared radiation that has by-passed the sample in some way.

Instrument/Software Effects

Even when all precautions have been taken to ensure that all the bands and the overall shape of the spectrum will be analytically representative of the sample in the context of the problem to be solved, and that the method of sample presentation will optimise the requisite information content, extra bands and anomalies may still occur within the recorded spectrum from instrument, experimental, or software artefacts. Close inspection may be needed to ascertain:

Is the band in the infrared spectrum near 2400 cm^{-1} a nitrile or an isocyanate, or in fact due to CO_2 imbalance?

Is the weak sharp band near 670 cm^{-1} in the infrared spectrum attributable to the sample or does it again arise from out-of-balance CO_2?

Are sharp bands in the 1800–1400 cm^{-1} and 3900–3600 cm^{-1} regions due to out-of-balance water vapour?

Does the spectrum really have a flat background or has a software flat routine removed scatter, and lost information?

Is the spectrum as strong as it appears? Check the scale, check for expansion routines. Has a smooth function been applied, which loses bands normally resolved? (Most data systems can be set to display and plot information on data manipulation. However, a lack of printed information does not mean manipulation hasn't occurred.)

Do broad bands in the Raman spectrum most likely arise from fluorescence or charring?

Could sharp bands in a Raman spectrum come from cosmic rays or room light emissions? The position in the spectrum of a 'nuisance' emission from a room light, such as that commonly encountered for the mercury emission from fluorescent tube lighting, will vary with the wavelength of the exciting line used for the Raman experiment!

Are sharp bands near 2330 Δcm^{-1} and 1555 Δcm^{-1} misconstrued, at a first distanced glance, as being due perhaps to a nitrile or a C=C moiety respectively? Or, are they merely due to N_2 and O_2 respectively; the latter pair arising perhaps from a poorly aligned sample, with the laser being focused just in front of the sample and exciting the air molecules? (Diatomics are Raman active.)

The Spectrum

Once all the information on the sample history has been noted, and all possible distortions and artefacts have also been noted, then interpretation of the spectrum and its band positions and strengths should begin, remembering that negative evidence is as important as positive features. A useful visual inspection practice for a material might proceed as follows:

1. Look at the total spectrum as a picture. Does it look as expected from the sample? Are the bands broad or sharp? Does it look organic or inorganic? Does it have the appearance of an aromatic or an aliphatic material? Does it look cyclic? Is the background sloping or flat? If it appears correct then continue with band position interpretation.
2. Starting at the high wavenumber end, in the 3600–3100 cm^{-1} region, are there any νOH or νNH bands? Refer to tables to determine the type, and look for confirmation bands in other parts of the spectrum, *e.g.* amides have carbonyl bands as well as νNH bands.
3. In the 3200–2700 cm^{-1} region are there aromatic or unsaturated or saturated aliphatic bands present; if saturated aliphatic, are they largely methyl, methylene, or methine groups? Again refer to the tables for confirmation by other bands, such as the wagging or rocking modes.
4. Check for bands in the cumulative bond (*e.g.* —N=C=O) region 2700–2000 cm^{-1}.
5. Check for bands in the double bond (*e.g.* νC=O, νC=C—) region 1800–1600 cm^{-1}, refer to the correlation tables, then look for confirmation bands in the rest of the spectrum.
6. Checks 1–5 should establish whether the specimen contains aliphatic, unsaturated or aromatic groups. Alcohols, phenols, amines, or carbonyl moities should have been established or dismissed. Multiple bond bands or carbonyl bands should also have been identified. Look at the rest of the spectrum for strong bands. Do they correspond to bands in the tables?
7. The region below 1600 cm^{-1} contains bands largely due to the fingerprint of the molecule. Structural information can be gained from this region, but many bands provide largely supporting evidence (*e.g.* deformation modes) for groups established in other regions via their stretching mode (ν) vibration bands. Functional groups with their ν bands in this region often arise from oxygenated organics, *e.g.* nitro, sulpho, or heavily halogenated moities, such as —C—Cl and —C—F.

8. If the 3200–2700 cm^{-1} region contains only very weak or no bands, then this negative information could suggest that a halogenated species might be present or that the sample is a non-hydrated inorganic. Oxygenated inorganics *e.g.* carbonates, chlorates, nitrates/nitrites, phosphates/phosphonates/phosphites, sulphates/sulphites, *etc.*, have bands in the 1500–900 cm^{-1} region.
9. Once it is known what functional groups may be present, it is possible to attempt to combine them into a molecule which might be expected, given the sample chemistry and history, together with a knowledge of possible impurities.

Whenever possible, cross-check the interpretation by visibly matching the analyte spectrum to a reference spectrum of the molecule or, failing that, one of very similar structure. If in doubt seek corroboratory evidence from another technique, such as NMR or MS. Never trust peak lists or computer search printouts without visually matching the spectra. If the general procedure outlined above is followed, then the maximum information will be obtained from the examination, and errors will be minimised.

6 Computer-aided Spectrum Interpretation

Computer aided spectrum interpretation can be essentially divided into two types. The most common is library searching or pattern-matching. The other is structural elucidation, which is sometimes part of a training or a library search package.

Library Search Systems

Most infrared instrument manufacturers now offer their own library search routines which contain a collection(s) of pre-recorded reference spectra, and which can be expanded with the users' own recorded spectra. Several can be enhanced with modern digitised versions of earlier well-known hard-copy libraries, such as the commercial Aldrich and Sadtler collections. The major library publishing companies also offer their own versions of electronic libraries as stand-alone library search systems. Several large industrial companies have attempted to build corporate libraries of spectra. SPECINFO® is a system originally devised by the German chemical company BASF, which combined an in-house, corporate, spectra collection with several libraries of published spectra. This has now been commercialised via Chemical Concepts, Weinheim, Germany, with the facility to search or add spectra over a network.

Structural Determination Aids

There have been many attempts to produce artificial intelligence or expert systems that can emulate the human thought process for spectral interpretation. Most have been successful for a limited range of similar chemical compounds or structures, but none has yet approached the full range attempted by humans. Many manufacturers are including software training packages in their range of offerings, which demonstrate very graphically the fundamental principles of interpretation. Several now include excellent 3-D graphical representations of band origins, with simultaneous twisting, bending, and stretching of bonds. These usually work very well for a limited or generic range or group of molecules, but cannot be added to by the spectroscopist.

Spectra Formats for Transfer and Exchange of Data

Vibrational spectroscopists very quickly realised the potential of computers to manipulate spectra for either quantitative or qualitative work. However, this initially required either using software supplied by the instrument manufacturer, or difficult and tedious manipulation of the spectra files for transfer from one computer to another. With the advent of PC work-stations and the establishing of large commercial databases, the demand grew for a universal format for data transfer. In 1987, the Joint Committee on Atomic and Molecular Physical Properties (JCAMP) proposed a format to be used internationally. This is known as JCAMP-DX. The format was intended to represent all data in a series of labelled ASCII fields of variable length. Very quickly, the major spectrometer manufacturers provided software to convert their spectra to and from JCAMP format. Unfortunately, whilst the data format was clearly specified, the file header format was less tightly specified. As a result, commas, spaces, *etc*., were used in different ways as delimiters. The effect was that each manufacturer supplied a slightly different JCAMP file. However, a number of commercial spectrum file converters are now available, which allow for the transfer of these JCAMP files, or unconverted spectra, to and from most spectrometers into third-party commercial data handling and manipulation software packages.

The Internet

We are probably at the beginning of what could become an explosion in the use of the Internet for spectroscopic information exchange and assistance. Most of the main instrument manufacturers have established home pages on the *Net*. These are mainly used for promotional material, but many have plans to provide access to applications, notes, and give details of training courses or seminars. The Society for Applied Spectroscopy (SAS) journal has monthly article listings. Recent editions have articles available in abstract form. Advance notice of meetings and events are also available. The Internet *Journal of Vibrational Spectroscopy* is a freely available source of 'how to do it' articles, as well as for more formal journal articles.

7 Instrument Performance Checks

If they are to carry out qualitative interpretations or quantitative measurements, spectroscopists have to assure themselves that their spectrometer is working correctly and reproducibly. If databases are going to be searched we need to know that a sample examined on any or all spectrometers will produce the 'same', searchable spectrum. Questions are also being asked with increasing perceptiveness by Regulatory Auditors, such as, 'How do you know your instrument is working properly? Is your instrument calibrated?'. The most likely response, especially from those industrial spectroscopists who have migrated from dispersive to Fourier Transform infrared instruments, is that a spectrum of a polystyrene film is recorded periodically and that provided this looks right the instrument is working adequately. Raman spectroscopists tend to give even more vague answers!

For many infrared spectroscopists this approach was seen as the simplest way of satisfying the questioner – but is the approach right? One is left with the uncomfortable feeling that we, as spectroscopists, are complying with Regulators but perhaps not really practising true instrument calibration. Serious doubts regarding the applicability and stability of polystyrene films are voiced every now and then, with variable composition, variable thickness, and traceability being the major concerns.

The ICI/Zeneca Polystyrene '*Round-robin*' Trial

At the end of 1990, members of the Vibrational Spectroscopy Panel of the then ICI plc Analytical Chemists' Committee were aware of the increasing pressure from Regulatory Data Authorities to establish calibration procedures for FTIR spectrometers. There was much discussion within the Panel on how such calibrations could best be carried out. Absolute calibration generally involved complex, time-consuming procedures. The majority of standard compounds recommended by authorities and academia *e.g.* benzene, halogenated hydrocarbons, are difficult to prepare and/or not easily handled in a laboratory because of Health and Safety regulations. In addition, these samples require fixed path-length cells, specific temperature control, *etc.* Even if the results of implementing such calibration methods indicate that the instrument is out of calibration there is very little that spectroscopists can do, other than call in a trained instrument engineer to make any necessary changes.

The consensus reached by the Panel in its deliberations was that industrial spectroscopists are more concerned with the reproducibility of the instrument than with its absolute calibration. 'Will the instrument give the same spectrum today and tomorrow as the one it gave yesterday?' This is a particular concern directly following a service engineer's visit! Therefore a straightforward performance check carried out regularly is essential. The problem is to establish a 'user-friendly' performance check – one which is simple, robust, easily transportable, and requires the minimum of sample preparation and handling. In

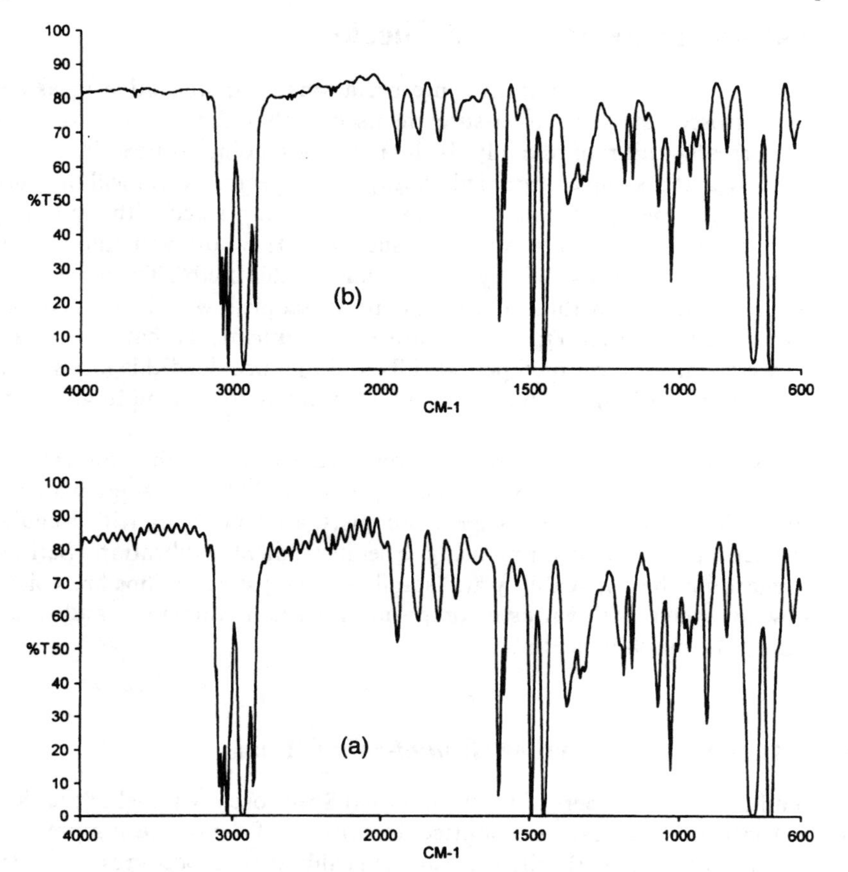

Figure 9 (a) *Transmission infrared spectrum of polystyrene film showing interference fringes,* (b) *Transmission infrared spectrum of matt finish polystyrene film, thickness* $\sim 30\ \mu m$

effect the IR equivalent of the Holmium Oxide filter used in UV-VIS spectroscopy.

Initially polystyrene film was disregarded as a suitable material, mainly because of its unknown thickness but also because the infrared spectrum often contained interference fringes, see Figure 9(a). However, after much debate the Panel decided that a polystyrene film possessed most of the 'user-friendly' attributes needed, and that it would undertake trials to establish whether some of the reservations expressed did significantly affect its use as a performance indicator.

Nicolet (UK) Ltd kindly donated to the Panel three matt-finished polystyrene films that had the advantage of being free of interference fringes, see Figure 9(b). They were also, of course, lightweight, easily transportable without any safety considerations, and required no special sample preparation. It was agreed to undertake a 'round-robin' type inter-laboratory trial, encompassing as many different FTIR instruments as possible.

The main objective of the initial trial was to compare the infrared spectra from the different instruments, defining only the number of scans (16), and the resolution setting (2 cm^{-1}). All other parameters (*e.g.* gain, apodisation, flushing time, *etc.*) were set by the manufacturer's software according to their normal/default operating conditions. Although the specific infrared bands to be measured were defined (3060, 3001, and 906 cm^{-1}), the method of measurement, whether by automatic software assessment or manual calculation, was left to the operator's discretion.

From this initial trial it soon became apparent that there were several inconsistencies between the various instruments from different manufacturers. 'The resolution parameter setting' meant different things on different instruments, *e.g.* collecting at one data point per wavenumber is translated into 2 cm^{-1} or 4 cm^{-1} resolution depending on the particular manufacturer's software package. Although the wavenumber reading should be consistent, as this is always referenced to the internal laser wavelength, it was noted that in some cases printed peak tables produced by the software recorded band positions to the nearest integer, while others reported them to 3 or 4 decimal places, using interpolation procedures to improve band profile (see pp. 93–94), which can affect peak height values. As FTIR spectra are single beam with a background spectrum removed, the shape of this background might affect the absorbance scale reading. The polystyrene peaks to be measured were chosen on account of their absorbance readings (<1 AU, absorbance unit) and they were in regions of the spectrum away from the influences of small changes in water vapour and carbon dioxide effects. Their peak heights were found to be extremely variable, being very dependent on the baseline chosen, some software producing very peculiar results. However the variation in reproducibility of peak height position was found to be negligible to the nearest wavenumber.

A standard operating procedure was then drawn up which took into account, and hopefully minimised, the effects of the noted inconsistencies. For example, the 'resolution' was defined as data collected at one data point per wavenumber, and peak height values as those generated by computer calculated peak tables. The variation in manufacturers' methods was not considered to be significant enough to produce significant variations in these trials. These results, presented in Table 3, have been separated according to the type of detector used. Included in this dataset were spectra from the instrument used in a long-term reproducibility test. The results of this single film on a single instrument trial over a four month period are given in Table 4, together with the ratio of the intensities of the 3060 cm^{-1} and 3001 cm^{-1} bands to the 906 cm^{-1} band.

The main conclusions reached at this point from the inter-lab trial were:

1. The polystyrene matt-finished film supplied had remained in a stable (invariant) condition when kept under ambient conditions for a period of at least two years.

Table 3 *Polystyrene band ratios from two films: Average absorbance values and associated errors*

	IR Peak	Average Absorbance Values TGS	MCT	Both
FILM 01	3060	0.98 ± 0.09	0.97 ± 0.03	0.98 ± 0.09
	3001	0.45 ± 0.06	0.45 ± 0.03	0.45 ± 0.06
	906	0.38 ± 0.04	0.39 ± 0.04	0.38 ± 0.03
	3060/906	2.57 ± 0.18	2.53 ± 0.24	2.56 ± 0.21
	3001/906	1.18 ± 0.05	1.18 ± 0.07	1.18 ± 0.06
FILM 02	3060	0.94 ± 0.05	0.92 ± 0.06	0.94 ± 0.06
	3001	0.44 ± 0.04	0.42 ± 0.05	0.43 ± 0.04
	906	0.36 ± 0.02	0.37 ± 0.02	0.36 ± 0.02
	3060/906	2.66 ± 0.21	2.51 ± 0.12	2.63 ± 0.25
	3001/906	1.22 ± 0.09	1.16 ± 0.08	1.21 ± 0.12

Table 4 *Polystyrene band ratios for a single film: Reproducibility test on a single instrument*

IR Peak	*Average Absorbance Values TGS*
3060	0.997 ± 0.009
3001	0.460 ± 0.003
906	0.391 ± 0.003
3060/906	2.55 ± 0.02
3001/906	1.18 ± 0.01

2. The variation found with a given film on a single instrument for bands with an absorbance <1 AU was better than 10 mAU. (Typical *S*/*N* values of 100/0.05 as quoted by the manufacturer would indicate that this is what might be expected!)
3. Transferring two films around several laboratories in the UK over a period of six months did not appear to degrade the spectral qualities of either film.
4. With the exception of one instrument, the variation of results in the inter-lab trial could be explained in terms of random errors alone.
5. There were no statistically significant differences at the 95% confidence level between:
   (a) instruments made by different manufacturers,
   (b) different models made by the same manufacturer,
   (c) same models with different detectors (DTGS and MCT, see pp. 69–72).
6. Typical variation found for the 0.9 AU band (3060 cm^{-1}) was between 90 and 50 mAU (at least a factor of 5 worse than the long-term reproducibility test).

Table 5 *Polystyrene film band comparisons on strong bands*

	Instrument A		*Instrument B*		*Instrument C*
IR Peak	*MCT*	*TGS*	*MCT*	*TGS*	*TGS*
3027	1.79	1.97	1.60	1.84	>2
2920	1.85	2.10	2.00	2.10	>2
1601	0.79	0.85	0.78	0.82	0.85
1493	2.50	2.10	2.00	2.10	>2
750	1.70	1.67	>2	1.65	>2

Polystyrene film of this type was therefore regarded by the ICI Panel as being suitable as a performance check material for bands less than 1 AU, but could it be used in a quantitative way for higher absorbance readings?

The strongest bands in all the spectra were also investigated. Those studied were at 3027, 2920, 1601, 1493, and 750 cm^{-1}. The results from two similar instruments (A and B) at different geographical locations with MCT and DTGS detectors are shown in Table 5. For comparison the results from an instrument (C) of a second manufacturer fitted with a DTGS detector only are also included. There appeared to be no direct correlation between detectors, background shape, or the previous groupings. A study of band strength led to the conclusion that <1.5 AU the band will fall into the groupings previously stated, between 1.5 and 1.8 AU the band strength variation increases, and above 1.8 AU the instruments will return a figure between 1.8 and 10 AU for the same band. This general comment applies whether the band occurs in the low or high energy regions of the spectrum.

The conclusions of the survey were that matt polystyrene film is a robust, easy to handle material, which can be used for instrument performance checks, under manufacturers' standard settings. A single film will produce the 'same' spectrum on a variety of manufacturers' spectrometers, and models within a range, but not all. Changing between MCT and DTGS detectors does not qualitatively affect a spectrum. Quantitatively, the spectra were in groups for bands up to 1 absorbance unit. Above 1.5 AU the spread increases, and above 1.8 AU the variation is too wide to be comparable. It was recommended that for a performance check a single polystyrene film could be used as a secondary standard, but that the search for a robust, easy to handle, primary standard *should continue*.

During the period of the survey, a number of instrument manufacturers became aware of our activities and responded with varying degrees of interest. The discussions with manufacturers raised the topic of how they test that an instrument is working. In many cases the signal-to-noise is set within specification, and a spectrum, if recorded, is only a visual aid. However, as many detectors and beam-splitters do not have a flat response, the spectra are not necessarily the same. Some manufacturers are now including polystyrene films and/or transmission filters built into their spectrometers as part of a validation package.

Commercially Available Traceable Standards for Infrared

Regulatory Authorities are rarely satisfied that a performance check is carried out; they also require that the standards used are certificated and traceable. At the time of the ICI trial traceable calibrated polystyrene films did not exist. There are now several commercial sources available, established since the survey was carried out.

Towards the end of the survey the National Physical Laboratory (NPL) in the UK produced their own standard polystyrene films and method. The calibration covers wavenumber position only and the band list stops at 1028 cm^{-1}. Smoothing, and a cubic interpolation is recommended in some cases, ideally run on a different computer than the instrument! The film calibration is carried out on a dispersive instrument. The 1997 current cost is approximately £20 plus £450 for calibration, with a recommended 2 year recalibration period. NPL also provide glasses for transmittance standards. These are for the region 4000–2000 cm^{-1}; the glass cuts off at the lower limit. The comment is made that 'for most purposes it is sufficient to establish linearity of an instrument in one wavelength region'!

The National Institute of Standards (NIST) in the USA have also produced a polystyrene film traceable standard. Again the wavenumber position only is calibrated, but covers the range to 545 cm^{-1}, with uncertainties quoted from 0.06 to 12.29. The calibration was carried out under vacuum at 0.5 cm^{-1} resolution with a centre-of-gravity-based apodisation function. Not a measurement easily reproduced on an industrial laboratory spectrometer! Five films are supplied for approximately US$200, with no end date for recalibration.

Both the NPL and NIST calibrations give very similar wavenumber positions, but these do not agree with some quoted listings in various Pharmacopoeias. This leaves the spectroscopist in the unenviable position of quoting reproducibility against traceable films or variation against published standards!

Other recent attempts at producing standards include polystyrene solutions from the Laboratory of the Government Chemist (LGC) in the UK and an IUPAC book of tables of toluene, benzene, chlorobenzene, and dichlormethane peak positions. With the LGC solutions, spectroscopists have to cast their own films. With the IUPAC book, a software package is supplied which allows comparison of the published data with the spectroscopist's own data.

This is clearly an area where a single meeting of minds of academe, manufacturers, regulators' and industrial spectroscopists has yet to occur.

Raman Spectroscopy Calibration Standards

The infrared industrial community is gradually, if somewhat reluctantly, addressing the issue of calibration standards. The Raman spectroscopy community is in a different position. Indene and sulphur have well-known band positions as measured on dispersive instruments, but calibrating relative peak heights is a rarely mentioned field. For NIR (near-infrared) FT-Raman spectrometers the situation is worse: while the sulphur spectrum maintains relative strengths, indene bands vary greatly in relative intensity with laser power. A number of compounds with absorption bands in their NIR spectra that occur near to the laser line at 1064 cm^{-1} show this effect. This appears also to particularly affect compounds with aliphatic hydrocarbon groups that have absorption frequencies near that of the exciting line. The νCH bands, and others near to the limits of the detector range, are also frequently strongly attenuated compared to spectra excited with a visible light laser, and bands are likely to be severely attenuated when they occur close to the cut-off edges of the filter used to block the elastically scattered radiation occurring at the exciting line frequency.

Halogenated dienes have been suggested as possible Raman standards. Alternatively, a neon lamp on the beam axis can provide a wavelength calibration standard. Very recently, a simple, practical calibration standard for instrument response correction, based on the use of luminescent standards (fluorophores), has been proposed by Ray and McCreery (see Bibliography and Figure 22 of Chapter 3).

8 Quantification

The procedures discussed so far have led only to qualitative interpretations, but vibrational spectroscopy is also used extensively as a quantitative tool in industry. Quantitative analysis is considered in detail in Chapter 5.

9 Closing Remarks

In this chapter we have tried to give a brief insight into many of the basic principles and practices necessary for successful application of vibrational spectroscopy in the industrial laboratory. We have also discussed numerous factors which can affect the spectra being produced. Vibrational spectroscopy can be carried out at many levels, from the simple identification test to an in-depth, full spectrum, qualitative and quantitative analysis. In all cases the simple basic considerations and practices laid out should be observed. Although we have been able to touch only briefly on several concepts in this opening chapter, many will be explored and discussed in more detail in subsequent chapters. A number will be used in each area of sample type, preparation, instrumentation, and quantification to exemplify the successes that can be gained or the traps fallen into, especially by the unwary, if basic principles, considerations, and practices are forgotten, short-cut or ignored.

10 Bibliography

There is an ever increasing *ocean* of literature on vibrational spectroscopy. Listed below are a number of publications which the authors have found useful for general reference, for theoretical and practical aspects of vibrational spectroscopy, and for aids in interpretation. A growing number of software packages are available which are intended for similar functions. A few of these are also listed:

Fourier Transform Infrared Spectroscopy, P.R. Griffiths and J.A. de Haseth, J. Wiley, New York, USA, 1986.

Laboratory Methods in Vibrational Spectroscopy, 3rd Edn, ed. H.A. Willis, J.H. van der Maas and R.G.J. Miller, J. Wiley, Chichester, UK, 1987.

Vibrational Spectroscopy: Methods and Applications, A. Fadini and F.-M. Schnepel, Ellis Horwood, Chichester, UK, 1989.

Introduction to Infrared and Raman Spectroscopy, 3rd Edn, N.B. Colthrup, L.H. Daly and S.E. Wiberley, Academic Press, San Diego, USA, 1990.

Infrared and Raman Spectroscopy. Methods and Applications, ed. B.H. Schrader, VCH, Weinheim, Germany, 1995.

Infrared Spectra of Complex Molecules, L.J. Bellamy, 3rd Edn, Chapman and Hall, London, UK, 1975.

Advances in Infrared Spectra of Complex Molecules, L.J. Bellamy, 2nd Edn, Chapman and Hall, London, UK, 1980.

Tabulation of Infrared Spectral Data, D. Dolphin and A. Wick, J. Wiley, New York, USA, 1987.

Infrared and Raman Spectra of Inorganic and Co-ordination Compounds, 4th Edn, K. Nakamato, J. Wiley, New York, USA, 1986.

Metal-Ligands and Related Vibrations, D.M. Adams, Edward Arnold, London, UK, 1967.

FT Raman Spectroscopy, P. Hendra, C. Jones and G. Warnes, Ellis Horwood, Chichester, UK, 1991.

The Handbook of Infrared and Raman Characteristic Frequencies of Organic Molecules, D. Lin-Vien, N.B. Colthrup, W.G. Fateley and J.G. Grasselli, J. Wiley, New York, USA, 1991.

Infrared Characteristic Group Frequencies, 2nd Edn, G. Socrates, J. Wiley, New York, USA, 1994.

Analytical Raman Spectroscopy, ed. J.G. Grasselli and B.J. Bulkin, J. Wiley, New York, USA, 1991.

Fourier Transform Raman Spectroscopy, From Concept to Experiment, ed. D.B. Chase and J.F. Rabolt, Academic Press, San Diego, USA, 1994.

Introductory Raman Spectroscopy, J.R. Ferraro and K. Nakamoto, Academic Press, San Diego, USA, 1994.

Practical Sampling Techniques for Infrared Analysis, ed. P.B. Coleman, CRC Press, Boca Raton, USA, 1993.

Advances in Applied Fourier Transform Spectroscopy, ed. M.W. Mackenzie, J. Wiley, Chichester, UK, 1988.

Infrared and Raman Spectroscopy of Polymers, H.W. Siesler and K. Holland-Moritz, Marcel Dekker, New York, USA, 1980.

Handbook of Near-Infrared Analysis, ed. D.A. Burns and E.W. Ciurczak, Marcel Dekker, New York, USA, 1992.

Resources and References for Spectral Interpretation. Part 1: Infrared and Raman, J.P. Coates, *Spectroscopy*, 1995, **10**, 7, 14 – contains a very useful list of reference works for IR/Raman Spectral Interpretation.

Fifty categories of ordinate error in Fourier transform spectroscopy, J.R. Birch and F.J.J. Clarke, *Spectroscopy Europe*, 1995, **7**, 4, 16.

Tables of Intensities for the Calibration of Infrared Spectroscopic Measurements in the Liquid Phase, prepared for publication by J.E. Bertie, C.D. Keefe and R.N. Jones, International Union of Pure and Applied Chemistry (IUPAC), Chemical Data Series No. 40, Blackwell Scientific, Oxford, UK, 1995.

Simplified Calibration of Instrument Response Function for Raman Spectrometers Based on Luminescent Intensity Standards, K.G. Ray and R.L. McCreery, *Appl. Spectrosc.*, 1997, **57**, 1, 108.

Standards for Infrared Spectroscopy, R.N. Jones, *European Spectroscopy News*, 1987, **75**, 28.

Units! Units! Units!, D.W. Ball, *Spectroscopy*, 1995, **10**, 8, 44.

Encyclopedia of Analytical Science, ed.-in-chief A. Townshend, Vol. 4, *Infrared Spectroscopy*, pp. 2153–2240, Vol. 7, *Raman Spectroscopy*, pp. 4369–4429, Academic Press, London, UK, 1995.

1995 Annual Book of ASTM Standards, Vol. 03.06, ASTM, Philadelphia, USA, © 1995, American Society for Testing and Materials:

Designation E 131–94: *Standard Terminology Relating to Molecular Spectroscopy.*

Designation E 932–89: *Standard Practice for Describing and Measuring Performance of Dispersive Infrared Spectrometers.*

Designation E 1421–94: *Standard Practice for Describing and Measuring Performance of Fourier Transform (FT-IR) Spectrometers: Level Zero and Level One Tests.*

Designation E 1683–95: *Standard Practice for Testing the Performance of Scanning Raman Spectrometers.*

Hard-copy Spectra Collections/Sources

1. *The Aldrich Library of FT-IR Spectra*, Edition 1, ed. C.J. Pouchert, Aldrich Chemical Company, Milwaukee, USA, 1985.
2. Biorad Laboratories Ltd., Sadtler Division: Multiple IR Spectra Databases/Collections, Handbooks and Software collections.
3. The Coblentz Society, Kirkwood, Mo, USA, Various Spectra Collections and Handbooks of Spectra.
4. Raman/Infrared Atlas of Organic Compounds, 2nd Edn, ed. B. Schrader, VCH, New York, USA, 1989.
5. Handbook of Infrared and Raman Spectra of Inorganic Compounds and Organic Salts, R.A. Nyquist, R.O. Kagel, C.L. Putzig and M.A. Leugers, Academic Press, London, UK, 1996.
6. Atlas of Polymer and Plastics Analysis, 3rd Edn (4 vols), D.O. Hummel, VCH, Weinheim, Germany, 1991.

Software Interpretation Tools and Databases, Internet Sites

1. Perkin-Elmer Ltd.: *IR Tutor*®.
2. Bio-Rad Laboratories, Sadtler Division: *IR Mentor*®.
3. Chemical Concepts: *Specinfo*® spectral databases and *SpecTool*®.
4. Internet *Journal of Vibrational Spectroscopy* (http://www.teamworks.co.uk/ijvs).
5. Society for Applied Spectroscopy. (http://esther.la.asu.eau/sas/journal.mtml).
6. *Internet Sites for Infrared and Near-Infrared Spectrometry. Part 1: On-line instruction and direct communication*, E.G. Kraemer and R.A. Lodder, *Spectroscopy*, 1996, **11**, 7, 24.

Data Transfer/Exchange/Processing

1. JCAMP-DX: *A Standard Form for Exchange of Infrared Spectra in Computer Readable Form*, R.S. McDonald and P.A. Wilks, *Appl. Spectrosc.*, 1998, **42**, 1, 151.
2. GRAMS/32®, Galactic Industries Corporation, Salem, USA.

CHAPTER 2

Sample Types and Analyses

1 Introduction

As stated in the previous chapter, vibrational spectroscopy is used to study a very wide range of sample types in the industrial laboratory. Although the techniques may not be as sensitive as other spectroscopies, microscopies, or some separation sciences, in their range of applications they are second to none. The sample types analysed and the information gained are as varied as the laboratories that use them. This chapter can only give a flavour of an ever expanding field.

The most common uses in industry for vibrational spectroscopy are simple identification, reaction following, and simple quantification. In the main, samples are provided as powders or liquids, which, whilst being very important and the staple diet for many spectroscopists, are not altogether challenging, unless they are very small, unstable, or very nasty (environmentally unfriendly). There are many other types of samples, which by their very nature require skilful preparation/presentation and/or the use of sophisticated accessories. Often these are complex mixtures or formulations, or manufactured products subject to customer complaints. Crisp bags, stained cloth, natural and man-made fibres, window-sills, toy soldiers, swimming pool filters, swimming costumes, elasticated underwear, roof panels, ski boots, bath panels, printing rollers, inks, ink jets, paper, CD disks, semiconductors, pharmaceutical tablets, spark plugs, skin lotion, hair, human body implants, tin cans, car seats, car panels, hot air balloon structures, Old Masters, air-borne pollutants, and rocks, are a few examples of the types of sample that have passed through the industrial laboratories in which the authors work. Applications are ever increasing; for example, recent advances facilitating the study of water-wet samples have led to a growth in publications, particularly in the area of protein analyses.

Many of the *interesting* samples listed above may be challenging in their sample preparation, handling, and/or the instrumental techniques involved, particularly if both chemical and physical characteristics are sought. However, various types of information can be obtained from spectra recorded even from samples using simple preparation procedures; structural chemistry is an area where even *simple* single compound samples can be challenging, especially if their spectra are being used to study polymorphism, hydrogen bonding, tautomerism, or racemates.

2 Structural Information

Many samples are examined purely for structural information, either to identify an unknown sample, or to monitor the products of a chemical synthesis.

With unknowns, if pattern-matching fails, the generic type/class of chemical can often be determined by identifying the chemical groups present. By comparing the major band peak positions with those in correlation tables, such as Tables 2(a)–2(d) in Chapter 1, a progressive identification can be attempted using the procedure set out in Chapter 1, Section 5. For example, does the spectrum contain bands due to organic or inorganic chemicals? If it appears to be inorganic are carbonate, nitrate, sulphate, or phosphate bands present? If the spectrum appears to contain bands due to organic species, are aliphatic, unsaturation or aromatic bands present? Is there a carbonyl band? If so, is it a carboxylic acid, ester, aldehyde, or ketone? Are other NH or OH bands present? Could it be an alcohol, amine, or amide? Are there unusual bands such as nitrile or isocyanate? The answers to all these questions should progressively lead to identifying or typing an unknown substance or suspected impurity.

Reactions may be followed or products analysed to determine whether an expected synthesis has taken place. For example, alcohols react with carboxylic acids to form esters. During this condensation reaction, infrared spectroscopy will show the νOH band of the alcohol to decrease, the intense acidic νOH band will also decrease, and the carbonyl band (νC=O) will move to a higher wavenumber as acid converts to ester. Many reactions can be followed in similar ways or by simply monitoring the intensity of a band in the spectrum that is related to the chemical moiety of interest. The study of reactions by following the appearance or disappearance of a particular band is quite common, since rate of reaction information and kinetic studies are of particular importance in plant design and process development. However, if a reaction is complex, full spectrum coverage can reveal information on any by-products or intermediates formed. In some instances the product formed may not have the expected spectrum, even though its chemical identity is correct. This can occur when the physical form of the product, or molecular interactions, affect the spectrum.

Polymorphism

The definition of polymorphism is a subject of much discussion in its own right.[1] For our purposes McCrone's definition of a polymorph as '*a solid crystalline phase of a given compound resulting from at least two crystalline arrangements of*

[1] Threlfall T.L., *Analyst*, 1995, **120**, 2435.

the molecules of that compound in the solid state' will suffice.[2] The study and identification of polymorphic forms are key to many industries. In the pharmaceutical industry, drugs in different polymorphic forms have differing solubilities, and therefore potentially great differences in the way in which they interact with the body.[3,4] Some doubt has been expressed as to the regularity of this occurrence,[3] but much effort is demanded from Regulatory Authorities and Patent Attorneys to determine whether polymorphs exist. Apart from solubility, however, other manufacturing and tablet problems can arise.[5–9] In the pigment, electrographic, and printing industries, differing polymorphic forms create dispersion and shade variations.[10–18] Products containing natural fats, waxes, soaps, petroleum products, sugars, and polysaccharides can all change behaviour when polymorphism occurs.[19–30] Many spectroscopic and thermal techniques have been employed to study polymorphism, and these are discussed and described in a review by Threlfall.[1]

Vibrational spectroscopy is a key tool in characterising polymorphs, since it can be used to distinguish between forms and quantitatively assess mixtures. For example, when sampled in an appropriate way, the alpha and beta polymorphic forms of copper phthalocyanine can be readily distinguished, see Figure 1 and refer to Chapter 4, Section 2.

[2] McCrone W.C., in *Physics and Chemistry of the Organic Solid State*, Vol. II, ed. Fox D., Labes M.M. and Weissberger A., Interscience, New York, USA, 1965, p. 725.
[3] Burger A., in *Topics in Pharmaceutical Sciences*, ed. Bremer D.D. and Speiser P., Elsevier, Amsterdam, 1983, p. 347.
[4] Halebian J.K. and McCrone W.C., *J. Pharm. Sci.*, 1969, **58**, 911.
[5] Nakamachi H., Yamaoka T., Wada Y., and Miyaka F., *Chem. Pharm. Bull.*, 1982, **30**, 3685.
[6] Chan H.K. and Doelker E., *Drug. Dev. Ind. Pharm.*, 1985, **11**, 315.
[7] Macle C.H.G. and Grant D.J.W., *Pharm. Int.*, September 1986, 233.
[8] Thoma K. and Serno P., *Deut. Apotek. Z.*, 1986, **124**, 43, 2162.
[9] Liversidge G.G., Grant D.J.W., and Padfield J.M., *Anal. Proc.*, 1982, **19**, 12, 549.
[10] Kendall D.N., *Anal. Chem.*, 1952, **24**, 382.
[11] Susich G., *Anal. Chem.*, 1950, **22**, 425.
[12] Ebert A.A. and Gottlieb H.B., *J. Am. Chem. Soc.*, 1952, **74**, 2806.
[13] Whitaker A., *J. Soc. Dyers Colour*, 1992, **108**, 282.
[14] Thomas A. and Ghode P.M., *Paintindia*, 1989, **39**, 6, 25.
[15] Khunert-Brandstätter M. and Riedmann M., *Mikrochim. Acta.*, 1989, **I**, 373.
[16] Warwicker J.O., *J. Text. Inst.*, 1959, **50**, T443.
[17] Etter M.C., Kress R.B., Bernstein J., and Cash D.J., *J. Am. Chem. Soc.*, 1984, **106**, 6921.
[18] Enokida T., Hirohashi R., and Nakamura T., *J. Imag. Sci.*, 1990, **6**, 234.
[19] *The Physical Chemistry of Lipids*, ed. Small D., Plenum, New York, USA, 1986.
[20] *Crystallisation and Polymorphism of Fats and Fatty Acids*, ed. Garti N. and Sato K., Marcel Dekker, New York, USA, 1988.
[21] Gupta S., *J. Am. Oil Chem. Soc.*, 1991, **94**, 50.
[22] Srivastara S.P., Handoo J., Agrawal K.M., and Joshi G.C., *J. Phys. Chem. Solids*, 1993, **54**, 639.
[23] Ungar G., *J. Phys. Chem.*, 1983, **87**, 689.
[24] Cebula D.J. and Ziegleder G., *Fett Wiss. Technol.*, 1993, **95**, 340.
[25] deMan L., Shen C.F., and deMan J.F., *J. Am. Oil Chem. Soc.*, 1991, **68**, 70.
[26] Saltmarsh M. and Labuza T.P., *J. Food Sci.*, 1980, **45**, 1231.
[27] Roos Y. and Karel M., *Int. J. Food Sci. Technol.*, 1991, **26**, 55.
[28] Herrington T.M. and Branfield A.C., *J. Food Technol.*, 1984, **19**, 427.
[29] Cammenga H.K. and Steppuhun I.D., *Thermochim. Acta.*, 1993, **229**, 253.
[30] Imberty A., Buleon A., Vihn T., and Perez S., *Starch/Staerke*, 1991, **43**, 375.

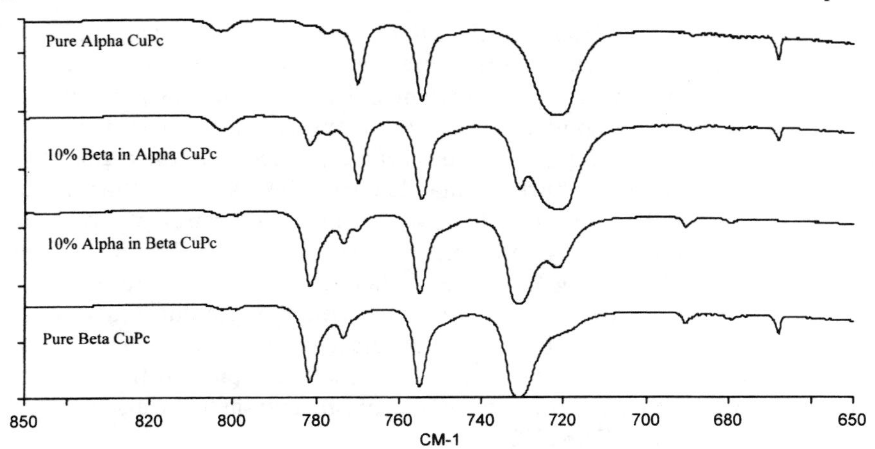

Figure 1 *Transmission infrared (Nujol™ mull) spectra of copper phthalocyanine (CuPc) alpha and beta polymorphs and wt/wt mixtures of the polymorphs. Spectra offset for clarity*

Choice of a correct sample preparation/presentation technique is imperative if the polymorphic state is not to be affected or changed. In solution, forms can convert, as they may do when prepared as an alkali-halide disk. Mulling techniques are better, generally being gentler treatments. This is of particular relevance with explosives such as cyclotetramethylenetetranitramine (HMX),[31] for which the different states have greatly varying sensitivity to detonation. Four different polymorphs can be measured quantitatively in samples prepared as mulls.

DRIFT (diffuse reflectance infrared Fourier transform spectroscopy) and PAS (photoacoustic spectroscopy) techniques, see Chapter 4, are more recent sampling approaches making headway for polymorphism studies, since they require minimal or no sample preparation. However, in both cases, particle size can affect reproducibility for qualitative and quantitative measurements. Raman spectroscopy, with its minimal sample handling requirements, is also a favoured technique,[32] although its application is limited when visible lasers are used because of fluorescence effects in colourless powders or burning/charring of strongly coloured pigments or dyes. Near-infrared (NIR) FT-Raman has extended the range of samples susceptible to examination,[33] but some problems still remain,[34] and some uncertainty exists around quantitative aspects of the technique, which are still the subject of several investigations.

[31] Corbett J.K., in *Laboratory Methods in Vibrational Spectroscopy*, 3rd Edn, ed. Willis H.A, van der Maas J.H. and Miller R.G.J., J. Wiley, Chichester, UK, 1987, ch. 17.
[32] Neville G.A., Beckstead H.D., and Shurvell H.F, *J. Pharm. Sci.*, 1992, **81**, 1141.
[33] Dent G., *Spectrochim. Acta Part A*, 1995, **51**, 1975.
[34] Asselin K.J. and Chase B., *Appl. Spectrosc.*, 1994, **48**, 699.

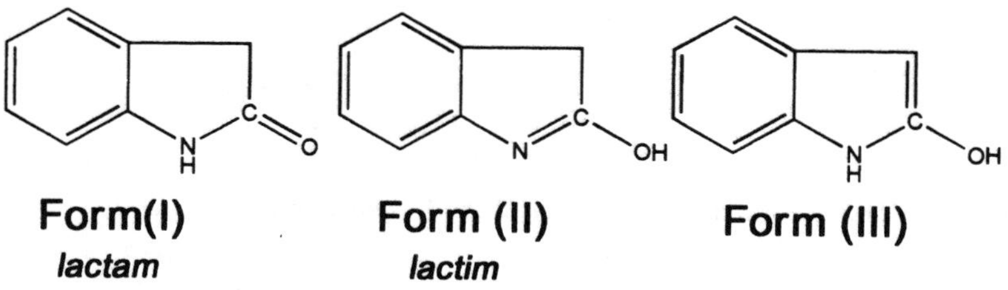

Form(I) *lactam*　　Form (II) *lactim*　　Form (III)

Figure 2 *Oxindole molecule, tautomeric structures*

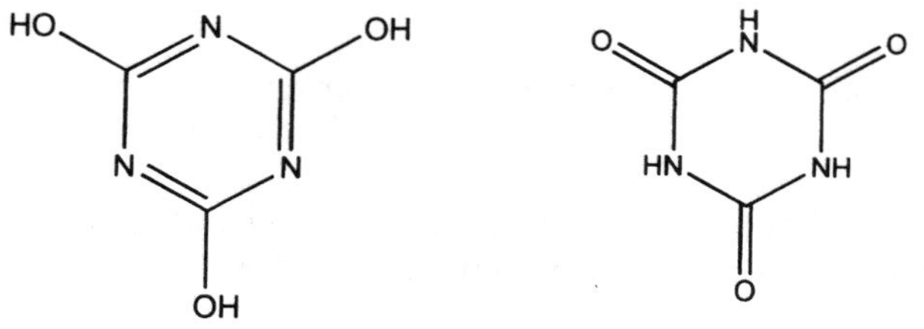

Figure 3 *Cyanuric acid, tautomeric structures*

Tautomerism

Another equilibrium state which can also be studied by vibrational spectroscopy is tautomerism. NMR is more sensitive for quantitative studies, but vibrational spectroscopy can easily and sensitively study the solid state. Typical of these are *keto–enol*, *lactam–lactim*, and *thio–thioamide* tautomers. These all contain characteristic groups, with well-defined bands which make the predominant species fairly clear. An awareness of tautomerism will be key in many studies, and to illustrate this four examples of tautomerism are discussed briefly below.

Figure 2 shows the oxindole molecule, which exhibits both *lactam* and *lactim* forms. *Lactam–lactim* systems, as with other *keto–enol* systems, may be characterised by the >C═O group of the *lactam* (*keto*) form and the —N═C—OH group in the *lactim* (*enol*) tautomer. Oxindole also exists in a third form, which can be distinguished readily through bands characteristic of its —C═C—OH entity, since, the νOH groups have characteristic wavenumber positions. Infrared spectroscopy has been used to show that in the solid state and in chloroform solution oxindole exists entirely in the *keto* form.[35]

Cyanuric acid is another well-known compound which exhibits (*amido–imidol*) tautomerism, see Figure 3. Several reference spectra in commercial

[35] Kellie A.E., O'Sullivan D.G., and Sadler P.W., *J. Chem. Soc.*, 1956, **2202**, 3809.

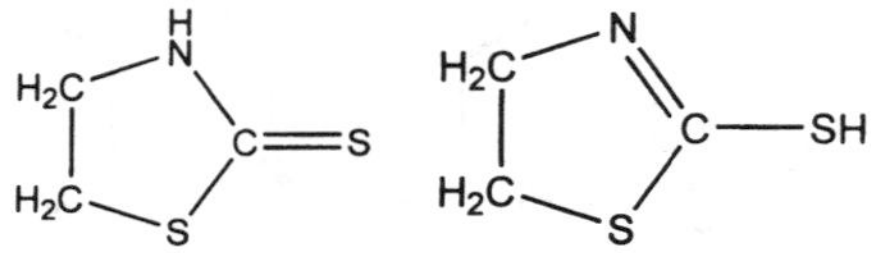

Figure 4 *Mercaptothiazoline (thiazoline-2-thiol), tautomeric structures*

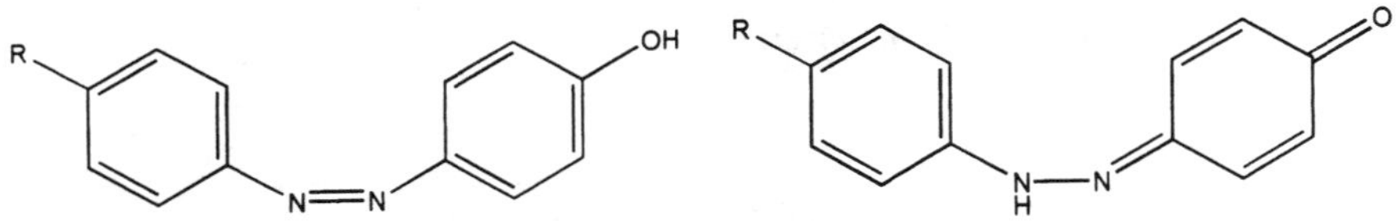

Figure 5 *Tautomeric structures of an azobenzene derivative*

collections[36–38] are dominated by a strong carbonyl band at 1740 cm^{-1} and NH absorptions, yet many reactions in solution depend on the presence of the —OH groups!

As with *keto–enol* forms, *thio* equivalents show similar effects, see Figure 4, although the >C=S and —SH group bands are less intense than those of the equivalent oxygenated species. The >C=S group has bands near 1140 cm^{-1} with the νSH band occurring near 2580 cm^{-1}. This latter band can be very weak in the infrared spectrum, and absence of the band is not always indicative of absence of the group. In solution spectra a strong νNH band appears around 3400 cm^{-1} supporting the existence of the *thio-ketone* tautomer.[39]

The fourth example of tautomerism concerns *azo–hydrazone* tautomerism, see Figure 5. The *azo* group, (—N=N—), a common moiety in many dyes, being symmetrical, is very weakly absorbing in the infrared spectrum but can give rise to a strong band in Raman spectra,[40] occurring at approximately 1440 Δcm^{-1}. When the *hydrazo* form is made, the —C=N— band can be seen in the Raman spectrum at approximately 1605 Δcm^{-1}.

Hydrogen Bonding

Infrared spectroscopy can provide much information on molecular interactions involving hydrogen bonding. Many methods are used to study hydrogen bonding;[41] infrared has the advantage of being direct and sensitive. Pioneering

[36] The Aldrich Library of FT-IR Spectra, Edn 1, Vol. 2, ed. Pouchert C.J., Aldrich Chemical Company, Milwaukee, USA, 1985, p. 849A.
[37] Spectrum No. 1403, *Infrared Spectra Collection*, The Coblenz Society, Kirkwood, Mo, USA.
[38] Spectrum No. 18149, *Infrared Grating Spectra Collection*, Sadtler Division, Bio-Rad Laboratories, Hemel Hempstead, UK.
[39] Flett M.St.C., *J. Chem. Soc.*, 1953, 347.
[40] Armstrong D.R., Clarkson J., and Smith W.E., *J. Phys. Chem.*, 1995, **99**, 51, 17825.
[41] Vinogradov S.N. and Linell R.H., *Hydrogen Bonding*, Van Nostrand Reinhold, New York, USA, 1971.

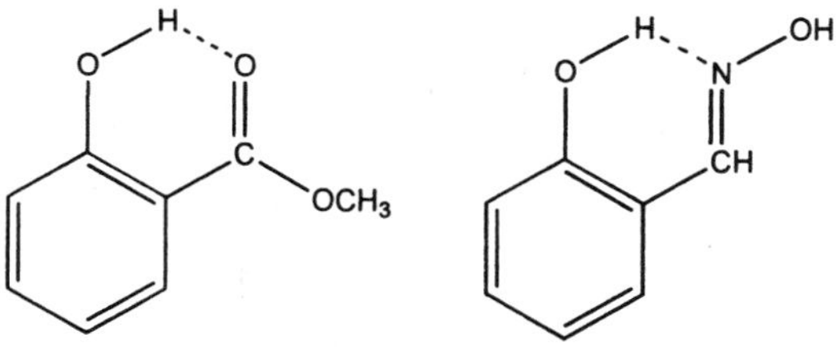

Figure 6 *Methyl salicylate and salicylaldoxime structures*

work in this area was carried out by Pauling.[42] In the vapour state, or in dilute solution in inert solvents, the —OH and —NH groups clearly show well-defined characteristic bands at high wavenumbers. As these groups form bonds, the bands broaden and move to lower wavenumbers. Many theories have been put forward as to the relationship between hydrogen bond length, strength, and/or formation. A very simplistic comment is that within a series stronger bonds give rise to bands that are broader and move further, but this is strictly limited in application.

The hydrogen bond most clearly recognised and easily studied is the intermolecular bond; less easy is the intramolecular bond. In the liquid or solid phase it is difficult to distinguish between inter- and intra-molecular hydrogen bonding. However, in weak solutions in inert solvents such as carbon tetrachloride the molecules may be separated far enough that intermolecular effects can be eliminated. The most obvious bondings are the —O—H···O and the —O—H···N groups. Electronic and steric factors affect a bond being formed, but the 6-membered ring is highly favoured, methyl salicylate and salicylaldoxime being good examples, see Figure 6.

The methyl salicylate spectrum shows strong effects on both the νOH and νC=O bands, especially in the neat liquid state, see Figure 7. Usually the νOH band would be expected to be centred at 3400–3300 cm^{-1} and the carbonyl at 1710 cm^{-1}. The effects of the hydrogen bonding can be seen quite clearly in the shifts to lower wavenumbers. The intramolecular hydrogen-bond band is broader and weaker than the free νOH ($\sim$3600 cm^{-1}) would be to such an extent that in some compounds, such as *o*-dihydroxyazobenzene, the —OH bonded band is not detected readily in the IR spectrum. However, Raman spectra clearly show bands from the azo group with shifts due to bonding.

Intermolecular hydrogen bonding occurs much more commonly, and is more readily observed, though little understood, in most spectra. Again the —OH and —NH groups are most commonly involved in intermolecular hydrogen bonding. As with the intramolecular bonding, intermolecular hydrogen bonds are best studied in weak solutions using inert solvents. Some workers[43] claim that this is the only way to study these effects. In an ideal world this would be

[42] Pauling L., *The Nature of the Chemical Bond*, 3rd Edn., Oxford University Press, London, UK, 1960.

[43] Taylor P., ICI Pharmaceuticals, Private Communication, 1980.

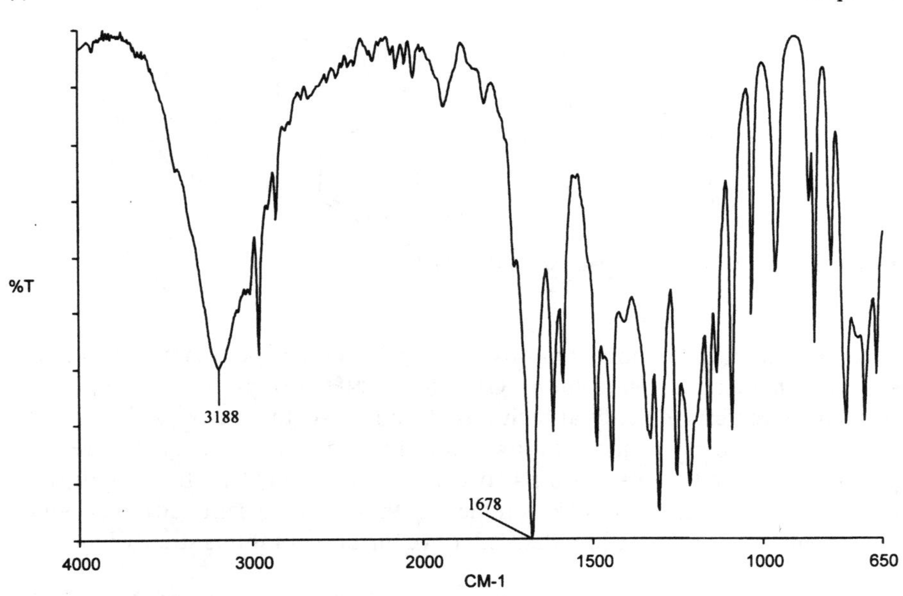

Figure 7 *Infrared spectrum (capillary layer film) of methyl salicylate, showing hydrogen-bonded νOH (*3188 cm^{-1}*) and shifted ν$C{=}O$ (*1678 cm^{-1}*)*

true, especially if quantitative studies of bond lengths and energies are to be made. This is not always possible, however, as many compounds are not soluble enough in inert solvents, and/or if bonding in crystals is to be studied, the act of dissolution destroys the bond of interest. Inert solvents are preferred whenever possible, since many bonds are quite weak, and sample–solvent bonding can be relatively strong. Toluene is a good solvent, but bonding to the ring does occur. In the 3100–2800 cm^{-1} region, aromatic and aliphatic bands can obscure the hydrogen-bond bands. Aromatic ring overtones in the 3600–3100 cm^{-1} region can also overlap with the hydrogen-bond bands. The free νOH band occurs at approximately 3500 cm^{-1} or above for most compounds. As two molecules form a bond to become a dimer, a weak broader band appears at a lower wavenumber. Increasing the concentration of the solution, causes this band to move further to lower wavenumbers, as trimerisation and polymerisation through the bond builds up, see the oxime spectrum of Figure 8. In a study of these effects, one of the authors made quantitative studies that related bond area to relative amounts of free dimer, trimer or polymerised material.

Varying the path-length, concentration, or temperature affects the band position and strength. Changing from an inert, non-polar solvent to solvents with weak interactions, such as ether or toluene, also affected the band position and strength.[44] *N*-Methylpyrrolidone may be used as a probe. By adding small

[44] Dalton R.J. and Seward G.W., in *Reagents in the Minerals Industry*, Institute of Mines and Metallurgy, London, UK, 1984, p. 107.

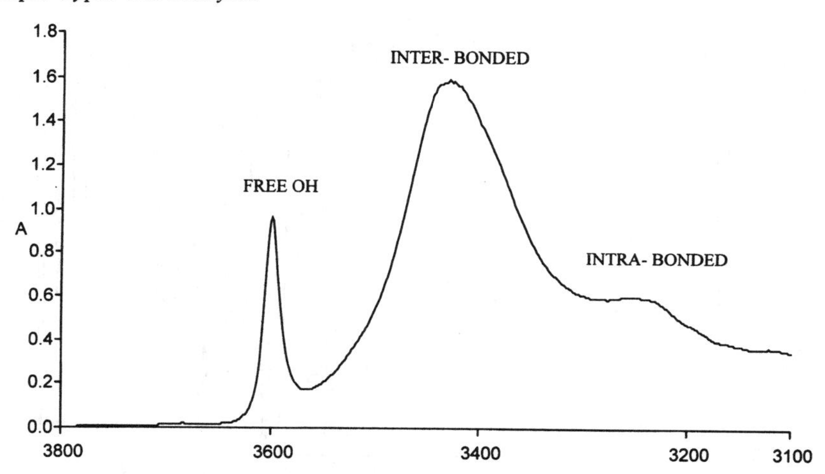

Figure 8 *Salicylaldoxime (5% hexane solution;* 0.5 mm *path-length cell) infrared spectrum, showing free, inter- and intramolecular bonded* νOH *bands*

amounts to the solution the position of the pyrrolidone carbonyl band moves as molecules preferentially bond to the carbonyl group rather than to each other. This is sometimes a useful way of testing bond strength.[45] By mixing molecules that bond to themselves and each other in varying concentrations, *e.g.* diphenylamine and phenol, some quite complex and fascinating hydrogen-bonding studies can be carried out.

Neat liquids and solids generally show strong intermolecular hydrogen bonding. Moving the molecules apart by diluting in weaker and weaker solutions, usually produces the free νOH and νNH bands. Two exceptions to this are carboxylic acids and sterically hindered compounds. Carboxylic acids form very strong intermolecular hydrogen bonds; a band at 960 cm^{-1} due to the dimer often appears in the spectra. Even in the vapour phase many carboxylic acids still appear in the dimer state. At the other extreme, sterically hindered phenols, such as 2,6 di-t-butyl phenol, see Figure 9, show a strong free νOH band at approx. 3600 cm^{-1} in the solid state due to the inability of the OH group to physically bond with another molecule.

Racemates

There are two principal types of crystalline racemates.[46,47] Conglomerates are simple juxtapositions of crystals of two enantiomers (optical isomers), and are mixtures separable by physical means. Much more common are racemic

[45] Morris J.J., ICI Pharmaceuticals, Private Communication, 1980.
[46] *Chirality in Industry*, ed. Collins A.N., Sheldrake G.N., and Crosby J., J. Wiley, Chichester, UK, 1992.
[47] Jacques J., Collet A., and Wilen S.H., *Enantiomers, Racemates, and Resolutions*, J. Wiley, New York, USA, 1981.

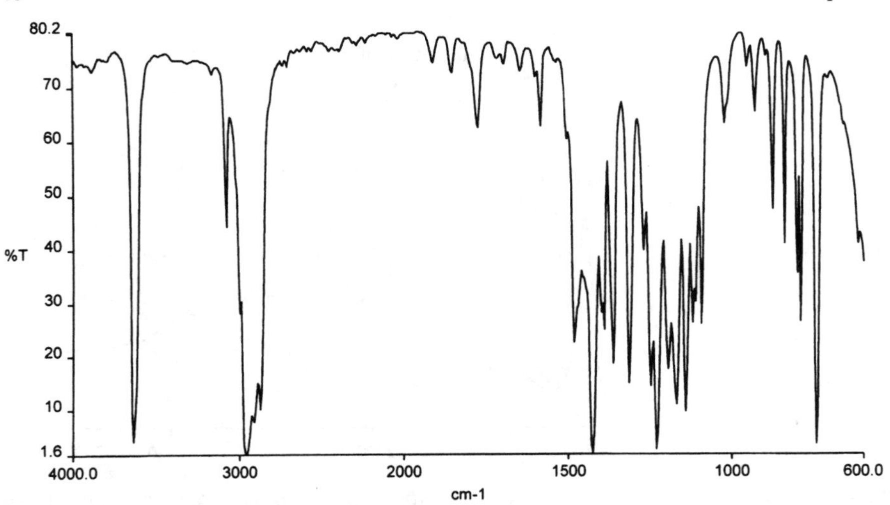

Figure 9 *Infrared spectrum of 2,6 di-t-butyl phenol*

compounds, whose crystals contain two enantiomers, in equal numbers, which comprise a unit cell. The production of pure enantiomers of drugs is of growing importance in the pharmaceuticals industry. In 1985, it was estimated that 85% of drugs with an asymmetric carbon were in a racemic form.[46] This has since been reduced to approximately 25%. The infrared spectra of enantiomers in conglomerates are directly superposable. In racemic compounds the resultant spectrum will be significantly different from that of the two enantiomers, see example of mandelic acid (phenylglycollic acid, $C_6H_5CHOH.CO_2H$) in Figure 10.[47] Not all optical isomers have detectable differences in their infrared spectra.

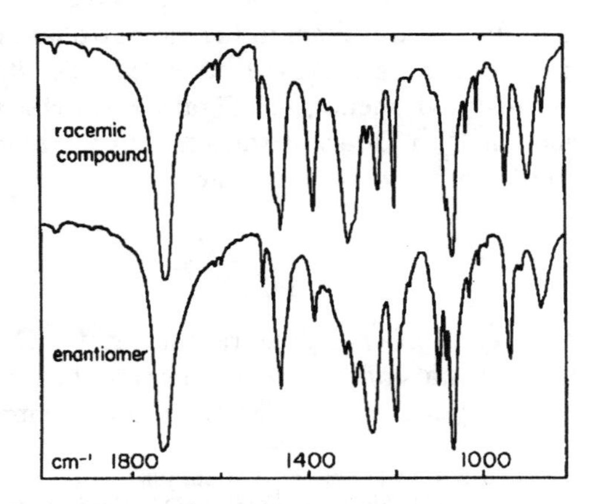

Figure 10 *Infrared spectra of mandelic acid*
(Reprinted from ref. 47 by kind permission of John Wiley & Sons Inc. © 1981)

Consideration must also be given to the appropriate use of other techniques, such as X-ray diffraction, to check that racemates have been formed. Spectral differences in the fingerprint region could be due to polymorphism, or other crystal effects.

3 Formulations

Many of the studies mentioned so far have significance in the pharmaceutical industry, where the physical form of the drug can greatly affect the metabolisation rate. These studies are carried out primarily on the drug itself, that is, the *active agent*. A very large proportion of chemicals sold are not pure chemicals but formulations which render the active agent most effective. Drugs in tablet form are diluted with inorganics and edible starches. Implants with time delayed drug release are becoming common. More widely used are those agricultural products with granules for slow release of lawn feed and weedkiller. Dyes and pigments are usually sold as solutions with dispersing agents or dedusting agents present. In the printing industry inks have to be very precisely applied, and have to be formulated with the correct rheological properties, as do paints, particularly the non-drip type. Surfactants and detergents are frequently multi-component formulations. In all these areas vibrational spectroscopy can contribute by identifying individual components, sometimes following from chromatographic separation (see Chapter 7), or by pattern-matching the formulations for both Quality Control and Technical Service concerns. However, a complete complex mixture analysis is rarely solvable by one particular technique, and NMR studies coupled with MS investigations tend to be the prime tools for many industrial 'deformulation' investigations of the organic components of an 'unknown' formulated product. These are often applied after a solvent or other treatment, perhaps to separate inorganics from organics, or ionics from non-ionics.

4 Polymers

Vibrational spectroscopy is a major contributor to the study and characterisation of samples from the petrochemical, polymer[48,49] and related industries. Samples may include raw materials from the cracker plant or from refining, catalysts, elastomers, and additives, with polymer and copolymer in various forms, including powder, granules, films, fibres, mouldings, latexes, dispersions and laminates.

There is nothing very peculiar about homopolymers when it comes to interpreting their spectra. One can simply view polymers as sub-structural units, for instance, as a regular sequence of the repeat unit, and apply the normal rules to correlate the major characteristic bands with functional groups.

[48] Haslam J., Willis H.A., and Squirrell D.C.M., *Identification and Analysis of Plastics*, 2nd Edn, Butterworth, London, UK, 1972.

[49] Chalmers J.M. and Everall N.J., in *Polymer Characterisation*, ed. Hunt B.J. and James M.I., Blackie Academic, Glasgow, UK, 1993, ch. 4.

One can thus work out, at least generically, what the material is, *e.g.* is it an aliphatic or aromatic hydrocarbon polymer, a polyester, a polycarbonate, a polyamide, a polyimide, *etc.*, remembering that negative evidence can be as important as positive features. Full identification is usually achieved through familiarity or pattern-matching a spectrum to that of a standard reference spectrum, such as one contained in a commercial collection. Remember that many commercial polymeric materials/products are formulations, and may contain additives, such as antioxidants and UV stabilisers, processing aids, such as plasticisers and lubricants, and fillers, for fire-retardancy or mechanical strength, as well as catalyst residues. These may often add significant intensity to the spectrum being interrogated. One other point worth emphasising is that polymers in themselves are not single entities. They have a molecular weight distribution, and for copolymers there may be a wide distribution of comonomer compositions, which may or may not vary with molecular weight. (Different molecular weights may also exhibit different solubilities, so if a solvent/solution treatment is employed, one may need to ensure that the sample examined is representative of the bulk average.)

The fine detail of homopolymer spectra can supply important second-order information, when viewed objectively and with proper consideration.[48] For instance, different nylons, (secondary amide units joined together with differing sequences of contiguous $—CH_2—$ groups), have distinctive fingerprint spectra. There are many sharp, weak bands in the region 1500–900 cm^{-1} which appear to be characteristic of particular nylon types. They are indeed characteristic, but unfortunately (from an unequivocal identification point of view), they are strongly dependent on and characteristic of the crystalline form of a nylon. There are different crystalline forms of many nylons, each of which may have a significantly different spectrum in this region. So identifying a nylon type solely from its infrared spectrum is sometimes fraught with danger, especially as many commercial materials may exist as nylon copolymers or blends.

In other circumstances, this fine detail can be put to extremely good use. This is particularly so for polyethylene types. Fingerprint spectra of aliphatic hydrocarbon polymers, defined as those spectra for which the absorbance band maxima of all the bands are evident, are of minimal use, and it is better to record the spectrum of a thicker specimen (typically 0.1 mm or greater thickness) in order to increase the contrast of bands other than the νCH near 2900 cm^{-1} and δCH near 1460 cm^{-1}. Low density polyethylenes (LDPEs) are characterised by a relatively high $—CH_3$ content, which may be recognised by the intensity at 1378 cm^{-1} of the $\delta_{sym}C—CH_3$. Accompanying this may be a particular pattern of bands attributable to unsaturated moities in the polymer chain or at an end, namely chain ($—CH{=}CH—$) at 965 cm^{-1}, vinyl ($—CH{=}CH_2$) at 910 cm^{-1}, and vinylidene (pendant methylene) ($>C{=}CH_2$) at 888 cm^{-1}, the relative intensities of which will differ according to the manufacturing route (catalyst used); a common type has the 888 cm^{-1} as the most prominent. By contrast, high density polyethylenes (HDPEs) are characterised by a very much lower $—CH_3$ content, and high vinyl unsaturation, which is evidenced readily by both the band at 910 cm^{-1} and its weaker counterpart at 990 cm^{-1}.

However, fingerprinting PEs by IR can still be complicated, since linear low density polyethylenes (LLDPEs), of which there are several types, are another important sub-class of commercial products. LLDPEs may be considered essentially as a HDPE modified (copolymerised) with a low level of another alkene, such as propylene, butene, hexene, or octene, some of which give rise to more distinctive spectra than others, but all of which give rise to an increase in $—CH_3$ content; this appears particularly enhanced for the propylene modification, since the δCH_3 absorption coefficient of the

$$\begin{array}{l} —CH—CH_2— \\ \;\;\;\,| \\ \;\;CH_3 \end{array}$$

group is much greater than for the other alkane side groups. Again, it should be noted that many commercial products are/contain blends of polyethylenes. Infrared (or Raman) *cannot* readily distinguish blends from homopolymers. Complementary information from NMR or DSC is a must if a commercial material is to be characterised unequivocally.

In HDPE the methyl groups are present essentially as end groups as opposed to being mainly part of side-chain branches, as in LDPE. In many polymers, particularly polycondensation-type polymers, the functionality of an end group or the end group balance in the finished material can be a key property in determining polymer thermal/processing stability. Excess of one particular monomer may be used in the synthesis to try to secure a particular end group in the final product, or, in some circumstances, chain transfer or chain terminating agents may be employed. Characterising and quantifying the concentration of particular end groups in certain polymers is an important analytical use in industry of vibrational spectroscopy. In this respect, a key requirement for the successful analysis of the presence and/or concentration of any hydroxyl end groups is that the specimen examined is dry. Polymers, for instance PET, may exist in equilibrium with significant amounts of adsorbed water, as much as $\sim 2\%$ in some cases, which will interfere with any measurement of hydroxyl end group, and must therefore be removed (*e.g.* by vacuum drying) before the analysis is undertaken. (The presence of adsorbed H_2O, can often be ascertained by merely leaving a polymer film in the sample beam of a dry, purged FTIR spectrometer and noting the decrease of appropriate absorption bands with time!).

Isomerism can be seen clearly to have a considerable effect on the vibrational spectra of a polymer. For instance, the three aliphatic hydrocarbon polymers, polyethylene, polypropylene, and polyisobutene, are chemical isomers, but each has very distinctive spectra. Distinct differences are also observed in the case of isomerism about double bonds, especially for the well-known out-of-plane deformation modes of hydrogen atoms attached to the carbon atoms of double bonds. This is well demonstrated in the spectra of various polybutadienes, in which appropriate characteristic bands may be evident for the *trans* (965 cm^{-1}) and *cis* ($\sim 680 \text{ cm}^{-1}$) 1,4 forms and the 1,2 form (910 cm^{-1} and 990 cm^{-1}).

For 1,2 polybutadiene, two different chain structures are possible, which yield different spectra. The polymer molecule can exist in two different stereoisomeric forms, the isotactic and syndiotactic species. The structural groups are the same in both spectra; the differences arise from the change from isotactic to syndiotactic. It would appear therefore that it is possible to distinguish isotactic and syndiotactic forms of polymers by vibrational spectroscopy.

This point requires clarification and understanding. If we draw a polypropylene molecule as a straight chain with all the methyl groups on one side of the main chain, then this is the *isotactic* species, and that most commonly found in commercial materials. If the pendent groups alternate from one side to the other of the main chain, this is the *syndiotactic* species, which is not sold commercially. Another less regular species, known as *atactic*, which is sometimes sold commercially, has a spectrum (and properties) somewhat similar to that of an amorphous polypropylene.

isotactic polypropylene:

$$\begin{array}{llllllllllllll}
(\text{—CH} & \text{—CH}_2\text{—} & \text{CH} & \text{—CH}_2\text{—} & \text{CH} & \text{—CH}_2\text{—} & \text{CH} & \text{—CH}_2\text{—} & \text{CH} & \text{—CH}_2\text{—} & \text{CH} & \text{—CH}_2\text{—} & \text{CH} & \text{—CH}_2\text{—})_n \\
\quad | & & | & & | & & | & & | & & | & & | & \\
\text{CH}_3 & & \text{CH}_3 & & \text{CH}_3 & & \text{CH}_3 & & \text{CH}_3 & & \text{CH}_3 & & \text{CH}_3 &
\end{array}$$

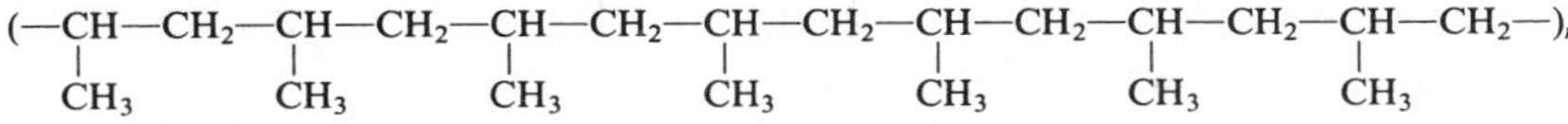

syndiotactic polypropylene:

$$\begin{array}{llllllllllllll}
 & & \text{CH}_3 & & & & \text{CH}_3 & & & & \text{CH}_3 & & & \\
 & & | & & & & | & & & & | & & & \\
(\text{—CH} & \text{—CH}_2\text{—} & \text{CH} & \text{—CH}_2\text{—} & \text{CH} & \text{—CH}_2\text{—} & \text{CH} & \text{—CH}_2\text{—} & \text{CH} & \text{—CH}_2\text{—} & \text{CH} & \text{—CH}_2\text{—} & \text{CH} & \text{—CH}_2\text{—})_n \\
\quad | & & & & | & & & & | & & & & | & \\
\text{CH}_3 & & & & \text{CH}_3 & & & & \text{CH}_3 & & & & \text{CH}_3 &
\end{array}$$

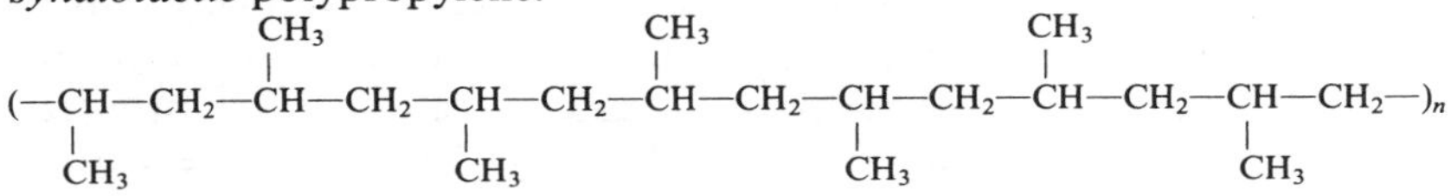

Like most isotactic polymers, isotactic polypropylene coils up into a helix. The isotactic polypropylene helix is a 3:1 helix. Viewed from the end of the chain, the methyl groups are arranged at the corners of an equilateral triangle, because in three monomer (repeat) units the structure has completed one turn of the helix. In the crystalline form this material has a rigid, rod-like structure that does not allow any rotation about single bonds. The bond angles are known with high precision and the force constants of the bonds are also well known, so a calculated spectrum is a good fit with the experimental spectrum. The calculated molecular vibrations break down into two distinct groups: 'A' modes, the parallel modes, in which during the vibrations of the molecule the direction of the dipole moment change is essentially parallel to the helix axis; and, 'E' modes, in which the direction of the transition moment (dipole moment change) is essentially perpendicular to the helix axis.

Polarised radiation can therefore be used on drawn (stretched) polymer samples to interrogate the directional properties of vibrational modes and assign bands.[49] Alternatively it can be used to examine the degree of *molecular orientation* of oriented specimens, see Figure 11, such as commercial polymer films, many of which are biaxially oriented. If the spectrum of a one-way stretched sample of polypropylene is first measured with the electric vector of the radiation along (parallel to) the direction of draw, then we get the 'parallel' spectrum, in which the parallel bands feature most strongly, whereas the perpendicular modes are weak or non-existent. If the electric vector of the

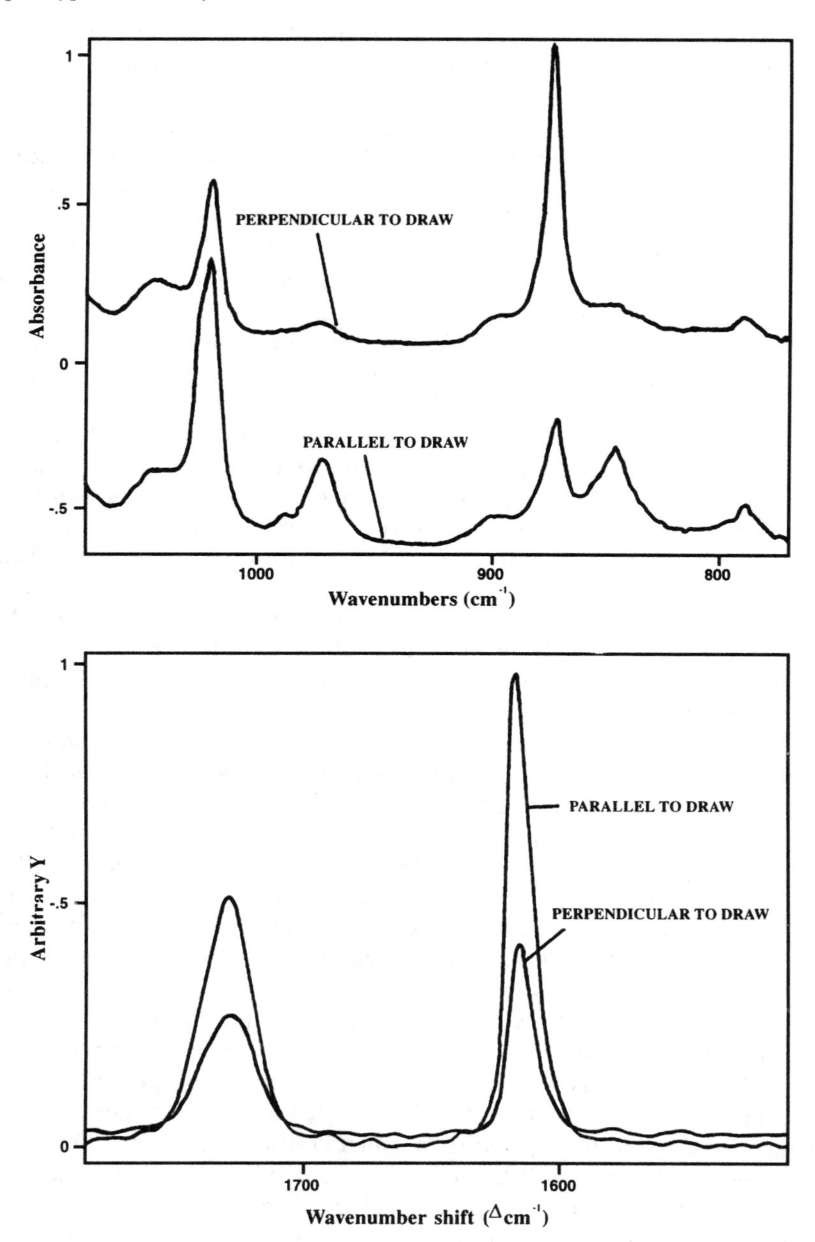

Figure 11 *Comparison of the dichroic infrared and Raman spectra of an oriented PET film.* Top: *infrared spectra measured with the electric vector parallel and perpendicular to the draw axis.* Bottom: *Raman spectra with laser polarised parallel and perpendicular to the draw axis. The spectra show the* 1615 Δcm^{-1} *(Raman) and* 1018 cm^{-1} *(IR) bands to have parallel character, while the* 875 cm^{-1} *(IR) and* $\nu C{=}O$ *(Raman) are shown to have perpendicular character with respect to the polymer draw axis*

polarised radiation is now rotated through 90°, the perpendicular modes become dominant, while the parallel bands are now comparatively much weaker or disappear.

The fingerprint spectra, infrared or Raman, of crystalline isotactic, syndiotactic and atactic polypropylene are very different. The distinct differences are due to *conformational isomerism*, and not directly to tacticity or crystallinity. The spectrum of crystalline isotactic polypropylene arises because it has a particular shape in space, and does not directly relate to the fact that it happens to be isotactic. Vibrational spectroscopy does not tell us anything directly about tacticity, only implicitly, (whereas NMR gives direct information). For instance, syndiotactic polypropylene can, by special means, be prepared in both helical and planar zigzag arrangements; these have very different spectra. Another well-documented example is that of isotactic polybutene (polybut-1-ene). This has two common crystalline forms, one having a three-fold (3_1) helix-like polypropylene, and the other a four-fold (4_1) helix, that is four repeat units for one complete turn. The fingerprint spectra of these two crystalline forms are again significantly different due to change of molecular conformation. (In fact, when a film is prepared from the melt, the Type II form is the first modification formed. This is the tetragonal 4_1 helix form, which is only stable at high temperatures. This slowly reverts to the higher density Type I form, which has the rhombohedral 3_1 helix crystalline arrangement, and is stable at room temperature. This process can be readily observed over a few days by infrared spectroscopy.)

Conformational isomerism must not be confused with crystalline/amorphous changes. The spectral differences observed in the fingerprint region between crystalline and amorphous regions of polymers mostly arise not directly from crystallinity, but from the presence of different conformations in crystalline and amorphous regions. In PET (polyethylene terephthalate), for example, different conformations can be made by rotation about the C—C bond of the alkyl portion of the molecule. In amorphous PET, both the *trans* and *gauche* isomers exist. Stretching a piece of amorphous film increases the *trans* content at the expense of the *gauche*, with no change in the crystallinity, but with a corresponding change in the vibrational spectrum. The most likely consequence of a change (increase) of crystallinity will be a reduction (narrowing) in the half-band width of some bands in the conformer in the crystalline phase. Indeed, the half-band width of the Raman νC=O band of PET has been correlated with polymer density, and by implication polymer crystallinity, see pp. 305–306 and Figure 39(b) of Chapter 6. If the infrared spectra of an essentially amorphous, an essentially oriented amorphous, and a bi-axially oriented, crystalline PET are compared, many differences can be observed associated with conformational changes and molecular ordering and packing.

In some circumstances, there are peaks within the spectrum of a polymer that are quite definitely associated with crystalline regions, because they originate from the interaction of two or more chains in the crystallographic unit cell. Polyethylene is perhaps the most quoted example. In polyethylene, because the —CH_2—CH_2— unit has a centre of symmetry, the rule of mutual exclusion applies; the modes which are active in the infrared are absent in the Raman spectrum, and vice-versa. As most frequently encountered, polyethylene crystallises into an orthorhombic arrangement, with two chains per unit cell. The various vibrations of these two chains can occur symmetrically or antisymmetrically, and there is a slight energy difference between these two, so there is splitting of the bands into pairs. This is particularly easy to see in the case of the $—CH_2—$ rocking mode, for which two bands (a doublet) occur at essentially 720 cm^{-1} and 730 cm^{-1}. Commercial samples of some polymers, *e.g.* polyethylene, may cover a wide range of crystallinities (and densities); consequently relative band intensities and widths may differ appreciably between spectra. This is particularly so for the ratio of the pair of bands at 720 cm^{-1} and 730 cm^{-1} for polyethylene. Structural defects along a polymer chain will hinder good lateral packing, so as chain branching increases crystallinity decreases. Consequently HDPE features a more intense, sharper doublet than does LDPE. The chains in the crystalline regions of polymers adopt a more or less fixed conformational arrangement, usually an all-*trans* arrangement for alkyl units, as part of a planar zigzag or regular helical structure. On the other hand, chains in amorphous regions are conformationally irregular and unordered, with many *gauche* bands in the case of alkyl chains, and a variety of conformational structures occur over a distance of a few repeat units. The vibrational frequencies of each of these conformational units differ slightly and when a range is present the overall effect is to broaden the band. Polymorphism is not uncommon among polymers that are largely crystalline. For example, polyoxymethylene normally occurs as a hexagonal system and individual chains have nine units in five turns, *i.e.* a 9_5 helix. This can be shown to have 33 spectroscopically active modes. A second form can be obtained from some types of polymerisation, in which the chains are planar zigzag and which has an orthorhombic unit cell. This arrangement gives 20 spectroscopically active modes.

So far we have considered essentially only homopolymers, yet many commercial materials are co- or ter-polymers. If one compares the spectra of two homopolymers, which are amorphous or of very low crystallinity, then the bands in both spectra relate largely to the polymer molecular functional groups. The spectrum of a copolymer may therefore be virtually indistinguishable from the appropriate summation of the spectra of the two homopolymers. Consequently, it may be extremely difficult to distinguish between a blend and a copolymer. So *beware of over-interpretation; seek complementary evidence.* Sequencing, normally the realm of NMR, can influence spectra and sometimes be observed. Propylene/ethylene copolymers are very common and are commercially available in a number of different forms, and there are significant differences in their infrared spectra according to the way the copolymers are made. These are observed in the CH_2 rocking mode region and relate to the environments of the copolymerised ethylene units, varying from the crystalline 'polyethylene' doublet in a block copolymer through to a single band at 733 cm^{-1} due to three contiguous CH_2 groups, *i.e.* an isolated ethylene unit. If one compares closely the infrared spectra of 1% VA and 28% VA ethylene/vinyl acetate copolymers (EVA), then the bandwidths of the νC=O and νC—O increase significantly with increasing VA content, probably due to bunching or adjacency of the VA groups. (The half-widths for 28% VA EVA copolymer blended with PE remain the same as for the undiluted EVA!).

Other key areas of polymer studies of industrial importance by vibrational spectroscopy include monitoring polymerisation rates and resin cure and cross-linking, and understanding and monitoring the mechanisms by which polymers deteriorate, such as the processes and molecular species involved in thermal, thermal-oxidative, photo, photo-oxidative, or radiation-induced damage.[49]

5 Fibres

Optical microscopists have long been identifying fibres by their shape, general appearance, birefringence, reponse to solvents, melting behaviour, *etc.* Vibrational spectroscopists have tried many ways to record spectra from fibres and coatings on fibres, because of their industrial importance. Identifying man-made or natural fibres is often critical in tracing sources of contamination, and has always been to the fore in forensic science, but their study is crucial to many industries. Fibres may need to be examined individually or as woven cloth, prepared for deliberate dyeing or staining. The dyes, pigments, and formulating agents all have differing surface interactions and effects that may be studied. The printing industry is becoming ever more demanding, producing intricate precise patterns for high-technology applications, particularly in the photocopying field. The action of cleaning agents or cosmetic treatments on human hair has direct relevance to our feelings of well-being. Inorganic fibres, such as carbon, have particular importance as reinforcing agents in composite materials used in demanding applications, for example in aerospace or military environments.

Because of the simplicity of sampling, Raman microscopy has been widely used in fibre studies for many years, with Resonance Raman micro-techniques now proving very sensitive for identifying dyes on fibres.[50] For a long time the most effective way of examining fibres by infrared spectroscopy was to wrap a bundle of fibres around an ATR crystal (ATR is attenuated total reflectance, see Chapter 4). The advent of the FTIR-microscope along with the ATR lens, or the use of a diamond window compression cell, has now made fibre fingerprinting by infrared spectroscopy a relatively much easier task, see Chapter 6. PAS has also been used as a tool for identifying surface coatings on fibres.[51]

6 Surface Analysis

Surface analysis in an industrial context encompasses a wide and diverse arena of application studies and problem-solving work, such as effect treatments, weathering/degradation behaviour, diffusion, corrosion, additive bleed, adhesion, stratification, printability, *etc.* 'Surface' also means different things to

[50] White P.C., Munro C.H., and Smith W.E., *Analyst*, 1996, **121**, 835.
[51] Urban M.W., Chatzi E.G., Perry B.C., and Koenig J.L., *Appl. Spectrosc.*, 1986, **40**, 8, 1103.

different investigators, and occasional ambiguities arise as to whether one is strictly characterising a surface or analysing a surface layer! In large industrial laboratories Surface Analysis *per se* often comes under the domain of ESCA, SIMS, *etc.*, which require large instruments, operating under high vacuum[52] – (ESCA, Electron Spectroscopy for Chemical Analysis; SIMS, Secondary Ion Mass Spectroscopy). These high-cost techniques provide surface specific elemental or molecular spectra over the top 50–100 Å. However, vibrational spectroscopy has been and will continue to be used for many years as a front-line, cost-effective tool in industry to supply surface or surface layer information, albeit, for other than specialised approaches, providing information from layer thicknesses of the order of 0.5 μm to approximately 20 μm, depending on the sampling technique used. Vibrational spectroscopy is also in many ways the only technique which can provide molecular interaction details.

Much of the pioneering work using infrared spectroscopy for surface characterisations was carried out in the 1950s,[53–57] with major reviews published in the 1960s.[58,59] A landmark in its industrial use came with the development and acceptance of internal reflection spectroscopy (IRS) techniques,[60] often more commonly referred to as ATR spectroscopy, see pp. 144–153. IRS accessories have now been employed routinely in many laboratories for many years, without any thought of it being a specialist technique. FTIR-microscopes, particularly with the ATR and grazing-incidence lenses, and modern Raman microprobes have recently brought surface analysis as a routine tool to many more industrial laboratories (see Chapter 6). Also, nowadays, Langmuir–Blodgett films are studied[61] and monomolecular layers examined more readily with approaches such as Reflection Absorption Infrared Spectroscopy (RAIRS),[62] Surface Enhanced Raman Spectroscopy (SERS)[63] and Surface Enhanced Resonance Raman (SERRS)[64] techniques.

Typical among the applications are many studies of polymer and metal surfaces. For example, modified polymer surfaces are examined to ascertain their chemical functionality in developing products with good adhesion or printing properties. The outermost layers of laminated polymer structures can be discriminated and fingerprinted. Metal surfaces are commonly examined for contamination and corrosion, while painted external or resin-coated internal surfaces are also easily checked for impurities or coating depth. The most

[52] Briggs D. and Seah M.P., *Practical Surface Analysis*, 2nd Edn, Vols. 1 and 2, J. Wiley, Chichester, UK, 1990.
[53] Terenin A.N., *Zh. Fiz. Khim.*, 1940, **14**, 1362.
[54] deBoer J.H., *Z. Physik. Chem. B.*, 1932, **16**, 397.
[55] Eischens R.P., Francis S.A.A., and Pliskin W.A., *J. Phys. Chem.*, 1956, **66**, 194.
[56] Eischens R.P. and Pliskin W.A., *J. Phys. Chem.*, 1956, **24**, 482.
[57] Eischens R.P. and Pliskin W.A., *Adv. Catalysis*, 1958, **10**, 1.
[58] Little L.H., *Infrared Spectra of Adsorbed Species*, Academic, London, UK, 1966.
[59] Hair M.L., *Infrared Spectroscopy in Surface Chemistry*, Marcel Dekker, New York, USA, 1967.
[60] Harrick N.J., *Internal Reflection Spectroscopy*, J. Wiley, New York, USA, 1967.
[61] Tani M. and Yamada H., *Surf. Sci.*, 1982, **119**, 266.
[62] Greenler R.G., *J. Chem. Phys.*, 1966, **44**, 310; ibid. 1969, **50**, 1963.
[63] Fleischmann M., Hendra P.J., and McQuillan, A.J., *Chem. Phys. Lett.*, 1973, **26**, 163.
[64] Rodger C., Smith W.E., Dent G., and Edmondson M., *J. Chem. Soc. Dalton Trans.*, 1996, **5**, 791.

common powders undergoing surface studies by vibrational spectroscopy are catalysts, but these generally require the use of more complex accessories.[65] Many products are sold as formulations or composites, which contain powders that are often coated, or added to resins, as fillers for strengthening. Information on this type of surface interaction is continually being sought using vibrational spectroscopy.

7 Inorganics, Metal Co-ordination Compounds, and Semi-conductors

Most of the sample types discussed so far have been organic but vibrational spectroscopy also makes a valuable contribution to structural analyses in co-ordination and inorganic chemistry. Many chemists seem to have the view that inorganic groups do not absorb in the infrared region. This is based on the spectroscopist's use of salt windows, KBr disks, *etc.*, but even these have absorption features outside the mid-IR region. Certain ionic crystals have a high reflectivity over a fairly narrow wavelength range in the far-infrared region. The position of this reflection is known as the *reststrahlen* frequency. For KBr, this lattice mode occurs near 125 cm^{-1}; for the NaCl crystal lattice this occurs near 185 cm^{-1}. (At 4 mm thick, these alkali halides have long wavelength transmission cut-offs at about 40 μm (250 cm^{-1}) and 25 μm (400 cm^{-1}), respectively.) Many vibrations key to characterising inorganics give rise to bands at low wavenumbers, and, ideally, for the study of inorganics the spectrometer should be capable of recording spectra down to 50 cm^{-1}. This is not problematic for Raman systems, but while many of the research-grade dispersive infrared spectrometers scan into the far-infrared region to 200 cm^{-1} or slightly lower, presently for FTIR systems it is necessary to change some of the optical components (detector, beam-splitter) normally used for mid-IR work in order to cover this region.

Many metal oxide bands occur below 800 cm^{-1}. For example, hydrated Fe_2O_3 (rust) has strong absorption bands between 600 cm^{-1} and 400 cm^{-1} and Sb_2O_3, a flame-retardant used in the polymer industry, has an intense absorption band at ~740 cm^{-1}. Titanium dioxide, a common whitener and filler, although absorbing strongly at 650 cm^{-1} and below, is much more specifically characterised into its three different crystal modifications (anatase, rutile, brookite) by Raman spectroscopy. The two most commonly encountered forms have very distinct Raman spectra: anatase has two intense bands at ~610 Δcm^{-1} and ~450 Δcm^{-1} and a weaker band at ~230 Δcm^{-1}; rutile is characterised by a sharp, very intense band at ~145 Δcm^{-1}, with much weaker features at ~395 Δcm^{-1}, 515 Δcm^{-1} and 640 Δcm^{-1}. Many other common fillers, such as calcium carbonate, barium sulphate, silica, talc, mica, and other minerals, have characteristic intense mid-IR spectra. It is also possible to distinguish between filler aids, such as sand, talc, quartz, and the various forms

[65] Parkyns N.D. and Bradshaw D.I., in *Laboratory Methods in Vibrational Spectroscopy*, 3rd Edn, ed. Willis H.A., van der Maas J.H., and Miller R.G.J., J. Wiley, Chichester, UK, 1987, ch. 16.

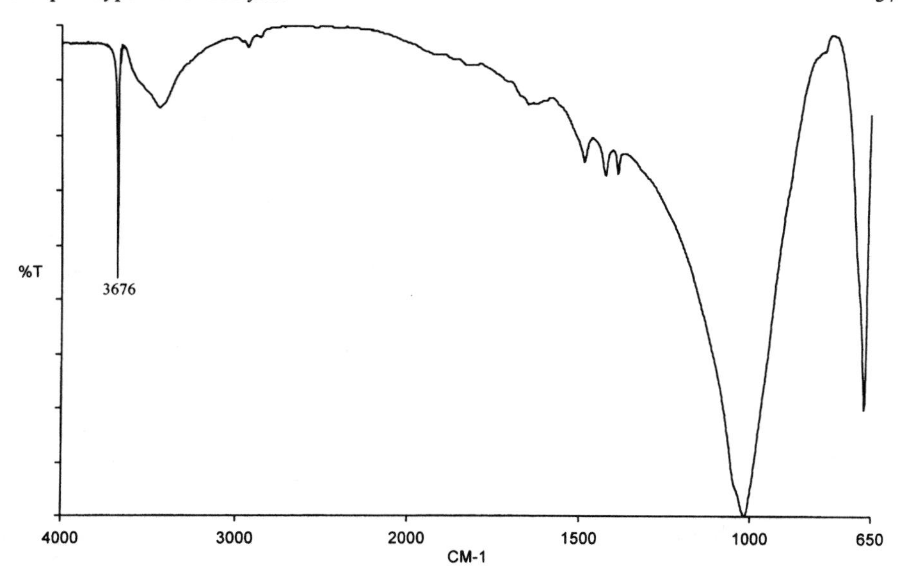

Figure 12 *Infrared spectrum (KBr disk) of an inorganic (magnesium silicate) showing hydrate bands*

of asbestos. The silicates have characteristic band shapes and positions. Many oxygen containing inorganics (carbonates, nitrates(ites), phosphates(ites), sulphates(ites), *etc.*) also have intense absorption bands, typically in the mid-IR region. These are also often incorporated as fillers into organic materials, plastics, paints, resins, *etc.*

Inorganic compounds can exist as hydrates, see for example, Figure 12. As well as changes to the main absorption bands, characteristic features are seen in the hydrated water bands. These generally appear above 3500 cm^{-1} and are often very sharp. For example, a common pattern *readily spotted once seen* is associated with the hydrated alumino-silicate kaolinite (sometimes referred to as china clay), which has a recognisable pattern of four sharp bands between 3700 cm^{-1} and 3600 cm^{-1}. Characteristic patterns like these can be used to distinguish between hydrates. Another 'interesting' spectrum worth acquainting oneself with is α-$Al(OH)_3$ (gibbsite), which has $\sim$5 bands between 3750 cm^{-1} and 3350 cm^{-1}.

Many aqueous solutions on being evaporated to dryness show a characteristically sharp nitrate band near 1350 cm^{-1}, while caustic solutions on being evaporated to dryness have spectra which are predominantly inorganic carbonate. Whilst X-ray examination can be more specific in identifying crystalline inorganic compounds, vibrational spectroscopy, particularly infrared, has the advantage that it can also readily study amorphous materials.

Direct study of inorganics *per se* is a limited and select area of industrial vibrational spectroscopy, but the subject is usually included in vibrational

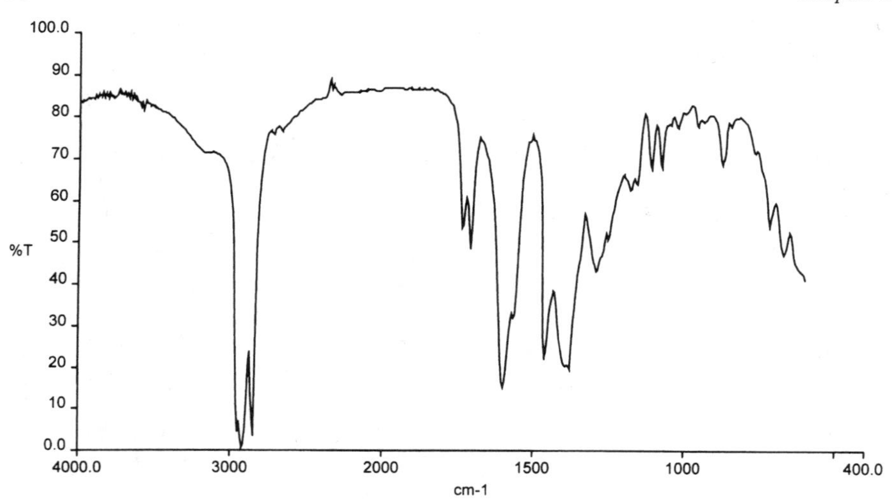

Figure 13 *Infrared spectrum, capillary film, of di-n-butyl-tin-dilaurate*

spectroscopy publications and is well covered by Russell[66] and several useful collections of inorganic reference spectra exist, including combined infrared and Raman inorganic spectra.[67] Much early Raman work was in the field of inorganics,[68–71] and, with the increasing installation of Raman spectrometers in industrial laboratories, its utility in this application area should grow considerably, particularly since inorganic bands generally tend to be narrower in Raman spectra than in their infrared counterparts.

Metal co-ordination compounds also tend to have most of their characteristic bands below 600 cm^{-1}; there are a number of books with peak tables and band positions which demonstrate the applicability of this region to structural determination.[72,73] However, unlike the mid-IR region with which we are usually mainly concerned, bands in this region can often be assigned from a known structure rather than the reverse. Some co-ordination bands will show strongly in the mid-IR wavenumber region; for example, typically the catalyst dibutyl-tin-dilaurate, has strong bands in the 1600 cm^{-1} region arising from the organic ligand, see Figure 13. The carbonyl band moves to lower wavenumber

66 Russell J.D., in *Laboratory Methods in Vibrational Spectroscopy*, 3rd Edn, ed. Willis H.A., van der Maas H.J., and Miller R.G.J., J. Wiley, Chichester, UK, 1987, ch. 18.
67 *Handbook of Infrared and Raman Spectra of Inorganic Compounds and Organic Salts*, ed. Nyquist R.A., Kagel R.O., Putzig C.L., and Leugers M.A., Academic Press, London, UK, 1996.
68 Dhamelincourt P., Wallart F., Leclercq M., Nguyen A.T., and Landon D.O., *Anal. Chem.*, 1979, **51**, 414A.
69 Rosasco G.J., Roedder E., and Simmons J.H., *Science*, 1975, **190**, 557.
70 Rosasco G.J., *Proc. 6th Int. Conf. Raman Spectroscopy*, Heyden, London, UK, 1978, **1**, 389.
71 Dele-Dubois M.L., Dhamelincourt P., and Schubnel H.J., *L'actualité chimique*, 1980, April, 39.
72 Adams D.M., *Metal-Ligands and Related Vibrations*, Edward Arnold, London, UK, 1967.
73 Nakamato K., *Infrared and Raman Spectra of Inorganic and Co-ordination Compounds*, 4th Edn, J. Wiley, New York, USA, 1986.

from ester through free acid to salt. The size of the metal atom on the salt pulls the band to lower wavenumber. The co-ordinated tin in this respect acts as a pseudo-salt. Metal carbonyls, terminal and bridging, are well studied in the infrared region between about 1850 cm^{-1} and 1700 cm^{-1}; the stretching vibrations of bridging carbonyls occur at substantially lower wavenumbers than those of a terminal group. Metal co-ordination chemistry has always been important in catalysis work and with some dyes, and since the mid 1970s it has come to the fore in metals recovery,[74] and, in each field, spectroscopic investigations have played a significant role in structural evaluation and mechanism elucidation.

One of the modern growth industries that also brought challenges for vibrational spectroscopy is the Semiconductor Industry.[75,76] The crystals themselves can be checked for purity and doping levels by studying the band cut-off points, and with the advent of the FTIR and Raman microscopes, the circuit boards can also be checked for delamination and impurities. Further examples are illustrated in more detail in Chapter 6.

8 Liquids

Many of the sample types described above have been solids or solids in solution. Vibrational spectroscopy is employed in examining and characterising liquids in many areas, though usually the challenge comes in handling the samples rather than from complex interpretation due to form changes. Apart from laboratory solution work, many liquids as raw materials are simply identified (pattern-matched) by infrared or Raman for QC or QA purposes. This has always been a simple practical measurement, but increasingly Regulatory bodies are asking for proof that chemicals in containers are what the label states. Raman is increasingly used in this area and NIR fingerprinting is growing in popularity, particularly in the Food and Pharmaceuticals industries; both techniques offer the advantage of examining samples directly in glass containers, such as vials and bottles. Another growth area is the examination of biological liquids, particularly protein analysis and characterisation. When liquids are mentioned, organic and aqueous systems are generally treated separately. Emulsions, latices and dispersions are important products of many industries. Paint, food (milk & butter), cosmetics, and resins are amongst the more common. Vibrational spectroscopy is employed both on plant and in the laboratory for fingerprinting of the formulation, for characterisation of the components or for researching interfacial phenomena.

[74] Laskorin B.N., Yaksin V.V., Ul'yanov U.S., and Mirokhin A.M., *Proc. Int. Solvent Extr. Conf.*, 2, 1775, ed. Jeffreys G.V., Society of Chemical Industry, London, UK, 1974.
[75] Needham C., *L'actualité chimique*, 1980, April, 43.
[76] Kuzmany H., in *Infrared and Raman Spectroscopy. Methods and Applications*, ed. Schrader B, VCH, Weinheim, Germany 1995, Sec. 4.8.

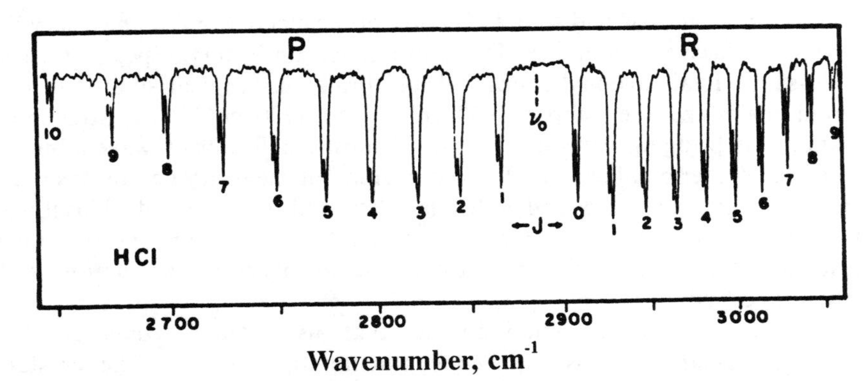

Figure 14 *High resolution, gas phase spectrum of hydrogen chloride, showing vibrational–rotational bands, and splitting of rotational lines into doublets, since the chlorine atom has two naturally occurring isotopes. The peaks for HCl^{37} occur at slightly lower wavenumbers than the peaks for HCl^{35}*
(Reproduced with permission from *Introduction to Infared and Raman Spectroscopy*, Colthup N.B., Daly L.H., and Wiberley S.E., Academic Press, New York, 1964, p. 21)

9 Gases and Vapours

In many chemical reactions vapours are produced either deliberately or from side reactions. The evolved gases are examined by vibrational spectroscopy either to gain further information on reaction paths or on plant to monitor reaction rates. Individual gases can be monitored by single bands with simple, single-filter instruments. The composition of complex mixtures can be monitored with long path-length gas cells[77] (see pp. 244–245 and Figure 2 of Chapter 6), multi-filter instruments or FTIR spectrometers, and chemometric analysis of the spectra. Emissions from chimneys, cooling towers, leaks, or across-a-site can be monitored with modern instrumentation; specific gases may be monitored with narrow-band tuneable infrared laser source spectrometers or multi-species with FTIR spectrometers.[78] Vapours and gases from small molecules *e.g.* HCl, CO, CO_2, SO_2, H_2O have sharp bands which are very sensitive, in strength and shape, to both temperature and pressure. These factors can also be utilised in reaction monitoring or plant control. The fine structure (rotation bands) of HCl can be used to determine the Cl^{35} to Cl^{37} isotope ratio,[79] see Figure 14.

[77] White J.U., *J. Opt. Soc. Am.*, 1942, **32**, 285.
[78] Herget W.F., in *Fourier Transform Infrared Spectroscopy: Applications to Chemical Systems*, Vol. 2, ed. Ferraro J.R. and Basile L.J., Academic, Orlando, USA, 1979.
[79] Colthrup N.B., Daly L.H., and Wiberley S.E., *Introduction to Infrared and Raman Spectroscopy*, 3rd Edn, Academic, New York, USA, 1990.

10 Closing Remarks

In a brief chapter such as this it is not possible to give an exhaustive description of all the uses of vibrational spectroscopy in industry, and only a limited number of references have been included; the list would be far too large for a full citation covering all sample types, analyses, and industries. In this overview we have attempted to show the diverse and largely unrestricted areas of application. One area hardly mentioned is the biological area, and in particular blood and cell analysis, as this is not often considered strictly an industrial application. Increasingly industry not only analyses the products as made, but also attempts to study the effects generated. Studying the 'chemicals' *in situ* and in use is an area where vibrational spectroscopy is now being extensively applied. Often people know what is present but are trying to elicit molecular and structural information on long-term processes, such as migration, toxicity, degradation, and stability. Along with these come investigations associated with product recycling, which sometimes involves studies related to so-called 'reverse-engineering' processes, such as polymer pyrolysis/incineration to monomers.

CHAPTER 3

Instrumentation

1 Introduction

This chapter will discuss briefly the modes of operation of spectrometers used to measure vibrational spectra, and the principal components which comprise these instruments. The basic building blocks of all spectrometers, whether infrared or Raman, for laboratory, plant, or field use, are a source of radiation, some collection optics, a means to discriminate between wavelengths, and a detector to record the intensity of these selected spectral elements. Nowadays, a fifth key requisite for most spectrometer systems is an integral data processing and control computer unit.

Sources for infrared are most commonly broad-band emitters, thermal for the mid-infrared, with the sample itself sometimes acting as the radiation source. Lasers are used as sources for more specialised or limited-range infrared applications, and lasers are probably now the only sources used for Raman measurements.

The collection optics system will attempt to optimise both radiation collection and its refocusing onto either the detector or the wavelength discriminator, depending on the order of the optical train. Most commonly they will involve mirrors and/or lenses, which may, in more specialised applications, be optically coupled to a prism, a fibre-optic or optical wave-guide, or perhaps an integrating sphere, or other optical component.

Selection, dispersion, or separation of analytical wavelengths (wavenumbers) may be achieved in a variety or combination of ways, depending on the instrument design application, through the utilisation of filters, prisms, gratings, monochromators, and interferometers.

Single channel detectors respond consecutively to the number of wavelength-resolved, wavenumber-resolved or time-resolved incident photons. In a multi-channel detection system the dispersed radiation is spread across an array detector: a one- or two-dimensional assembly of radiation responsive elements. In a multiplex system, all of the radiation passing through the spectrometer is incident on a single element detector; the recorded signal is then subsequently decoded (separated) into the individual spectral resolution elements which comprise the spectrum.

The computer, sometimes referred to as the data-station, may, as in some lower cost spectrometers, be incorporated within the body of the spectrometer, but is more often a separate module. It may range from a specific of the spectrometer manufacturer to a commercial desk-top PC (Personal Computer) running a mixture of manufacturer-specific spectrometer control software and third-party data manipulation and analysis routines; PC control and spectral processing have now become the standard capability.

2 Infrared Spectrometers

The dominant commercial infrared instrument now manufactured and purchased for general industrial laboratory use is the Fourier Transform infrared (FTIR) spectrometer. These instruments initially came to prominence as high performance research tools[1,2] in industry in the late 1970s, since which time advances in technology have led to a wide range of FTIR instruments now being available,[3–5] from low cost (<US$20,000) bench-top spectrometers, to dedicated process analysers, through to versatile, multi-range, multi-sampling-port spectrometers, to systems optimised and interfaced as dedicated detectors to other analytical techniques, *e.g.* microscopy and gas chromatography (GC). A 4000 to 400 cm^{-1}, 4 cm^{-1} resolution infrared spectrum may be recorded in less than a second with a FTIR spectrometer. FTIR spectrometers have largely supplanted dispersive spectrometers for mid-infrared applications;[3–6] the foremost dispersive infrared spectrometer from the mid-1970s was the ratio-recording, double-beam instrument.[7,8] At the present time, although specific near-infrared interferometric instruments have recently become available,[9] dispersive or selective-filter designs for near-infrared analysers (NIRA) remain those most commonly in use.[10]

3 Raman Spectrometers

Prior to the mid-1980s, relative to infrared, Raman installations within industry were comparatively rare. These dispersive spectrometer systems tended only to be found in the laboratories of major companies or very specialist application

[1] Griffiths P.R., Sloane H.J., and Hannah R.W., *Appl. Spectrosc.*, 1977, **31**, 6, 485.
[2] Chenery D.H. and Sheppard N., *Appl. Spectrosc.*, 1978, **32**, 1, 79.
[3] Griffiths P.R. and de Haseth J.A., *Fourier Transform Infrared Spectrometry*, Wiley-Interscience, New York, USA, 1986.
[4] Johnston S.F., *Fourier Transform Infrared. A constantly evolving technology*, Ellis Horwood, Chichester, UK, 1991.
[5] Griffiths P.R., *Anal. Chem.*, 1992, **64**, 18, 868A.
[6] Griffiths P.R., in *Laboratory Methods in Vibrational Spectroscopy*, 3rd Edn, ed. Willis H.A., van der Maas J.H., and Miller R.G.J., J. Wiley, Chichester, UK, 1987, ch. 6.
[7] Steer I.A. and Miller R.G.J., in *Laboratory Methods in Vibrational Spectroscopy*, 3rd Edn, ed. Willis H.A., van der Maas J.H., and Miller R.G.J., J. Wiley, Chichester, UK, 1987, ch. 5.
[8] Sheppard N., *Anal. Chem.*, 1992, **64**, 18, 877A.
[9] (a) Bomem MB160, Bomem Inc., Hartmann & Braun, Quebec, Canada; (b) InfraProver, Brann + Luebbe, Norderstedt, Germany; (c) IFS 28/N, Bruker Analytische Messtechnik GMBH, Karlsruhe, Germany.
[10] *Handbook of Near-Infrared Analysis*, ed. Burns D.A. and Ciurczak E.W., Practical Spectroscopy Series, Vol. 13, Marcel Dekker, New York, USA, 1992.

laboratories, even though the microprobe advantages of the technique were widely appreciated.[11] While infrared still remains the dominant partner, since about 1986 there has been a steady and growing increase in the popularity of Raman spectroscopy among industrial spectroscopists.[12] Key among the early motivators to this development was the introduction of near-infrared laser excitation FT-Raman spectroscopy,[13–16] which enabled the certain, routine study of many more industrial materials and products by circumventing fluorescence,[17] which for so long had been the principal difficulty in the application of the Raman technique to industrial problem-solving and research. FT-Raman systems also brought a simplicity of sample-positioning that infrared practioners had grown accustomed to.[18] They are commercially available as stand-alone bench-top units or as adjuncts to FTIR spectrometers.

CCD (charge-coupled device) detectors[19,20] also came to prominence in the late 1980s, and are now being extensively used in Raman point, line-focus, and imaging applications. In the early 1990s, holographic optical elements – notch filters and gratings – had a major impact on Raman instrumentation.[21–25] The notch filters provided an optimum means for rejection of the Rayleigh scattered light. Holographic gratings and notch filters used in combination with a CCD detector and diode-laser source are now the prime components of compact, robust, fast, Raman spectrometers, which, integrated with a fibre-optic and probe head, are well suited to both routine laboratory sampling and process monitoring.[26–30] A single-shot 3 cm^{-1} resolution spectrum over the range about

[11] Louden J.D., in *Practical Raman Spectroscopy*, ed. Gardiner D.J. and Graves P.R., Springer-Verlag, Berlin Heidelberg, Germany, 1989.
[12] Chase B., *Appl. Spectrosc.*, 1994, **48**, 7, 14A.
[13] Hirschfeld T. and Chase B., *Appl. Spectrosc.*, 1986, **40**, 2, 133.
[14] Chase D.B., *J. Am. Chem. Soc.*, 1986, **108**, 7485.
[15] Hallmark V.M., Zimba C.G., Swalen J.D., and Rabolt J.F., *Spectroscopy*, 1987, **2**, 6, 40.
[16] Parker S.F., Williams K.P.J., Hendra P.J., and Turner A.J., *Appl. Spectrosc.*, 1988, **42**, 5, 796.
[17] (a) Authors various in *Applications of Fourier Transform Raman Spectroscopy*, ed. Hendra P.J., special edition *Spectrochimica Acta*, 1990, **46A**, No. 2; (b) Authors various in *Applications of Fourier Transform Raman Spectroscopy – II*, ed. Hendra P.J., special edition *Spectrochimica Acta*, 1991, **47A**, No. 9/10; (c) Authors various in *Applications of Fourier Transform Raman Spectroscopy – III*, ed. Hendra P.J., special edition *Spectrochimica Acta*, 1993, **49A**, No. 5/6; (d) Authors various in *Applications of Fourier Transform Raman Spectroscopy – IV*, ed. Hendra P.J., special edition *Spectrochimica Acta*, 1994, **50A**, No. 11.
[18] Hendra P.J., Jones C., and Warnes G., *Fourier Transform Raman Spectroscopy. Instrumentation and Chemical Applications*, Ellis Horwood, London, UK, 1991.
[19] Falkin D. and Vosloo M., *Spectroscopy Europe*, 1993, **5**, 5, 16.
[20] Epperson P.M., Sweedler J.V., Bilhorn R.B., Sims G.R., and Denton M.B., *Anal. Chem.*, 1988, **60**, 5, 327A.
[21] Carraba M.M., Spencer K.M., Rich C., and Rauh D., *Appl. Spectrosc.*, 1990, **44**, 9, 1558.
[22] Yang B., Morris M.D., and Owen H. *Appl. Spectrosc.*, 1991, **45**, 9, 1533.
[23] Pelletier M.J. and Reeder R.C., *Appl. Spectrosc.*, 1991, **45**, 5, 765.
[24] Everall N., *Appl. Spectrosc.*, 1992, **46**, 5, 746.
[25] Tedesco J.M., Owen H., Pallister D.M., and Morris M.D., *Anal. Chem.*, 1993, **65**, 441A.
[26] Newman C.D., Bret G.G., and McCreery R.L., *Appl. Spectrosc.*, 1992, **46**, 2, 262.
[27] Battey D.E., Slater J.B., Wludyka R., Owen H., Pallister D.M., and Morris M.D., *Appl. Spectrosc.*, 1993, **47**, 11, 1913.
[28] Everall N.J., Owen H., and Slater J., *Appl. Spectrosc.*, 1995, **49**, 5, 610.
[29] Everall N.J., in *An Introduction to Laser Spectroscopy*, ed. Andrews D.L. and Demidov A., Plenum Press, New York, USA, 1995, p. 115.
[30] Pelletier M. and Davis K., *Int. Lab.*, 1996, May, 11D.

50 to 1000 Δcm^{-1} may be acquired in about 3 sec through a 100 m fibre probe.[28] Instruments are of a comparable price to medium-to-high-performance FTIR spectrometers.

4 Sources

Figure 1 shows some theoretical black-body emission (Planck radiation equation) curves over a range of temperatures; their absolute wavenumber dependences are compared in terms of both thermal and photon radiation.

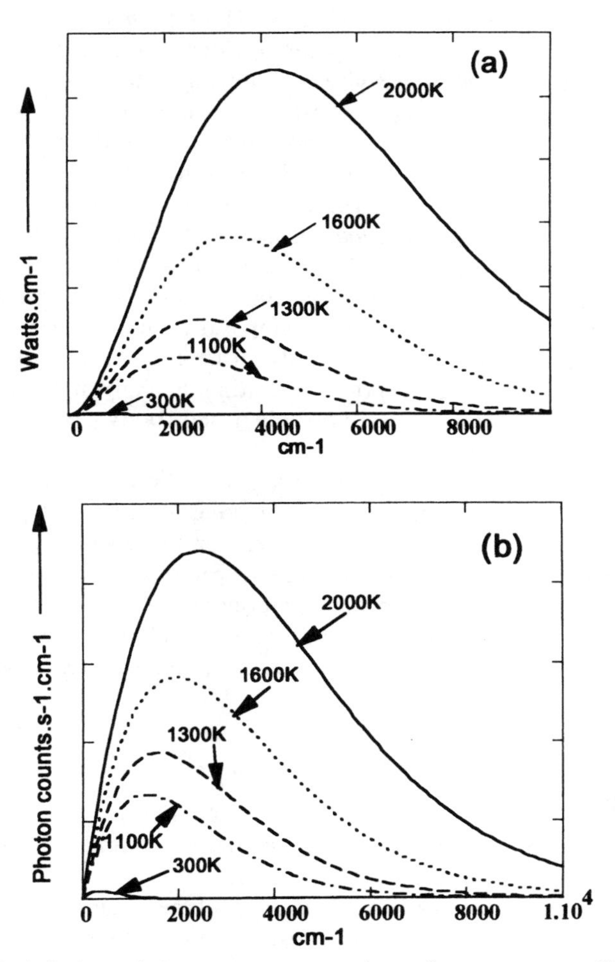

Figure 1 *Black-body emission curves at a variety of temperatures. Abscissa scale in absolute wavenumbers. Ordinate scales:* (a) *watts per wavenumber,* (b) *photon counts per second per absolute wavenumber. (Note: the Nd^{3+}:YAG laser operating at* 1.064 μm, *emits at* 9398.5 *wavenumbers)*

Infrared Sources

Broad-band sources for mid-infrared spectrometers, whether dispersive or FT, are usually rod-like elements of a refractory material. The element is heated to a sufficient temperature to emit continuous radiation across the mid-infrared region.[3,6,7,31] They are pseudo-black-body radiators. Ideally, sources need to be stable, long-lived, preferably rugged, and of high emissivity.[31] *Nichrome coils* wound round a ceramic which are heated to incandescence (~1100 K) are inexpensive, low-intensity, durable sources. *Nernst glowers* are cylinders of rare earth oxides which can be operated at temperatures up to 2000 K. They have the disadvantages that they are fragile and require secondary heating to pre-heat the element to initiate electrical conduction. Nernst glowers are not used as sources in commercial FTIR spectrometers. The *Glowbar* (or *Globar*) silicon carbide rod source has a high intensity and a wider useful range than a Nernst glower, with relatively high emissivity to both high (>2000 cm^{-1}) and low (<200 cm^{-1}) wavenumbers; its spectral output is about 80% of that of a theoretical black-body radiator.[31] In some instruments the Glowbar source requires water-cooling. For the far-infrared, beyond 200 cm^{-1} and certainly below ~100 cm^{-1}, the long wavelength radiation from the plasma created within a *high-pressure mercury arc lamp* is used as the source. For work in the near-infrared region specifically, ~12,500 cm^{-1} to 4000 cm^{-1} (~800 nm to 2500 nm), an *incandescent lamp source*, usually a *tungsten filament lamp*, is used.

Glowbar or other proprietary ceramic, temperature stabilised, frequently air-cooled, sources are those most commonly installed in a commercial FTIR spectrometer for mid-infrared purposes, operating typically in the range 1300 K to 1600 K. Other proprietary sources include a blackbody cavity source and a black coated hot-wire source.

Tuneable diode lasers emit high intensity radiation over a very narrow band of wavelengths. They are normally operated at cryogenic temperatures, and tuned by varying either the diode current or the operating temperature. Their most common use is in non-dispersive process analysers, required to monitor a small number of specific components. The *tuneable CO_2 laser* emits ~100 closely spaced discrete lines over the 1100 to 900 cm^{-1} range. The source is typically used in long path-length (many metres) environmental monitoring.

Raman Sources

The source for a Raman measurement is essentially a collimated, narrow cross-section monochromatic line output from a laser.

Lasers may be considered essentially as of two types: continuous wave (CW) and pulsed. *CW lasers* are by far the most commonly used type in industrial laboratories, being both intrinsically safer and less expensive. *Pulsed lasers* tend

[31] Ciurczak E.W., *Spectroscopy*, 1993, **8**, 9, 12.

to be reserved for the more specialist applications, such as time-resolved studies, as tuneable ultra-violet sources for resonance Raman observations, or for non-linear Raman methods.[32–34]

A visible line (single-wavelength output) from a CW gas ion laser was for a long time the primary source used with commercial Raman dispersive spectrometers. The two most widely used gas ion lasers were the *argon-ion* and the *krypton-ion*, which together offered various discrete lines over the wavelength range 324 nm to 799 nm, although the most intense were those to which they were most commonly tuned as Raman excitation sources. These are the Ar^+ laser lines at 488.0 nm (blue) and 514.5 nm (green) and the Kr^+ laser line at 647.1 nm (red), which are significantly more intense than any of the other lines. The more powerful of these lasers require water cooling, although any modern air-cooled argon ion laser will readily deliver > 100 mW in each of its two most intense lines, which is plenty for most studies.[32,33] In industrial laboratories utilising dispersive spectrometers the Ar^+ laser remains a prominent source. It is used either directly for short wavelength (blue/green) visible excitation, or at higher powers to provide the pump laser beam for either a dye laser or a titanium–sapphire diode laser, both of which will then emit at longer wavelengths. The *helium–neon* (He–Ne) laser, which emits a red line at 632.8 nm, although relatively a much weaker source than the Kr^+ laser has advantages in its robustness and low cost, and is air cooled. The output from a He–Ne laser is sufficient for use in combination with modern CCD detectors,[35–37] and as a consequence has made somewhat of a 'comeback' as a Raman excitation source for use in modern Raman imaging microscopes, which also use holographic optical elements.[37–38]

Dye lasers, which utilise a dye solution, are continuously tuneable over a range of wavelengths and typically have a useful output for Raman excitation over about 100 nm range. For example, useful wavelength selectable emission ranges of $\sim$570–640 nm and $\sim$690–790 nm are available from Ar^+ pumped Rhodamine 6G and Pyridine 1 dye lasers respectively. The solid-state *Ti–sapphire* laser is continuously tunable over a much wider wavelength range; an Ar^+ pumped Ti–sapphire laser will yield 640–900 nm radiation. Single-line outputs of greater than 200 mW may be readily achieved with both dye and Ti–sapphire systems. Consequently, where a higher power Ar^+ laser (*e.g.* $\geqslant 0.5$ W) is available, then one of these pumped laser systems provides a more flexible arrangement than the limited number of discrete lines selectable from

[32] Gerrard D.L. and Bowley H.J., in *Practical Raman Spectroscopy*, ed. Gardiner D.J. and Graves P.R., Springer-Verlag, Berlin Heidelberg, Germany, 1989, ch. 3.
[33] Chase D.B., in *Analytical Raman Spectroscopy*, ed. Grasselli J.G. and Bulkin B.J., J. Wiley, New York, USA, 1991, ch. 2.
[34] Carey P., in *Laboratory Methods in Vibrational Spectroscopy*, 3rd Edn, ed. Willis H.A., van der Maas J.H., and Miller R.G.J., J. Wiley, Chichester, UK, 1987, ch. 21.
[35] Pemberton J.E. and Sobocinski R.L., *J. Am. Chem. Soc.*, 1989, **111**, 2, 432.
[36] Pallister D.M., Liu K-L., Govil A., Morris M.D., Owen H., and Harrison T.R., *Appl. Spectrosc.*, 1992, **46**, 10, 1469.
[37] Garton A., Batchelder D.N., and Cheng C., *Appl. Spectrosc.*, 1993, **47**, 7, 922.
[38] Purcell F., *Laser Focus World*, 1993, **29**, 10, 135.

Table 1 *Continuous Wave Raman Sources*

Laser type	*Wavelength* (nm)	*Wavenumber* (cm^{-1})	*Notes*
Ar^{+}	454.4	22 007.0	Lines listed are some of the more intense.
	457.9	21 838.8	The 488.0 nm and 514.5 nm are the two most intense, and most widely used.
	465.8	21 468.4	
	472.7	21 155.1	
	476.5	20 986.4	
	488.0	20 491.8	
	496.5	20 141.0	
	501.7	19 932.2	
	514.5	19 436.3	
Kr^{+}	413.1	24 207.2	Lines listed are the principal lines.
	468.0	21 367.5	The most commonly used and most intense is the 647.1 nm line.
	476.2	20 999.6	
	482.5	20 725.4	
	520.8	19 201.2	
	530.9	18 835.9	
	568.2	17 599.4	
	647.1	15 453.6	
	676.4	14 784.1	
	752.5	13 289.0	
	799.3	12 510.9	
He–Ne	632.8	15 802.8	
Dye	~450–860	~22 222–11 628	Ar^{+} pumped. Several dye solutions required to cover the whole region.
Ti-sapphire	~640–900	~15 625–11 111	Ar^{+} pumped.
Nd^{3+}:YAG	1064	9398.5	Yields a useful line at 532 nm when frequency-doubled, see process instrument example in ref. 38.

a Kr^{+} laser. The Ti–sapphire option is much more convenient to use, and does not require the use or changing of relatively short-lifetime dye solutions, many of which are carcinogenic; it is also less costly, and has become the preferred system as a long wavelength tuneable visible excitation Raman source.

The near-infrared (NIR) emission at 1.064 μm from a *neodymium-doped yttrium aluminium garnet* (Nd^{3+}:YAG) solid-state continuous wave laser is that most commonly used for FT-Raman measurements. The models used are typically capable of delivering between 1 and 5 W in single mode.

Table 1 summarises some of the more intense and most commonly used Raman excitation wavelengths from continuous wave lasers. (See also Section 6 of Chapter 6.)

5 Detectors

Infrared Detectors

Infrared detectors are essentially of two types: thermal or quantum;[3,6,7,39] photoacoustic detection is a special case and is considered separately in Chapter 4, Section 7.

Thermal detectors sense and respond to a change in temperature. They are usually operated at room temperature and have a wide wavelength coverage, but compared to quantum detectors are less sensitive and relatively slow, which limits the frequency by which the signal may be modulated. *Thermocouple* detectors contain two junctions between two dissimilar metals, one of which, the receiver, is blackened to make it absorb effectively. Infrared radiation incident on this junction causes it to heat up and its potential increases above that of the other junction. This potential difference (voltage), thermal electromotive force, is proportional to the intensity of the radiation. A *thermopile* detector is essentially a series of thermocouples; it has a typical response time of around 30 ms. The output from a *thermistor* detector is caused by a change in electrical resistance. These bolometers are manufactured from sintered oxides of manganese, cobalt, and nickel, and have a high temperature coefficient of resistance ($\sim$4% per °C). A *pyroelectric* detector contains a ferroelectric element. Below their Curie point these materials exhibit a large spontaneous polarisation (alignment of electric dipole moments). The output from a pyroelectric detector is an electrical signal generated by a change in the degree of polarisation as a consequence of a temperature change caused by incident infrared radiation. Common pyroelectric detectors are TGS (triglycine sulphate) and DTGS (deuterated TGS); other examples contain $LiTaO_3$ and $LiNbO_3$. The Curie point of TGS is between 45 °C and 49 °C, but its operating range is increased by L-alanine doping. Pyroelectric bolometers have fast response times, of the order of 1 ms, and hence are used in FT spectrometers as well as in dispersive instruments. The sensitivity of TGS pyroelectric detectors is essentially independent of wavenumber, see Figure 2.

Quantum (or *photon*) detectors are both sensitive and fast. They depend on the interaction between quanta of the incident radiation and the electrons in a solid, which for most infrared purposes is a semiconductor. The incident photons must be able to excite electrons in the semiconductor from one state to another, thereby raising it from a non-conducting state to a conducting state. The excited electrons in the conduction band and the holes in the valence band can then carry current – the photoelectric effect. The sensitivity of these detectors tends to fall off rapidly close to their long wavelength limit, since each photon has an energy (E) which is directly proportional to its frequency (ν). ($E = h\nu = h(c/\lambda)$, where: h is Planck's constant; c the speed of light in vacuo, and λ a photon's wavelength). For operation or optimum performance the elements of many of these infrared photon detectors must be cooled; for some, simple thermoelectrical cooling is sufficient, others need to be maintained at

[39] Ciurczak E.W., *Spectroscopy*, 1993, **8**, 8, 12.

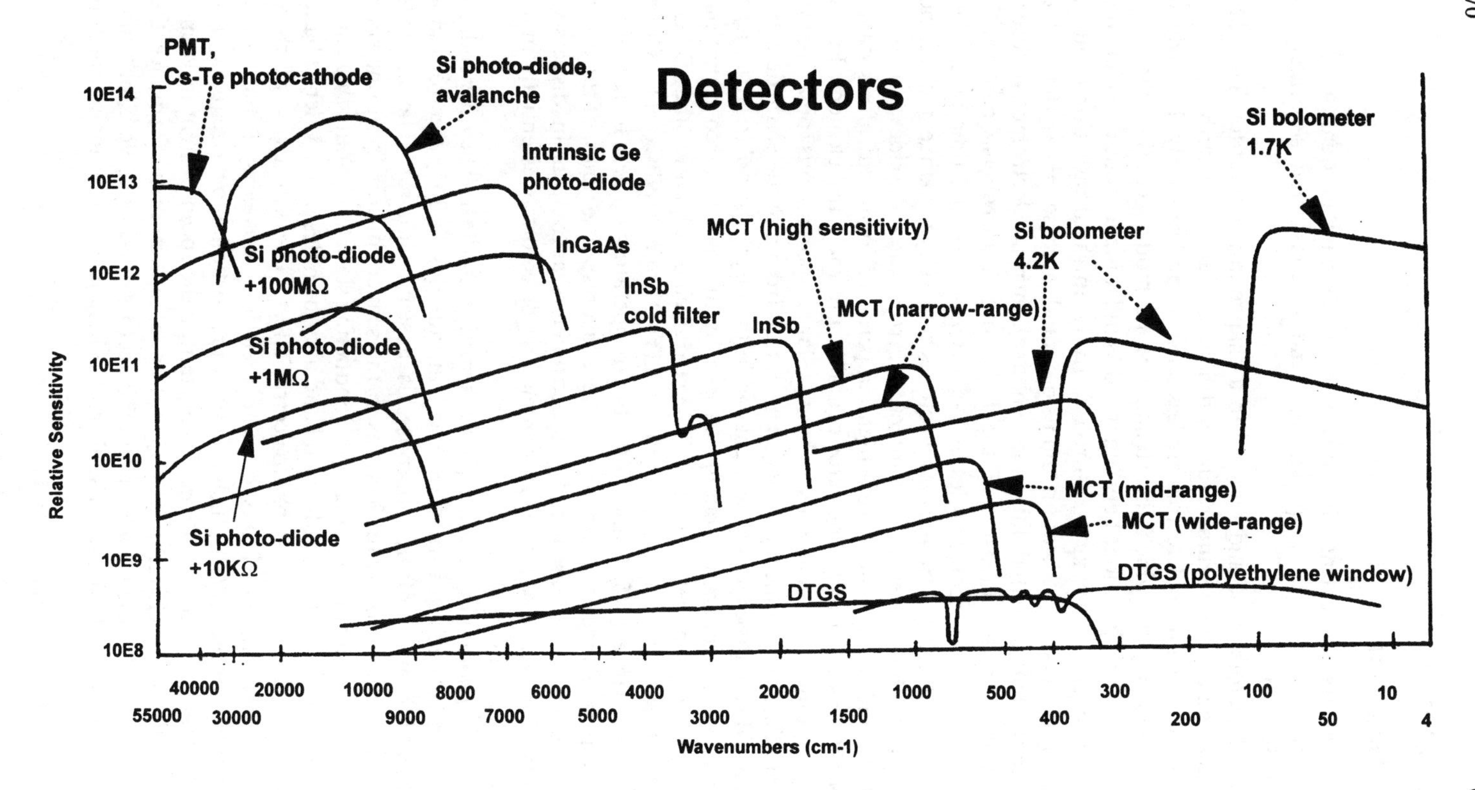

Figure 2 *The relative sensitivities of a range of detectors used with infrared and Raman spectroscopy*
(Figure adapted with permission from publicity literature of Bomem, Inc., Québec, Québec, Canada)

liquid nitrogen temperatures (77 K). In a *photoconductive* detector the element is a *n*-type (homogeneous) or doped *n*-type chip, the electrical resistance of which is lowered and monitored following electron excitation to the conduction band. In a *photovoltaic* detector, radiation striking the *p*-surface of a *p–n* junction in a semiconductor will create hole-electron pairs, which are separated by the internal field existing at the junction. This generates a small voltage which is amplified to provide a measure of detection. Examples of quantum detectors are PbS, PbSe, InGaAs, InSb, PbSnTe and CdHgTe. The latter is most commonly referred to as the MCT (mercury cadmium telluride) detector, and has a response time as high as 20 ns. During operation it must be maintained at liquid nitrogen temperatures (77 K).

At present the most commonly used mid-infrared detectors are DTGS, for both dispersive and FT spectrometers, and MCT, for more sensitive and/or rapid FT applications, with $LiTaO_3$ detectors installed in some of the lower cost spectrometers. The first commercial FTIR spectrometer fitted with an array detector was launched in 1996.[40] The InSb focal plane array detector has 128 × 128 active pixels, optimised for operation over the nominal spectral bandpass range 3950–1975 cm^{-1}, and the instrument is capable of collecting over 16 000 interferograms in about 16 s.

The most commonly used detector for the far-infrared (<400 cm^{-1}) is a TGS pyroelectric bolometer, in which the mid-infrared transparent window, usually of ZnSe, has been replaced by a polyethylene (PE) window. For the very far-infrared (<100–50 cm^{-1}) or high-sensitivity far-infrared studies a liquid-helium-cooled detector may be required, such as a germanium or silicon bolometer or InSb hot-electron detector.[3] For the far-infrared below about 200 cm^{-1} a *Golay* detector has been used.[3,7] In a Golay cell infrared radiation passing through a transmitting window is absorbed by a blackened diaphragm which forms one end of a gas-filled cell; the gas is usually a noble gas such as xenon or krypton. This causes the temperature of the gas in the cell to rise and it expands. This expansion distorts a more delicate membrane at the other end of the cell. This remote diaphragm is silvered on the outside and acts as a mirror reflecting light from a source onto a photo-electric cell. Changes of intensity of light reaching the photo-electric cell will change the measured photo-current strength, which may then be related to the intensity of the infrared radiation. Golay detectors are slow and cannot be used with rapid scan FT systems.

For high sensitivity at high wavenumbers (>5000–2000 cm^{-1}) an InSb detector is frequently used.

Table 2 summarises some information for the detectors most commonly found in industrial laboratories. Figure 2 illustrates the relative sensitivity of a wide range of the detectors available for use with the higher performance, research-grade FT instruments.

[40] Bio-Rad FTS Stingray 6000, Bio-Rad Digilab, Cambridge, USA.

Table 2 *IR Detectors*

Detector	*Operating temperature*	*Instrument*	*Approximate range** (cm^{-1})
InGaAs	Room	FT	11 000–6350
	Liquid nitrogen	FT	11 000–6000
PbS	Room	FT	> 10 000–3300
InAs	Liquid nitrogen	FT	15 000–3300
PbSe	Room	FT	11 000–2000
InSb	Liquid nitrogen	FT	10 000–1850
$LiTaO_3$	Room	Dispersive or FT	> 8000–350
DTGS	Room	Dispersive or FT	> 7500–350 (KBr window)
			> 6400–225 (CsI window)
MCT (narrow band)	Liquid nitrogen	FT	10 000–700
MCT (medium band)	Liquid nitrogen	FT	10 000–580
MCT (wide band)	Liquid nitrogen	FT	10 000–400
DTGS fitted with a PE window	Room	Dispersive or FT	650–20
He-cooled bolometer	1.7 K for maximum sensitivity	FT	400–10

* range values (and sensitivities) quoted will vary for different model numbers and from different manufacturers; specification performance is often a compromise between maximum sensitivity and spectral range

Raman Detectors

For many years the optimum system for single-channel detection has been a *photomultiplier tube* (PMT).[18,32,33,41] These relatively low cost detectors have high sensitivity and dynamic range and low background count (dark current), particularly when cooled. A photon striking the photocathode of the PMT undergoes a photoelectric collision, generating an electron which is then accelerated by a potential difference towards a dynode. On collision with this dynode the high kinetic energy electron causes the release of multiple (two or more) electrons by secondary emission. These electrons are then accelerated towards a second dynode, and the multiplication process continues over many dynodes, typically 10–14, to produce an amplified current which is essentially proportional to the detected (incident) radiation intensity. The preferred display is photon-counting. This is achieved by an electronic gating technique which sorts signal pulses from noise pulses, which are mainly of higher or lower energy, in order to produce an output which is proportional to the number of photons per second arriving at the detector. Since the spectral responsivity of photomultipliers for visible detection falls off rapidly towards their long wavelength limit, usually in the extreme visible (red-end) or near-infrared, *solid-state* detectors such as silicon photodiodes have been used, although these generally exhibit higher dark currents.[18] Cooled PMTs with photocathodes such as the Ga–As or Na–K–Cs–Sb (S-20), have long wavelength response beyond 800 nm.[18,33]

Multichannel detection systems have included image-intensified closed circuit TV cameras (vidicon) and diode arrays.[17,18,32,42] The former was the choice for early Raman microprobe spectrometers.[11,42,43] A *photodiode array* comprises a linear arrangement of many individual diodes (elements or pixels),[32] which typically might be 512 or 1024, with a pixel size of about 25 μm × 25 μm. The array is aligned along an exit focal plane of a Raman spectrometer, and a portion of the Raman spectrum is dispersed along the array, so that successive array elements detect successive elements of the Raman spectrum, although these will not be exactly equal in wavenumber resolution, see below. The diode array detector is especially useful for time-dependent studies, since a Raman spectrum of the dispersed spectral region can be recorded in a few milliseconds.

The primary function of the intensifier is to multiply (not improve) the signal-to-noise of the dispersed Raman signal to a value at which the quantum noise of the photocathode becomes negligible compared to the Raman signal intensity.[43,44] Cost and engineering restraints usually limit the diameter of the microchannel plate of the intensifier to less than that required to fill a 1024 element array length, about 800 pixels typically being covered. For a grating with a groove density of 1800 grooves/mm, the wavelength dispersion is 1.5 nm/mm,

[41] Gilson T.R. and Hendra P.J., *Laser Raman Spectroscopy*, Wiley-Interscience, London, UK, 1970.
[42] Campion A. and Wodruff W.H., *Anal. Chem.*, 1987, **59**, 22, 1299A.
[43] Delhaye M. and Dhamelincourt P., *J. Raman Spectrosc.*, 1975, **3**, 33.
[44] Everall N.J., PhD thesis, University of Durham, UK, 1986.

which, spread along a 800 × 25 μm array, yields a wavelength range of about 30 nm. If the system is set to record from 600 Δcm^{-1}, then using the 488.0 nm (20 491.8 cm^{-1}) line of an Ar^+ laser the array will register from 600 Δcm^{-1} (19 891.8 cm^{-1}, 502.7 nm) to about 1720 Δcm^{-1} (18 772.3 cm^{-1}, 532.7 nm), implying a resolution of ~ 1.4 cm^{-1} per pixel. However the resolution is linear with wavelength (~ 30 nm/800 pixels ≈ 0.0375 nm/pixel), and consequently will be about 1.48 cm^{-1} at ~ 600 Δcm^{-1}, whereas at ~ 1720 Δcm^{-1} it will be about 1.3 cm^{-1}. If the dispersion is spread from 2500 Δcm^{-1} (555.8 nm) to ~ 3421 Δcm^{-1} (585.8 nm), the resolution along the array increases from ~ 1.2 cm^{-1} to ~ 1.1 cm^{-1}. For a spectrum generated using 647.1 nm excitation, equivalent coverage from 600 Δcm^{-1} will be to ~ 1233 Δcm^{-1}, *i.e.* a smaller wavelength range but at higher resolution.

In recent years the *CCD (charge coupled device)* detector has become perhaps the most popular multichannel system for Raman purposes.[18,19] CCDs are high sensitivity, low dark current (signal), large dynamic range, low read-out-noise detectors.[19,20] They may be thermoelectrically (Peltier) cooled, or, for maximum sensitivity, low dark current cooled to about − 133°C (140 K) using a liquid nitrogen Dewar. CCD detectors have high overall quantum efficiency, which extends well into the red. They are two-dimensional array detectors, comprising a matrix of semiconductor detectors (elements, pixels), which are available in a variety of sizes. A '1/2 inch' (~ 1.25 cm) CCD chip may have about a 512 × 512 pixel format; larger arrays are available, such as '1 inch' chips with 1024 × 1024 pixels, with an individual pixel size of 20 μm × 20 μm. A charge is generated on a pixel by the photoelectric effect when light is incident upon it. Changes in electric potential may then be used to transport this charge to the bottom row (output register) of the CCD array and then along the bottom row to a corner read-out, where it is amplified and converted to an electrical signal. A CCD detector may be operated in several modes.[19,20] For rapid (a few milliseconds upwards) kinetic studies, successive spectra may be imaged (dispersed) along the row furthest from the output register and consecutively driven to it and read out. The CCD may also be segmented to simultaneously image signals from multiple sources onto the array; each row corresponds to a separate Raman spectrum from each source. Figure 3(a) shows the the spectra recorded from simultaneous excitation of four liquids; the Raman scattered radiation from each was collected by a fibre-optic and imaged onto a different section of the CCD array.

Line profile or 2D images may be built up from illuminations using a Raman-microprobe/microscope system, see Chapter 6. The signal-to-noise ratio (S/N) characteristic of a spectrum will increase with the signal integration time. The signal on each pixel may be read out individually, or the signals in adjoining pixels may be summed ('binned'). The S/N characteristic of a single dispersed spectrum can also be rapidly enhanced by co-adding ('binning') the signals in a column; the S/N is increased by the number of rows binned, while the noise rises with the square root of the number of pixels binned.[19] In a recent Raman imaging holographic spectrograph[38,45] wide spectral coverage is achieved in a

[45] HoloSpec/Holoprobe systems, Kaiser Optical Systems, Inc., Ann Arbor, USA.

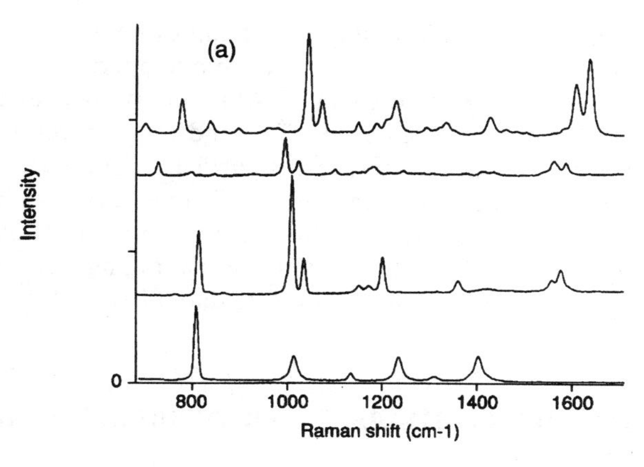

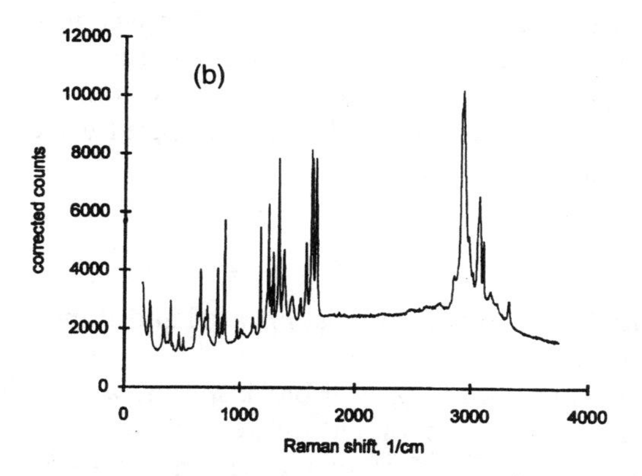

Figure 3 (a) *Four simultaneously acquired Raman spectra taken using fibre optic inputs and setting up four read-out regions on the CCD chip. The Raman system was a SPEX 270M, a* 0.25 m *imaging single spectrograph, with a holographic edge filter. The four liquid samples were held in test tubes and a different fibre optic collected the scattered light from each. From top to bottom the liquids are: styrene, t-butyl benzene, toluene and cyclohexane*
(Reproduced by kind permission of Instruments S.A. (UK) Ltd.)
(b) *In situ Raman spectrum of Tylenol*^{TM} *(pain-reliever) in a blister pack obtained with a Holoprobe 532 system (*5 mW, 532 nm *radiation at the sample) through a* 100 m *fibre probe, acquistion time* 15 s
(Reproduced from ref. 38, by kind permission of Laser Focus World, Pennwell Publishing Company)

single shot by recording consecutive spectral ranges on different rows. Figure 3(b) shows a wide spectral range, high S/N Raman spectrum recorded in 15 s through a 100 m fibre optic, with a spectrometer fitted with a CCD detector.

FT-Raman spectrometers are usually fitted with an InGaAs detector, which can be operated at ambient or liquid nitrogen temperatures, or with the more sensitive liquid nitrogen cooled Ge detector. Although offering higher sensitivity when cooled, the InGaAs detector has reduced spectral coverage at large wavenumber Raman shifts (Stokes), imposing a limit of about 3000 Δcm^{-1} at 77 K compared to about 3300–3500 Δcm^{-1} at room temperature.

The CCD detector and those used in FT-Raman spectrometers are both susceptible to interference by cosmic rays,[46–49] and must therefore, whenever possible, be shielded from such events; this is particularly important in the FT case, since a spike in the interferogram will transform as a sinusoidal variation in the spectrum. Software/hardware methods for spike removal/minimisation include an 'intensity-possibility' summation and averaging approach of two spectra recorded under identical conditions,[46] median filtering,[47,48] and electronic filtering circuits.[49]

6 Filters, Prisms, Gratings, Monochromators, and Interferometers

Spectrometers for general laboratory qualitative applications use will normally be required to interrogate as much of the spectral range as possible, whereas analysers or process monitors may only be required to measure at selective or over a narrow band of wavelengths. Discrete infrared wavelength selection is only possible with a laser source.

Filters

Filters are essential components of most infrared and Raman instruments, and can have several functions.

Optical filters may be used to pass or block specific wavelength ranges, either as narrow band or as cut-on or cut-off respectively, or to totally remove unwanted radiation.[3,4,7,11,19,32,33,50] In FT systems, high-, low- and band-pass frequency electronic filtering may also be used[3,4,19] to restrict the spectral coverage. (Digital (numerical) filtering by software algorithms may also be used to specify a function which acts on the recorded data to limit the frequency response).[3,4,19] The two most common types of optical filters are absorption and interference.[50] Holographic optical filters are now common components in new Raman instrumentation.[26,27] The principle underlying interference filters is that a narrow range of wavelengths of the incident radiation will constructively interfere and be passed (transmitted or reflected), while the remainder will destructively interfere and be blocked. Single-layer (Fabry–Perot) etalon interference filters comprise a transparent dielectric spacer placed between two thin sheets of glass. The centre wavelength passed depends on the refractive index of the spacer. It also depends on the angle of incidence of the incident radiation. Alternating layers of high and low refractive index materials are used to

[46] Takeuchi H., Hashimoto S., and Harada I., *Appl. Spectrosc.*, 1993, **47**, 1, 129.
[47] Treado P.J. and Morris M.D., *Appl. Spectrosc.*, 1990, **44**, 1, 1.
[48] Barraga J.J., Feld M.S., and Rava R.P., *Appl. Spectrosc*,. 1992, **46**, 2, 187.
[49] Chase D.B. and Rabolt J.F., in *Fourier Transform Raman Spectroscopy*, ed. Chase D.B. and Rabolt J.F., Academic Press, San Diego, USA, 1994, ch. 1.
[50] Ciurczak E.W., *Spectroscopy*, 1994, **9**, 1, 12.

construct multi-layer dielectric filters, which act as if they were a series of Fabry–Perot filters. They can have narrow bandpass half-widths (1–100 nm commercially available) and high light transmission. Wedge interference filters have formed the operating basis of scanning tilting-filter NIR instruments.[50,51]

Some mid-infrared range spectrometers manufactured for process measurements or environmental gas monitoring make use of a *circular variable* filter (CVF) to scan consecutively through the wavelengths, *e.g.* the mid-infrared MIRAN Analyser.[52,53] (The MIRAN analysers have very recently been superseded by instruments using a linear variable filter.)[54] The infrared beam is incident on the CVF, which is mounted around and close to the circumference of a disc. Rotation of the disc about its axis causes successive narrow wavelength regions to be transmitted through the filter. Such a filter was also incorporated into a dedicated dispersive infrared microscope system.[55,56] Circular variable filters have low spectral resolution.

A recent development is the *acousto-optical tuneable* filter (AOTF), a solid-state electronically tuneable spectral bandpass filter, which relies on the acoustic diffraction of light in an anisotropic medium.[57–59] AOTFs will provide fast discrete wavelength selection over a broad range. The AOTF comprises a birefringent crystal such as TeO_2 bonded to a piezoelectric transducer.[57–59] When a rf (radio-frequency) signal is applied to the transducer, the propagating acoustic wave produces a periodic modulation of the index of refraction within the crystal. This creates a moving phase grating that diffracts portions of incident radiation. They are currently available for coverage of the wavelength range from ~200 nm through to ~4500 nm, although individual filters only cover ranges of between about 150 nm and 1500 nm, offering resolutions of between 2 nm (50 cm^{-1}) at the low wavelength end[59,60] to about 20 nm at the high wavelength end.[60] An AOTF monochromator is at the heart of a novel NIR fibre-probe spectrometer;[61] it is capable of acquiring 4000 data points per second over the range 840 nm to 1700 nm (standard configuration). AOTF technology in combination with a CCD detector has also been employed recently to produce a high-fidelity, no-moving-parts, fast Raman imaging spectrometer.[58] Both AOTFs and LCTFs (Liquid Crystal Tuneable Filters) have been evaluated and compared as to their suitability for Raman imaging microscopy.[62]

[51] Workman J.J. and Burns D.A., in *Handbook of Near-Infrared Analysis*, ed. Burns D.A. and Ciurczak E.W., Practical Spectroscopy Series, Vol. 13, Marcel Dekker, New York, USA, ch. 4.
[52] Miran™ Infrared Analyzers, The Foxboro Company, MA, USA.
[53] Laboratory Application Data: LAD 001–002; LAD 001–013; LAD 001–014; LAD 001–030; LAD 001–032, The Foxboro Company, MA, USA.
[54] Sapphire™, The Foxboro Company, MA, USA.
[55] Scott R.M. and Ramsey J.N., *Microbeam Anal.*, 1982, **17**, 239.
[56] NanoSpec™/20IR, Nanometrics Inc., Sunnyvale, CA, USA.
[57] Wang X., *Laser Focus World*, 1992, **28**, 5, 173.
[58] Treado P.J., Levin I.W., and Lewis E.N., *Appl. Spectrosc.*, 1992, **46**, 8, 1211.
[59] Eilert A.J., *NIR News*, 1991, **2**, 3, 6.
[60] AOTFs, Brimrose Corporation, Baltimore, MD, USA.
[61] The Luminar 2000, Brimrose Corporation, Baltimore, MD, USA.
[62] Morris H.R., Hoyt C.C., and Treado P.J., *Appl. Spectrosc.*, 1994, **48**, 7, 857.

In the early 1990s holographic notch filters rapidly became the choice for a laser rejection filter in FT-Raman systems, replacing various interference filter arrangements. They have high reflectivity (>99.99%, equivalent to an optical density of >4) to the laser wavelength and steep band edges, which enables efficient rejection of Rayleigh scattering in FT and other Raman systems.[12,21–25,27,49] Their high transmission (>75%) affords much better throughput than triple monochromator Raman systems, although current technology limits observations to Raman shifts of greater than about 50 Δcm^{-1}.

Prisms and Gratings

Prisms and gratings, although slower than AOTFs, both generally have superior operational bandwidths. *Prisms*, the original dispersing elements in commercial mid-infrared spectrometers, with their frequently inherent hygroscopic nature and relatively poor transmission ranges, are no longer used as monochromators.

Reflection *diffraction gratings* have been the most widely used dispersive element, offering a higher resolving power than a prism. A grating may be either rotated, as in a monochromator to focus consecutive wavelength elements on to a mono-channel detector, or fixed, as in a spectrograph with the dispersed radiation being spread across a multi-channel detector. While both configurations find considerable use in Raman instrumentation, see later, the former was employed in conventional commercial dispersive infrared spectrometers. (IR linear array spectrometers have been developed for specialised time-resolved applications. In these instruments the InSb linear array detector is placed at the focal plane of the monochromator exit slit.)[63,64]

A diffraction grating for the infrared consists of a series of close, evenly spaced, lines (the pitch) 'ruled' onto an optical flat, usually aluminised glass. Each line (groove) has a tilted sawtooth profile to give it a blaze angle. This blaze angle disperses the radiation incident on it and directs it at a specific angle from the normal to the grating surface. A grating's wavelength dispersion is approximately constant over a wide angular rotation (*e.g.* 0° to 30°).[7] For coverage of the mid-infrared spectral region at least two gratings are normally required; for instance, the region above 2000 cm^{-1} may use a grating with between 300 and 100 lines per mm, while below 600 cm^{-1} a grating with between 25 and 9 lines per mm may be used. Filters are used in combination with the gratings to 'order sort' overlapping spectral information; integer multiples of a radiation frequency are diffracted to the same angle. Replica ruled diffraction gratings for the infrared are mostly cast from a master.

For good dispersion of the shorter wavelength visible Raman radiation, higher pitch gratings must be employed. Typically 1200 and 1800 grooves per

[63] Alawi S.M., Krug T., and Richardson H.H., *Appl. Spectrosc.*, 1993, **47**, 10, 1626.
[64] Richardson H.H., Pabst V.W., and Butcher J.R., *Appl. Spectrosc.*, 1990, **44**, 5, 822.

mm used in combination will allow wide spectral coverage and good resolution. Holographic rather than ruled gratings are now extensively used to disperse Raman radiation.

The angular dispersion rad nm^{-1} of a grating is given by:[65]

$$\frac{d\theta}{d\lambda} = \frac{pn10^{-6}}{\cos\theta_p} \quad (1)$$

where, $d\theta$ is the angular separation (radians) between two spectral wavelengths λ nm and $\lambda + d\lambda$ nm, and n is the grating groove density (grooves/mm) and θ_p the diffraction angle for the pth order (integer value).

The linear dispersion is expressed as the extent to which a spectral interval is spread across a focal plane; the smaller the interval (*i.e.* the greater the angular dispersion), the higher the dispersion, and the greater the potential for resolving fine spectral detail,[65] see Equation 2. In a given order θ_p increases as λ increases, that is the dispersion is greatest towards the red end.

Monochromators (Dispersive Spectrometers)

The components which comprise the optical train of a monochromator are essentially: an entrance slit; a collimating mirror to produce a parallel beam of light; the dispersing element – a grating; a second mirror, if the grating is plane, to re-focus the radiation; and an exit slit. The entrance and exit slits of a monochromator are frequently coupled and usually maintained at equal width, or, ideally, the exit slit-width should be equal to the image of the entrance slit-width. This slit-width and the dispersion of the grating primarily govern the resolution and spectral bandpass of the monochromator. Traditionally, multiple monochromator systems with intermediate slits were necessary to discriminate the weak Raman scatter from the strong Rayleigh scatter. High resolution (3 cm^{-1}), high throughput Raman spectrograph systems are now available, which utilise two volume-phase hologram gratings with a groove density of 2400 grooves/mm combined in a single optical element, each one differently centred to give wide spectral coverage.[38] This transmission grating simultaneously disperses the Raman radiation, splits it in half, and spatially displaces the two segments across two rows of a CCD detector. Thus a single spectrum from about 100 to 4000 Δcm^{-1} may be obtained in a short exposure time.[27,38] A recent system[45] extends single spectral coverage from the anti-Stokes (~ -1800 Δcm^{-1}) through the filtered exciting line to $\sim$4000 Δcm^{-1} on the Stokes side. The resolution of a monochromator system may be considered as its ability to resolve overlapping bands, remembering that the higher the resolution (*i.e.* the narrower the slits) the lower the throughput (*i.e.* the poorer the signal-to-noise characteristics) of the spectrum.

[65] (a) Turner J.M. and Thevenon A., *The Optics of Spectroscopy. A Tutorial V2.0*, J-Y Optical Systems, Instruments SA Inc., Longjumeau, France, 1988; (b) *Guide for Spectroscopy*, Instruments S.A., Inc., Jobin Yvon/Spex Division, Longjumeau, France, 1994.

If for a monochromator, f_x is the effective exit focal length, *i.e.* the distance between the exit slit and either the focusing mirror (for a plane grating system) or a concave grating, and z the unit interval in mm,[65] then the linear dispersion (nm/mm) perpendicular to the diffracted beam at a central wavelength is given by:[65]

$$\frac{\mathrm{d}\lambda}{\mathrm{d}z} = \frac{10^6 \cos\theta_p}{pnf_x} \qquad (2)$$

Equation 2 shows that the linear dispersion is inversely proportional to the exit focal length, the diffraction order and the grating groove density; it varies directly as the cosine of the angle of diffraction.

For a collimated beam incident on a grating at an angle of incidence α, then the grating equation gives:

$$10^{-6}pn\lambda = \sin\alpha + \sin\theta_p \qquad (3)$$

The wavelength of the radiation focused on the exit slit of a monochromator is determined by the angular position of the grating. In most spectrometers a grating is rotated about its centre, so that:

$$[\theta_p - \alpha] = \text{a constant} \qquad (4)$$

Blazed grating groove profiles are calculated for the Littrow condition,[65] when the incident and reflected rays are in autocollimation, *i.e.* they propagate along the same axis so that $\alpha = \theta_p$. The grating equation then becomes:

$$10^{-6}pn\lambda = 2\sin\theta_p \qquad (5)$$

(At the 'blaze' wavelength λ_B, the blaze angle $\omega = \alpha = \theta_p$.)[65]

For a grating operating in first order with all other orders filtered out, and for small collimated angles of incidence, Equation 3 gives:

$$\sin\alpha + \sin\theta_p \approx \text{a constant} \qquad (6)$$

Thus the grating equation is often simplified and quoted in the form:

$$p\lambda = 2d\sin\theta_p \qquad (7)$$

where, d the groove spacing in centimetres equals $(1/n \times 10^{-6})$, with the wavelength now measured in centimetres, such that its wavenumber, $\tilde{\nu}$ (cm^{-1}), equivalent becomes:

$$\tilde{\nu} = \frac{p}{2d}\operatorname{cosec}\theta_p \qquad (8)$$

The bandpass is the spread of wavelengths (or wavenumbers) which pass through the exit slit; it depends on both the linear dispersion and the exit (or image of the entrance) slit width, and may, assuming perfectly matched slits and minimal system aberrations, be defined as:

$$\text{bandpass} \approx \text{linear dispersion} \times \text{slit width.}$$

(Note: dispersion varies as cos θ, but the value of θ depends on the angle of incidence, so bandpass varies as cos α.)[65]

Dispersive Mid-infrared Spectrometers: Ratio-recording, Double-beam

As already stated, few dispersive spectrometers are now produced commercially. Those that are tend to be intended more for educational purposes than for industrial analytical spectroscopy or academic research, being of very low cost and limited performance. Nevertheless, dispersive infrared double-beam spectrometers still occupy space on many laboratory benches; these will most likely be of the ratio-recording type, that being the latest technology at the time of their decline. In a double-beam optical arrangement, the source radiation is divided into two beams, a sample beam and a reference beam.

Figure 4 shows schematically two different monochromator designs for infrared dispersive spectrometers. A schematic layout of a double-beam, ratio-recording instruments is shown in Figure 5. It features pre-sample chopping in addition to post-sample ratio-recording modulation. This additional modulation prevents the detection electronics from measuring infrared emission which might emanate directly from the sample as a consequence of its temperature being raised through absorption processes. [If measured, this emitted radiation would have a similar effect on an absorption spectrum to that of stray-light (see p. 128, effect 5).]

Broadband radiation, most of it in the infrared, emitted from a heated (incandescent) ceramic tube source is divided into two beams, which are alternately blocked by the pre-sample chopper mirror, such that pulses of radiation are produced. These pulses then pass alternately through the sample compartment and reference beam paths. The radiation pulses which constitute the sample beam will be attenuated by the absorption characteristics of any sample placed in its path and reduced by other loss processes, such as reflection from accessory mirrors. After the sample compartment the beams are recombined in space by the action of a second beam switch (rotating sector, chopper) mirror, which is synchronised with the pre-sample chopper. The beam, consisting of alternate pulses of radiation emanating from the sample and reference beams, is then passed through a pupil baffle, perhaps a common-beam polariser if required by the experiment, and an optical filter before being focused onto the entrance slit of the monochromator. (The optical filter, selected from a set (7 interference, 1 transmission, for the system shown in Figure 5) on a filter-wheel for the spectral region being scanned, rejects higher order radiation that would

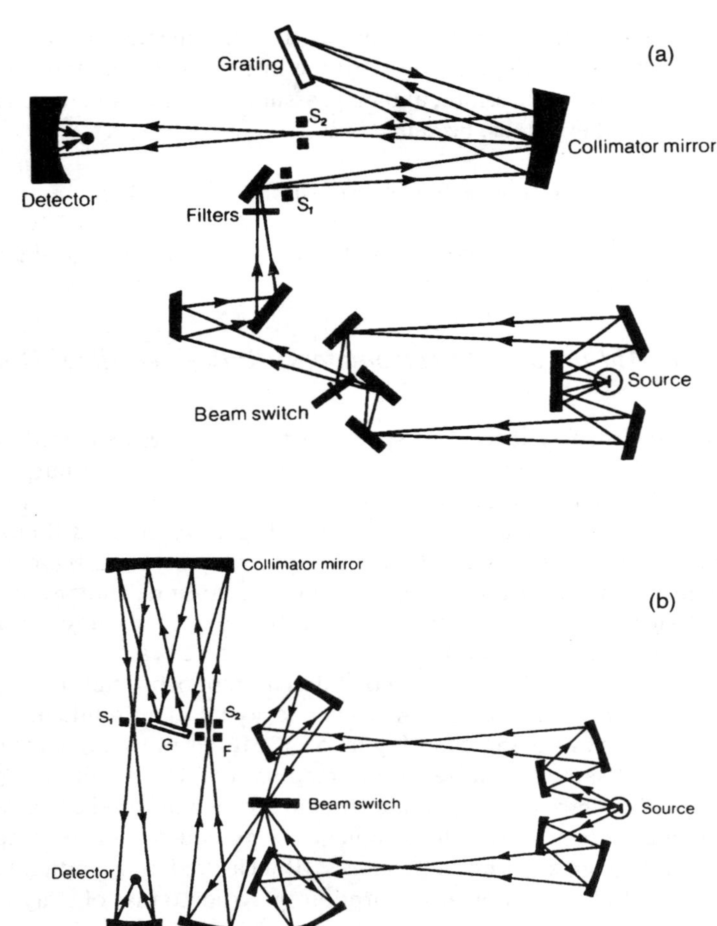

Figure 4 *Schematics of two infrared dispersive spectrometers:* (a) *employing a Littrow, monochromator;* (b) *employing an Ebert monochromator*
(Reproduced with kind permission of Thermo Unicam, from *Principles and Practices of Infrared Spectroscopy*, 2nd Edition, by R.C.J. Osland, published by Pye Unicam Ltd., Cambridge, UK. (1985))

be diffracted from the grating in the monochromator at the same angle as the desired wavenumber element.)

The monochromator depicted in Figure 5 actually contains four automatically selectable gratings, all used in their first order. The ranges they cover are 4000–1980 cm^{-1}, 1980–630 cm^{-1}, 630–330 cm^{-1} and 330–180 cm^{-1}, although the exact limits are variable by about 30 cm^{-1} if one needs to operate close to a 'change-over' wavenumber. (The grating rulings per mm are 288, 96, 25 and 9 respectively.) The combined pulse beam passing through the monochromator is dispersed by the selected grating, which is rotated such that the dispersed spectrum scans across the exit slit of the monochromator. The width of this slit

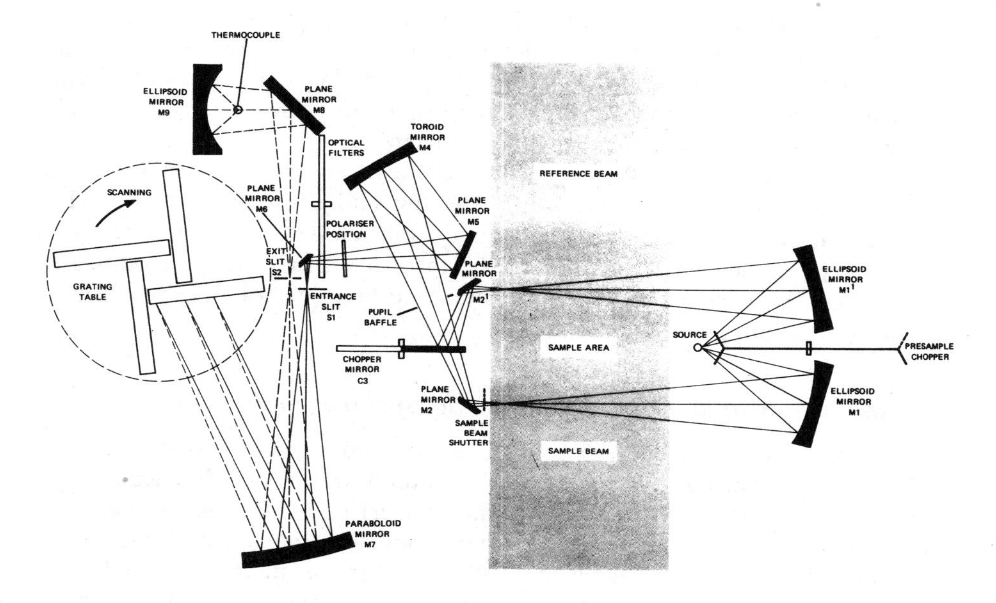

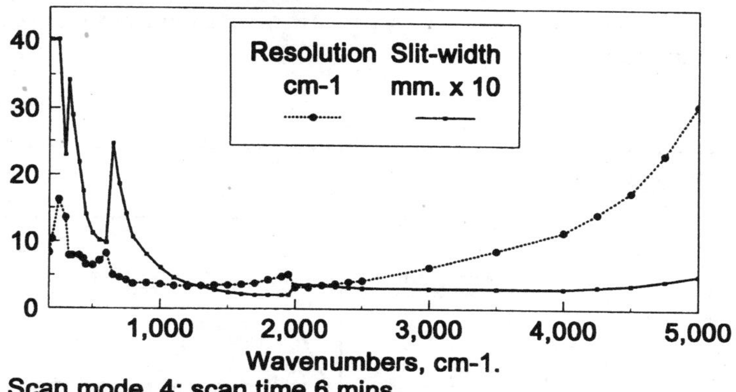

Figure 5 Top: *Optical system design of the Perkin-Elmer 983 ratio-recording dispersive infrared spectrometer*
(Reproduced by kind permission of Perkin-Elmer Ltd, Beaconsfield, UK)
Bottom: *Variation of resolution and slit-width of the Perkin-Elmer 983 infrared spectrometer scanning at scan mode 4 (scan time* 6 min*). The values plotted were read from the instrument display panel*

(maintained equal to the entrance slit by bilateral coupling), which determines the width of the wavenumber band emerging from the monochromator (*i.e.* the spectral resolution), is usually varied in order to maintain approximately constant energy at the detector, see example in Figure 5. Thus the spectral resolution will vary across a spectrum; it will be relatively very narrow in regions of high source energy (see Figure 1), increasing in width as the source energy reduces. As each grating is switched into use, the slits will be reset to maintain a near constant energy profile. A fixed slit-width would yield a constant spectral resolution spectrum, but of vastly different signal-to-noise characteristics at its extremes compared to those near 2000–1000 cm^{-1}.

The consecutive wavenumber resolved elements are then focused and arrive at the thermocouple detector for processing and subsequent display. The amplified, demodulated alternating signal from the detector yields the signal (intensity) of the two beams separately; these intensities are then ratioed to determine the attenuation (*e.g.* absorption and loss) of radiation by a sample placed in the sample beam.

Dispersive Raman Spectrometers and Spectrographs

In contrast to infrared, dispersive instrumentation remains to the fore for Raman spectrometers. Two modes of operation are commonly encountered: monochannel (spectrometer, spectrophotometer) and multichannel (spectrograph), both of which may be available in a single instrument. With the advent of holographic optical elements and CCD detectors, single mode multichannel systems with their high speed and sensitivity are rapidly becoming the most appropriate dispersive Raman instrument for many industrial applications.[26,27,29,38] Figure 6 is an optical schematic of an axial transmissive spectrograph configuration.

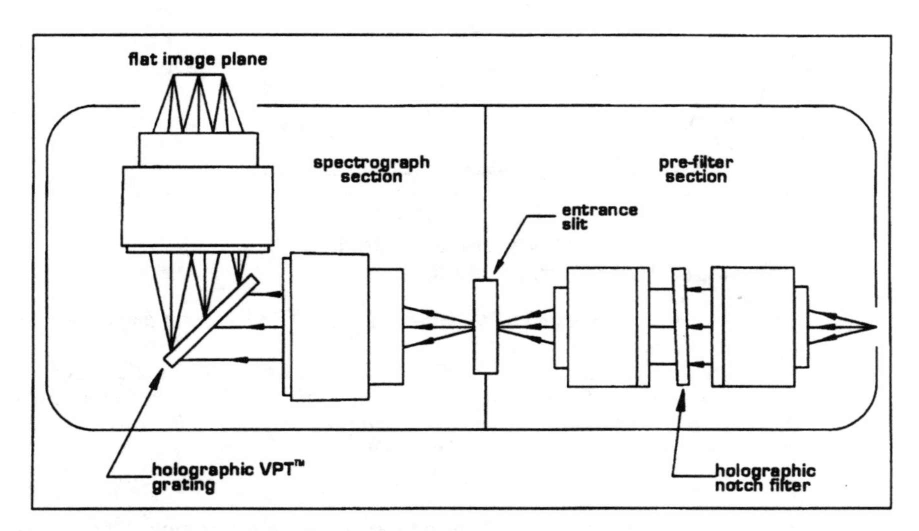

Figure 6 *Optical schematic of the axial transmissive spectrograph used in the HoloSpec f/1.8i Raman system*
(Reproduced by kind permission of Kaiser Optical Systems, Inc., Ann Arbor, MI, USA)

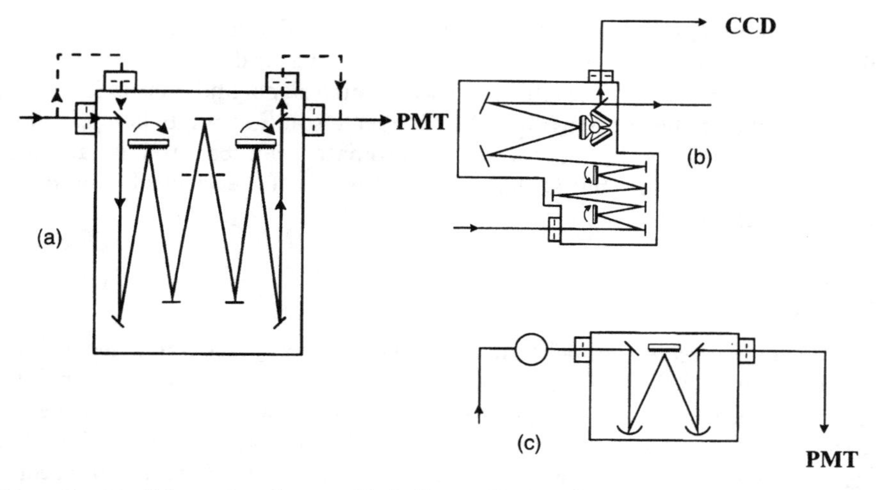

Figure 7 (a) *Schematic of a modified Czerny-Turner f/7.8 aperture,* 0.85 m *double monochromator Raman spectrometer (SPEX 1403);* (b) *Schematic of the SPEX 1877 triple spectrograph for multichannel detection. A zero-dispersion double spectrometer operating in subtractive mode to eliminate elastically scattered radiation before directing the Raman light into the spectrograph stage, where it may be dispersed by one of three gratings with different groove densities;* (c) *Schematic of single-grating spectrometer (SPEX M series) fitted with a holographic notch filter in front of its entrance slit. (PMT = photo-multiplier detector)*
(Diagrams reproduced by kind permission of Instruments S.A. (UK) Ltd)

Until the advent of holographic optical elements, because of the need for high rejection of the intense elastically Rayleigh scattered radiation, single monochromator systems were largely inappropriate; double or triple monochromator instruments were needed for high dispersion (wavenumber resolution) and good stray-light rejection. The principal advantage of triple monochomator systems are their very high stray-light rejection, and they remain, perhaps, the optimal configuration for work at wavenumber shifts very close to the exciting line ($< \pm\Delta 30$ cm^{-1}), but at a cost. They have relatively low throughput, and are relatively expensive to manufacture, but have wide capabilities.

A common configuration found in many Raman instruments in industrial analytical laboratories is based around a double monochromator system, see Figure 7. The double monochromator, which uses two gratings, has two modes of operation; these are referred to as the double additive and double subtractive cases. In the former, the second monochromator improves the stray-light rejection and enhances the resolution for single channel operation, see Figure 7(a). (In some instruments there is an option to add another monochromator and extend the train to a triple additive configuration.) In the subtractive mode, the double monochromator acts essentially as a fore-monochromator, to minimise the elastically scattered light, before directing the Raman radiation onto the grating of a spectrograph for multichannel detection, see Figure 7(b).

The two monochromators act in reverse and in combination to provide spatial filtering of the Rayleigh light; the first monochromator disperses the radiation, while the second recombines the dispersed Raman signal and focuses it onto the entrance slit of the spectrograph. For higher resolution multichannel studies with the spectrograph, the fore-monochromator can be used in a double additive configuration, but at the expense of width of the spectral band passed to the spectrograph.

Interferometers

Unlike a dispersive monochromator, which spatially separates the spectral radiation into individual components for successive measurement, an interferometer measures all of the spectral radiation simultaneously;[3,4,6,18] a spectrometer based on an interferometer is a multiplex detection system. This offers very distinct advantages over a wide range of applications for infrared spectrometers used within analytical and other industrial laboratories; for Raman, the principal justification is very specific, namely, avoidance of fluorescence emission in the visible region.[13–18,32,33,66] In assessing the FT-Raman case, its value needs to be carefully compared with the development and performance of modern spectrograph systems using long wavelength visible laser excitation, such as the Ti-sapphire tuned to 760 nm (or lower power solid-state diode lasers, for example, operating at 785 nm ($\sim$250 mW), see Section 6 of Chapter 6), particularly where microsampling is a consideration,[67,68] although the spectrograph approach may prove a disadvantage if fluorescence is of particular concern![69,70]

The Michelson Interferometer

The most common interferometer used in commercial vibrational spectrometers is a Michelson type, or a derivation thereof.[3,4,6,18,71,72] Its principle of operation is best illustrated initially by considering its action with a monochromatic (*i.e.* single wavelength, λ) radiation source. The components of a Michelson interferometer, illustrated in Figure 8, comprise a beam-splitter and two mirrors, one fixed in position, while the other is scanned back and forth. Collimated monochromatic radiation incident on an ideal beamsplitter will be

[66] Authors various in *Fourier Transform Raman Spectroscopy*, ed. Chase D.B. and Rabolt J.F., Academic Press, San Diego, USA, 1994.
[67] Williams K.P.J., Dixon N.M., and Mason S.M., *Proc. 13th Int. Conf. Raman Spectrosc., Wurzburg, Germany*, 1992, J. Wiley, Chichester, UK, 1992.
[68] Mason S.M., Conroy N., Dixon N.M., and Williams K.P.J., *Spectrochim. Acta*, 1993, **49A**, 5/6, 633.
[69] Hendra P.J., Pellow-Jarman M.V., and Bennett R., *Vib. Spectrosc.*, 1993, **5**, 311.
[70] Wilson H.M.M., Pellow-Jarman M.V., Bennett B., and Hendra P.J., *Vib. Spectrosc.*, 1996, **10**, 89.
[71] Perkins W.D., *J. Chem. Ed.*, 1986, **63**, 1, A5.
[72] Geick R., *Fresenius Z. Anal. Chem.*, 1977, **288**, 1.

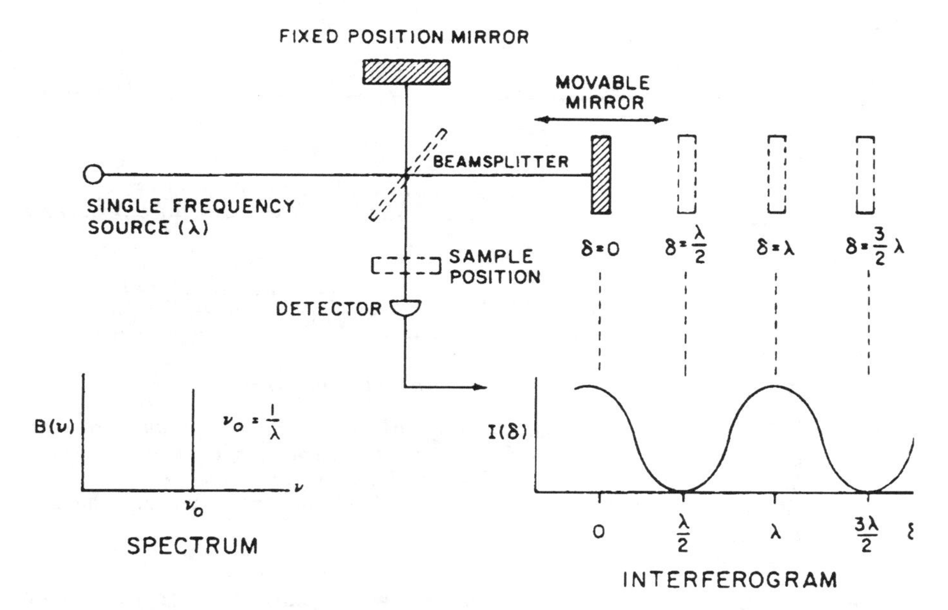

Figure 8 *Schematic of a Michelson interferometer. The figure also shows the spectrum of an infinitely narrow line source and how its cosine function interferogram is generated as the moving mirror is translated*
(Reprinted with permission from ref. 71. © (1986) American Chemical Society)

divided into two equal intensity beams, such that 50% is transmitted to one of the mirrors, while 50% is reflected to the other mirror. These two beams are reflected back and returned to the beamsplitter, where they are recombined and 50% is sent to the detector via the sample compartment, while the other 50% is essentially lost, since it is returned towards the source. As the moving mirror is scanned, the path difference between the two recombined beams will be varied. When this path difference is an integral number of wavelengths, then the beams will be in phase and constructive interference will occur between the recombined beams, *i.e.* their intensities will be additive. Destructive interference results when the optical path difference between the two beams is $\lambda/2$, *i.e.* the moving mirror has moved (been displaced) a distance equivalent to $\lambda/4$ from the in-phase position. The optical path difference is sometimes referred to as the retardation (or optical retardation). If the moving mirror is scanned at constant velocity, the detector will measure a sinusoidally varying signal as the beams move in and out of phase. The intensity measured by the detector is therefore a function of the moving mirror displacement.

The Interferogram. As can be seen from Figure 8, as the moving mirror travels away from the position of zero path difference between the two recombined interferometer beams the intensity of radiation reaching the detector varies as a

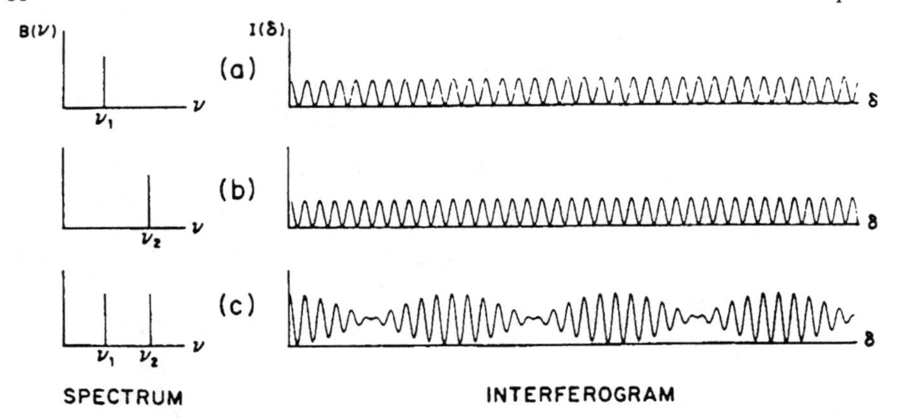

Figure 9 *Schematic showing the summation of the two cosine function interferograms generated from a two-frequency source. The periodicity of each cosine function interferogram varies with the frequency of the emitting source*
(Reprinted with permission from ref. 71. © (1986) American Chemical Society)

cosine function of the optical retardation, which may be expressed as an equation of the form:[3,4,18,71–73]

$$I(\delta) = B(\tilde{\nu})\cos(2\pi\delta/\lambda) = B(\tilde{\nu})\cos(2\pi\delta\tilde{\nu}) \tag{9}$$

where, $I(\delta)$ is the intensity of radiation at the detector, δ is the optical retardation and $B(\tilde{\nu})$ represents the single-beam spectrum intensity at wavenumber $\tilde{\nu}$. $I(\delta)$ is the cosine Fourier transform of $B(\tilde{\nu})$. The single-beam spectrum $B(\tilde{\nu})$ is calculated by computing the cosine Fourier transform of $I(\delta)$.

The monochromatic source of wavelength λ (the spectrum), and the cosine wave (its interferogram), represent a Fourier transform pair.

Figure 9 illustrates the situation for a two wavelength source.

For a polychromatic source, such as a broad band infrared source or a Raman spectrum, the interferogram represents the summation of all the individual cosine functions corresponding to each of the wavelengths (wavenumbers) in the source. These will only all be in phase at the position of zero path difference, hence the strong interferogram centre-burst, see Figure 10.

For a continuum source equation (9) becomes,[3,4,18,71,73]

$$I(\delta) = \int_{-\infty}^{+\infty} B(\tilde{\nu})\cos(2\pi\delta\tilde{\nu})\mathrm{d}\tilde{\nu} \tag{10}$$

and its spectrum can be calculated from,

$$B(\tilde{\nu}) = \int_{-\infty}^{+\infty} I(\delta)\cos(2\pi\tilde{\nu}\delta)\mathrm{d}\delta \tag{11}$$

[73] Bell R.J., *Introductory Fourier Transform Spectroscopy*, Academic Press, New York, USA, 1972.

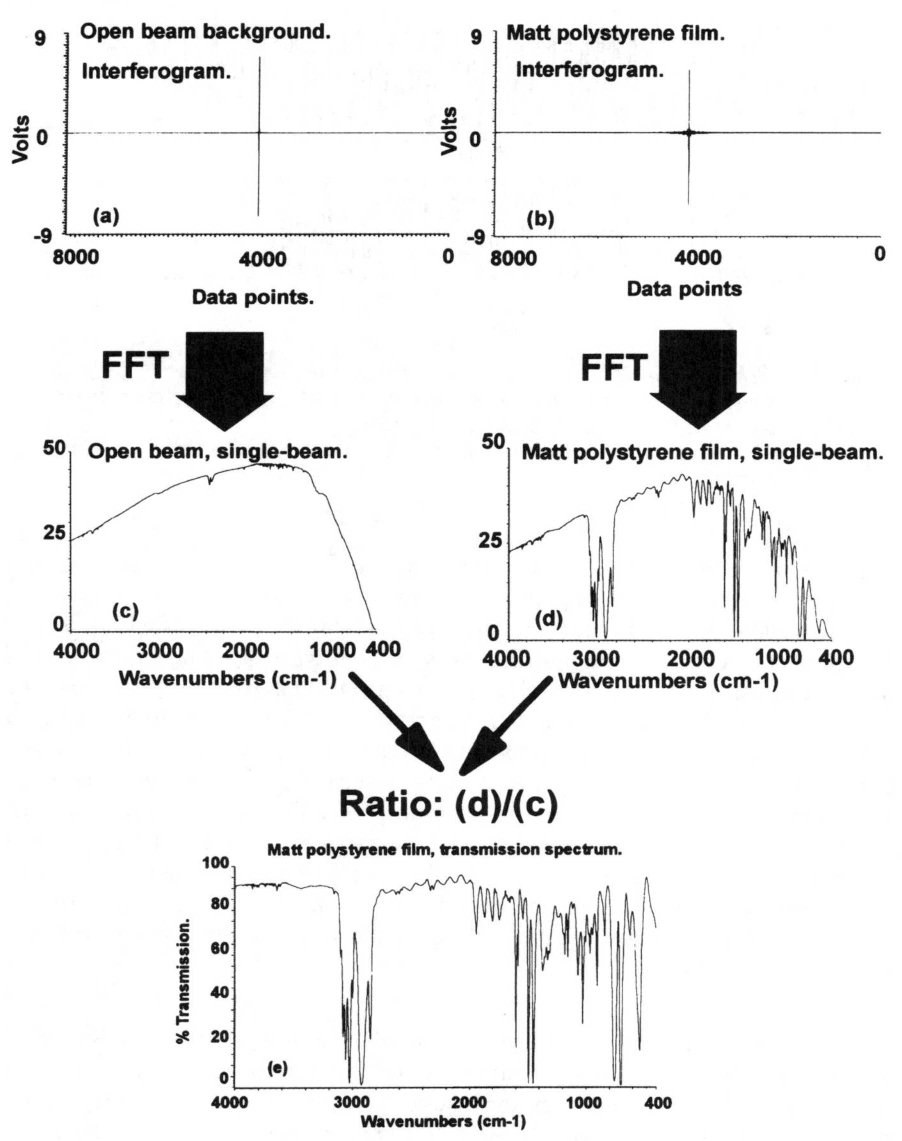

Figure 10 *Summary diagram of the principal steps involved in generating a FTIR transmission spectrum from a sample, in this case a matt polystyrene film. (FFT stands for fast Fourier transform)*

Since the signal magnitude of the centre-burst is large compared to that in the wings of the interferogram, where, comparatively, the signal-to-noise may be low but much of the spectrally important information is encoded, the ADC (analogue-to-digital converter) of the detection system must have a large dynamic range, typically 10^{16} or better.[3,18]

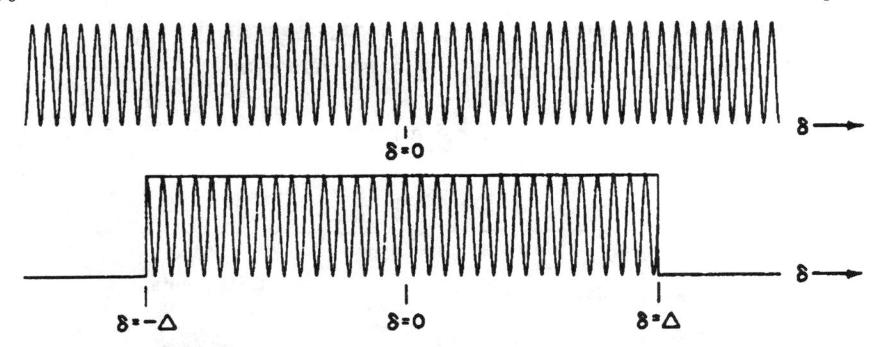

Figure 11 *Schematic of Boxcar truncation of the interferogram for a single frequency source. The interferogram is recorded for optical retardation δ cm between − Δ and + Δ, δ = 1/2 × optical path difference*
(Reprinted with permission from ref. 71. © (1986) American Chemical Society)

Apodisation. The moving mirror only moves a finite distance, since it cannot travel from minus infinity to plus infinity, so the interferogram is truncated by a boxcar function, see Figure 11, the Fourier transform of which is a sinc function. The effect of convoluting the spectrum with this sinc function is to introduce positive and negative side lobes to spectral bands. These lobes may be minimised (or effectively eliminated) to greater or lesser extents (depending on the true bandshape), at the expense of (increased) band width and (decreased) band intensity, by replacing the boxcar truncation function with another algebraic (weighting) function,[3,4,18] see examples in Figure 12. These functions are known as *apodisation* functions. They may be operator selective. The computer default option selected by the instrument manufacturers is not always the same, and while they may represent the best compromise for the 'average' sample, special experiments (*e.g.* high resolution) may benefit from the use of an alternative.[3]

Spectral Resolution. The spectral resolution of a Fourier transform spectrum is implicit in how far the moving mirror of the interferometer has travelled. To resolve two bands in a spectrum, the extent of travel of the moving mirror must equal or exceed that necessary to register one complete beat pattern generated between the two cosine waves in the interferogram that represent the wavenumber positions of the two bands. For a conventional Michelson interferometer, the spectral resolution (cm^{-1}) is given approximately by:

$$\Delta\tilde{\nu} = 1/\delta_{max} \tag{12}$$

where δ_{max} is the maximum optical retardation, equal to twice the distance travelled by the moving mirror from its position at zero path-length difference (ZPD) between the two beams. Thus:

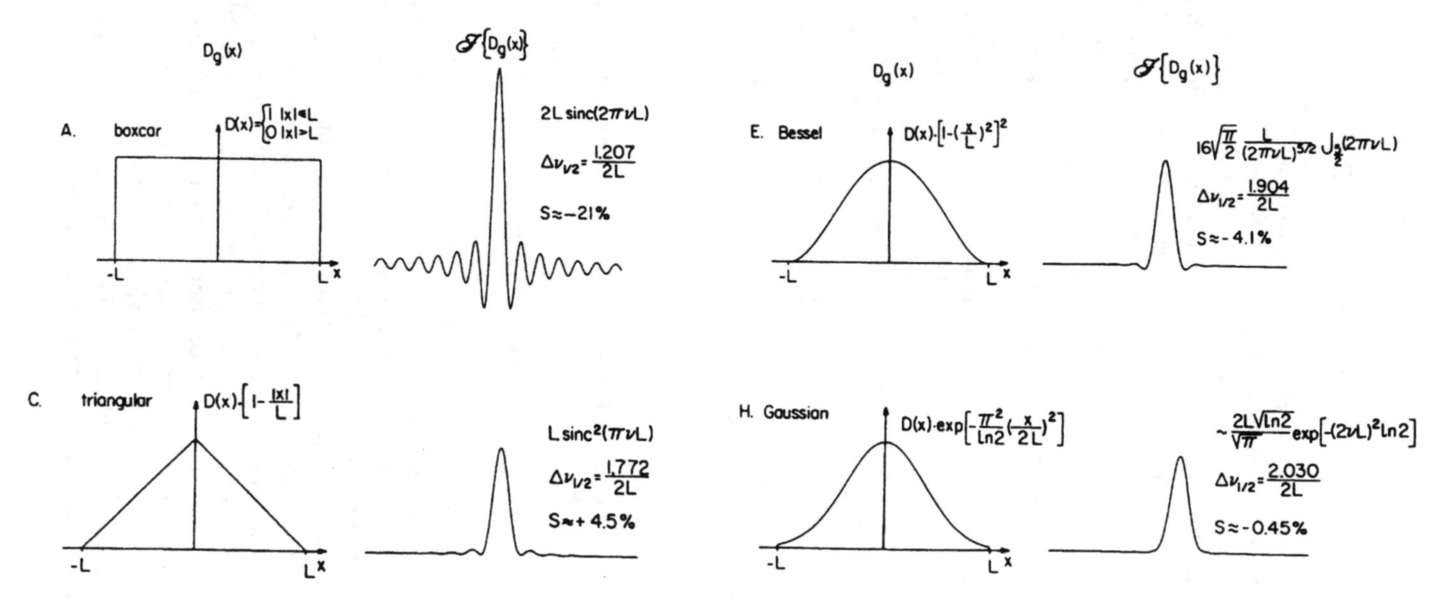

Figure 12 *Some apodisation functions $D_g(x)$ and their corresponding line shape functions $\mathcal{F}\{D_g(x)\}$. The full widths at half-height, $\Delta\nu_{1/2} = a/2L$, of the line shape function and the relative magnitude S (%) of the strongest sidelobe of the line shape function are listed; see original reference for full details*
(Reproduced from Kaupinnen J.K., Moffat D.J., Cameron D.G., and Mantsch H.H., *Applied Optics*, 1981, **20**, 10, 1866, by kind permission of The Optical Society of America, Inc.)

a mirror travel from ZPD of ~0.25 cm will yield a spectrum of 2 cm^{-1} resolution,

a mirror travel from ZPD of ~2 cm will yield a spectrum of 0.25 cm^{-1} resolution.

The recorded interferogram is a set of equally spaced digitised values of signal intensity. It may be recorded as either a single-sided or a double-sided interferogram.[3,4] The Nyquist Criterion (theorem) may be stated as, there must be at least two recorded data points per resolution element to unambiguously define a sinusoidal waveform. The sampling trigger for recording the interferogram data points is provided by a second interferogram generated within the interferometer from a low power reference He–Ne laser. The zero-crossings of this sinusoidal signal (interferogram) from the 632.8 nm wavelength He–Ne emission line are used to trigger sampling and define the sampling interval.

Spectral Bandwidth and Interferogram Data Points. If a data point is recorded every other zero-crossing of the He–Ne laser interferogram, then the data point spacing will be 632.8 nm. The maximum wavelength bandwidth that may be covered is then from 0 to 1265.6 nm, which corresponds to wavenumber bandwidth from 0 to 7902 cm^{-1}. This adequately covers the mid-infrared region; optical and electronic filters may be employed to restrict the bandwidth detected.

For a mirror travel of 0.25 cm from ZPD, the number of data points for a single-sided interferogram would be:

$$(2 \times 0.25 \times 10^7)/632.8 = 7901. \quad (13)$$

For a mirror travel of 2 cm from ZPD, the number of data points for a single-sided interferogram would be:

$$(2 \times 2 \times 10^7)/632.8 = 63\,211. \quad (14)$$

In practice, the number of data points is taken from the binary scale, so the above two values would be extended to become 8192 (2^{13}) and 65 536 (2^{16}), respectively; this is done to optimise utilisation of the Cooley–Tukey algorithm[74] for fast Fourier transformation.[3,4,18,73]

The absolute wavenumber range of a 100 Δcm^{-1} to 4000 Δcm^{-1} Stokes shifted Raman spectrum generated through 1.064 μm excitation is 9298 cm^{-1} to 5398 cm^{-1}, which is equivalent to approximately 1.08 μm to 1.85 μm. This is beyond the maximum wavelength bandpass covered if a data point is only taken every other zero crossing of the reference laser. If a data point is taken every zero crossing of the reference laser then the data point spacing becomes 316.4 nm and the maximum wavelength bandwidth, according to the Nyquist requirement, decreases from 0 to 632.8 nm (*i.e.* 0.6328 μm). This equates to a maximum wavenumber within the near-infrared of 15 803 cm^{-1}.

[74] Cooley J.W. and Tukey J.W., *Math. Comput.*, 1965, **19**, 297.

For a mirror travel of 0.25 cm from ZPD, the number of data points for a single-sided interferogram would be:

$$(2 \times 0.25 \times 10^7)/316.4 = 15\,803, \tag{15}$$

extended to the binary value 16 384 (2^{14}).

The FFT (Fast Fourier Transform), Zero-filling and Spectral Data Point Density. The interferogram in commercial interferometers is usually processed by an algorithm commonly referred to as the fast Fourier transform (or FFT) method,[3,4,18,71–76] which was first realised by Cooley and Tukey[74] and promoted for Fourier spectroscopy by Forman.[75] This will produce one (real spectrum) data point per resolution element. When plotted directly this will produce a disjointed spectrum composed of a series of points connected by straight lines, see Figure 13. Clearly the point density of such a display is insufficient for measuring peak positions to any reasonable degree of precision, and it is likely that exact coincidences between spectrum data points and absorption band maxima will be very rare. A mathematically justifiable method[3,4,18,76] of increasing the number of spectral data points is to zero-fill the interferogram. It is a very common pratice to zero-fill to twice the number of interferogram data points, so that the number of transform points corresponding to Equations (13) and (14) above would become 16 384 (2^{14}) and 131 072 (2^{17}) respectively. The number of transform points must be an integer multiplication of the number of interferogram data points, and higher levels of zero-filling are used, but at the expense of computation time.[3,4] Further interpolations to increase point density may also be applied by, for example, taking blocks of consecutive data points and fitting a polynomial through these. Increasing the number of spectral data points in these ways may be essential to many qualitative and quantitative analytical processes, but it must be remembered that the spectral resolution is implicit and defined in the interferogram (before zero-filling) and *not* by the data point interval in the spectrum; the interpolation routines are primarily aesthetic functions implemented to improve band contours.

Signal Modulation. The conventional mode of operation of commercial FT spectrometers is the rapid-scanning mode, (step-scan operation is discussed later, see p. 112). The moving mirror is scanned at constant velocity, in order to maintain constant frequency of the laser reference signal. Signal modulation is an intrinsic property of rapid-scan FT systems. Each source wavelength results in an interference pattern that is sinusoidal, and which is modulated by a frequency determined by the velocity of the moving mirror (V cm s^{-1}); that is, the interferometer modulates each wavelength at a frequency, f (Hz), which is directly proportional to wavenumber.[3,4,6,72,73]

$$f_{\tilde{\nu}} = 2V\tilde{\nu} \tag{16}$$

[75] Forman M.L., *J. Opt. Soc. Am.*, 1966, **56**, 978.
[76] Brigham E.O., *The Fast Fourier Transform*, Prentice Hall, New Jersey, USA, 1974.

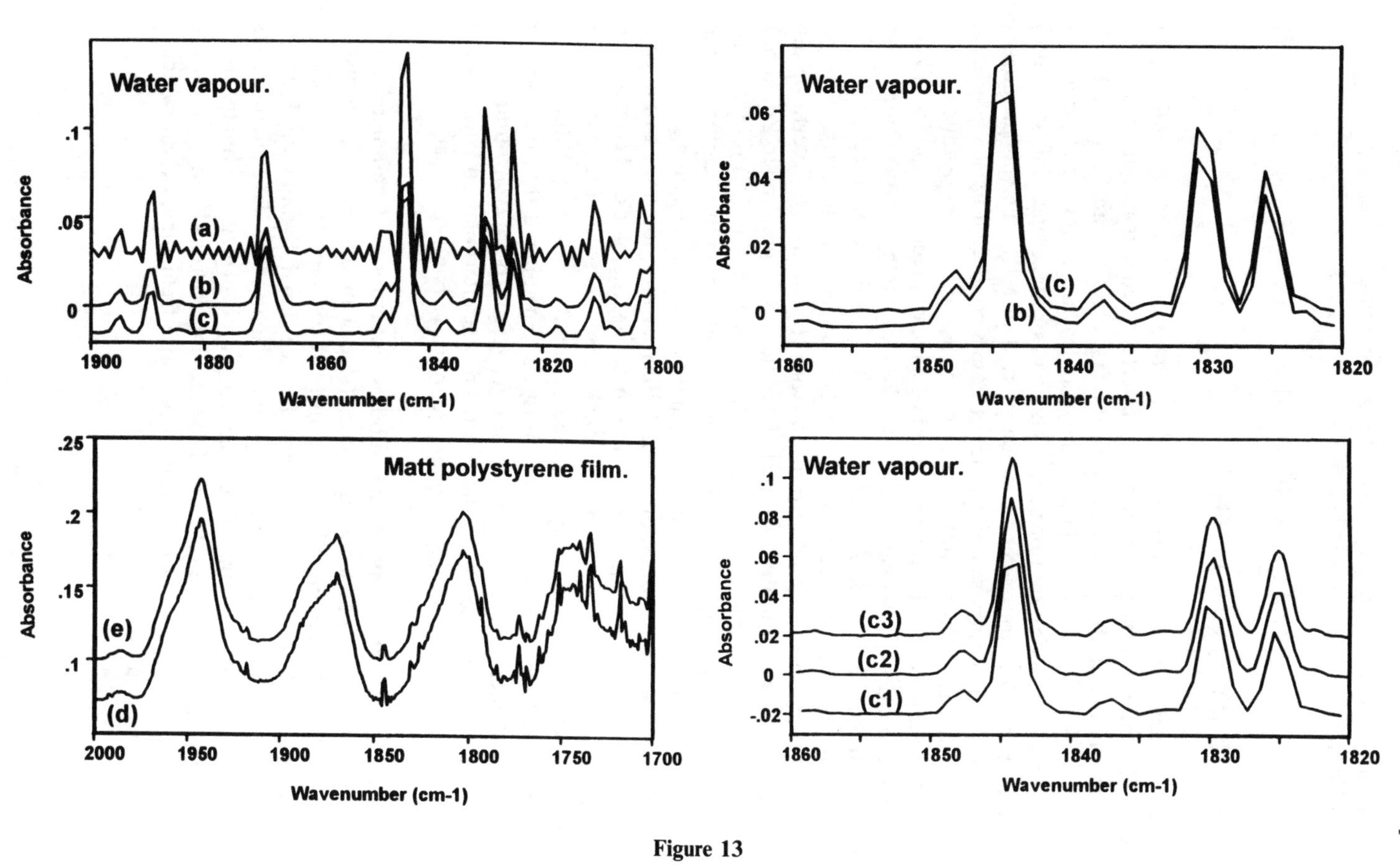
Water vapour.
(a)
(b)
(c)
Absorbance
.1
.05
0
1900
1880
1860
1840
1820
1800
Wavenumber (cm-1)
Water vapour.
(c)
(b)
Absorbance
.06
.04
.02
0
1860
1850
1840
1830
1820
Wavenumber (cm-1)
Matt polystyrene film.
(e)
(d)
Absorbance
.25
.2
.15
.1
2000
1950
1900
1850
1800
1750
1700
Wavenumber (cm-1)
Water vapour.
(c3)
(c2)
(c1)
Absorbance
.1
.08
.06
.04
.02
0
-.02
1860
1850
1840
1830
1820
Wavenumber (cm-1)

Figure 13

Scan Speed. Mirror velocities used with infrared detectors are typically ~2 mm s^{-1} (DTGS detector) and ~25 mm s^{-1} (MCT detector), while for Raman ~1 mm s^{-1} (InGaAs detector) is used. The modulation frequencies for a mid-range spectrum would then be approximately:

4000 cm^{-1}: 1600 Hz (DTGS) 20 240 Hz (MCT)
400 cm^{-1}: 160 Hz (DTGS) 2240 Hz (MCT)

and for the shorter wavelength Raman using 1.064 μm (9398 cm^{-1}) excitation:

4000 Δcm^{-1}: 10 796 Hz (InGaAs)
400 Δcm^{-1}: 17 996 Hz (InGaAs)

These are essentially in the audio-frequency range.

At these scan speeds, it is clear that the interferogram itself is recorded very rapidly.

For a mirror travel of 0.25 cm:
at 2 mm s^{-1} this would take 1.25 s;
at 25 mm s^{-1} this would take 100 ms.

(Clearly, to this time must be added processing time, display time, and maybe mirror fly-back time, *etc.* to get the true single-beam spectrum acquisition time interval.)

Potential Advantages of Interferometric Measurements of Vibrational Spectra

Three important and often quoted primary advantages of Fourier transform infrared (FTIR) spectrometers over dispersive infrared spectrometers are frequently referred to by the names of the pioneer scientists who first defined or were closely associated with the principles.[3,4,6,18,73] They are known as the Fellgett advantage, the Jacquinot advantage, and the Connes' advantage. These

---

Figure 13 Top left: *Open beam water vapour absorbance spectrum,* 1900–1800 cm^{-1}, 2 cm^{-1} *resolution,* 100 scans: (a) *Boxcar apodisation, no zero-filling;* (b) *Triangular apodisation, no zero-filling;* (c) *Happ-Genzel apodisation, no zero-filling. Note the reduction in band intensity on going from* (a) *to* (b) *and* (c)
Top right: *Scale-expanded spectra of* (b) *and* (c). *Note the slightly lesser reduction in band intensity in* (c) *compared to* (b)
Bottom right: (c1) ≡ (c), *Happ–Genzel apodisation, no zero-filling;* (c2) *Happ–Genzel apodisation, 1 level of zero-filling;* (c3) *Happ–Genzel apodisation, 2 levels of zero-filling. Note the apparent improvements in band profile and definition of peak maxima on going from* (c1) *to* (c2) *to* (c3)
Bottom left: *Matt finish polystyrene film plus water vapour absorbance spectrum,* 2000–1700 cm^{-1}, 2 cm^{-1} *resolution,* 100 scans; (d) *Boxcar apodisation, no zero-filling;* (e) *Happ–Genzel apodisation, no zero-filling. Note the greater significance of the apodisation function to the intrinsically narrower water vapour bands compared with the broader bands of the solid polymer.*
All these spectra were recorded on a Nicolet 750 FTIR spectrometer, equipped with a DTGS detector. A 'dry' open beam background interferogram was Fourier processed in a manner equivalent to each of the spectra cases above in order to provide appropriate single-beam backgrounds. Some of the spectra have been offset for clarity

are concerned with the multiplex, throughput, and wavenumber precision aspects of interferometric measurements, respectively.

The Fellgett, or Multiplex, Advantage. The multiplex advantage of an interferometer was first described by Peter Fellgett in his thesis and later published as ref. 77.

In a monochromator dispersive spectrometer the single-channel detector only observes one resolution element at a time, scanning successively through consecutive wavelength/wavenumber elements until it has built up a spectrum; typically this may take 10–20 min to cover the range 4000 cm^{-1} to 400 cm^{-1} or 4000 Δcm^{-1} to 400 Δcm^{-1}. In an interferometer all of the wavelengths/wavenumbers of the signal radiation are being detected simultaneously. Typically, for an equivalent wavenumber range spectrum, this may take 1 s or less. Thus, all other parameters being equal (*e.g.* source, detector, throughput, *etc.*), for equivalent signal-to-noise ratio (S/N) spectra there is clearly a significant time advantage in using an interferometer over a scanning dispersive spectrometer. This is put to considerable use in infrared kinetic studies and time-dependent observations (*e.g.* real-time GC-FTIR, see Chapter 7). (Rapid multiplexed spectra may also be recorded with a spectrograph, although there may have to be a trade-off between spectral bandwidth and resolution; modern Raman spectrographs using holographic optical elements are capable of producing good S/N, single scan, wide bandwidth, 3 cm^{-1} resolution spectra within a few seconds acquisition time.)[38]

The Fellgett or multiplex advantage may be considered as follows. In a system which is limited by detector noise (*i.e.* essentially random fluctuating noise), then, providing precise data point triggering is retained between successive scans (see Connes' advantage, below), successive interferograms may be co-added (superimposed) in order to improve the S/N characteristics of the resultant spectrum. If the spectrum recorded from a scanning dispersive single-channel spectrometer is composed of M resolution elements, then the spectrum at the same resolution transformed from a set of co-added interferograms recorded on an interferometer over the same time period will show a $\sqrt{M}$ better S/N, all other parameters being equal. This advantage is crucial to many modern applications involving mid-infrared spectroscopy, for which the S/N may be considered to improve as $\sqrt{\text{number of co-added scans}}$, up to the limit allowable by the dynamic range of the ADC. For Raman studies undertaken using visible or UV excitation, where detectors are shot-noise limited, this advantage is not realisable.[12,78,79] PMTs are photon shot-noise limited, and the noise increases as $\sqrt{\text{signal level}}$.

In the array detector of a spectrograph, signal levels may be increased either by increasing the integration (measurement) times or by co-adding successive signals, and they therefore exhibit a multiplex or Fellgett-like advantage.[19] For

77 Fellgett P., *J. Phys. Radium*, 1958, **19**, 187.
78 Hirschfeld T., *Appl. Spectrosc.*, 1976, **30**, 1, 68.
79 Everall N.J. and Howard J., *Appl. Spectrosc.*, 1989, **43**, 5, 778.

a diode array the former choice is usually preferable in order to minimise array read-out noise. This is generally not such a significant problem with CCDs and either or both methods may be used to good advantage, although again the former is preferable.[46] For optimum S/N, signal integration levels should be kept somewhat below the level at which signal saturation sets in.

The Connes's, or Wavenumber Repeatability, Advantage. In Connes's earlier instruments, the optical path difference was determined accurately using a monochromatic lamp beam (*e.g.* cadmium or mercury) as a reference channel.[4,80] This advantage nowadays ensues from the use of the He–Ne reference laser to monitor the position of the moving mirror and trigger data acquisition. This ensures an exact correspondence between consecutive data points in successive interferograms. Also, since the wavelength of the laser is known very precisely, the wavenumber positions of the data points in the Fourier transformed spectrum are known very accurately. This precision in data point repeatability and wavenumber position are critical to successful co-addition of interferograms, and the subsequent superimposition of spectra. They are invaluable spectral properties for the application of spectral subtraction processes and prerequisites for many analyses involving the comparisons of closely matched spectra. However, it must be emphasised that it is the position (wavenumber) of the data points which is repeatable and accurately known, not necessarily the position of spectral features. These will be affected by nuances between sample and reference preparation and presentation, which should be kept to a minimum for the realisation of optimum difference spectroscopy.

The abscissa precision of a data point in a FT spectrum recorded on a commercial spectrometer is usually quoted as better than 0.01 cm^{-1}.

The Jacquinot, or Throughput, Advantage. The Jacquinot advantage[81–84] is contained within the fact that in an interferometer the *etendue* (ability to accept radiation) is determined by the collimating area of the interferometer mirrors or the diameter of the interferometer circular entrance aperture, the so-called Jacquinot stop. The area of the Jacquinot stop is generally considerably greater than that of slits within the optical train of a monochromator system, and hence the etendue and thus the throughput are much greater in an interferometer. The extent of the gain will be both wavenumber and resolution dependent, but improvements in the order of 50–100 times are typical mid-range ($\sim$2000 cm^{-1} or $\sim$2000 Δcm^{-1}) values.[3,4,18,49,73,85]

Throughput may be defined as the usable photon flux at the exit of the instrument. Etendue is proportional to the product of the area of a beam at a

[80] Connes J. and Connes P., *J. Opt. Soc. Am.*, 1966, **56**, 7, 896.
[81] Jacquinot P. and Dufour P., *J. Rech. CNRS*, 1948, **6**, 91.
[82] Jacquinot P., *J. Opt. Soc. Am.*, 1954, **44**, 10, 761.
[83] Jacquinot P., *Rep. Prog. Phys.*, 1960, **23**, 267.
[84] Jacquinot P. and Girard A., in *Advanced Optical Techniques*, ed. Van Hell A.C.S., North Holland, Amsterdam, The Netherlands, 1967, ch. 3.
[85] (a) Geick R., *Topics in Current Chemistry*, 1975, **58**, 75; (b) Geick R., *Fresenius Z. Anal. Chem.*, 1977, **288**, 1.

focus (source, slit, or aperture) and the solid angle into which it propagates or the solid angle the source subtends with the area of a collimating mirror,[3,18,49,73,81–85] see schematics in Figure 14.

The definitions below are taken or adapted from refs. 65 and 86.

(1) The system flux for a spectrometer is given by the photons s^{-1} (or watts) emitted from a source or through a slit or aperture of given area into the solid angle at a given wavelength or bandpass.
(2) Intensity is the flux at a given wavelength or bandpass per unit solid angle (*e.g.* watts/steradian).
(3) Luminance (Brightness) is the luminous flux emitted from a surface per unit solid angle per unit of area, projected onto a plane normal to the direction of propagation. It is the intensity per surface area of the source, *e.g.* watts $steradian^{-1}$ cm^{-2}.
(4) Geometric etendue characterises the ability of a system to accept radiation. It is a function of the area of the emitting source and the solid angle into which it propagates. Etendue is a constant of the optical system and is determined by the least optimised segment of the entire optical train.

References 1 and 87 compare several performance characteristics of infrared dispersive and interferometric spectrometers.

FTIR, Fourier Transform Infrared Spectrometers

A wide range of FTIR spectrometers are now available, either separately or incorporated as specific detectors in hybrid systems or process analysers. Commercial spectrometers for laboratory use are all two-beam interferometers. They vary in optical design, performance, and flexibility, with the Michelson-type being the most common optical configuration, and, while dual beam optical null FTIR spectrometers have been produced,[88] the vast majority operate in a single beam mode. (For sole use in the far-infrared region, particularly at very low wavenumbers, other interferometer designs are more appropriate.[3])

The sequence of events leading to display of a mid-infrared absorption spectrum were summarised in Figure 10. The single beam background energy profile is a consequence of a combination of many factors, including source emission (see p. 66), detector response (see p. 69), beamsplitter properties, and residual atmospheric absorptions due to water vapour and carbon dioxide, see Figure 7 of Chapter 1. These atmospheric absorptions are a significant nuisance to many studies, and are usually minimised or eliminated. Many recent instruments contain sealed and/or desiccated optics, which serve two purposes;

[86] *The Photonics Dictionary*: Book 4 of The Photonics Directory, 1994 Edition, Laurin Publishing Company, Pittsfield, USA.
[87] Sheppard N., Greenler R.G., and Griffiths P.R., *Appl. Spectrosc.*, 1977, **31**, 5, 448.
[88] Galaxy™ 8020, Mattson Instruments, Madison, USA.

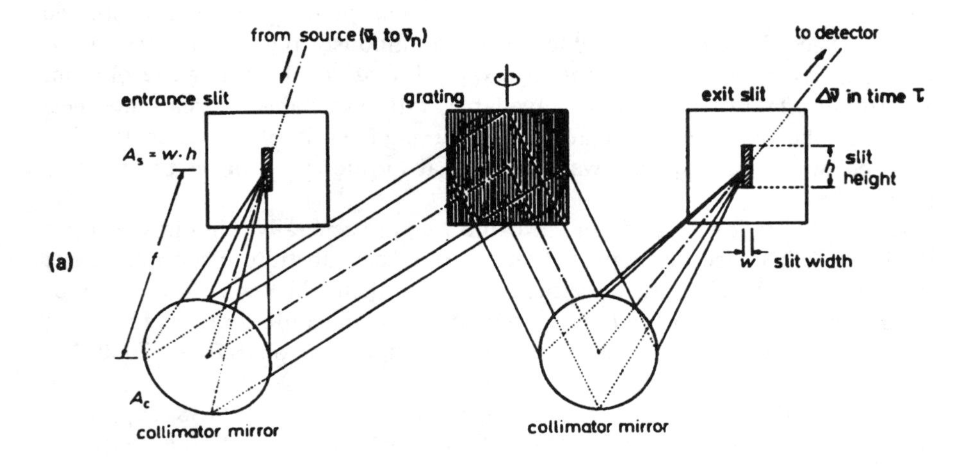

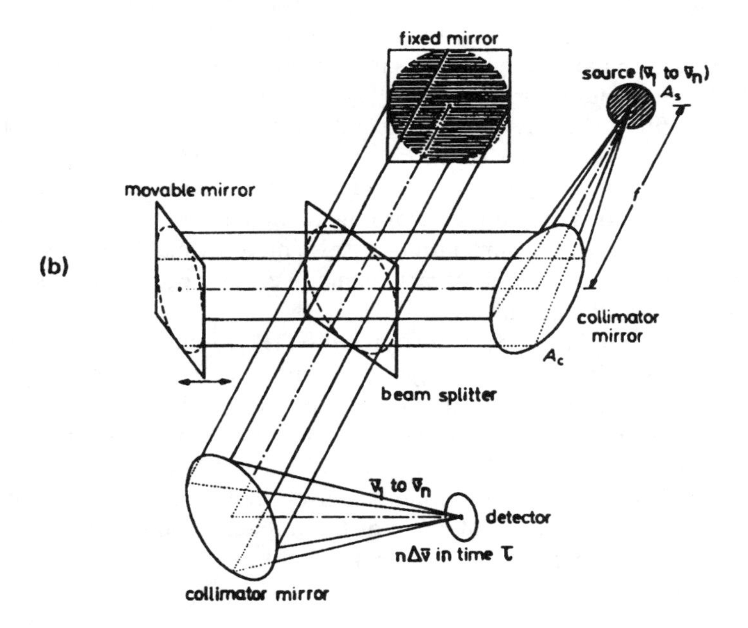

Figure 14 *Schematic representations of* (a) *a grating monochromator and* (b) *a Michelson interferometer. The figures illustrate the factors which determine the multiplex and throughput advantages. A_S, source area; A_C, collimator mirror area; f, focal length*
(Reproduced with permission from ref. 85b, *Fourier Methods in Analytical Chemistry* Geick R., *Fresenius Z. Anal. Chem.*, **288**, 1–18 (1977). Copyright (1977). Springer-Verlag GmbH & Co. KG)

the first is to better protect hygroscopic components such as a KBr beamsplitter substrate; the second is to reduce the system path open to atmospheric interferences. It is common practice to continuously purge with dry air or nitrogen any optical path potentially exposed to atmosphere. High resolution, research-grade instruments are available with evacuable optical modules; these can be particularly advantageous for high sensitivity work in the far-infrared, where atmospheric water vapour absorptions can prove particularly aggravating.

Beamsplitters for the mid- and near-infrared are generally thin films (<1 μm thickness) of a dielectric supported on a transparent substrate. Their reflectance will be dependent on their refractive index, their thickness, the angle of incidence of the beam, and the wavenumber of the radiation.[3,73] The most common mid-infrared beamsplitters are Ge supported on KBr ($\sim$7000–350 cm^{-1}) or, for extended range, supported on CsI ($\sim$6400–200 cm^{-1}). For the near-infrared, beamsplitters are materials such as Si, Fe_2O_3 or Sb_2S_3 supported on CaF_2 ($\sim$10 000–1200 cm^{-1}) or thin films of Al, TiO_2, Si or Fe_2O_3 on quartz ($\sim$50 000–4000 cm^{-1}, the range depending on the coating material). To compensate for dispersion by the beamsplitter substrate plate, a same material compensator plate of exactly the same thickness is placed on the other side of the reflective film. For the far-infrared, typical beamsplitters are different thickness films of poly(ethylene terephthalate), frequently referred to by a manufacturers tradename of this material, *i.e.* Mylar®. Typical thicknesses are in the range 12–100 μm, the interference fringe patterns generated offering different operating windows according to film thickness, (see pp. 183–186). Metal mesh/wire grid beamsplitters are also made available for some regions by some spectrometer manufacturers, as may be other proprietary beamsplitters, particularly for the far-infrared.

Since one might expect commercial FTIR spectrometers on delivery to perform to specification or designated purpose, any detailed discussion of the finer points, merits, and limitations of particular designs or components is somewhat academic, and outside the scope of this book; references 3 and 4, published in 1986 and 1991 respectively, are good sources of more detailed background information on commercial developments and instrumentation. However, a brief mention of some of the differences and specifications is still relevant to the discussions and comparisons of this chapter.

Figure 15 *Schematics of the optical lay-outs of two early commercial rapid scan FTIR spectrometers:* (a) *Digilab FTS 14, introduced in 1969*
(Reproduced from ref. 2, Chenery and Sheppard, *Appl. Spectrosc.*, **32**, 1, 79–89 (1978), by kind permission of the Society for Applied Spectroscopy)
(b) *Nicolet 7199, introduced in 1976. (*S1, *source;* D2, *detector;* L2, *reference laser;* L1, *alignment laser;* BS1, *beamsplitter and compensator;* BS2, *white light beamsplitter;* WS1, *white light source;* WD1, *white light detector;* M4, *fixed mirror;* M7, *flat mirror to external port;* M6, *multi-position mirror)*
(Reproduced with kind permission of Nicolet Instruments Corporation, Madison, WI, USA)

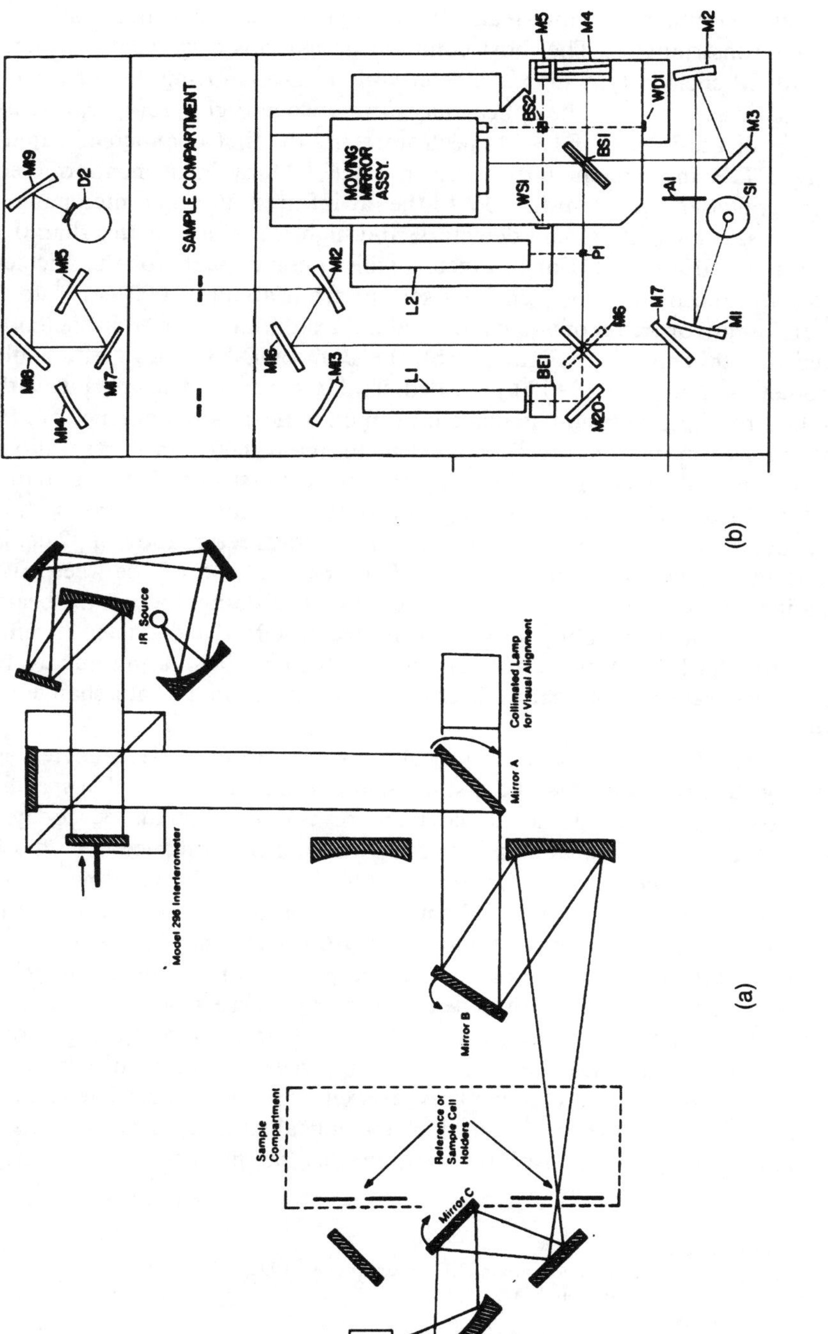

Model 296 Interferometer
IR Source
Collimated Lamp for Visual Alignment
Mirror A
Mirror B
Mirror C
Sample Compartment
Reference or Sample Cell Holders
TGS
(a)
SAMPLE COMPARTMENT
MOVING MIRROR ASSY.
M5
M4
M2
M3
WDI
BS2
BS1
AI
SI
WSI
PI
L2
M6
M7
MI
M20
BEI
LI
M13
M16
M12
M15
M19
D2
M14
M18
M17
(b)

Figure 15

The early commercial rapid-scan FTIR spectrometers tended to be large stand-alone instruments. The most common design was a conventional flat-mirror 90° Michelson-type interferometer with a linear moving mirror travel, see schematic as Figure 8. The optical diagram shown in Figure 15(a) represents the layout of the Digilab FTS-14 spectrometer,[89] the first commercial rapid-scanning FTIR instrument, introduced in 1969.[90] Many instruments offered wide spectral coverage, from the UV to the far-infrared, through interchangeable sources, beamsplitters and detectors, and high resolution, better than 0.5 cm^{-1} in the mid-infrared. For instance, auxiliary components for the Nicolet 7199[91] (introduced in 1976[90]) enabled set-ups from about 50 000 cm^{-1} to 5 cm^{-1} and a mirror travel of ~8 cm offered a maximum resolution in the mid-infrared of 0.06 cm^{-1}, see Figure 15(b); the Bomem DA3 series[92] evacuable spectrometers (launched 1980[90]) provided similar spectral coverage, but with the highest resolution version having a maximum optical path difference of 250 cm (about 1.25 m mirror travel!) equivalent to about 0.003 cm^{-1} resolution. Another evacuable wide spectral range spectrometer first introduced in about 1976,[90] the Bruker IFS 113v,[93] offering a maximum resolution of 0.03 cm^{-1}, is based around a Genzel-type interferometer. The optical retardation of 32 cm is achieved by traversing a double-sided mirror back and forth between two collimating mirrors, see schematic in Figure 16(a). This design focuses the beam onto the beamsplitter enabling it to be much smaller (effective diameter 10 mm) than in standard Michelson interferometers, thereby facilitating relatively simple incorporation of a beamsplitter carousel for automatic interchange of beamsplitters.[3,4,93]

The industrial need for such high resolution mid-infrared spectrometers is limited, *e.g.* to specialised gas-phase studies, and although comparable specification instruments are still available from several manufacturers, today's industrial market tends to be dominated by medium performance (*e.g.* $\leqslant 0.1$ cm^{-1} resolution), wide spectral coverage spectrometers and much lower cost compact, lower resolution (*e.g.* ~1–2 cm^{-1} as standard maximum resolution), limited spectral range, bench top routine analytical instruments. Many variations on the Michelson interferometer are now incorporated in commercial FTIR spectrometers. Some, with cube corner retroreflectors, were introduced essentially to eliminate the need for tight control or monitoring of mirror tilt.[3,4,73,94] Other innovations and modifications were driven by design constraints for robustness and compactness, as well as reduced cost. The refractively scanned interferometer,[3,4,95] in which the optical retardation is introduced by a scanning wedged beamsplitter, was another design also used in early

[89] Bio-Rad Digilab, Cambridge, USA.
[90] Ferraro J.F., Jarnutowski R., and Lankin D.C., *Spectroscopy*, 1992, 7, 2, 30.
[91] Nicolet Instruments Corporation, Madison, USA.
[92] Bomem, Inc., Quebec, Canada.
[93] Bruker Analytische Messtechnik GMBH, Karlsruhe, Germany.
[94] Crabb N.C. and King P.W.B., in *Process Analytical Chemistry*, ed. McLennan F. and Kowalski B., Blackie Academic, Glasgow, UK, 1995, ch. 6.
[95] Doyle W.M., McIntosh B.C., and Clarke W.L., *Appl. Spectrosc.*, 1980, **34**, 5, 599.

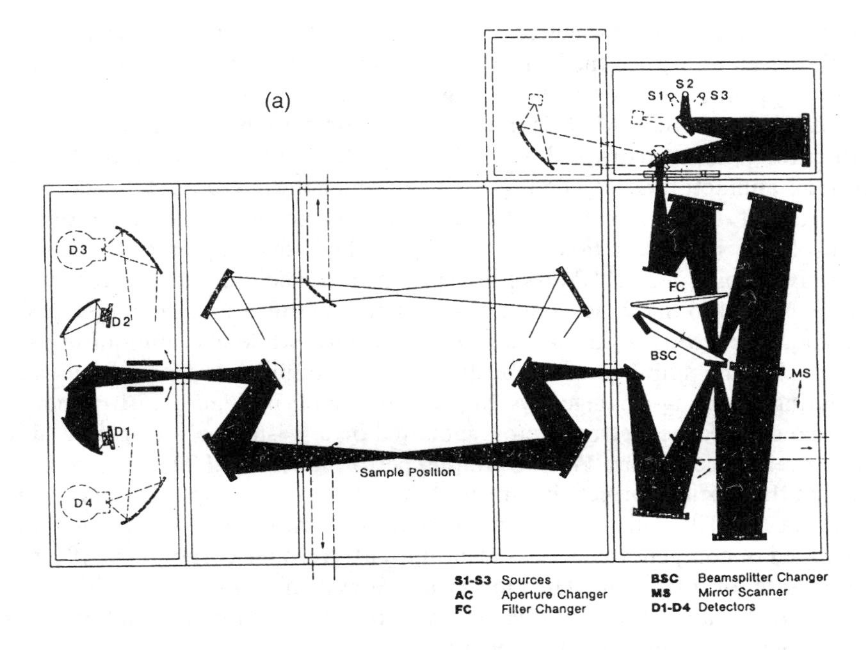

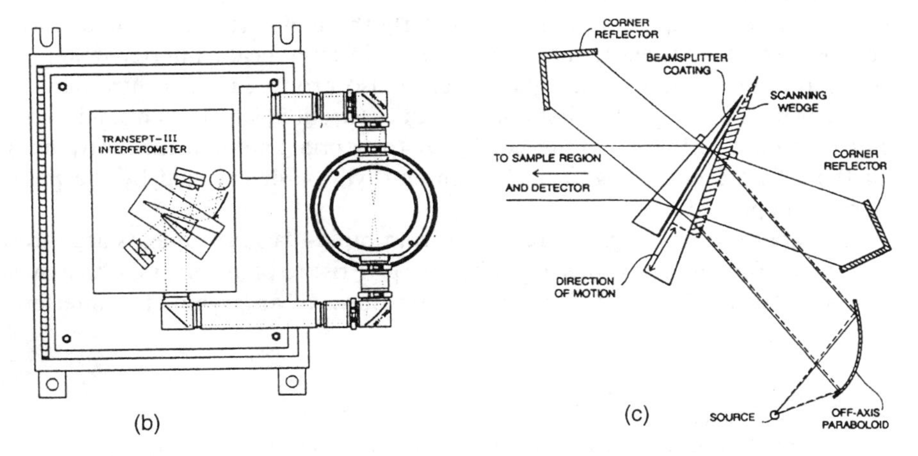

Figure 16 *Schematics of the optical lay-outs of two commercial rapid scan FTIR spectrometers:* (a) *Bruker IFS 113v, employing a Genzel-type interferometer*
(Reproduced by kind permission of Bruker Spectrospin Ltd., Coventry, UK)
(b) *Schematic of a refractively scanned interferometer used as part of a high pressure, high sensitivity gas monitor*
(Reproduced by kind permission of Analect Instruments, Inc., Irvine, CA, USA)
(c) *Schematic of a refractively scanned interferometer*
(Reproduced from ref. 95, Doyle et al., *Appl. Spectrosc.*, **34**, 5, 599–603 (1980), by kind permission of the Society for Applied Spectroscopy; © 1980)

commercial FTIR spectrometers, see schematic in Figure 16(c); interferometers of this design now form the heart of a series of process measurement and plant control systems,[96] see, for example, Figure 16(b).

Not all commercial Michelson-type interferometers have been or are built with orthogonal mirrors and a 45° beamsplitter configuration; models from several manufacturers use a configuration in which the mirrors are angled at 60° with respect to each other. This gives a 30° angle of incidence on the beamsplitter, see schematic of Figure 17(a), which gives a greater throughput for a given solid angle.[3,4] Market expansion into medium and low resolution models has led to many design innovations and modifications being used so as to achieve optical retardation more economically, while also minimising size and maximising stability and robustness. For example, one variation uses an oscillating 'rotating-carriage' system, see schematic in Figure 17(b); another uses two cube corner retroreflectors mounted on a 'wishbone' swing on a 'flex-pivot' bearing, see Figure 17(c). Optimum performance of a FTIR spectrometer relies on there being correct alignment of the interferometer optical components during a scan. Tilt, roll, yaw, shear, *etc.* of mirrors must therefore be essentially eliminated or compensated for; as mentioned above, cube corner retroreflectors compensate for mirror tilt. The interferometer system[97] shown schematically in Figure 17(d) is an example of one designed to be free (self-compensating) from tilt and shear dynamic alignment errors.

In many FTIR spectrometers, particularly those of early design, in addition to the infrared interferogram and the He–Ne laser beam interferogram, an interferogram from a white-light source is also produced, the centre-burst of which is used to trigger the first point of data collection, which for a single-sided interferogram will be set to precede by an appropriate amount the centre-burst of the infrared interferogram. This amount will be determined by the phase correction procedure.

In practice, interferograms recorded from rapid-scan instruments are rarely symmetric functions; various imperfections give rise to phase errors, which must be corrected for.[3,4,18,72,98] The Fourier transform of an asymmetric interfero-

[96] Laser Precision Analytical, Irvine, USA.
[97] Perkin-Elmer Limited, Beaconsfield, UK.
[98] Forman M.L., Steel W.H., and Vanasse G.A., *J. Opt. Soc. Am.*, 1966, **56**, 1, 59.

Figure 17 *Schematic optical diagrams of some other commercial FTIR interferometers:*
(a) *Michelson type with 30°* incident angle
(Reproduced by kind permission of Shimadzu Corporation, Tokyo, Japan)
(b) *'Rotating-carriage' design*
(Reproduced by kind permission of Perkin-Elmer Ltd., Beaconsfield, Bucks., UK)
(c) *'Rocking-wishbone' design*
(Reproduced by kind permission of Bomem, Inc., Québec, Québec, Canada)
(d) *'Periscope' design*
(Reproduced by kind permission of Perkin-Elmer Ltd., Beaconsfield, Bucks., UK)

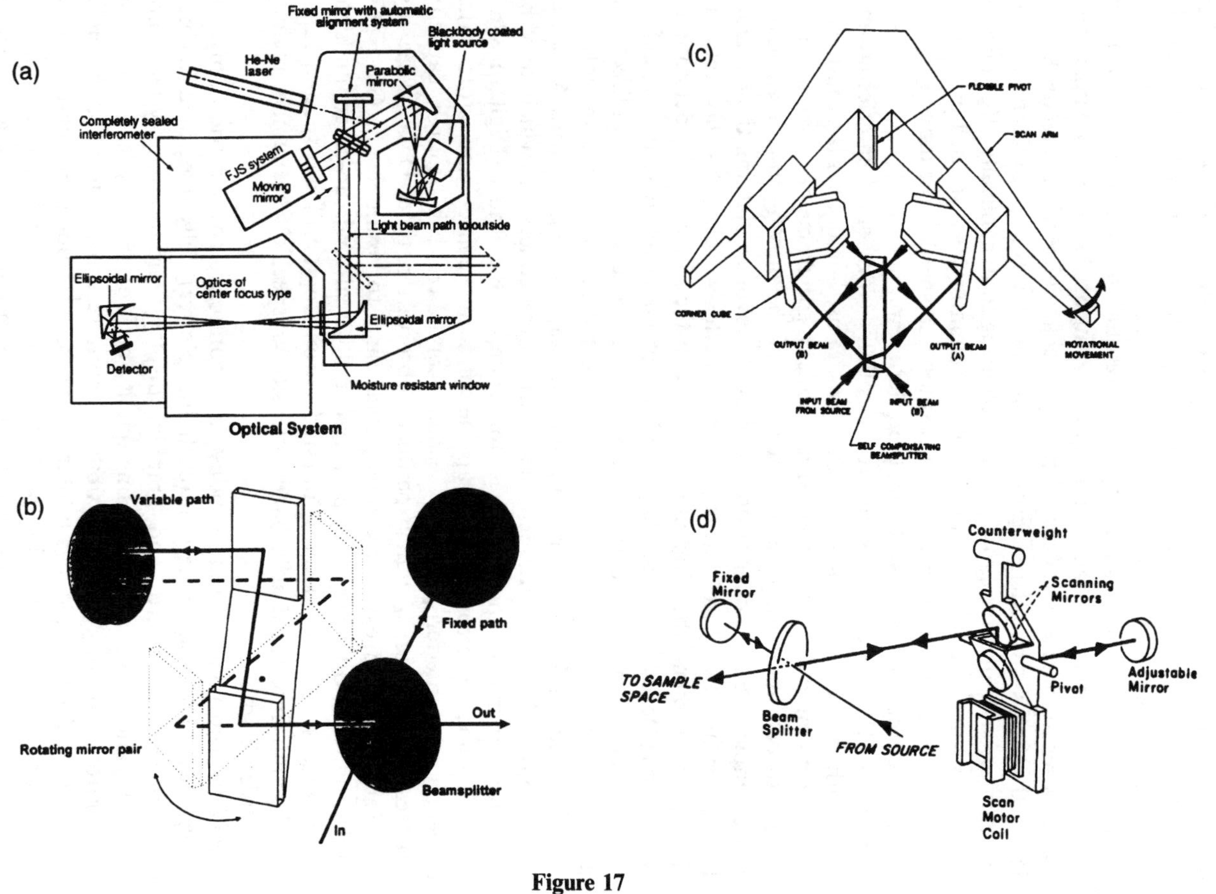
(a)
Fixed mirror with automatic alignment system
Blackbody coated light source
He-Ne laser
Parabolic mirror
Completely sealed interferometer
FJS system
Moving mirror
Light beam path to outside
Ellipsoidal mirror
Optics of center focus type
Ellipsoidal mirror
Detector
Moisture resistant window
Optical System
(b)
Variable path
Fixed path
Out
Rotating mirror pair
Beamsplitter
In
(c)
FLEXIBLE PIVOT
SCAN ARM
CORNER CUBE
OUTPUT BEAM (B)
OUTPUT BEAM (A)
ROTATIONAL MOVEMENT
INPUT BEAM FROM SOURCE
INPUT BEAM (B)
SELF COMPENSATING BEAMSPLITTER
(d)
Counterweight
Fixed Mirror
Scanning Mirrors
TO SAMPLE SPACE
Pivot
Adjustable Mirror
Beam Splitter
FROM SOURCE
Scan Motor Coil

Figure 17

gram would contain both a real and an imaginary part. Fortunately, the phase errors do not vary rapidly with wavenumber, and may therefore be readily computed from a short double-sided interferogram.[97] (Hence, even when essentially recording a single-sided interferogram, a short extension must be recorded the other side of ZPD (the centre-burst).)

Many of the instruments offering higher resolution capabilities, particularly those of earlier manufacture, use linear air-bearing drives for the moving mirror travel, while more recent designs use purely mechanical bearings, *e.g.* the Bruker IFS 120 HR,[93] which offers a greater than 5 m optical path difference. Many systems use laser interferogram phase referencing to dynamically align mirrors during a scan.[3,4] Laser 'fringe monitoring or counting' is also used in many of the recent instruments to measure the position and speed of the moving mirror; this negates the need for a white-light interferogram.

Some FTIR spectrometers record essentially single-sided interferograms; while others record double-sided interferograms; some offer an option of either form of data collection. Most modern spectrometers also use laser fringes to determine the direction of travel of the moving mirror assembly. For example, the laser quadrature technique[3,4] involves dividing the reference laser beam into two parts and imparting a quarter-wavelength delay (retardation) between them; the direction of travel can then be noted by determining which signal leads and which lags. (Other techniques are described in refs. 3 and 4). As a consequence many recent spectrometers operate bidirectional data collection, *i.e.* interferograms are collected and co-added in both the forward and backward scan directions of the moving mirror. At lower resolutions, this allows for extremely rapid data collection, which may be important to many kinetic studies; for example, Bio-Rad's (Digilab) 896[89] research-grade spectrometer is capable of scan, collect and store rates exceeding 80 interferograms a second; computer disk storage capacity imposes the limit on overall experiment observation time. Data processing techniques are also extremely rapid, such that, for example, in GC-FTIR mode, systems will visually display essentially real-time, up-dated multiple absorbance spectra, or several 'infrared chromatograms', *i.e.* specific functional group (integrated absorbance intensity *vs.* specific wavenumber region) plots (see Chapter 7).

FTIR spectrometers used in industrial laboratories cover a wide range of flexibilities. Low cost, single unit (excluding plotter), fully integrated (including software, key-board, and graphics display) mid-infrared systems are available, and well-suited to routine quality control or analytical tasks. For a laboratory requiring greater flexibility, a more versatile system may be more appropriate; for example, a modular system based around an interferometer with dual computer selectable sources, interchangeable beamsplitters, and four exit ports. These might be configured to service a GC-IR accessory, a FTIR microscope, and standard sample compartments for mid-infrared and near-infrared work. The associated computer (today's standard being a PC-type) will offer multi-tasking operation and will be programmed with search and multivariate quantitative analysis software. Auxiliary or 'external' modules with their own dedicated detector are available for most medium performance instruments. Research-grade spectrometers may also offer step-scan operation (see pp. 112–117), in addition to the normal continuous, rapid-scan mode.

A perfectly collimated beam, as implied in Figure 8, can only be generated from a point (infinitely small) source. In fact a FTIR spectrometer uses an extended source; therefore one must consider rays that are both on-axis and oblique to the optical axis.[3,4,73] The off-axis rays will experience a reduced path difference compared to those parallel to the optical axis. This divergence leads to a small shift to lower wavenumbers from the true value in the Fourier transformed spectrum, and in addition a wavenumber spread. It is therefore important in many circumstances to minimise this beam divergence, particularly for medium and high resolution studies. This is achieved by the Jacquinot stop (J-stop), a limiting circular aperture sited at a beam focus usually between the source and the Michelson interferometer. This optimises the beam solid angle prior to collimation for a particular resolution and wavenumber range by limiting the spread of rays entering the interferometer. It may be either a computer controlled variable iris diaphragm or computer selectable from a range of diameter apertures on a rotatable wheel. Other apertures may also be employed in the optical train to control beam convergence.[3,4] Spectrometers built for low resolution, limited performance applications may have an essentially fixed J-stop.

Sample compartments vary between instruments, in that for some the sample point focus is to one side ('side-focusing'), while in others it is located at the mid-point ('centre-focusing'). Maximum (or default/standard operating condition) sample beam diameters at this focal plane also vary; typical values lie between 5 and 13 mm; the centres of beams may also be located at different heights from the spectrometer base-plate. Consequently, accessories are not always readily interchangeable between instruments of different manufacture, and accessories which incorporate focusing optics need to be specified accordingly, and/or an appropriate base-plate purchased.

Without a sample in place, the throughput of a Michelson interferometer is dependent on the diameters of both the J-stop and the beamsplitter. The sample beam optics, usually after the interferometer, are characterised by their focus diameter and convergence, the former being usually an image of the source or J-stop. The F-number ($f/\#$) denotes the ratio of the equivalent focal length of a lens/mirror to the diameter of its entrance pupil.[86] (The *aperture ratio* is the ratio of the lens/mirror aperture to its focal length ($1/f\#$).)[86] The specifications given in manufacturers' brochures usually refer to the value for the collimator optics; they are typically in the range $\sim f/3$ to $f/5$, with slower optics (higher $f/\#$s required for high resolution models, *e.g.* the Bruker 120FS with $f/6.5$), with nominal maximum collimated beam diameters for Michelson-type interferometers varying over a wide range according to performance, for example values of 34 mm to 80 mm may be found in manufacturers' brochures.

FT-Raman, Fourier Transform Raman Spectrometers

For the industrial spectroscopist the advent of the modern FT-Raman technique largely satisfied a long recognised need – the capability to record analytical Raman spectra that were essentially free from fluorescence from a wide range of industrially relevant samples and products. While it is not a panacea for all occurrences, it did make circumvention of visible region fluorescence a practical and routine reality through the use of a longer wavelength, near-infrared laser as the excitation source.[13–17,99] For this reason the technique is sometimes referred to as NIR FT-Raman. As previously stated, the current standard source is the 1.064 μm (9398.5 cm^{-1}) line from a diode-pumped Nd:YAG laser, which means that in absolute terms the Stokes-shifted Raman spectrum will appear approximately in the region 9398 cm^{-1} to 5398 cm^{-1}, *i.e.* equivalent to about 0 to 4000 Δcm^{-1}. Typical single mode maximum powers available at the sample are quoted as between 1W and 4W.

Two types of commercial FT-Raman systems are presently produced. Systems are available essentially as add-on modules to FTIR spectrometers. These accessories or adjuncts, which were the first type to appear, are linked to the Michelson-type interferometers through an external port of the FTIR system. They offer the opportunity for recording both infrared and Raman spectra from a sample in a single instrument, see example in Figure 18. As the technique became established, manufacturers began to produce optimised dedicated spectrometers, see example in Figure 19. Currently some manufacturers offer both alternatives. The principal detector choice is between Indium Gallium Arsenide (InGaAs) and Ge, see p. 75, with some spectrometers accommodating both options. There may also be a choice for the Rayleigh scattered radiation rejection filter, which may be either a dielectric filter offering a typical cut-on of about 130 Δcm^{-1}, or a notch or holographic notch filter, transmitting Stokes shifted radiation from closer to the exciting line, typically cutting-on between 50 to 100 Δcm^{-1}; anti-Stokes observations may typically extend from between -150 to -300 Δcm^{-1} to between -1200 to -2000 Δcm^{-1}. Since commercial FT-Raman systems are operating in the near-infrared region, non-hygroscopic components may be used, and the interferometer beam-splitters are typically either quartz coated with Fe_2O_3 or TiO_2, or CaF_2 coated with silicon or Sb_2S_3. However, to obtain improved sensitivity, reflective mirrors in dedicated spectrometers may be gold coated for high reflectivity, as opposed to the aluminium coated mirrors frequently used in mid-infrared FTIR systems.

Efficient minimisation of the intense Rayleigh line by filtering is essential to the operation of FT-Raman spectroscopy, otherwise the overwhelming shot-noise associated with the Rayleigh line would be redistributed (spread) throughout the spectrum derived from the FT measurement[13,14,100] and Fellget's advantage would not be realised. Also, the relative intensity of the radiation

[99] Schrader B., Hoffmann A., Simon A., Podschadlowski R., and Tischer M., *J. Mol. Struct.*, 1990, **217**, 207.
[100] Cutler D.J., pp 123–129 in reference 17(a).

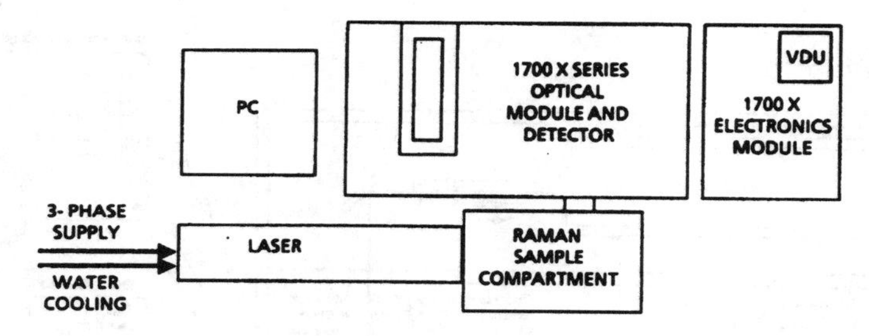

Schematic Diagram of the Raman System

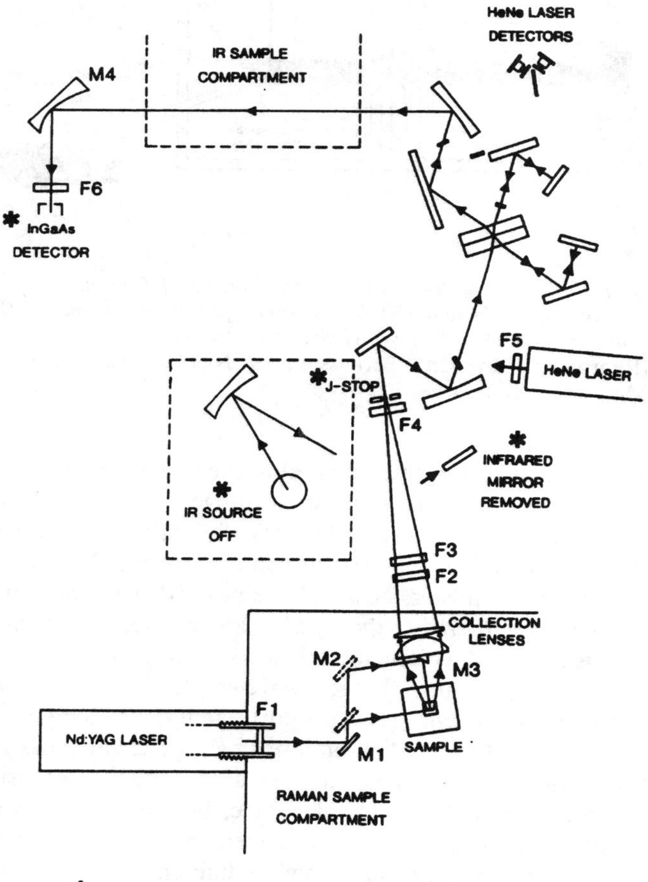

Optical Path Through the Raman Sample Compartment and Infrared Spectrometer

Figure 18 *Schematic diagram and optical path diagram of a Perkin-Elmer 1760 combined FTIR/FT-Raman spectrometer*
(Reproduced by kind permission of Perkin-Elmer Ltd., Beaconsfield, UK)

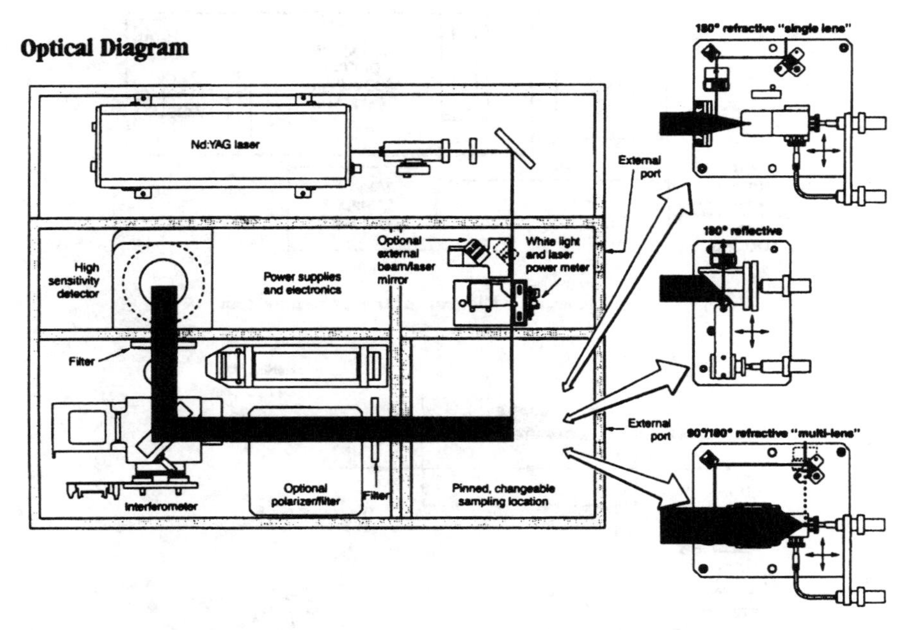

Figure 19 *Optical diagram of Nicolet 900 series dedicated FT-Raman system. Available sample modules include 180° backscattering refractive geometry, 180° reflective geometry and a 90°/180° switchable refractive system*
(Reproduced by kind permission of Nicolet Instrument Corporation, Madison, USA)

scattered at the laser wavelength compared to the much weaker Raman signals would pose a serious (very large) dynamic range problem for the detection system.[13,15] Provided that the intense inelastically scattered light is eliminated, the multiplex advantage may be used to compensate for the lower sensitivity (higher noise equivalent power, NEP) of the near-IR detector and its accompanying electronics (compared to the visible Raman detector systems). The FT-measurement also helps in enhancing the weaker Raman signals emitted as a consequence of using a much longer wavelength excitation laser; Raman intensities have a ν^4 dependence ($1/\lambda^4$ proportionality), so changing from 514.5 nm to 1.064 μm will lead to a reduction in Raman scattering cross-section by a factor of about 18.[18] A benefit accrued as a consequence of using the FT multiplex detection method should, of course, be the inherent wavenumber precision of each data point in the transformed spectrum – a prerequisite for optimal application of comparisons involving difference spectroscopy (spectral subtraction) or multivariate analysis methods. However, while the FT approach does bring good wavenumber shift reproducibility to Raman spectra, it does *not* guarantee reproducible absolute wavenumber values. The source, the Raman scatter, is effectively the sample itself, which unlike in a conventional FTIR spectrometer must therefore precede the interferometer in the optical train. Consequently, reproducible illumination of the interferometer is very dependent both on proper sample alignment, and on the illumination falling centrally and filling the Jacquinot stop or limiting aperture. Otherwise, shifts in absolute wavenumber scale may occur, which readily manifest themselves in an inability to undertake satisfactory spectral subtraction processes, giving rise to 'derivative-like' features in the difference spectrum.

As with the single-beam FTIR spectrum, a FT-Raman spectrum as recorded represents the convolution of several effects – two of these being the vibrational spectrum (in this case the Raman radiation emitted) and the instrumental response profile.[18] As with a FTIR spectrometer, the latter is far from constant (or linear) with wavelength and contains a summation of the characteristics of the optical components and the detector response. For any form of comparative analytical spectroscopy, whether qualitative or quantitative, as with single-beam FTIR spectra, FT-Raman spectra should be corrected, that is, the instrument response profile must be ratioed out, particularly if any meaningful comparisons are to be made between spectra recorded on different instruments. To 'background correct', a FT-Raman spectrum for instrument wavelength sensitivity an appropriate (*i.e.* equivalent optical configuration, spectral resolution) instrument response curve must be generated. This correction curve is ideally determined by measuring the spectral response to a blackbody or near-blackbody emitter, such as a cavity furnace source, which is then divided point-by-point into a computer-generated theoretical blackbody curve normalised to the same temperature. The resultant correction function is then used to normalise the recorded FT-Raman spectrum of the sample to produce the 'corrected' spectrum.[18,101] A more convenient routine laboratory practice is to use a secondary standard such as a large area element lamp, the corrected output spectrum of which has been determined against a furnace source. The colour-temperature of the lamp needs to be known accurately, and be stable with time. Some commercial systems are provided with a calibrated reference source which mounts in the sample compartment and is used to record an instrument response function for correcting FT-Raman spectra.[18] Others feature automatic correction through the use of an 'integrated' white light source and blackbody generating software,[18] while others provide computer correction routines.[18] Failure to correct FT-Raman spectra for the varying instrument response sensitivity will leave spectra in which the relative intensities of the Raman bands may be far from the truth and extremely misleading, see Section 7.

A measured and corrected FT-Raman spectrum may also contain a thermal emission contribution that emanates from the sample itself. Thermal emission from the sample as a consequence of its becoming heated through absorption of source radiation is a particular problem in FT-Raman experiments, as the sample is placed before the interferometer so any thermal emission from it will be modulated and detected along with its Raman signal. (In most conventional FTIR laboratory experiments the sample is placed after the interferometer and any thermal radiation emanating from it will not be modulated and is therefore

[101] Petty C.J., Warnes G.M., Hendra P.J., and Judkins M., pp 1179–1187 in reference 17(b).

not convoluted with the single-beam spectrum; the infrared source emission other than that absorbed or reflected is, of course, ratioed out.) To aesthetically improve a Raman spectrum, any thermal emission background may be minimised by a baseline-flattening data manipulation routine; this may be necessary for an optimal comparison search against a spectrum library. At least one manufacturer supplies a thermal background correction kit in the form of a variable temperature source.[97]

FT-spectrometers: Step-scan Mode (and 2D Spectroscopy)

Step-scan operation is now an option available on many research-grade FTIR spectrometers.[102] Although not a new principle, since the late 1980s step-scan operation has increasingly become the preferred method for several areas of application, particularly in the mid-infrared region.[102,103] In a commercial interferometer operated in the more conventional rapid, continuous-scan mode, data are collected during the constant velocity scan of the moving mirror, and it is control of the retardation velocity that is important. However, as pointed out earlier (p. 93), this leads to each wavenumber in the resultant spectrum being modulated at a different frequency, which increases with increasing wavenumber; this 'Fourier frequency'[104,105] was defined in Equation (16). Also, since the interferogram itself takes a finite time to record (typically >100 ms, see p. 95), uncoupling temporal domain information from the spectral multiplexing is complicated. These characteristics are removed in step-scan operation,[102–106] which benefits certain studies such as photoacoustic depth profiling and observations on time-resolved or time-dependent phenomena.

In step-scan operation, the interferogram is built up point by point essentially by stopping the moving mirror (velocity is equal to zero) at each data collection point; the required signal-to-noise ratio (S/N) is then achieved by collecting data for an appropriate time at each retardation position; successive interferograms may also be added to further enhance S/N. Thus, in contrast to continuous-scan operation, step-scan operation depends on precise control of mirror position.[102,103,107,108] This may be achieved using opto-electronic feedback circuitry to monitor interferogram(s) (interference records) generated by the same He–Ne laser interferogram used to control sampling interval in the continuous-scan

[102] Palmer R.A., *Spectroscopy*, 1993, **8**, 2, 26.
[103] Palmer R.A., Chao J.L., Dittmar R.M., Gregoriou V.G., and Plunkett S.E., *Appl. Spectrosc.*, 1993, **47**, 9, 1297.
[104] Budevska C.J., Manning C.J., Griffiths P.R., and Roginski R.T., *Appl. Spectrosc.*, 1993, **47**, 11, 1843.
[105] Dittmar R.M., Chao J.L., and Palmer R.A., *Appl. Spectrosc.*, 1991, **45**, 7, 1104.
[106] Palmer R.A., Manning C.J., Rzepiela J.A., Widder J.M., and Chao J.L., *Appl. Spectrosc.*, 1989, **43**, 2, 193.
[107] Manning C.J. and Palmer R.A., *Proc. 7th Int. Conf. Four. Trans. Spectrosc.*, SPIE, 1989, **1145**, 577.
[108] Manning C.J., Palmer R.A., and Chao J.L., *Rev. Sci. Instrum.*, 1991, **62**, 5, 1219.

mode.[102,103,107,109,110] One method of generating a control signal is by *phase modulation* (PM) of the laser signal, which is fed to a phase-sensitive detector,[102,103,105,108] such as a lock-in amplifier. This modulation may be created by rapidly vibrating one of the interferometer mirrors back and forth across each 'stopped' step position – this leads to no net change in retardation. This path-difference modulation at each stepped mirror position is often referred to as 'dithering' or 'jittering' of the mirror. Commercial step-scan interferometers now use digital control of both mirror stepping and positioning to an uncertainty of about ± 1 nm,[102,103,111] with phase modulation as an option.

Modulated infrared radiation signals for detection may be generated in one of two ways.[102,103,111] External modulation, that is *amplitude modulation* (AM), can be achieved either by placing a rotating mechanical chopper (50% transmitting) into the infrared beam or by inducing sample perturbations, which change its spectrum. Alternatively, internal modulation can be provided by path-difference modulation such as moving mirror 'dithering', that is, by PM. The latter is usually preferred because it offers a S/N advantage.[102,103,108,112]

For photoacoustic spectroscopy (PAS), a straightforward benefit of using step-scan FTIR is that each wavenumber in the PA spectrum can be subjected to the same single (constant) modulation frequency. Thus, in the absence of photoacoustic saturation, a uniform sampling depth will apply to the entire spectral range scanned, which can then be changed by altering the modulation frequency for subsequent spectra recorded from the same sample, see Figure 20(a).[105,113] Alternatively, since the thermal diffusion process creates phase lags (time dependencies) in PA responses from different depths within a sample, more depth specific information may be obtained (phase separated) easily by extracting, for example, the in-quadrature (surface) and in-phase (bulk, substrate) signals using a lock-in amplifier(s), see Figure 20(b).[105,113] Phase rotation, together with measurements recorded at different modulation frequencies, offers the potential for spectral depth profiling of multilayer products, such as polymer film laminates. However, for regions where the surface layer exhibits intense absorption bands there will probably be some obscuration of absorption features characteristic of the submerged layer through absorption by the surface layer, see Figure 20(b). Higher depth resolution and more layer-discriminatory information is potentially available by analysing the signal phase directly.[102,103,114–116]

The step-scan PAS measurement is an example of a synchronous modulation–demodulation ('phase-resolved' or 'frequency-domain') experiment.[102,103]

[109] Manning C.J. and Griffiths P.R., *Appl. Spectrosc.*, 1993, **47**, 9, 1345.
[110] Smith M.J., Manning C.J., Palmer R.A., and Chao J.L., *Appl. Spectrosc.*, 1988, **42**, 4, 546.
[111] Palmer R.A., *Proc. 9th Int. Conf. Four. Trans. Spectrosc.*, SPIE, 1993, **2089**, 53.
[112] Nicolet FT-IR Technical Note TN-9253, Nicolet Instruments Corp., Madison, USA.
[113] Gregoriou V.G., Daun M., Schauer M.W., Chao J.L., and Palmer R.A., *Appl. Spectrosc.*, 1993, **47**, 9, 1311.
[114] Jiang E.Y., Palmer R.A., and Chao J.L., *J. Appl. Phys.*, 1995, **78**, 1, 460.
[115] Palmer R.A., Jiang E.Y., and Chao J.L., *Proc. 9th Int. Conf. Four. Trans. Spectrosc.*, SPIE, 1993, **2089**, 250.
[116] Palmer R.A., Jiang E.Y., and Chao J.L., *Progress in Fourier transform spectroscopy: 10th Int. Conf. Four. Trans. Spectrosc., Mikrochim. Acta [Suppl.]*, 1997, **14**, 591.

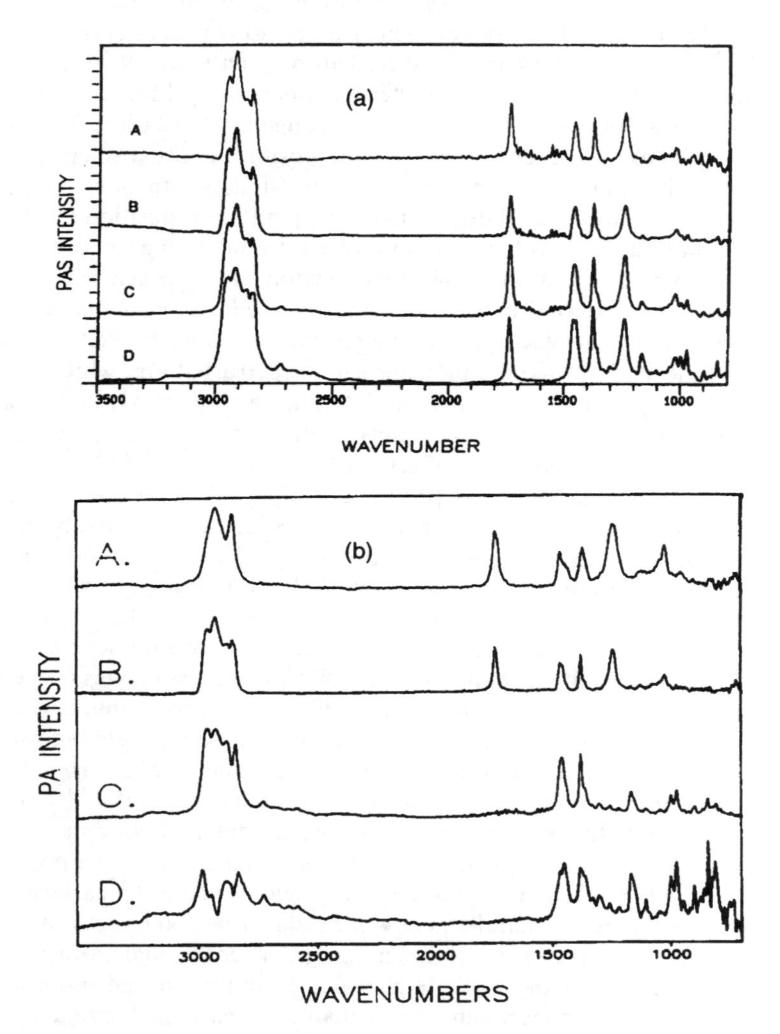

Figure 20 (a) *FTIR step-scan PAS magnitude spectra of an ethylene–vinyl acetate copolymer (EVA),* ~12 μm *thick, film on a* ~60 μm *thick polypropylene (PP) substrate. Spectra collected with phase modulation frequencies: (A) 800 Hz; (B) 600 Hz; (C) 400 Hz; (D) 100 Hz;* (b) *Step-scan FTIR phase-separated PAS spectra of EVA/PP sample, compared to separately determined spectra of EVA and PP (A and C respectively). The phase difference between vector components B and D is 135°. All measurements were made with a phase modulation frequency of 100 Hz. See ref. 105 for a fuller explanation of experimental details*

(Reproduced from ref. 105, Dittmar et al., *Appl. Spectrosc.*, **45**, 7, 1104–1110 (1991), by kind permission of the Society for Applied Spectroscopy; © 1991)

In frequency-domain step-scan measurements, 'simultaneous' data at multiple modulation frequencies can be recorded, and then sequentially demodulated by lock-in amplifiers,[103] or, as has been introduced recently into commercial spectrometers,[117] sorted more conveniently by digital signal processing (DSP) methods.[109,118] The multiplex and throughput attributes inherent in FTIR instrumentation, coupled with the advantages of modulation–demodulation step-scan measurements, are well-suited to studying certain systems undergoing dynamic change.[102,103,119] In these experiments it is an essential requirement that the dynamic perturbation *e.g.* a mechanical, electrical, optical, thermal, chemical, or acoustic stimulus, is repeatable reproducibly and that the induced spectral changes are reversible and respond linearly to the perturbation.[102,119,120] The time-dependent spectral fluctuations (*e.g.* band position, absorption intensity, molecular orientation/reorientation, molecular conformation, sample path-length) may then be retrieved with respect to the applied stimulus. To enhance observation of the information content and gain much greater insight into the connectivity and interactions among various functional groups, a two-dimensional infrared (2D-IR) spectral display may be generated by a pair-wise correlation of the dynamic variations of the spectral signals induced by the external perturbation.[119–124] The 2D-IR approach was developed as a novel infrared concept in the mid-1980s, primarily for interpreting dynamic infrared dichroism data,[120,122] initially recorded using dispersive instrumentation.

A 2D-IR spectral plane, a spectrum defined by two independent wavenumbers, is constructed by applying a correlation analysis procedure to the time-dependent infrared signals. The position of peaks on this plane, which can be plotted either as a fishnet or, perhaps more usefully, as a contour representation, describe the connectivities and interactions among functional groups associated with the infrared bands,[120] see Figure 21. The degree of coherence between the dynamically fluctuating infrared signals is characterised by the intensities in the synchronous correlation plot, *i.e.* the synchronous spectrum is a representation of two infrared signals that are changing in-phase with each other, see Figure 21(b). The peaks on the diagonal, referred to as *autopeaks*, correspond to the autocorrelation of perturbation-induced dynamic variations. Their intensities indicate the susceptibilities of the corresponding absorbance bands to the external perturbation.[120,122] Off-diagonal peaks at two different wavenumbers,

[117] Various oral and poster presentations at the *2nd Int. Symp. Adv. Infrared Spectrosc.*, June 17–19, 1996, Duke University, Durham, USA.

[118] Manning C.J., Pariente G.L., Lerner B.D., Perkins J.H., Jackson R.S., and Griffiths P.R., in *Computer Assisted Analytical Spectroscopy*, ed. Brown S.B., J. Wiley, Chichester, UK, 1996, ch. 1.

[119] Palmer R.A., Manning C.J., Chao J.L., Noda I., Dowrey A.E., and Marcott C., *Appl. Spectrosc.*, 1991, **45**, 1, 12.

[120] Noda I., *J. Am. Chem. Soc.*, 1989, **111**, 21, 8116.

[121] Noda I., *Appl. Spectrosc.*, 1993, **47**, 9, 1329.

[122] Noda I., *Appl. Spectrosc.*, 1990, **44**, 4, 550.

[123] Palmer R.A., Manning C.J., Rzepiela J.A., Widder J.M., Thomas P.J., Chao J.L., Marcott C., and Noda I., *Proc. 7th Int. Conf. Four. Trans. Spectrosc.*, SPIE, 1989, **1145**, 277.

[124] Marcott C., Dowrey A.E., and Noda I., *Anal. Chem.*, 1994, **66**, 21, 1065A.

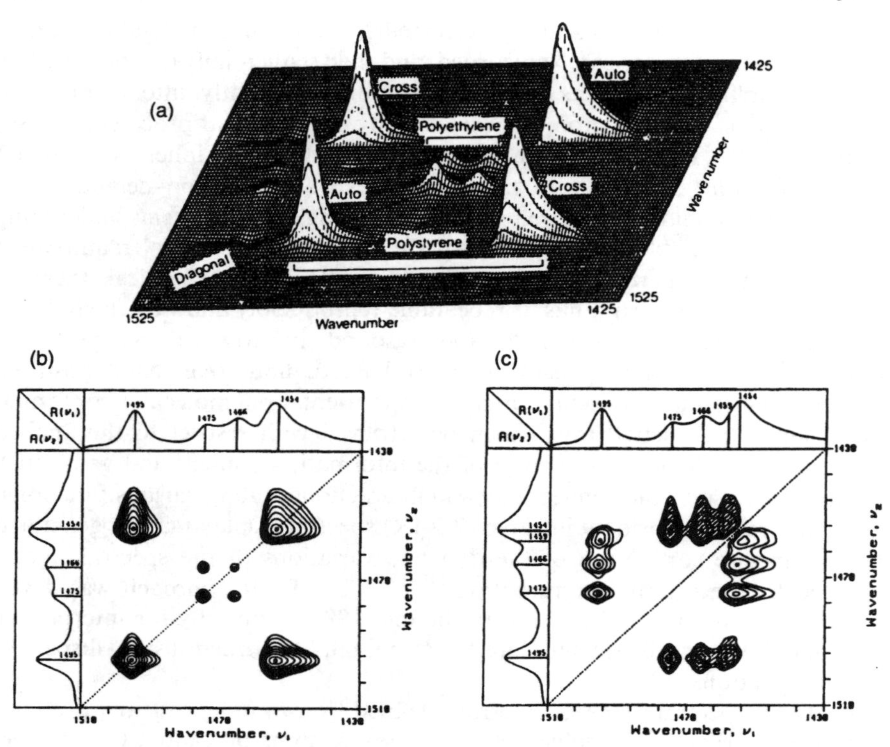

Figure 21 (a) *Fishnet representation of the synchronous 2D-IR correlation spectra of a film made from a mixture of atactic polystyrene and low-density polyethylene;* (b) *Contour map representation of the synchronous 2D-IR correlation spectrum. Conventional absorbance spectra are displayed along axes for reference purposes;* (c) *Contour map representation of the asynchronous 2D-IR correlation spectrum. Shaded areas represent negative intensity regions. See ref. 120 for a fuller description*
(Reprinted from ref. 120. Copyright (1989). The American Chemical Society)

referred to as *cross peaks*, appear when the dynamic fluctuations are correlated or anticorrelated (*i.e.* opposite sign) to each other. In contrast, intensity in the asynchronous plot represents independent and uncoordinated fluctuations of the infrared signals. That is, the information is complementary, with cross peaks appearing if the infrared fluctuating signals are not completely synchronised,[120,122] see Figure 21(c).

Examples of step-scan 2D-FTIR correlation analyses include several rheo-optical studies of polymer films. In these, in order to stay in a regime which has a reversible, linear response, small tensile-strain induced fluctuations are introduced into a polymer film sample by a low-amplitude (typically 0.1%–0.3%), low-frequency (typically <30 Hz) oscillatory (sinusoidal) perturbation with a dynamic mechanical stretcher. Such studies have been carried out on isotactic[104,123] and atactic[122,125] polypropylene, a blend of polystyrene with a low density polyethylene,[120] a melt-crystallised Nylon 11[126] and uniaxially drawn PET.[127] The electric-field induced submolecular reorientation dynamics of nematic liquid crystals have also been the subject of several studies,[102,103,128–130] including a polymer-dispersed liquid crystal system.[131] In the liquid crystal studies, the modulation (perturbation) was supplied by an ac signal generator (usually sinusoidal) operating in the frequency range 0–15 Hz, with about 5 V peak-to-peak amplitude modulation in the range 0–10 V.

Another mechanism by which step-scan spectra are recorded is by the impulse–response experiment.[102,103] In these time-domain measurements, the impulse is repeated at each stepped position and the data are collected as explicit functions of time and time intervals after each impulse. The perturbation may also be repeated at each step until the required S/N is achieved. The recorded data are then sorted to give time-resolved spectra. Studies of this kind have, for example, included heme protein sub-μs photodynamics.[102,103,132,133]

Time-resolved and 2D-IR measurements are, of course, not limited to step-scan operation, and many excellent applications of both using continuous-scan FTIR data are still being reported. Indeed, for vibrational circular dichroism (VCD) measurements, the approach has not yet developed sufficiently to offer any significant advantage over fast- and slow-continuous-scan approaches.[134]

As yet, few step-scan FT-Raman studies have been reported; one such is the picosecond time-resolved spectrum of the first excited state of the highly fluorescent chromophore 9,10-diphenylanthracene in cyclohexane and ethanol solutions.[135,136]

7 Closing Remarks

In this chapter we have sought to introduce and describe briefly the major components and concepts associated with infrared and Raman spectrometers, which are found most commonly in industrial analytical and research laboratories. Advances in both techniques continue to keep pace with the introduction

[125] Noda I., Dowrey A.E., and Marcott C., *Mikrochim. Acta [Wien]*, 1988, **1**, 101.
[126] Singhal A. and Fina L.J., *Appl. Spectrosc.*, 1995, **49**, 8, 1073.
[127] Sonoyama M., Shoda K., Katagiri G., and Ishida H., *Appl. Spectrosc.*, 1996, **50**, 3, 377.
[128] Nakano T., Yokoyama T., and Toriumi H., *Appl. Spectrosc.*, 1993, **47**, 9, 1354.
[129] Sasaki H., Ishibashi M., Tanaka A., Shibuya N., and Hasegawa R., *Appl. Spectrosc.*, 1993, **47**, 9, 1390.
[130] Gregoriou V.G., Chao J.L., Toriumi H., and Palmer R.A., *Chem. Phys. Lett.*, 1991, **179**, 5/6, 491.
[131] Hasegawa R., Sakamoto M., and Sasaki H., *Appl. Spectrosc.*, 1993, **47**, 9, 1386.
[132] Palmer R.A., Plunkett S.E., Dyer R.B., Schoonover J.S., Meyer T.J., and Chao J.L., *Proc. 9th Int. Conf. Four. Trans. Spectrosc.*, SPIE, 1993, **2089**, 488.
[133] Plunkett S.E., Chao J.L., Tague T.J., and Palmer R.A., *Appl. Spectrosc.*, 1995, **49**, 6, 702.
[134] Wang B. and Keiderling T.A., *Appl. Spectrosc.*, 1995, **49**, 9, 1347.
[135] Jas G.S., Wan C., and Johnson C.K., *Appl. Spectrosc.*, 1995, **49**, 5, 645.
[136] Jas G.S., Wan C., and Johnson C.K., *Spectrochim. Acta*, 1994, **50A**, 11, 1825.

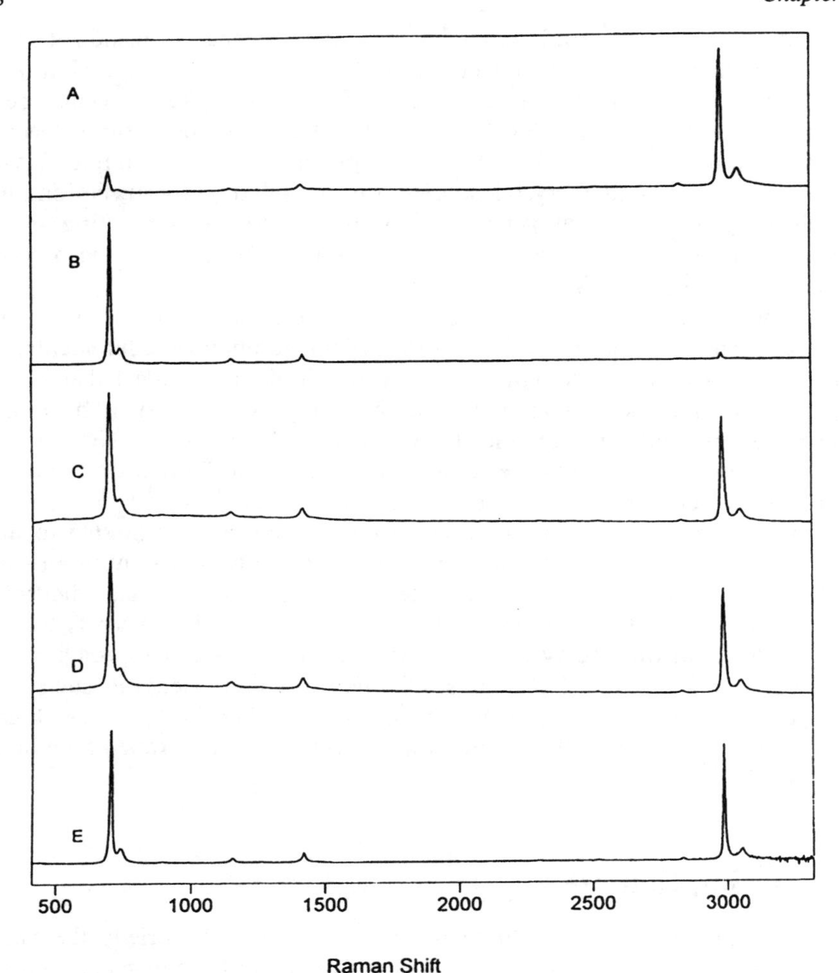

Figure 22 *Raman spectra of CH_2Cl_2 obtained as:* A, *uncorrected,* 514.5 nm; B, *uncorrected,* 785 nm; C, 514.5 nm, *corrected with reference to a tungsten bulb;* D, 514.5 nm, *corrected with reference to coumarin standard;* E, 785 nm, *corrected with reference to 2412 glass. See original reference for full details* (Reproduced from ref. 137, by kind permission of the Society for Applied Spectroscopy; © 1997)

of new, or commercial release of novel, technologies, and there is much promise for the future.

Step-scan FTIR applications are beginning to emerge, which DSP techniques make easier to use in everyday industrial situations. The industrial potential of infrared imaging is just beginning to be explored, while new solid-state long wavelength visible lasers are making an impact on Raman investigations in the industrial laboratory, where Raman instrumentation is becoming a common partner alongside the more traditionally favoured infrared spectrometer. For quantification, the uncertainty in relative intensities is a problem general to all Raman spectra, because it is a single-beam technique. Recently a procedure has been proposed for dealing with the general problem of correcting for widely varying instrumental responses, based on using practicable, routine secondary standards.[137] Viable secondary standards, referenced to a primary standard tungsten source, made from fluorophores that produce broad band emissions when excited have been demonstrated, see Figure 22. For 514.5 nm and 785 nm light, these are based on a 0.084 mM coumarin 540a fluorescent laser dye in methanol, in a 1 mm cuvette, and a commercially available red filter semiconductor doped glass, respectively.[137] In the process and production areas, fibre-optic sampling and probe-head devices interfaced to both vibrational spectroscopy techniques are now commonplace, providing means for measurements for monitoring and feedback control. While synchrotron radiation, which has for a long time found favour as a source for its far-infrared properties, is now finding value as a source for mid-infrared FTIR-microscopy studies, the free electron laser (FEL) is just beginning to be publicised as a specialised source for infrared radiation.[117] For the mid-infrared one might anticipate the eventual development of infrared lasers fully tuneable over the fingerprint region.

[137] Ray K.G. and McCreery K.G., *Appl. Spectrosc.*, 1997, **51**, 1, 108.

CHAPTER 4

Sampling Techniques and Accessories

1 General Introduction

A prime consideration in sample presentation has to be the purpose of the study. In many circumstances, this can be the most challenging part of solving a problem. Optimum choice of which sample preparation method and/or accessory to use is vital, as the spectrum detail obtained may be very dependent upon this choice. This is particularly important to infrared investigations, which are mostly undertaken using transmission or reflection techniques. By contrast, in Raman spectroscopy, which analyses scattered radiation, many samples can be examined directly without preparation, sometimes without even removing them from their container. When dealing with a solid one needs to decide whether its chemical nature is of sole interest, or whether its physical characteristics are of equal importance. For example, if the purpose is to study polymorphism in drugs or minerals, or the requirement is to determine the degree of crystallinity or molecular orientation in polymers, then the integrity of these properties must be maintained throughout sample preparation, presentation and measurement. Methods which provide simple, rapid generic structural identification may prove inadequate for more detailed studies, such as compositional or conformational analyses. Also, the specimen or specimens examined must be representative of the sample in the context of the study purpose or problem definition, particularly if the sample is heterogeneous. Sampling techniques that are inherently surface sensitive may not yield spectra that are characteristic of the sample bulk.

Most organic materials and many inorganic materials are suitable for infrared spectroscopic analysis. They can be in the form of continuous or powdered solids, liquids, dispersions, or gases; although, in our experience, the majority of samples submitted for analysis to an industrial laboratory are powders or liquids, and may simply require examination using traditional infrared transmission techniques at room temperature. However, many will have to be prepared into a suitable form at a different temperature or be cast from a solution or require use of an accessory to obtain a reflection spectrum. For example, a polymer powder may need to be formed from the melt into a continuous film, or ground at liquid nitrogen temperatures in order to decrease the particle size to that appropriate for an alkali halide disk examination. There are a plethora of commercial accessories available for purchase for infrared and Raman studies, which enable sample examinations to be made under a wide temperature range from liquid N_2 or below to 900°C or above, from very low to very high pressures, from very long path-lengths (25 metres or greater) to the very thin (monomolecular layers), and the very small (5 μm diameter).

2 Basic Principles of Sample Preparation for Infrared Spectroscopy

Many accessories and techniques are available that allow samples to be examined by infrared with the minimum of handling or preparation, *e.g.* DRIFT, PAS, ATR, which are discussed in detail later. However many samples need to be contained in a cell or dispersed in a matrix, prior to examination in transmission in an infrared spectrometer. Raman samples are mostly examined in a suitable glass container or, in many cases, if solid, directly as received.

To record a good, relevant infrared spectrum, the same basic principles apply to all samples examined under normal conditions. Preparation of a sample is determined by its physical state, which may need to be selected, along with the experimental parameters for operating the spectrometer, according to the information required. For example, in Chapter 1 we stated that while, for simple identification, copper phthalocyanine (CuPc) pigments may be prepared as halide disks, to determine the alpha/beta polymorphic state of a CuPc requires a less rigorous treatment, such as preparation as a mull, see Figure 1 of Chapter 2. Figure 1 shows the infrared spectra of CuPc samples prepared by different methods. As can be seen, each sample preparation procedure produces a slightly different spectrum. The alkali halide disk spectrum exhibits a significantly sloping background, due to scatter from particles, and this will vary with particle size. The DRIFT spectrum shows different relative band strengths and a poorer spectral contrast, which apparently enhances the intensity of the weak overtone and combination bands. The mull spectrum is overlayed with the absorption features of the mulling agent. However, all the spectra of Figure 1, which were recorded at 4 cm^{-1} resolution, will identify the sample as being substantially a CuPc. To determine the alpha/beta polymorph ratio of a sample, its spectrum should be recorded from a mull at a higher spectral resolution (1 cm^{-1}), to give greater spectral point density, as was done with the spectra of Figure 1 of Chapter 2. Variation in sample grinding can cause differing levels of polymorph conversion; for example, preparation as an alkali halide (KBr) disk can convert one polymorph to the other (alpha to beta), if a mechanical grinding process is used.

To repeat our earlier statement, *all samples must be presented into the instrument's infrared beam in such a way that the total cross-section of the beam*

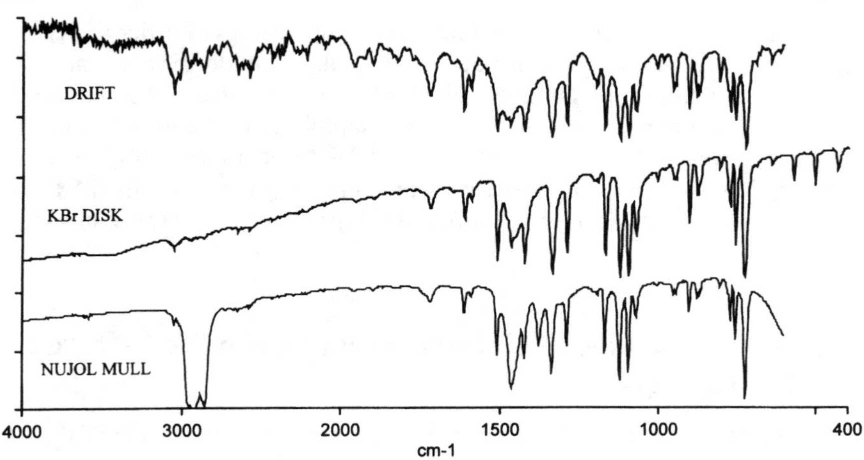

Figure 1 *Infrared copper phthalocyanine (CuPc) spectra* (4 cm^{-1} *resolution) from three different sample preparation techniques:* Top, *DRIFT* (1024 scans); Middle, *KBr disk* (8 scans); Bottom, *Nujol mull* (8 scans). *Spectra have been offset for clarity*

is covered. Test specimens for transmission measurements must have a cross-section greater than that of the sample beam, or be appropriately masked, and be without holes, bubbles, or cracks in the cross-section, which should also be essentially of even thickness. If the sample is dispersed, for example, in an alkali halide disk or mulling agent, it *must* be homogeneously dispersed, with a particle size small enough to reduce scatter loss and minimise absorption band distortions. Ideally, solvents used for examining samples in solution should be non-polar, so that there are little or no solvent–sample interactions, which can lead to band shifts. Carbon tetrachloride is, in this respect, spectroscopically an ideal solvent for solution work, having a spectrum which is minimally interfering except for a strong absorption band near 800 cm^{-1}. However, the use of this and other similar, previously favoured solvents has been discouraged by Health and Safety regulations, and their use is prohibited in many industrial laboratories, or, if allowed, they must be handled and used with extreme care. The thickness (path-length) of sample examined will depend on the information required, the sample, and its method of presentation. A transparent sample with truly parallel faces, may show interference fringes. In practice, very minor deviations, such as a small amount of surface roughness, minimise fringes and aid interpretation. For full range transmission spectra for qualitative purposes, the sample thickness should normally allow for 95–90% transmission at the baseline compared to the background, with the strongest band having a maximum between 5 and 15%T. For many aliphatics, however, thicker specimens are often more useful as the relatively stronger νCH bands are 'blacked-out' at 0%T, enhancing the fine detail in the rest of spectral region, see, for example, Chapter 2, Section 4.

3 Window Materials for Infrared Spectroscopy

Mid- and far-infrared radiation are strongly absorbed by normal glass (quartz) optical mirrors and lenses. This problem can be overcome by front silvering (or gold coating) of mirrors, but windows need to be manufactured out of different materials. Fortunately alkali metal halides such as sodium chloride (NaCl), potassium bromide (KBr) and caesium iodide (CsI), do not have absorption bands in the mid-IR region of the spectrum. These materials fuse under pressure and when polished give infrared and optically transparent 'glass-like' windows. Cells can be made with these materials as windows to hold liquids and vapours over wide temperature and pressure ranges. One of their disadvantages is that they are hygroscopic, and will fog if not handled and stored correctly. Also, since they are soluble in water, other materials are required for aqueous work or wet samples. A whole range of window materials are available commercially for infrared spectroscopy. These have differing properties; some, such as ZnSe, BaF_2 and Ge, are hard and may fracture under pressure, while others, such as AgCl, CsI and KRS-5, may cold flow. They also have different transmission ranges, see Table 1, and costs. [KRS-5 is a potassium thallium bromide (42 mol%) iodide (58 mole%) mixed crystal. It is toxic if ingested, and eye and skin contact may cause irritation.]

For the majority of applications, rock salt (NaCl) or KBr windows are used, with BaF_2 or ZnSe being frequently used for those studies involving aqueous or wet samples. The salt and KBr windows have transmission cut-offs at about 600 cm^{-1} and 400 cm^{-1} respectively, which is more than adequate for most applications. BaF_2 and ZnSe, being mechanically glass-like, need more careful handling. BaF_2 at 2 mm thickness has a useful range only to 1000 cm^{-1}, preventing observation of the =C—H out-of-plane deformation vibration bands in the region 900–675 cm^{-1}, which are often used to establish aromatic ring substitution patterns. ZnSe has a wider transmission window to about 450 cm^{-1} when used at a thickness of about 2 mm, but at this thickness and being highly polished with a high refractive index occasionally causes interference fringe problems. Diamond is popularly used as windows for high pressure compression cells, particularly those developed for examining small samples by FTIR-microscopy. KRS-5, ZnSe and Ge are commonly used materials for infrared internal reflection spectroscopy elements.

4 Preparation and Cleaning of Windows for Infrared Spectroscopy

The basis of all good preparation is to ensure that the windows are clean and fit for use. The advantages, apart from cost, of NaCl and KBr are the ease of forming, cleaning and polishing windows made from them. Originally these salts were sold in huge lumps, with spectroscopists having to cleave their own windows. Many polished pre-cut shapes and sizes can now be bought off-the-shelf, with others being prepared to special order, although there will probably be minimum thickness restrictions depending on the materials' properties. Matt

Table 1 *Common Window Materials for Infrared Spectroscopy*

Material	Approx. useful transmission range*	Refractive index†	Max. useful temp °C, in air	Remarks
NaCl	40 000–600	1.49	400	Very common, lowest cost, window material. H_2O soluble. Easily cleaved.
AgCl	25 000–400	1.98	200	H_2O insoluble. Darkens on standing in light. Cold flows easily.
KBr	40 000–400	1.53	300	Common, low cost window material. Hygroscopic. Easily cleaved.
CsI	33 000–200	1.74	200	Easily scratched, soft. Very hygroscopic.
BaF_2	60 000–800	1.42	500	H_2O insoluble, does not fog. Easily cracked by mechanical shock.
CaF_2	75 000–900	1.40 @ 2000 cm^{-1}	900	H_2O insoluble. Withstands high pressures.
KRS-5	20 000–250	2.37	200	Good ATR material, but cold flows/deforms under pressure. Toxic.
ZnSe	20 000–500	2.41	300	Irtran-4™. Good, common ATR material. Easily cracked.
ZnS	50 000–750	2.2	300	Irtran-2™. Hard. Good thermal properties.
Ge	5500–600	4	270	Good ATR material. Brittle.
Si	8300–1600 400–35	3.42	300	Good for far-IR region.
AMTIR‡	11 000–600	2.5	300	Brittle.
Sapphire (Al_2O_3)	50 000–1600	1.74	1700	Very hard. Inert.
Diamond, Type II	4500–2500 1600–35	2.4		Very hard. Inert.

* Cut-off (lowest useful wavenumber) will depend somewhat on window thickness. The values here are what might be expected for a window of a thickness of between about 2 and 5 mm.

† The values given here are approximate and will vary with wavenumber, for example, the refractive index of KBr at 22°C will vary from about 1.537 at 4000 cm^{-1} to about 1.463 at 400 cm^{-1}. Except where stated, the values are for ~1000 cm^{-1}, at about 25°C.

‡ AMTIR™, Amorphous Material for Transmitting Infrared Radiation, Amorphous Materials Inc., Texas, USA. AMTIRs are chalcogenide glasses based on Group IV, V, and VI elements of differing stoichiometries, *e.g.* AMTIR-1 is $Ge_{33}As_{12}Se_{55}$.

or fogged windows can be polished using a 50–50 mixture of ethanol and water, a little Jeweller's Rouge, and gentle rubbing on a Selvyt^{TM} cloth. If a salt or KBr window is heavily contaminated, the whole surface can often be rubbed down with a fine abrasive (emery) paper, then repolished. Flatness of the window must be checked after heavy polishing to ensure that a variable thickness (path-length) has not been created. Also, it is advisable to record the spectrum of a used plate which has been polished before it is used for an analysis; a transparent-looking, polished plate is not necessarily free from traces of contamination from previous uses; some may have become embedded in the window material. This repolishing technique is not recommended for cell windows used for quantitative work, where a high degree of parallelism is essential. For hard, water insoluble, materials like zinc selenide, surface damage is less common, although most suppliers do provide a repolishing service.

5 Transmission Techniques for Infrared Spectroscopy

Capillary Films

A transmission spectrum sufficient for chemical identification may often be recorded from a capillary layer of a non-volatile liquid. The neat liquid is usually prepared for examination as a capillary film sandwiched between two alkali halide polished flats; provided, of course, that they are resistant to attack by the liquid, otherwise another window material must be selected. Volatile, hazardous, or toxic liquids are more normally examined in sealed cells. Care must be taken at all times to avoid touching the faces of the flats or cell windows, as oils and moisture from fingerprints can damage the surface and/or interfere with the spectrum. The flats should be handled only by their edges, preferably whilst wearing thin plastic gloves. (Be careful in the choice of glove material, though, since some may exude additives that can be more problematic than finger contaminations – PVC gloves are likely to exude phthalates, which are commonly used as plasticisers and have intense characteristic infrared spectra).

Once flats of a suitable material have been chosen, their faces should be visually inspected to ensure that they are clean (if in doubt, run a spectrum as a check), polished, and free from scratches, pits, or lumps. One salt flat (plate) is placed, face up, on a clean, level surface. The liquid sample is checked to ensure that it is homogeneous. Using a capillary tube, a drop of the liquid is transferred to the centre of the flat. A second flat is then placed centrally over the drop. The flats are gently pressed together, producing a capillary film between them. The film should be free from holes, of uniform thickness and with an area greater than the cross-section of the spectrometer beam. If the film is holed or too small, the flats should be separated and cleaned and the procedure repeated using a larger drop. Once the quality of the sample is satisfactory, it is then mounted in the standard transmission mount in the spectrometer. The beam should pass through the centre of the film. Normal organic liquids should produce a spectrum with an essentially flat baseline at approximately 90%T. Ideally, the strongest absorption band should have its maximum at about 5%T, particularly if the spectrum is to be pattern-matched against a computerised library of reference spectra.

Cast Films

Samples which have been dissolved in a volatile solvent can be prepared by casting a thin film from the solution and allowing the solvent to evaporate. This should be undertaken in a fume cupboard or other appropriate containment environment, and due regard must be taken of any toxicological or other hazards associated with the solvent. A single clean, polished plate is prepared as described for preparing a capillary film. After a drop has been transferred to the centre of the flat and gently spread over the plate surface with the capillary pipette tip, the solvent is allowed to evaporate. The film may contract during evaporation and need to be gently spread across the flat, using the edge of the capillary pipette, to maintain its area. Particular care must be taken to produce an even film, which should have the same qualities as capillary films. The sample is mounted in the spectrometer in the same way as for a capillary layer. The aim should be to obtain a good quality spectrum, see Figure 2.

For less volatile solvents, evaporation may be facilitated by placing the flat on a hot-plate which is then turned on and warmed to a suitable temperature. This is normally quite hot to the touch, but not too hot or the flat may crack. Alternatively, the window may be warmed under an IR lamp. When all the solvent has evaporated and a film of the required thickness prepared, the film on the flat must be allowed to cool to room temperature prior to recording its infrared spectrum. Persistent solvent residues may need to be removed in a vacuum oven.

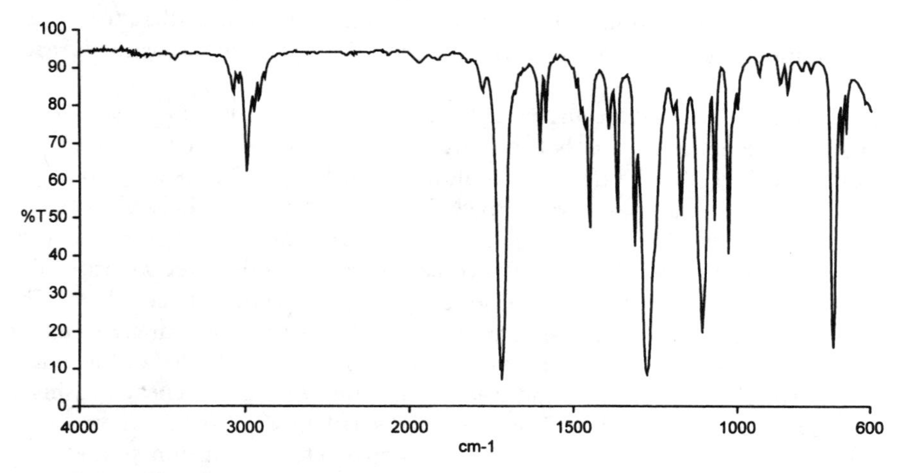

Figure 2 *'Good' example of a fingerprint transmission infrared spectrum from a cast film preparation*

Self-supporting films may be prepared by initially casting onto a support other than an IR-transmitting flat. A well-used example in the polymer industry is the casting on glass microscope slides of thin films of many polyamides (nylons) that have been dissolved in formic acid. After evaporation of the solvent, the sample is washed with water, wiped dry, peeled from the glass support, and mounted in an appropriate holder for sample presentation.

Spectrum Quality of Capillary and Cast Films

There are several possible causes for the production of a poor quality spectrum.

(1) Spectral contrast is poor. There are probably large variations in the film thickness.
(2) The baseline slopes. Either there is an instrument fault or, more likely, the sample was inhomogeneous, perhaps a dispersion containing a scattering component, such as a carbon or silica filler. Another frequent cause is crystallisation of the film after evaporation of the solvent with consequent scattering of the infrared beam. Scatter is wavelength dependent, being greatest at shorter wavelengths (higher wavenumbers), and so produces a sloping baseline. For highly crystalline materials, band shapes may also appear distorted due to refractive index dispersion effects.
(3) The baseline cannot be set to (dispersive measurement) or is not near (FT measurement) 90%T and/or peaks extend well below 5%T. The sample is probably too thick, see Figure 3(b). Repeating the sample preparation using a smaller drop should achieve a better result.

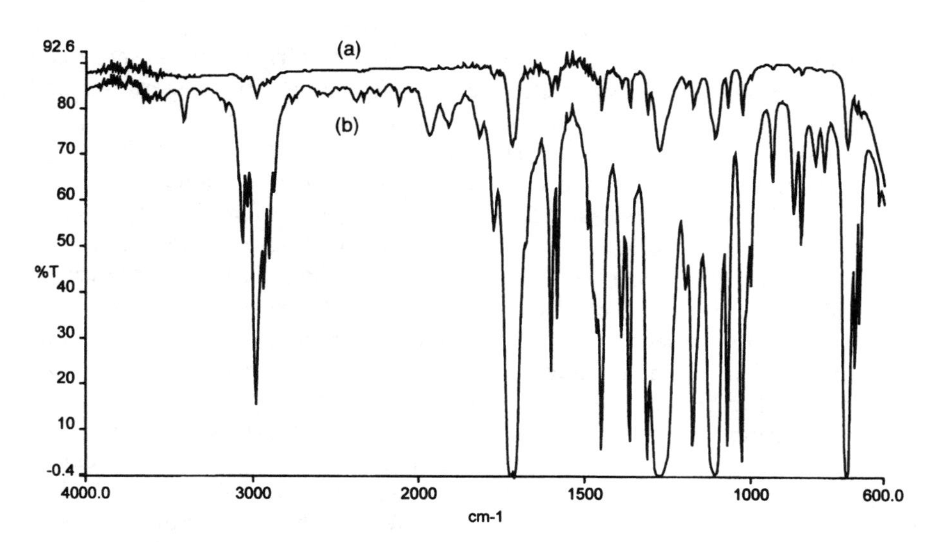

Figure 3 *'Poor' fingerprint transmission infrared spectra from cast film preparations:* (a) *too weak;* (b) *too strong*

(4) The peak maximum of the strongest band is well above about 5%T, see Figure 3(a). The spectrum is too weak, and should be measured again using more of the sample. This is much more appropriate than simply using 'excessive' ordinate scale expansion to enhance contrast artificially.
(5) Peaks are flattened at their maxima, and spectral contrast is poor/distorted. The film contains holes and/or some of the infrared beam is passing round the edges of the film and reaching the detector. Radiation reaching the detector without having passed through the sample is commonly referred to as stray-light.
(6) On a dispersive instrument, peaks become relatively weaker towards the end of the spectrum scan. The sample is too volatile and is evaporating during recording of the spectrum. In a FT measurement this effect will most likely manifest itself as a 'hole in the sample'.

Films from the Melt

Thin films from low-melting point, thermally stable solids can be prepared by heating and/or pressing. Solids with melting points just above room temperature may be prepared by gently heating until a liquid is formed. Alternatively, some solids may be melted or made to flow merely by squeezing it between a pair of infrared transparent plates. The purpose of pressure and softening/melting is to form an evenly thick film, which will cover the cross-section of the beam from a few mg of a sample placed at the centre of the plate sandwich. The comments on spectral quality given above also apply to this procedure. Heating may be accomplished by radiation from a heating lamp or use of a gradient temperature hot-plate. The windows should be heated gently to allow the sample to melt evenly and to prevent the windows cracking. After heating, the windows should be allowed to cool gradually. If they are picked up too soon the double disaster of burned fingers and cracked windows from thermal shock can occur! The potential effects on the spectrum of changes of temperature and physical form should not be forgotten. For example, the spectra of a polycaprolactone at 40°C (melt, liquid state) and room temperature (solid, crystalline state) show marked differences, as can be clearly seen in Figure 4.

The above approach is suitable for many amorphous materials such as waxes, tarry solids, hot-melt adhesives, as well as for low-melting-point solids. It is not recommended for materials which will crystallise on solidification, since scatter and molecular orientation effects will lead to both poor spectral contrast and irreproducibility. Rubbery materials may have a tendency to retract, since only hand pressure is normally used, and the plate–rubber–plate assembly may therefore need to be clamped at the edges.

For higher melting solids, such as polymers, windows clamped in spring clips may be heated on an electrically heated hot plate, or under an infrared heating lamp. Hot compression moulding from powders, granules or specimens taken from fabricated articles is one of the most effective and satisfactory means of preparing free-standing films from many thermoplastic resins for infrared transmission measurement, particularly for quantitative analysis, with the

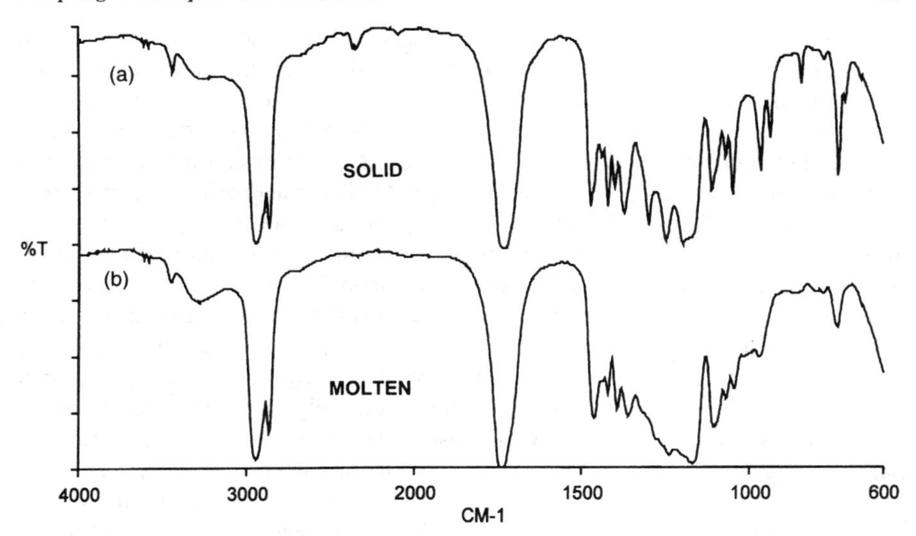

Figure 4 *Transmission infrared spectra of polycaprolactone:* (a) *solid;* (b) *molten. Sample prepared as a capillary film at* ~40°C, *and then allowed to solidify in the infrared beam*

proviso, of course, that the polymer is thermally stable at the pressing temperature. Hand-pumped, laboratory hydraulic presses are available commercially and are most appropriate for laboratories regularly producing such samples. They are capable of applying loads of up to 20 tons (~20 000 kg) on a ram of 4 in (10.16 cm) diameter. The platens are typically 8 in (~20 cm) square, and may be heated electrically to temperatures in excess of 350°C. Water flowing through the platens is used for rapid cooling, which is flushed out by air when at room or handleable temperature. Some presses incorporate control units for programmed heating and cooling cycles. Smaller constant thickness film-making kits are also available commercially, and may be better suited to the non-specialised laboratory; apparatus for the production of thin films is also available (water-cooled and heated, temperature controlled platens) for use with a manually-operated, hydraulic press usually used for making alkali halide disks.

With the larger laboratory presses, samples may be pressed between a pair of metal plates of the same size as the platens. We use polished stainless steel plates a few mm thick, which have been conditioned to prepare samples which are optimal for infrared examination. Their surfaces have been grit-blasted (25–30 mesh) and then spray-overcoated with a very thin layer of a PTFE (polytetrafluoroethylene), which acts as a mould-release agent. The introduced surface roughness is generally sufficient to minimise the appearance of an interference pattern in the prepared film's infrared spectrum. Overall, the prepared film should be of uniform thickness, which may be determined using a dial-gauge micrometer. If the film thickness may be determined precisely enough, its value may be used to normalise absorbance measurements in quantitative analyses. The thickness required will be dependent on the purpose of the study. Thin films suitable for a chemical structure survey may be moulded with or without the aid of feeler gauges inserted between the platens; thicker specimens better suited to a quantitative quality assurance application may be more easily prepared using a template. For fingerprinting purposes a film thickness of 0.1–0.2 mm (100–200 μm) is generally appropriate for aliphatic hydrocarbons, although this will be too thick to allow observation of the band maxima of the νCH bands, and probably some or all of the δCH bands near 1460 cm^{-1}. To allow clear observation of all the band maxima for a strongly absorbing material, such as an aromatic polyester, the film thickness may need to be less than 10 μm. The optimum conditions for pressing a particular material will need to be found empirically, although the compression temperature will most likely be close to the material's melt temperature. Samples need to be dry and essentially free from solvent or catalyst residues before hot-pressing, otherwise bubbles may form in the film, or hydrolysis or further chemical reaction take place, which will lead to the recording of erroneous spectral information. Pressing at too high a temperature or for too long may also give rise to bubbles or may cause thermal degradation, which, if excessive, may lead to discolouration. Too low a temperature or insufficient pressure may result in a sample which is not fully coalesced. If the sample is formed from a powder, this may give rise to scatter in a spectrum; if the film is formed from granules, it may separate readily at deformed granule boundaries; if formed from layered thinner films, it may delaminate. If a prepared film exhibits interference fringes in its infrared spectrum, these can often be reduced in intensity or eliminated by lightly scouring the film surfaces with wire wool or abrasive paper, but this is likely to produce a sloping baseline through added scatter.

For thin films, <25 μm thick, the procedure commonly used is to place a few mg of the sample between the press plates and then insert the assembly between the press platens, which have been preheated to the pressing temperature. The press platens are then closed onto the assembly, which is allowed to heat for 1–2 min before the pressure is applied for a further minute or so. The heating is then turned off and the press cooled according to the manufacturer's instructions. The sample is left to cool under pressure, after which the press plates are removed and separated to allow access to the film, which is then mounted in an appropriate mask or holder. Certain analyses may require that the hot-pressed compression-moulded film is 'frozen' into an amorphous state. This may be achieved by melting a few mg of sample between a pair of thin press plates or aluminium sheets, and then, using an appropriate safe handling procedure, quenching the assembly directly from the pressing temperature in iced water or solid carbon dioxide in methanol. Annealed (crystalline) specimens may be prepared by simply turning off the press at the pressing temperature with the sample assembly in it and allowing it to cool slowly unaided, maintaining the pressure until it is cool. More controlled morphologies may be better introduced through the use of a controller to set programmed temperature cycles.

Rubbery samples, which may contract after pressing, may be hot-pressed and then subsequently examined as a sandwich between two thin sheets of AgCl. Thin films of brittle materials may need to be similarly prepared. Films of high-temperature thermoplastics, with melt temperatures near to or above 300°C, can be prepared as a sandwich between aluminium foil. For thicker specimens the foil can be peeled away to reveal the sample, but for thin samples it can be dissolved away in sodium hydroxide solution. Such samples are pressed between uncoated plates, since PTFE coatings should not be heated to temperatures much in excess of 300°C.

Fixed and Variable Path-length Cells

Volatile liquids and solutions, prepared for quantitative work, are usually examined in fixed path-length cells, which are either permanently or semi-permanently sealed. The cells are assembled from two infrared transparent windows with a spacer between them. The spacers are usually made from metal, commonly lead but sometimes tin, or a polymer, most commonly PTFE (Teflon®), but sometimes PET (Mylar®), polystyrene, or a nylon. Spacer thicknesses for semi-permanent cells are available typically from 6 μm (tin) or 10 μm (PTFE) to many millimetres; permanent cells are sealed with a lead amalgam spacer, and have a similar thickness range, starting at about 15 μm. The upper window of the cell assembly has pre-drilled holes near each end for filling and emptying the cell. Gaskets are placed between the windows and cell surround plates, see Figure 5(a); the backplate is designed to mount directly in a standard infrared spectrometer sample holder.

A path-length (spacer thickness) of about 12 μm (accepted as being typical of a good capillary layer preparation, see above), will provide a sample thickness which is useful for surveying all the bands of many neat liquids, although, for many neat non-polar liquids, better spectral detail is likely to be achieved if a 0.1 mm path-length cell is used. As a solution is diluted, the path-length will need to be increased to maintain analyte bands at near equivalent intensity, *e.g.* a 10% solution of a polar molecule would typically be examined in a 0.1 or 0.2 mm cell. Path-length and concentration may also have to be compromised according to the viscosity of the liquid.

Permanent cells, in which the windows are sealed together, are the optimum choice for repetitive quantitative measurements, with clean homogeneous mixtures; they are also best suited to examining environmentally unfriendly materials, which must remain sealed in the cell. Cleaning these cells is difficult unless the samples can be flushed thoroughly. No traces of water can be allowed in the cells with salt or KBr windows, otherwise the windows will gradually dissolve, increasing the path-length. More commonly used are demountable cells, in which the spacers can be varied, and the windows gently cleaned if contaminated. Great care must be taken in cleaning these windows to maintain optical flatness and parallelism, particularly if the cells are being used for quantitative purposes; in such instances, rather than assuming constant thick-

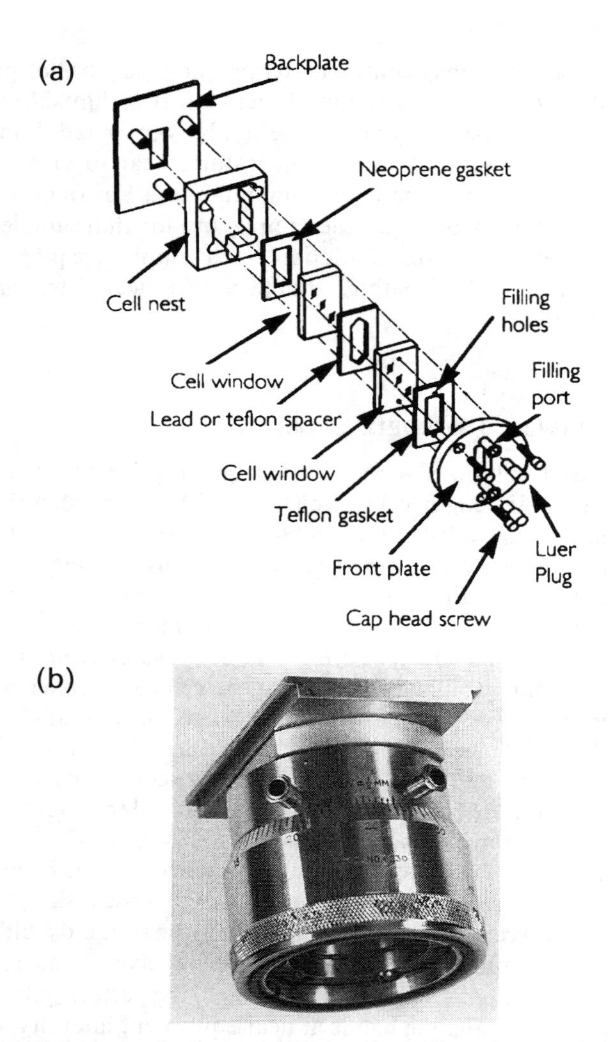

Figure 5 (a) *Schematic of a demountable (semi-permanent) transmission infrared liquid cell;* (b) *Photograph of a variable path-length liquid cell*
(Figures reproduced by kind permission of Graseby Specac Ltd., Orpington, UK)

ness, it is better practice to invoke an internal normalisation procedure, see p. 186.

Cells are filled by inserting a syringe or mini-pipette into the lower port of the cell and introducing the liquid so that it flows with an even wavefront between the windows. The liquid must fill all the cell, without air bubbles being present. If bubbles are introduced they may sometimes be removed by raising one end of the cell until they migrate out of the liquid (this rarely happens in practice), by injecting more liquid into the cell to force the bubbles out, or by gently applying vacuum to the upper cell port. The last method requires great care. The dangers are that a volatile solvent may evaporate too quickly and alter the concentration of the solution, or that the sample is too viscous for the path-length employed, in which case the cell windows may crack. Patience is a virtue when filling narrow path-length cells, particularly with viscous materials, since they often fill quite slowly by capillary attraction.

The path-length of a cell with flat, polished, parallel windows may be determined from a recording of the interference fringe pattern generated by the empty cell, see pp. 183–186. Maintaining and cleaning narrow path-length permanent cells can be troublesome, particularly since traces of viscous liquids may prove extremely difficult to remove. Continual disassembling and cleaning of demountable cells will also probably lead to degradation of the cell's transmission. Some of the problems associated with the routine use of thin transmission cells may be circumvented through the use of internal reflection accessories as described later in this chapter.

The diagram in Figure 5(a) shows a cell fitted with Luer plugs. Cells are also built with tubes instead of these ports, to enable flow-through (kinetic) studies to take place. Cells are also available with a constantly variable path-length, rather than a fixed spacer, see Figure 5(b). In this type of cell the path-length is changed by turning the front window assembly. The cell body contains a micrometer calibration, against which its path-length may be set accurately. A potential problem with the cell is that, at very narrow path-lengths, grit or contact can cause scratches on the opposing faces as the windows are rotated. More sophisticated cell designs are available in which, although the front mount and scale turn, the window moves in and out without turning. Variable path-length cells are very useful in dispersive instruments, where software spectral subtraction is unavailable, for balancing solvent bands, by placing a cell in the reference beam of a double-beam spectrometer.

A typical macro-sample liquid cell of path-length 0.1 mm will require about 0.3 ml to fill. Micro-volume cells are also produced for use when sample availability is restricted. Cavity cells of volumes as small as 0.2 μl are available commercially, although these are usually used in combination with a beam-condenser. Small volume flow-through liquid cells are also manufactured for use with HPLC-FTIR or GPC-FTIR applications. Hyphenated technique applications are described in Chapter 7. At the other extreme, where very dilute solutions are to be examined, for example in hydrogen bonding studies, cells of 1–10 cm are employed. These are usually tubular in construction, with the cell windows attached to the ends. Previously, matched pairs had to be made, but spectral subtraction routines have made this essentially unnecessary.

Powder Sampling Methods

For qualitative purposes (non-polymeric) powders are conventionally prepared for transmission examinations as fine particle solid suspensions dispersed in a suitable fluid, *e.g.* mulled in liquid paraffin (Nujol^{TM}), or in a disk of a transparent medium, *e.g.* KBr. Ideally, the diluent will not have absorption bands in the same regions of the spectrum as the sample. As stated earlier, the preparation method of a sample should be determined by its physical state, and the analytical information required. Some substances, such as base hydrochlorides, may exchange halogen with potassium bromide powder, in which case the mull method is preferable. Samples dispersed in alkali halide powder, or in a mull, must be homogeneously dispersed, and finely ground to a particle size which is preferably below that of the wavelength being used to interrogate the sample, *i.e.* small enough not to cause excessive radiation scatter. Suspension methods are best suited to crystalline, brittle or easily ground solids. They are widely applicable to monomeric organic and inorganic substances.

Mulls

Samples can be prepared as mulls for examination by transmission. The objective of this approach is to fill the interstices between the particles of a finely powdered solid with an infrared transparent or semi-transparent liquid medium, *i.e.* the solid is dispersed in a suitable liquid. This mixture is then sandwiched into a thin film between a pair of infrared-transparent windows. Ideally, the mulling agent should have a refractive index matching or close to that of the solid dispersed in it.

Most liquids have mid-infrared absorption bands. A liquid mulling agent is chosen for having suitable low-absorption, wide range transmission windows in its spectrum. Mineral oil (liquid paraffin, Nujol™) is the most commonly used mulling agent. This is a mixed chain length saturated aliphatic hydrocarbon and will therefore most likely obscure detail in the CH stretching band (3000–2800 cm^{-1}), CH deformation band (1500–1340 cm^{-1}), and CH rocking mode (near 720 cm^{-1}) regions. If hydrocarbon groups are of particular interest, a second mull may need to be prepared. In this case a halogenated oil mulling agent such as a perfluorohydrocarbon (Fluorolube™) or hexachlorobutadiene can be used. Together, the two mulls will allow almost unhindered observation of a sample's mid-infrared absorption bands.

To prepare a mull, transfer approximately 10–20 mg of powder into an agate or mullite mortar, and, if necessary, by using the pestle vigorously in a rotary movement, grind the powder to a fine particle size. Add one drop of mulling agent to the mortar, and, with the agate pestle, mull until the mixture has the consistency of a uniform paste, *i.e.* it is 'Vaseline®-like'. Transfer the mixture to a clean infrared transparent window (*e.g.* a NaCl, KBr, or CsI flat). Place a second window on top of the mull and squeeze the sandwich to form a thin film free from bubbles, *cf.* capillary film preparation described earlier. Mull samples often look translucent. If the mull does not produce an even capillary layer, then it is probably 'too thick', *i.e.* too concentrated. Repeat the preparation with less sample. Mulls can have too much or too little sample, but still produce an apparently good spectrum. Compare the mull spectrum with a reference spectrum of the mulling agent. If the strongest sample band is weaker than the strongest mull band, the mull is probably too weak. Repeat the preparation with more sample. It is a common fault to add too much mulling agent initially. Figure 6 shows a Nujol™ mull spectrum.

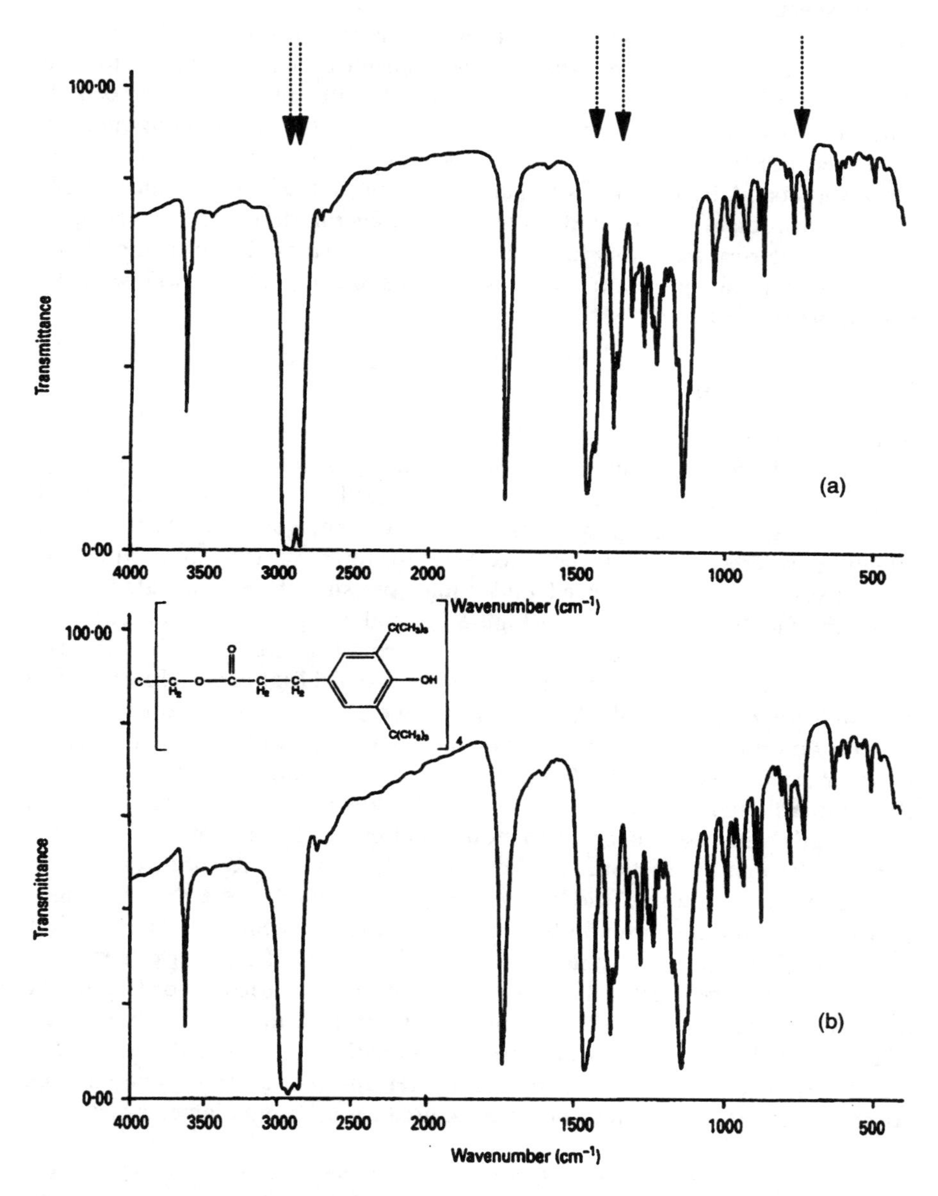

Figure 6 *Infrared Nujol mull transmission spectra of IRGANOX 1010 (Ciba-Geigy), a phenolic antioxidant for polymers:* (a) *a well-ground sample;* (b) *a poorly ground sample. Arrows indicate the regions where liquid paraffin (Nujol) has its principal absorption bands, which will overlap coincident absorption features of the analyte*

The spectrum of a mull prepared from an insufficiently ground sample exhibits a gradual decrease in background transmission on going from low to higher wavenumbers, since radiation scatter from particles increases towards shorter wavelengths. This is worse the greater the mismatch between the refractive index of the dispersed sample and that of the surrounding medium. Also, because the refractive index of a sample changes in the vicinity of an absorption band, becoming lower than the average to the high wavenumber side of an absorption maximum and higher than the average to the low wavenumber side, asymmetrical scattering loss will be superimposed on the absorption band. This results in the appearance of a distorted band profile, and is known as the Christiansen effect.

Alkali Halide Disks

This important sample presentation method involves making an intimate mixture of the finely ground solid sample and dry powdered alkali halide. The wt ratio of alkali halide to analyte is large, typically in the order of 200–100:1. The vast majority of samples prepared as alkali-halide disks use KBr powder. Finely powdered dry KBr will coalesce to form a clear disk with high transmission when it is pressed under high pressure in an evacuated die. A diagram of a KBr die is shown in Figure 7, see also Figure 3 of Chapter 6. KBr has a useful working range of 40 000 to 400 cm^{-1}, and can be obtained as a 'Spectroscopic Grade' from most commercial chemical suppliers. However, it is advisable to prepare a blank disk every time a new bottle is opened, as a precaution – interpreting contaminants not associated with the sample can be an infuriating waste of time! As with the mulling technique, the particle size of the sample should ideally be reduced to well below that of the shortest infrared wavelength being used, in order to reduce scatter loss and minimise absorption band distortions (Christiansen effect). The strength of the spectrum will be dependent on the amount and homogeneity of the sample dispersed in the KBr powder. The amounts given in the disk preparation procedure described below are for guidance only, the analyte bulk density or use of another diluent may require these to be varied. They will also need to be varied according to the diameter of the disk produced. Commercial dies in a range of diameters are available, including those which produce minidisks of 1 mm diameter. The weights quoted are for a 13 mm diameter disk, a commonly used size. Approximately twice as much needs to be used for a 16 mm diameter disk.

Recommended Method. Although preferred local procedures and practices will be developed and standardised, they should follow the general principles outlined. (They differ slightly in the laboratories in which we both work, because of the types of materials most commonly examined, the apparatus available, and local preferences). An optimal procedure, recommended by many experienced spectroscopists in order to achieve optimal spectrum quality and contrast, requires pregrinding of the sample prior to mixing it with the KBr to a homogeneous mixture, which is then used to make the disk. Many

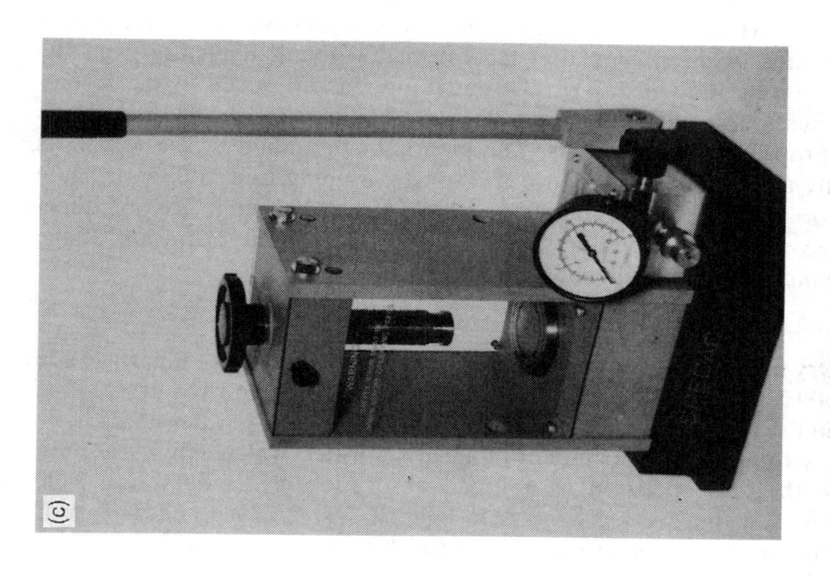

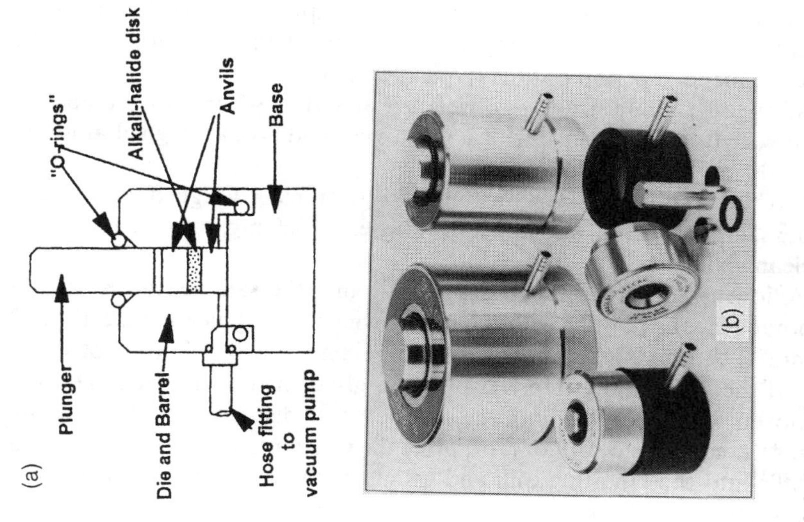

Figure 7 *Alkali-halide disk-making apparatus:* (a) *schematic of a die for preparing* 13 mm *and smaller diameter alkali-halide disks;* (b) *photograph of a range of evacuable pellet dies;* (c) *photograph of a manually operated hydraulic press for use with* KBr *disk sample preparations* (Figures (b) and (c) reproduced by kind permission of Graseby Specac Ltd., Orpington, UK)

expert practitioners consider that grinding the sample and KBr together initially is non-ideal practice, claiming that it is inefficient for reducing the particle size of the analyte. Also, any reduction in the particle size of the KBr powder creates more active surfaces, which means that it could adsorb water more rapidly. However, in our experience, the pre-grinding step is frequently short-circuited in many industrial laboratories, without detracting from the analytical results obtained. The alternative options are given in the recommended methodology below, and new users should select that which proves most appropriate to their needs. The pregrinding/grinding of the sample is commonly achieved by one of two methods – grinding in a mullite or agate mortar, or using a vibrating agate or stainless steel ball mill. The manufacturer's recommended procedures must be followed for the operation of the disk-making accessory, mill, and press.

(1) Dry a working quantity of 'IR-quality' KBr powder with a particle size of 100–200 mesh (~100 μm) overnight in a vacuum oven at ~150°C and store it in a desiccator ready for use. The KBr powder needs to be dry to minimise the amount of adsorbed water, which will reveal itself in the disk spectrum as a broad absorption centered around 3400 cm^{-1}, with a weaker band near 1640 cm^{-1}. The oven and desiccator should be kept free from other materials to avoid contamination of the KBr.

(2) Pre-grind about 10–20 mg of a sample in a mullite or agate mortar to a fine particle size, or place a few mg of the sample inside the clean agate or steel vial of a ball mill together with the balls and grind for the recommended period of time (typically 30 s).

(3) After pregrinding, transfer a weighed amount, ~2 mg, of the analyte powder to a clean mortar, or a clean vial, and add a weighed amount, ~300 mg, of dry KBr powder.

If the pregrinding step was missed out, transfer weighed amounts of the sample, ~2 mg, and dry KBr powder, ~300 mg, into the mortar or clean vial.

(4) Following the pregrinding step, gently mix the sample and KBr to a homogeneous powder with the pestle, which may typically take about 2 min. If the mixing is done in a vial the mixing time should be kept short.

If the pregrinding step was omitted, grind the powders together in a mortar with the pestle until the sample is well dispersed and the mixture has the consistency of flour, or, place the vial containing the mixture and balls into the vibrating mill and grind for the recommended period of time (typically 30 s).

(5) Transfer as much as possible of the KBr and sample mixture to a clean 13 mm die; fill and assemble the apparatus according to the manufacturer's instructions. Typically this involves:

(i) assembling the die with the lower pellet (anvil) in place, polished face uppermost,

(ii) transferring the ground mixture into the cylinder bore, so that it is evenly spread over the polished face of the lower pellet,

(iii) inserting the second pellet, polished face towards the mixture, into the bore, followed by the plunger, which should be pressed down.

(6) Place the die assembly into the hydraulic press, between the ram and the piston.
(7) Ensure that the die is firmly held in the press.
(8) Connect the die to a vacuum system and evacuate for about 2 min.
(9) Apply pressure and increase it slowly to the required pressure. A load of ~10–12 tons (~10 000–12 000 kg) is normally applied to the ram for preparing a 13 mm disk; follow the manufacturer's instructions for maximum pressure to be used with this and other diameter dies.
(10) Maintain the pressure for between 1 and 5 min, then slowly release it.
(11) Carefully release the vacuum, and remove the die from the press.
(12) Dismantle the die, and transfer the KBr disk to the disk holder. Avoid touching the faces of the disk.
(13) Check that the disk is translucent, intact, uncracked, and not fogged, and that the sample is homogeneously distributed in the disk.
(14) Mount the disk holder in the sample holder of the infrared spectrometer.

The spectrum quality will be affected by the quality of the disk. Figure 8 shows some disk preparation spectra. The flatness of the baseline depends on

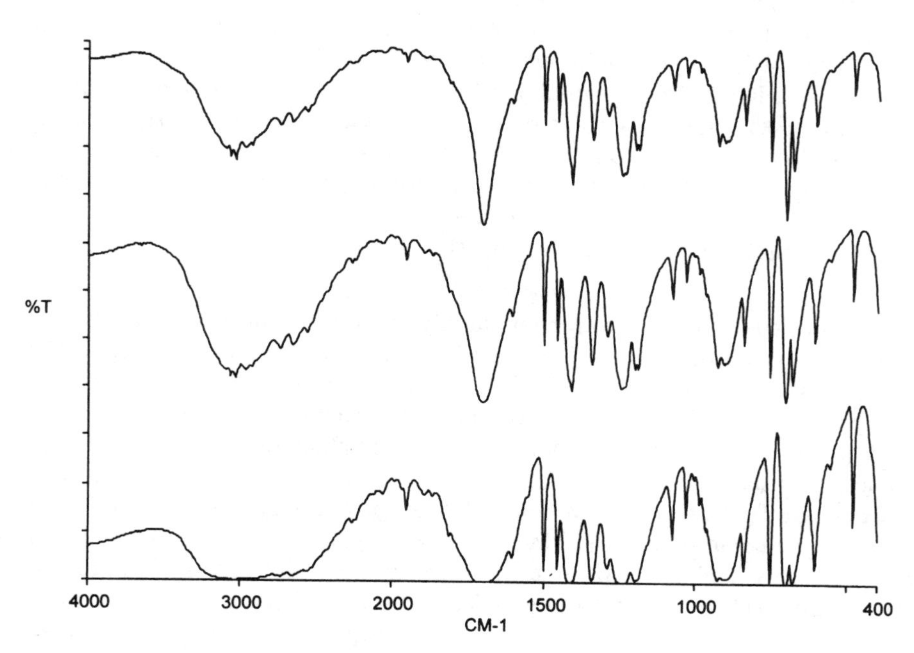

Figure 8 *Transmission infrared KBr disk spectra with increasing concentrations of the analyte phenyl acetic acid. Spectra have been offset for clarity*

the particle size and dispersion of the sample in the KBr powder. Check the disk and spectrum for the following faults:

(a) disk breaks on removal from the die. This indicates that the disk is probably too thin, caused by too little powder, or too much pressure for too long. Increase the sample load, and check the pressure used.
(b) disk is not translucent. This can have numerous causes:
 (i) uneven distribution of powder in die,
 (ii) too much sample,
 (iii) too much powder,
 (iv) poorly dispersed sample,
 (v) inadequate grinding of the analyte,
 (vi) water in the disk, from wet/moist sample or insufficiently dried KBr powder,
 (vii) disk pressed at too low a pressure or for too short a time.
 All except the last fault can be remedied by re-grinding and pressing with adjusted amounts. Small amounts of water are removed in the vacuum drying, step 8 in the procedure. Heating the disk in a clean oven, providing this does not degrade or adversely affect the sample, and re-pressing the disk should remove further residual water.
(c) disk turns brown or discolours.This could be due to the sample being an oxidising agent. Check the spectrum for halide degradation and re-examine as a mull if possible.
(d) bands in the spectrum are truncated or the spectral contrast is poor. This is caused by a poorly dispersed sample, a poorly ground sample or holes/cracks in the disk. Check the disk visually and repeat the preparation.
(e) the spectrum has a sloping baseline. This is also due to a poorly dispersed and/or a poorly ground sample. If necessary, use a mechanical or low temperature (liquid nitrogen cooled) grinding accessory. Some substances are too hard or intractable (*e.g.* many polymers) or too crystalline (*e.g.* anthraquinone) to disperse properly.

The faults listed above have been commonly found by the authors, sometimes in publications, but the list is not necessarily exhaustive. Poor sample preparation often leads to avoidable errors in interpreting the resultant spectrum, so time and patience spent in sample preparation will be well rewarded later.

While KBr is the most commonly used alkali halide, other alkali halides are sometimes used *e.g.* KCl, CsI. Caesium iodide, with a transmission cut-off at 200 cm^{-1}, offers extended wavenumber coverage, although it is softer and much more hygroscopic than KBr. Finely-powdered polyethylene can be used as a matrix for disks which need to be examined in the far-infrared spectral region. PTFE (polytetrafluoroethylene) powder and many of its copolymer powders will cold-coalesce at ambient temperature under pressure, and the alkali-halide die apparatus is very convenient for preparing analytically useful specimens for examination by infrared transmission spectroscopy. For a 13 mm disk, about

300 mg of neat powder pressed in a manner similar to that described, will form a translucent PTFE disk without the need of grinding and vacuum.

Gases and Vapours

While the majority of samples dealt with in industrial laboratories are either liquid or solid at room temperature, gases and vapours are also examined.They require special cells. The standard cell is a cylinder approximately 10 cm long, with a diameter of 3 cm. This holds a volume of approximately 70 cm^3. In use, a cell of these dimensions may typically be evacuated, connected to a vessel containing the vapour, the tap opened and a volume of gas sucked in. Flow-through versions of such cells are also available commercially. The cell body (pyrex glass, Perspex® and stainless steel are common) and window materials have to be considered. For example, water in a vapour may condense onto salt windows, and will introduce anomalous broad bands into the spectrum. Any other vapour that condenses will also mean that weak liquid-like transmission spectra are recorded rather than true vapour-phase spectra. Isocyanates have one of the strongest functional group absorption bands in the IR spectrum, but attempts at quantitative work often fail because the isocyanate has a strong affinity for the cell walls, rather than remaining in the vapour phase. Corrosivity also needs to be allowed for.

The strength of an infrared (or Raman) spectrum is ultimately dependent on the number of molecules in the beam. In a vapour the molecules are well dispersed, but also have greater motion. Figure 9 compares the vapour and

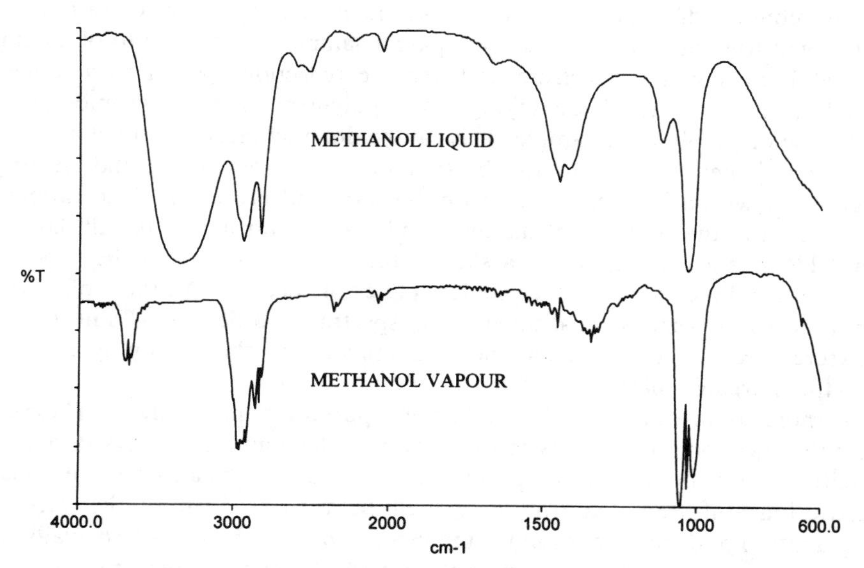

Figure 9 *Transmission infrared spectra of methanol:* Top, *capillary layer liquid film;* Bottom, *vapour in a 10 cm path-length cell at room temperature. Spectra offset for clarity*

liquid phase spectra of methanol. Low resolution (*e.g.* 4 cm^{-1}) vapour phase spectra, compared to their analogous condensed phase spectra, may show the rudiments of sharp bands due to rotational branching, (*cf.* also water and water vapour spectra). The pressure or partial pressure of a gas and its temperature will affect the shape and intensities of its absorption (or Raman) bands. At higher pressures band broadening occurs; molecular collisions will broaden the rotational fine structure of the gas-phase bands. If vapour spectra are to be used quantitatively, the working temperatures and pressures must be accurately measured.

For quantitative work or environmental studies, very low concentrations of vapours may have to be analysed. For this work very long path-length (typically, 25 or 50 m) gas cells are employed. The cells themselves are relatively low volume. The longer path is achieved by reflecting the beam several times backwards and forwards through the cell, see Figure 2 of Chapter 6. Typically a 20 m path-length is achieved with a cell cavity of about 0.5 m length.

Pyrolysis

Rubbers and polymer products can be examined as solids by transmission or by reflection techniques. Their spectra can be quite complex due to bands from various additives and agents in the polymer. Inorganic fillers present at high levels may have strong bands which dominate the spectra, overlapping key bands of interest – spectral subtraction techniques are rarely very useful in overcoming their presence. It is more likely that high concentrations of fillers will frustrate all attempts to prepare a sample appropriate to obtaining a useful transmission spectrum and degrade reflection spectra to a useless level! A simpler way of identifying many polymers in this case may be to volatilise or pyrolyse the sample. This can be done by placing a small amount of the polymer product in the bottom of a long test tube, and heating vigorously with a bunsen-burner, with due care and protection. The vapours will collect at the cool end of the tube, and may be transferred to salt flats or an ATR crystal. Alternatively, a shorter tube may be used and the vapours condensed directly onto the cooler plate or crystal. Neither of these approaches is very sophisticated. The spectra recorded may still be of mixtures, but are often simpler being dominated by the homopolymer and not the inorganic filler.

A more sophisticated method is to use a purpose-built pyrolyser. These fit into the injection port of a gas chromatograph. The sample is pyrolysed rapidly in-situ. The vapours are swept into the GC and subjected to the usual separation. Infrared detection may then be used to identify the various degradation or depolymerisation components. Direct pyrolysis is also available from accessories, in which a resistively heated platinum filament pyrolyser is housed in the base of a specially designed volatiles cell, which fits into the sample compartment of a FTIR spectrometer.

Variable Temperature and Pressure

Infrared and Raman spectroscopy are vibrational techniques. Band shapes, intensity, and to some extent positions are dependent on the physical state of the sample. These parameters are also temperature dependent. As seen earlier in this chapter, the comparison spectra of Figures 4 and 9 show significant differences. The bands in a spectrum arise from vibrations of the molecule, which are dependent on the degree of freedom and energy of the molecule. At lower temperatures the molecules have less energy, and may be held in a more solid and/or crystalline state. The molecules have less freedom to vibrate and the bands in the spectrum therefore tend to increase in intensity and appear much sharper. As the temperature is raised, molecules gain more energy, have more freedom and the bands tend to become broader, as intensity is spread into the wings of bands. At high pressures, vapour phase spectra often appear broader and more diffuse than their liquid state counterpart, although hydrogen-bonding effects will be non-existent or much reduced. Ultimately, in the vapour state at low pressure, molecules will have the greatest energy and degree of freedom and, at higher spectral resolutions, may feature many more sharp bands, with very specific wavenumbers of absorption, as the narrow bands associated with individual vibration–rotation energy level transitions become evident. For example, compare the spectrum of water vapour seen in the single beam spectrum of an unpurged FTIR instrument to that of liquid water.

The view that a substance is a frozen crystalline solid, a liquid when warm, and a vapour when hot is somewhat simplistic. Some room temperature liquids form glasses on freezing, rather than having a crystalline morphology. Their spectra at low and room temperatures are very similar. Solids at elevated temperatures may melt, vaporise, or degrade with consequent changes in their spectra. The vapour phase spectra recorded from above a solid at elevated temperature may show an increasing background slope, caused by scatter from subliming particles rather than absorption from the vapour.

It is to study spectral characteristics associated with changes in physical states, as well as studying the kinetics and pathways of chemical reactions, that variable temperature cells have been developed. The simplest, for elevated temperatures, are a metal block with a controlled heater, a thermocouple, and a slot to take thin films or KBr disks, while cell mounts with a flow-through coolant can be used for lower temperature experiments. Simple gas cells can have their bodies wrapped with heating or cooling jackets. More sophisticated variable temperature cells are available commercially which cover ranges from liquid N_2 temperatures to over 900°C; some of them may be temperature profile programmed. A limited-range type found in many multi-purpose industrial laboratories is shown in Figure 10. The sample is placed in an appropriate holder, which in turn is usually fixed at the end of a metal tube that can be filled with coolant. Alternatively the cell can be warmed with electrical heaters. The whole finger is then placed in a canister with windows; the space between may be evacuated like a vacuum flask to prevent condensation on the sample and

Figure 10 *Photograph of a variable temperature transmission infrared cell* (Figure reproduced by kind permission of Graseby Specac Ltd., Orpington, UK)

windows, or an inert gas may be passed through to minimise sample oxidation at higher temperatures.

High pressure cells are also available, although these tend to be used for more specialised applications.

6 Reflection Techniques for Infrared Spectroscopy

All the preparation techniques discussed up to this point have dealt with means by which a sample could be suitably presented to the spectrometer for measurement of its transmission spectrum, *i.e.* the radiation beam will usually strike the presented sample at essentially normal incidence, and pass through the sample to the spectrometer detector unless attenuated (absorbed) by the sample, or lost by other processes, such as scatter or back reflection from front surface. There are many types of samples and investigations for which this approach is not optimum, desirable or even practicable, *e.g.* aqueous solutions, urethane foams, polymer laminates, and surface coatings. To obtain spectra from these types of sample it is more usual to employ a reflection technique. Infrared radiation is directed at a sample surface, usually at an angle away from the normal, and the attenuated radiation, reflected back from that surface, is then detected. The most widely used reflection techniques in industrial laboratories are those based on internal reflection spectroscopy (IRS) and diffuse reflectance infrared Fourier transform spectroscopy (DRIFT), supported by specular reflectance and grazing/glancing angle reflection methods for more particular circumstances.

Internal Reflection Spectroscopy (IRS) Techniques

Common IRS method abbreviations are: ATR (Attenuated Total Reflectance) and MIR (Multiple Internal Reflectance). An alternative name sometimes used for ATR is Total Frustrated Reflection (TFR), (sometimes perhaps a more apt name, since one often gets *totally frustrated* by the quality of spectra produced by some users, who forget or ignore the basic criteria by which the method works, and then proceed to try to interpret spectra full of anomalies, or, perhaps worse still, perform quantitative measurements on them!). A basic appreciation of the principles and limitations of IRS is essential if it is to be applied effectively.

Looked at simply one might wonder why IRS works. 'Why should a beam reflected from an interface carry information on the absorbance characteristics of the surface layer of the sample below the interface?' It requires electromagnetic wave theory to describe why it does. However, it is not our purpose here to relate this, but merely to detail in a simple manner the key parameters necessary to a basic understanding. A rigorous mathematical explanation is included in the book *Internal Reflection Spectroscopy* by N.J. Harrick, which is probably the most quoted book on the IRS technique, and recommended reading/referencing for serious practitioners. While perusing the Harrick book, one should take particular note of the effects associated with using polarised infrared radiation, the differences between terms such as d_p (the depth of penetration) and d_e (the effective depth of penetration), and, importantly, the effects strongly absorbing bands can have on an IRS measurement.

From simple optical physics, if a beam of infrared radiation is incident on an interface at an angle greater than the critical angle, reflection rather than refraction will take place, provided that the material through which the radiation has passed has a higher refractive index than that of the medium the other side of the interface. The critical angle (α_c) is defined by: $\sin\alpha_c = n_2/n_1$, where, n_1 is the refractive index of the high refractive index medium, and n_2 is the refractive index of the low refractive index medium; n_2/n_1 is sometimes written as n_{21}. Simplistically, one may say that anything which interferes with these conditions destroys the effect.

The IRS technique relies on having a sample, the less dense medium, of refractive index n_2, in good optical contact with a surface of an essentially infrared transparent reflection element/prism of higher refractive index n_1, through which infrared radiation passes at an angle greater than the critical angle, so that total internal reflection occurs at the element–sample boundary. In the reflectance element a standing-wave pattern is established perpendicular to the surface of the element; within the denser medium (IRS element) there is a

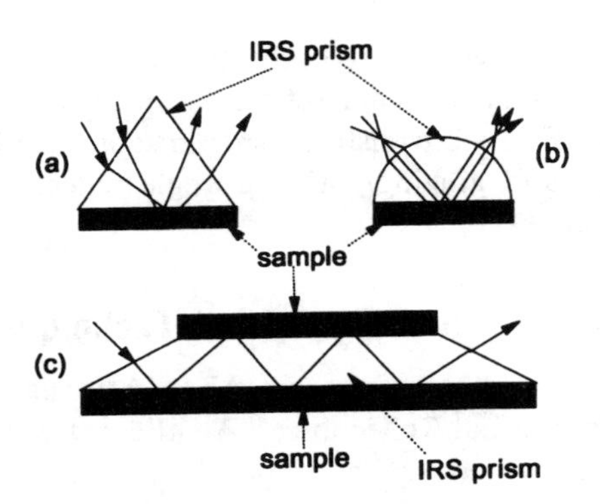

Figure 11 *Schematics of Internal Reflection Spectroscopy (IRS) Prisms and ray diagrams:* (a) *triangular ATR set-up;* (b) *hemi-spherical ATR element;* (c) *multiple internal reflection (MIR) schematic*

sinusoidal variation of electric field amplitude with distance, which, by suitable choice of incidence angle, can be made large at the boundary; whereas, within the rarer medium (the sample), the electric field amplitude falls off exponentially from its value at the boundary. This decaying amplitude, often referred to as the *evanescent wave*, will be attenuated by the absorption properties of the sample, resulting in a lower intensity of the reflected beam at the absorption wavelengths emerging from the internal reflectance element.

The above may be simply thought of as the infrared beam arriving at the prism–sample boundary, and penetrating the sample a short distance before returning back through the prism, carrying an absorption spectrum characteristic of the sample. Schematics of the ATR technique are shown in Figure 11. A depth of penetration (d_p) is most often defined as that at which the evanescent wave amplitude has fallen to 1/e, *i.e.* 37%, of its value at the interface, and is given by:

$$d_p = \frac{\lambda}{2\pi n_1(\sin^2\alpha - n_{21}^2)^{1/2}} \tag{1}$$

In Equation (1), frequently referred to as the Harrick equation, λ is the wavelength of the radiation, α its angle of incidence, *i.e.* the angle between the radiation beam and the normal to the surface, n_{21} is n_2/n_1, and $\alpha > \alpha_c$.

Equation (1) immediately tells us a lot about the appearance of an ATR-type measurement. It can be seen that the depth of penetration depends on the angle of incidence, the wavelength of the irradiating beam, and the relative refractive indices of the two media. Thus, in comparing an ATR spectrum with a transmission spectrum from the same material, while the two spectra may be very similar in that they both have bands at, or very near, the same wavenumber positions, the relative intensities of the bands within each spectrum are very different, the absorption bands in the ATR spectrum

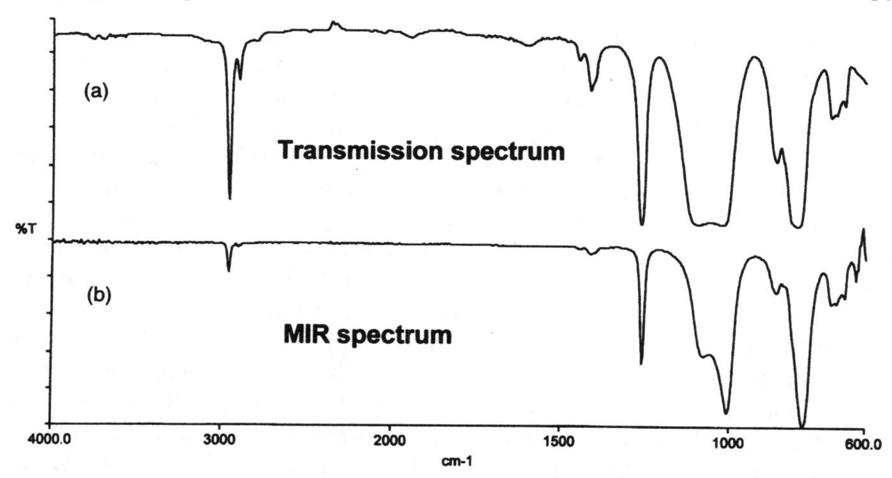

Figure 12 (a) *transmission infrared spectrum of a silicone oil smeared onto a ZnSe plate;* (b) *ZnSe MIR spectrum of the silicone oil*

becoming relatively stronger with decreasing wavenumber (increasing wavelength). Figure 12 compares the transmission and MIR spectra recorded from a silicone [poly(dimethylsiloxane)] oil. Equation (1) also tells us that d_p will decrease as the angle of incidence is increased, and that d_p will decrease if the refractive index of the IRS element is raised, conditions that may be to some extent experimentally selected to probe differing surface layers of a sample, remembering that d_p will be different at each λ. For large values of α, d_p is a fraction of λ; for values near to α_c, d_p increases to multiples of λ, although close to α_c the approximations used to derive d_p are no longer valid. For a sample of refractive index of 1.5, typical of many polymers, then at ~1725 cm^{-1} (λ = 5.8 μm) at an incidence angle of 45°, for a KRS-5 prism, refractive index 2.37, d_p calculates to 1.24 μm, while, for a Ge prism, refractive index 4.0, d_p calculates to 0.39 μm; for KRS-5 at 60° incidence angle then the equivalent d_p calculates to 0.66 μm. α_c for KRS-5 is 39°; for Ge it is 21.5°.

Among the most commonly used prism materials are KRS-5, Ge, and ZnSe. Of these, KRS-5, a mixed thallium bromide/iodide crystal (not *krypton pentasulphide*!), is perhaps the most generally used. It has a refractive index of 2.4, and will provide good spectra from many materials. A new, 25-reflection crystal properly aligned in a MIR accessory on a dispersive instrument will have ~60%T at 45° angle of incidence. For a regularly used crystal this may be expected to drop to ~20%T; crystals should probably be discarded if this falls much below ~10%T. For FTIR instruments, with their circular beams, thicker, shorter (10-reflection) prisms are sometimes preferred so as to increase the throughput. Depending on the beam diameter and focusing optics used, a new prism will probably yield between 10 and 25%T. Since ATR is very much a surface contact technique, particular care must be taken with KRS-5, which is relatively softer than most other IRS elements, not to use samples which may introduce tiny scratches onto the crystal surface, or to put too much pressure on the crystals, since they will cold flow; both effects will significantly degrade the performance of the ATR element. (One should also note carefully the hazard information supplied with the element by the manufacturers; KRS-5 is toxic if ingested). KRS-5, although not water soluble, will develop a surface bloom which makes it unsuitable for use with wet or aqueous samples. Germanium is another popularly used crystal, because of its higher refractive index of 4.0, which allows for thinner surface layer examinations. Ge has the advantage of being water resistant, but is very hard and brittle and can consequently be cracked fairly easily. Zinc selenide has a refractive index of 2.4, and is water resistant. Although the material is glass-like and may crack under pressure, many commercial accessories with fixed crystals are now using ZnSe. This material has the advantage that it is also resistant to many alkalis and acids, in our experience often out-performing manufacturers' specifications. The crystals are also very easy to clean, being washable under a tap if necessary. Other crystal materials and their refractive indices include: ZnS (2.2); AMTIR (2.5), mostly used as an ATR rod element for flow-through cells; Si (3.42); sapphire (1.74), and diamond (2.4), usually restricted to microscopy applications. They vary significantly in their mechanical properties, inertness/resistance to acids, alkalis, water and complexing agents, and have different regions in which they are transparent to infrared radiation, see Table 1.

The basic ATR accessory comprises a prism and holder, with a set of between 2 and 4 mirrors mounted on a base-plate to direct the spectrometer sample beam into and out of the prism. Early prisms gave a single reflection, but use of these has mostly been superseded by multiple reflection (MIR) prisms, see Figure 11.

A *key point to remember* in IRS experiments is that the refractive index of an absorbing sample is *not constant*. It varies in the vicinity of an absorption band, and may change markedly for strongly absorbing bands. This dispersion in the refractive index will cause distortions in the absorption bandshape and shifts in band maxima positions compared to those registered from a transmission measurement. Also, remember that the critical angle, which is $\sin^{-1}n_2/n_1$, will be affected by this dispersion. The refractive index falls below the average on the high wavenumber side of an absorption band, rising above the average on the low wavenumber side, the average being the value in the absence of any absorption. Thus d_p will be less on the high wavenumber (low wavelength) side of an absorption band and higher on the low wavenumber side. Bands will therefore narrow to the high wavenumber side, and may even appear to go slightly negative at the extreme, and broaden to the low wavenumber side, with their maxima shifted to lower wavenumbers. These effects increase at lower angles of incidence and the lower the refractive index of the prism, and will be particularly noticeable for strong absorption bands; compare the spectra shown in Figure 13. Therefore, if the distortion is marked, to record an IRS spectrum in which the bandshapes most closely resemble those in transmission, an internal reflection element with a higher refractive index should be used and/or the angle of incidence increased, although both will reduce the penetration depth and hence the overall intensity of the spectrum. However, this is not

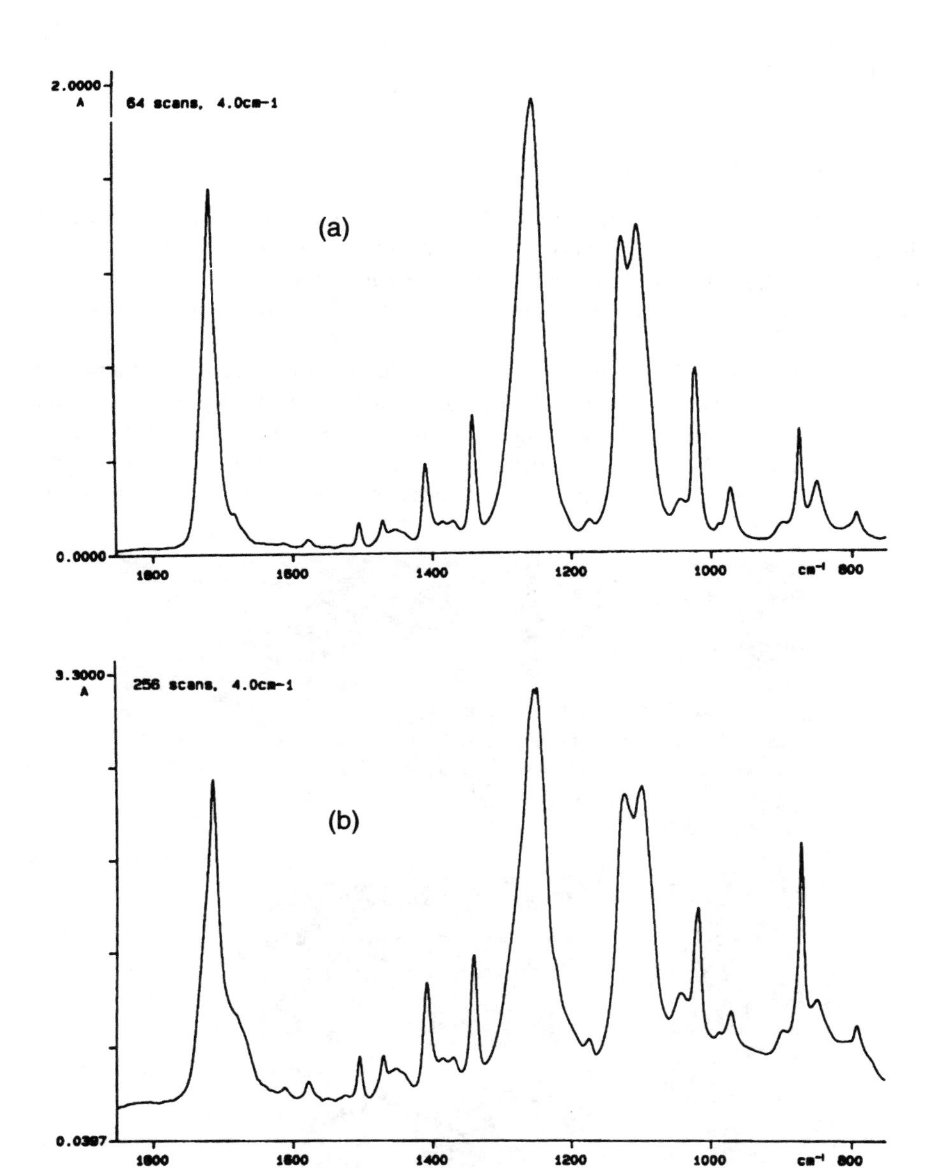

Figure 13 *KRS-5 MIR FTIR spectra of a PET (polyethylene terephthalate) film, 4 cm^{-1}* resolution: (a) *KRS-5 trapezoidal prism* (50 mm × 3 mm), *60° incidence angle; both sides of the prism covered with the film;* (b) *KRS-5 trapezoidal prism* (30 mm × 3 mm), *45° incidence angle; both sides of the prism covered with the film. Note the anomalous dispersion, particularly to the low wavenumber sides of the strong bands near* 1725 *and* 1250 cm^{-1}

usually a problem with FTIR instruments, since the signal-to-noise can be readily recouped by increasing the number of scans.

Horizontal ATR Accessories

This is sometimes abbreviated to H-ATR. There are many arrangements of these accessories. A popular model has a horizontal, top mounted, prism which allows solid, liquid, and paste samples to be easily introduced onto the IRS element, which is supported in a sampling plate and can be easily cleaned, see Figure 14.

This type of model has been adapted for *in situ*, non-destructive, skin analysis of fingers, arms, or any other part of the body which can be brought into contact with the crystal. Contact sampling is readily accessible to a wide range of viscosities and semi-solid compositions. Liquid samples are examined easily by pouring them into a trough plate. Cast films may be examined similarly, by

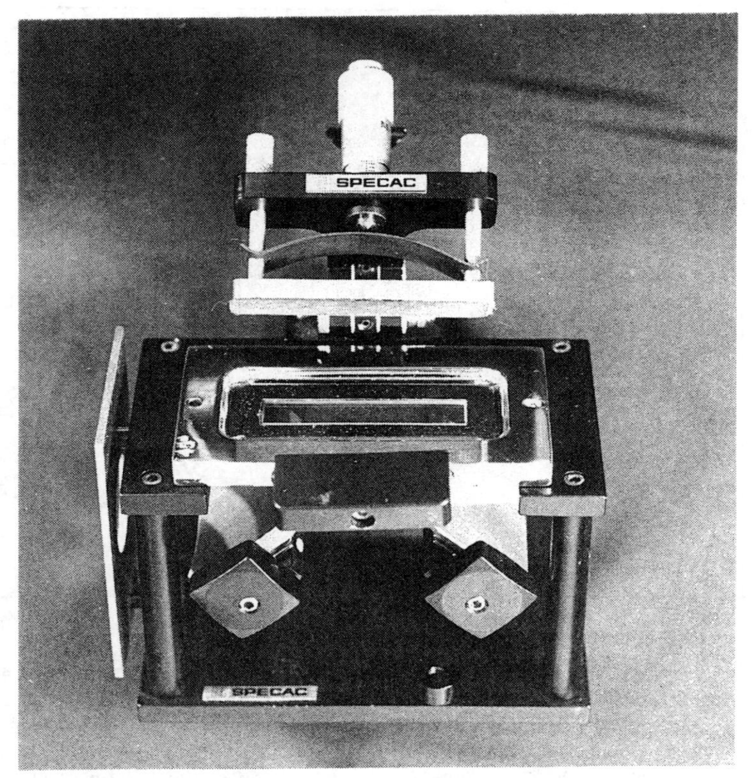

Figure 14 *Photograph of a horizontal ('overhead') ATR (MIR) accessory, fitted with a clamp, with a micrometer gauge pressure adjustment*
(Figure reproduced by kind permission of Graseby Specac Ltd., Orpington, UK)

spreading a few drops of solution onto the plate and allowing the solvent to evaporate. Pastes, oils, gels, foodstuffs, waxes, detergents, cosmetics, and the like may be carefully smeared or wiped across the prism surface. Dispersions and latices may be similarly handled, although care must be exercised, since it is not unknown for the dispersed material to preferentially migrate to the ATR prism surface.

In an industrial laboratory dealing with a high, routine workload of samples, H-ATR seems to offer many advantages (or perhaps more correctly many conveniences!), over transmission techniques for qualitative studies, although perhaps they are all not best practice! Since ATR is reliant on surface contact, in a correctly aligned and shielded set-up, stray-light phenomena are much less of a problem. H-ATR units do provide effective short constant path-length sampling, typically $\sim$25 μm for a ZnSe H-ATR unit, which is well-suited to routine liquid analyses. However, a liquid does not necessarily have to cover the whole beam (prism surface), or even be central on the prism, in order to record a reasonable quality fingerprint spectrum. Also, provided that the sample thickness is significantly greater than the penetration depth, it doesn't have to be of even thickness. This gives ATR cast film studies a sensitivity advantage, if the amount of material is too low for normal transmission work. The disadvantages lie with the possibility of evaporation causing skinning, and trapping solvent in the sample. A ZnSe prism set in a H-ATR trough sampling plate is a very useful support for examining the solute of aqueous solutions that can be allowed to evaporate. With liquids, pastes, and cast films, good surface contact is easily attained. Solids often do not have a flat surface. A clamp or pressure plate is used to press the material against the crystal. A wide variety of solids with not very flat surfaces, *e.g.* cloth, fibres, urethane foams, can be examined in this way. Many polymer films are the easiest type of solid to examine by H-ATR. By varying the pressure, a limited degree of empirical, qualitative depth profiling can sometimes be achieved! In some circumstances, the spectrum of a surface layer may be measured more cleanly with light pressure, while greater penetration into the bulk material will occur with increasing pressure. H-ATR can prove very effective for examining surface blooms on polymers and rubbers. After a polymer or rubber has been pressed against a crystal, and the sample is removed, a deposit of surface material may be left on the crystal. This deposit can then be examined without interference from the bulk. This is often useful in examining additive bleed or mould release agents on black rubber or polymer surfaces. When examining samples with hard or rough surfaces, or powders, care must be taken not to crack or scratch the crystal, for example, in the case of Ge or ZnSe. Some types of solid will not be flat even with pressure, *e.g.* rigid urethane foam. Contact can be improved by placing a high refractive index liquid between the crystal and the sample. This effectively extends the crystal surface onto the sample. A typical liquid is methylene chloride. This particular liquid is volatile, and a halogenated hydrocarbon. It is considered a health hazard, *so would have to be used with extreme care and due consideration of the hazards involved.* KRS-5, since it cold flows too much under pressure and can be easily indented, is not a standard H-ATR element offered by manufacturers of these units.

Vertical ATR Accessories

Many variations of holders for IRS are available. The earliest types of ATR holders held the crystals vertically, since they were primarily designed for use with dispersive spectrometers with slit-shaped beams. Although these have the disadvantage that it is not easy, without a special cell, to contain non-solid samples, they have the distinct advantage that units are available in which the angle of incidence may be easily varied, and consequently are still in very wide use, particularly in the polymer and related industries. We have found that uniform, higher, reproducible pressure may be applied much more easily, safely (without causing cracking of the prism), and routinely with these units than with H-ATR units, which is particularly important to quantitative studies on polymer films. KRS-5 prisms are readily available commercially for use with these units.

Samples are held in place with many designs of pressure plate. The simplest flat plates are for solids, but liquids can be examined by plates with cavities. A variation on the liquid cell is the flow-through cell, which has been used for kinetic studies or as an HPLC detector. Both solid and liquid cell platens can have electrical heaters attached for higher temperature work. The crystals lose transmission at high temperatures, dependent on the crystal material, see Table 1. It is common practice to aid sample/ATR prism contact and reduce the risk of prism damage, by inserting a backing material, such as a piece of a flexible rubber sheet, between a continuous polymer film sample and the clamp holder pressure plate. We have found that three thicknesses of freshly cut Whatman No.1 filter paper serve as an excellent material in this respect. For FTIR measurements, the single beam background should be recorded through the 'blank' prism, *i.e.* the prism with nothing contacting its sampling faces. Any backing material should be removed. (*Yes. We have received spectra where the single beam sample spectrum has been ratioed against the single beam spectrum of the backing material!*)

As the ATR spectrum is dependent on sample contact rather than sample thickness, crystals have been built into the sides of small reaction vessels, which can be mounted into the sampling compartment of a spectrometer. The same principle is also being increasingly applied in industrial plants, with rod-like probes based on the ATR principle being inserted into large reaction vessels and pipes, or used to make 'dip-in' probes for examining the contents of drums. ATR elements also feature as the method of spectrum generation at the tip of fibre-optic probes.

Micro-ATR

Many ATR designs are also reproduced for examining micro amounts of samples. This is particularly useful in the flow-through cell for HPLC. The best design minimises dead volume, and matches the cell volume to the chromatographic peak volume at half height. A micro-ATR accessory will tend to have a more limited number of reflections and employ some degree of beam-condensing to maximise the energy throughput. The ultimate in micro-ATR at present are the ATR objective lenses used with FTIR-microscopy, which is described in Chapter 6. With this, the single reflection technique is very much back at the forefront of the technology.

Rod-like ATR Units

In the examination of many liquids and solutions by transmission infrared spectroscopy, one of the disadvantages previously mentioned has been the requirement to have fairly small path-lengths. Many liquids will not flow easily through narrow transmission cells, and the infrared spectra of aqueous solutions are dominated by the water bands. With the ATR technique these problems can be circumvented due to the effective small path-length of the penetrating beam into the sample. Rod-like ATR accessories have been designed with the object of maximising the potential of this attribute. A relatively large cylindrical cell allows liquids to be pumped through, passing over a rod-shaped ATR element. Cells may be jacketed for heating or cooling. The spectra can be used to monitor the composition of the liquid. The elements are usually ZnSe, which allows a large range of solutions to be examined including acid or alkali aqueous systems; Ge and AMTIR are other popular element materials. Provided that the materials are homogeneous, emulsions such as milk can be examined in this cell. The shape of the rod maintains good flow characteristics and reduces problems with the deposition of solids from solution.

The first such unit to gain a high profile in industrial laboratories used a cylindrical ATR rod, see Figure 15. It is marketed by Spectra-Tech, Inc., under the trade-name of the CIRCLE® cell, and the approach is often loosely referred to by this name, although other variations are available, *e.g.* the SQUARE-COL® (Graseby Specac, Ltd) and the TUNNEL® cells (Axiom Analytical, Inc.). Cells of this type find uses in the quality assurance of a wide variety of beverages, such as beer, whiskey, fruit juices, and carbonated soft drinks.

Diffuse Reflectance Infrared Fourier Transform Spectroscopy (DRIFTS)

When infrared radiation falls on a surface, depending on the characteristics and environment of that surface, it may be absorbed, specularly (directly) reflected, internally reflected, or diffusely scattered over a wide area. It is the latter that, as the name implies, is essentially studied in diffuse reflection spectroscopy. The prime advantage of DRIFTS is that it enables infrared spectra to be recorded on diffusely scattering solids without the need for extensive sample preparation.

DRIFT spectra can be complex. They are strongly dependent upon the conditions under which they are obtained. They can exhibit both absorbance

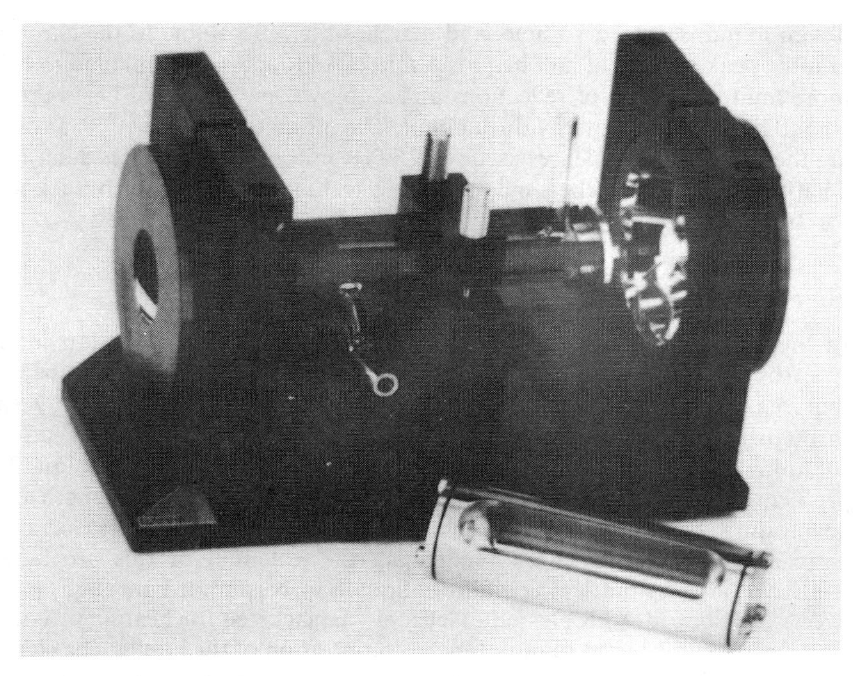

Figure 15 *Photograph showing a CIRCLE® cell, cylindrical internal reflectance unit, fitted with a flow cell. Also shown in the foreground is a trough/open boat cell* (Figure reproduced by kind permission of Spectra-Tech Inc., Shelton, CT, USA)

and reflectance features due to contributions from transmission, internal, and specular reflectance processes, as well as scattering phenomena in the collected radiation. An infrared beam focused onto a fine particulate material will interact with the particles in one of several ways, see Figure 16. Radiation may be reflected off the front surface of the particles. This represents true specular reflection, and is a function of the refractive index and absorptivity of the sample, see below pp. 160–162. The radiation may undergo multiple reflections off particle surfaces (all without penetrating the sample). This reflection, which is diffuse specular reflectance, will emerge from the sample at random angles relative to the incident beam. True diffuse reflectance results from the penetration of the incident beam into one or more sample particles and subsequent scatter from the sample matrix. This radiation will also emerge at any angle relative to the incoming beam. Since it has travelled (been transmitted) through the particles it will contain information about the absorption characteristics of the sample material. It is this latter information which we want to interrogate, since it should have some similarities to a transmission spectrum of the sample. Unfortunately, diffuse reflection cannot be separated optically from diffuse specular reflectance, but the contribution of the latter to the overall signal

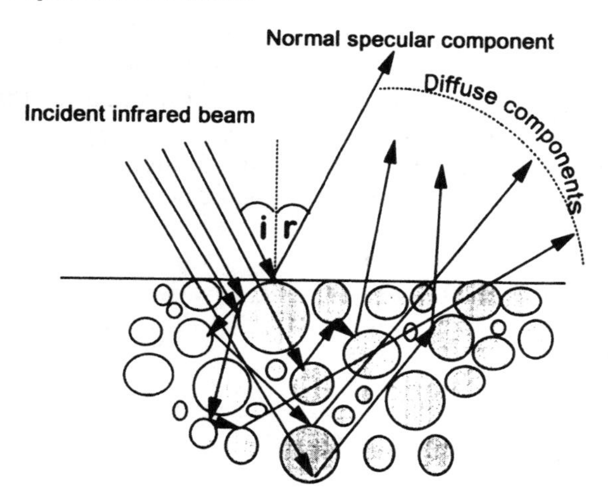

Figure 16 *Schematic of some processes leading to diffuse reflectance infrared spectra from a powder sample*

intensity can be minimised, see below. Notwithstanding, specular reflectance is a major cause of distortion in DRIFT spectra.

An integrating-sphere would be the ideal detection system for DRIFT measurements, and has indeed been employed, but it is much more common to compromise and use a sampling accessory with high-efficiency collecting optics, which will conveniently fit into the normal sample compartment of the spectrometer, see Figure 17. As can be seen from Figure 16, specular reflectance components will almost certainly be collected, which in severe cases will make interpretation of the spectrum almost impossible. Many DRIFT accessories employ a plate perpendicular to the sample surface, which is said/intended to

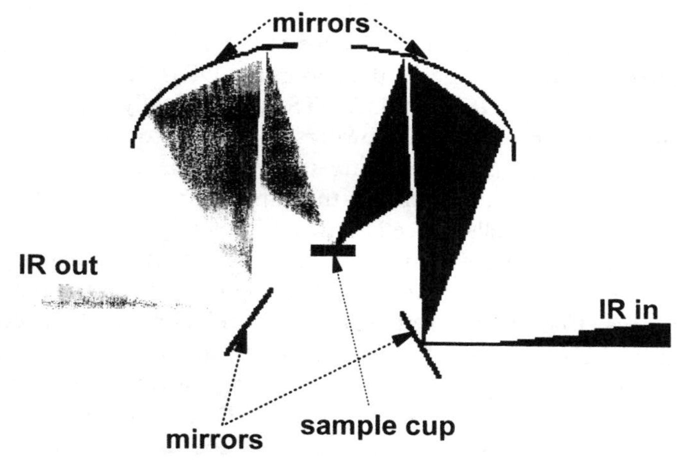

Figure 17 *Schematic of a ray diagram of a typical DRIFT accessory*

act as a blocker to the specular reflection component from the surface. It may well deal to some extent with some normally reflected specular components, but we have not yet examined a sample by DRIFTS for which using this has shown any significant improvement in spectral quality!

Several properties of the sample will exert significant influences on the quality of spectrum recorded from a DRIFT measurement, see Figure 18. These include:

- refractive index of the sample
- particle size
- packing density
- homogeneity
- concentration
- absorption coefficients

Specular reflectance components, particle size, and packing differences can produce pronounced changes in band shapes, relative peak intensities, and spectral contrast for organic samples, particularly for those with high absorption coefficients, which result in non-linearity of the relationship between concentration and band intensity. They similarly affect inorganic samples, and, for very strongly absorbing samples, can even produce inverted bands. Band intensities and widths are dramatically altered from those expected, *i.e.* by comparison with a conventional absorption spectrum, as the particle size increases (*this is also true for a KBr disc spectrum – try it sometime!*), and are more pronounced, of course, in highly absorbing samples. Grinding a sample to a smaller particle size will reduce the contributions of external reflectance from the larger particles.

DRIFTS of Powders

In theory, powdered samples require little or no preparation other than filling a cup to the requisite height, running the edge of a spatula across the top to even the surface height and then placing the cup in the DRIFTS accessory. Indeed, some early claims suggested that DRIFTS might totally replace KBr disk preparations. However, many neat powders absorb far too strongly, and need to be diluted if meaningful spectra are to be obtained. To record optimal DRIFT spectra from powders, so that they appear more similar to transmittance spectra than bulk reflectance spectra, it is mostly necessary to have the sample present in a fine particulate form and to dilute it in a finely powdered non-absorbing matrix, such as KCl or KBr. The former is favoured, because, although less easy to grind, it is less hygroscopic and more highly reflective. The single beam background for a DRIFT measurement should be made using a neat sample of the similarly ground non-absorbing matrix material, in order to provide an 'ideal' diffuse reflector for the FTIR analysis. Diluting ensures deeper penetration of the incident beam and less specular reflectance from the sample surfaces, thus increasing the relative contribution to the recorded

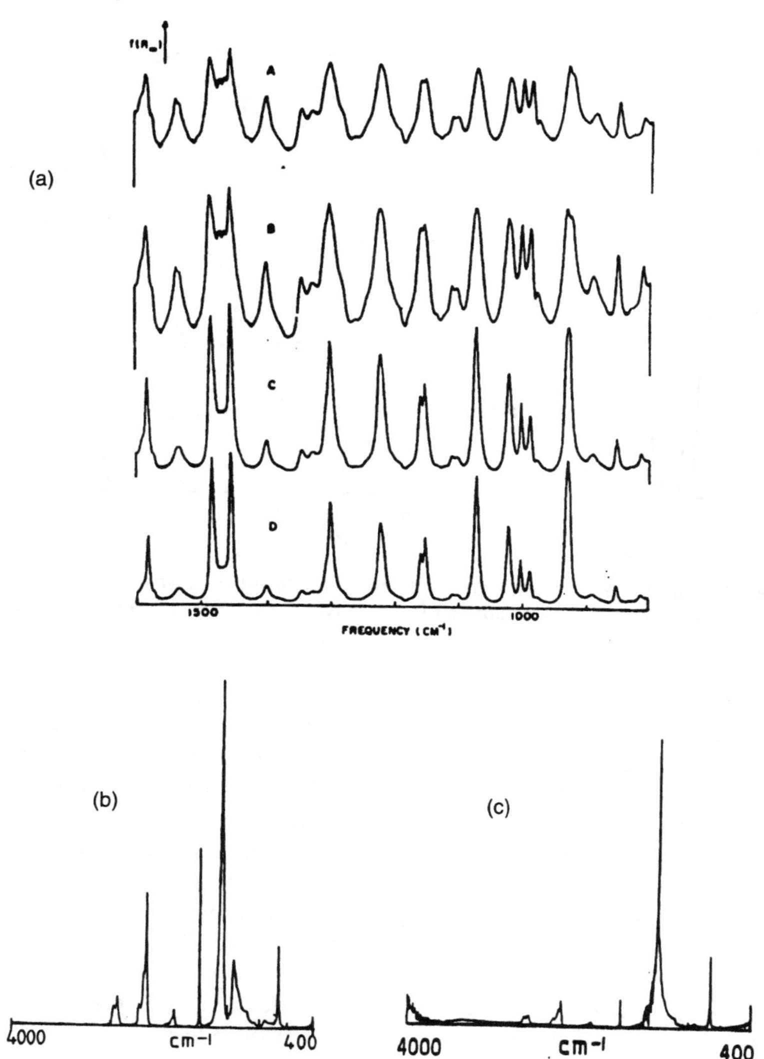

Figure 18 (a) *Effect of particle size (diameter, d) on the diffuse reflectance spectrum, plotted as* $f(R_\infty)$, of neat azobenzene *vs.* KCl reference: (A) $\bar{d} < 90\ \mu m$; (B) $75 < \bar{d} < 90\ \mu m$; (C) $10 < \bar{d} < 75\ \mu m$; (D) $\bar{d} < 10\ \mu m$;
(Reproduced from *Diffuse Reflectance Measurements by Infrared Fourier Transform Spectrometry*, Fuller M.P. and Griffiths P.R., *Anal. Chem.*, 1978, **50**, 13, 1906, by kind permission of the American Chemical Society, © 1978)
(b) *DRIFT spectrum of neat ball-milled* KNO_3 *at a compaction pressure of* ~520 kPa *vs. KCl reference;*
(c) *DRIFT spectrum of 1% wt/wt ball-milled* KNO_3 *in ball-milled KBr at a compaction pressure of* ~520 kPa *vs. KCl reference*
Note 'missing' ~1380 cm^{-1} KNO_3 *band in (b) due to the influence of front-surface (specular) reflectance*
(b) and (c) reproduced from *Infrared Diffuse Reflectance Spectroscopy* (R.N. Ibbett, PhD thesis, University of East Anglia, UK, 1988)

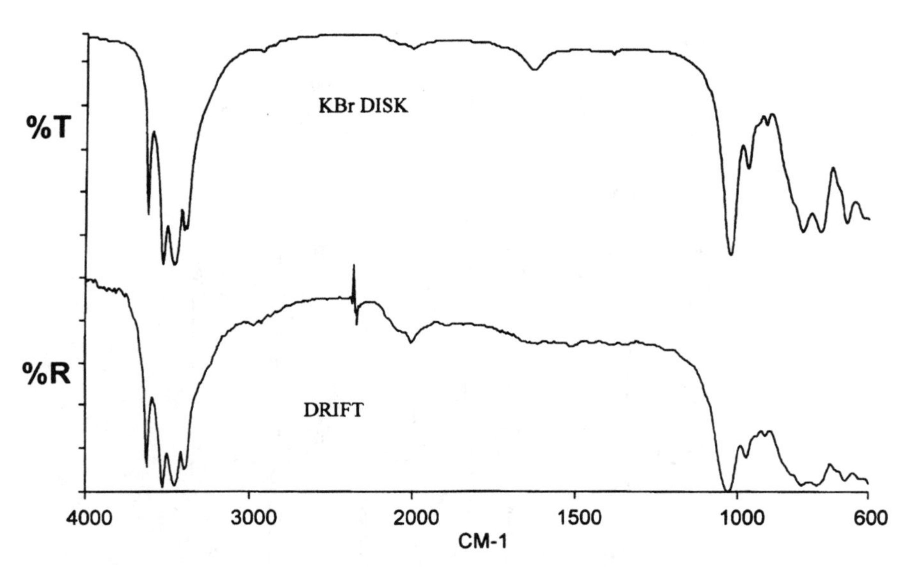

Figure 19 *Aluminium tri-hydroxide (ATH) powder infrared spectra:* Top, *KBr disk transmission spectrum;* Bottom, *DRIFT spectrum of ATH diluted (mixed with) with KBr powder*

spectrum of the components carrying absorbance characteristics. DRIFT spectra of finely divided powders very similar to KBr disk transmission spectra can be recorded at dilutions of approximately 10% wt/wt in KCl. Rather than grinding with the powders, as in some disk procedures (see above), the sample should be gently mixed, 'folded in', with the diluent powder. This preserves the physical form for polymorphic studies. Gentle mixing also preserves surface coatings on powders, a surface analysis application for which DRIFT has been shown to be useful in some circumstances. Figure 19 compares a DRIFT spectrum with a KBr disk spectrum.

Best results are achieved if the diluting powder and sample are ground to a similar particle size, typically less than 10 μm, prior to mixing. The alkali-halide diluent should be dried, stored and handled in a careful manner, similar to that for alkali-halide disk preparations. Quite acceptable spectra can be obtained for qualitative analysis, though some distortion can occur if the specular reflection component is not successfully blocked. DRIFTS also enhances weak bands preferentially to strong bands which again can lead to distortion. For quantitative studies two major problems are the non-linearity of band intensity with concentration and spectrum reproducibility. The former is normally addressed by transforming spectra with the Kubelka–Munk (K–M) function, see Equation (2), rather than the equivalent of transmission to absorbance conversion.

$$f(R_\infty) = \frac{(1 - R_\infty)^2}{2R_\infty} = \frac{k}{s} = \frac{2.303\text{ac}}{s} \tag{2}$$

where, R_{∞} is the ratio of the diffuse reflectance of the sample at 'infinite depth' (*i.e.* at a depth beyond which the signal does not change, usually 2–3 mm) to that of a selected standard (*e.g.* powdered, dried KCl), k is the absorption coefficient, c is the concentration of the sample, a is the molar absorptivity, and s is a scattering constant.

s is related to the particle size. Ideally, small regular, monodisperse spheres should be used for DRIFT quantitative analysis. Real-world samples for analysis rarely come in this form; they have a particle size distribution, tendency to aggregate, and are randomly orientated and irregularly shaped. Hence, reproducibility can be a problem. Other factors affecting reproducibility are packing density, sample surface flatness, and sample cup orientation. Careful sample handling procedures can reduce these effects and improve quantitative accuracy. Most spectrum handling packages on spectrometer PCs contain a software option which automatically converts reflectance spectra to a Kubelka–Munk scale. However, it is our general experience that DRIFT measurements are mostly limited to semi-quantitative work, and alternative methods should be used if high precision is required. Exceptions, of course, will be found, and, indeed, we have reported such for the direct compositional analysis of a series of tetrafluoroethylene/hexafluoropropylene (TFE/HFP) copolymer powders, when weak bands were used.

A useful approach for sampling intractable, composite, or gross objects is to abrade a fine powder from the article's surface with some silicon carbide or diamond powder abrasive paper. The abraded powder may then be examined *in situ* on the abrasive paper or removed and dispersed in dry potassium chloride powder. Abrasive pads/disks of ~ 13 mm diameter are sold commercially to facilitate such sampling. The DRIFT technique does provide a very convenient means of studying catalysts and catalytic processes. Special environmental, evacuable chambers are available, which allow for *in situ* conditioning of catalysts at elevated temperatures ($>700°C$), and monitoring of species adsorbed from gases flowing over the catalyst. High pressure (~ 50 at) and low temperature (liquid N_2 cooled) versions of these chambers are also manufactured. DRIFT spectra of samples deposited on KBr have been recorded, as a fraction collection apparatus for LC-FTIR examinations. This is described in more detail in Chapter 7.

DRIFTS of Polymers

Despite all the apparent difficulties, a DRIFT accessory can be a very sensitive and useful industrial tool. We have used a DRIFT accessory to effectively analyse quantitatively, resin coatings (thickness 0.5 μm), on aluminium foil although these were more truly transflectance spectra than diffuse reflectance, see next section. The DRIFT technique can be a quick, if less than rigorous, way of analysing hard polymers, which would normally require more tedious or specialised sample preparation. Notably, we have found it particularly well suited to the direct spectral 'fingerprinting' of polymer foams, whether rigid or flexible, open-structured foams. Polymer products are often shiny, with very reflective surfaces, so direct examination by DRIFTS would suffer extensively from specular reflection effects. These can be reduced by slightly roughening the surface with abrasive (emery) paper. If the sample is too large for the DRIFT accessory, the abrasive paper with a quantity of the transferred polymer on it can be examined in the DRIFT accessory, and often good quality spectra of the polymer can be recorded in this way. A blank sheet of abrasive paper is usually used as a background reference. As the polymer particles on the paper reduces its apparent surface roughness, the beam is scattered less by the sample than the reference. This can lead to spectra being recorded with baselines much greater than 100% R. Paint films can also be examined by this abrasive paper technique, thereby sometimes overcoming the problems of recording a spectrum with mixed specular reflection, diffuse reflection and 'transflectance' components, which can occur when some paint films on metal surfaces are examined directly.

Specular Reflectance

Specular reflectance is the condition under which incident radiation is reflected directly and collected, normally from a planar surface, according to the relationship, $\angle$incidence = $\angle$reflectance, see Figure 20(a). For a pure specular reflectance spectrum then only the component reflected normally from the front surface of a sample must be detected; back-surface reflected or diffuse radiation will give rise to absorption-like features or interference fringes. This requires that the sample area interrogated is essentially optically flat and that the depth is homogeneous and 'optically-thick'. The proportion of radiation reflected, R, at an angle of incidence near normal ($0° < \alpha < 30°$) from a sample in air is given by:

$$R = \frac{I}{I_0} = \frac{(n-1)^2 + n^2k^2}{(n+1)^2 + n^2k^2} \tag{3}$$

where, α is the angle of incidence, k is the absorption coefficient, n is the refractive index of the sample, and I and I_0 are the intensities of the reflected and incident radiation, respectively. However, as stated, the refractive index of a sample varies with wavenumber in the vicinity of an absorption band. The dependence of R on both n and k means that R falls on the high wavenumber side of an absorption band and rises on the low wavenumber side, see Figure 20(b), and consequently pure specular reflectance spectra have a first-derivative-like appearance. Specular reflectance spectra are therefore almost impossible to interpret in any great detail. Fortunately by applying software manipulation in the form of the Kramers–Kronig (K–K) transform, an absorption index spectrum can be extracted from the specular reflectance data. This algorithm is readily available within most FTIR data processing packages.

The parameters n and k are not independent of each other, and the overlay of dispersion and absorption, based on the Fresnel equation, may be separated by subjecting the measured specular reflectance spectrum to the K–K transforma-

(a) Specular Reflectance

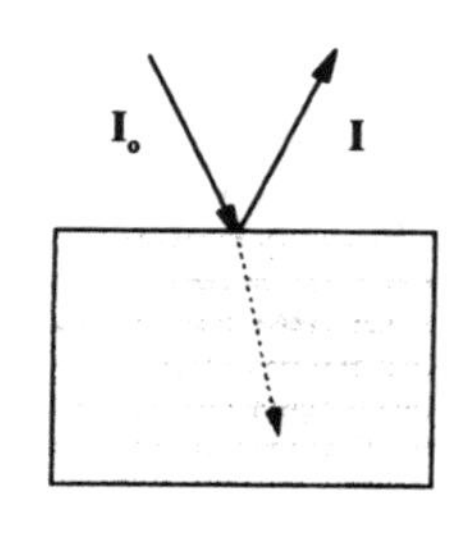

(c) "R = *f*(n,k)"

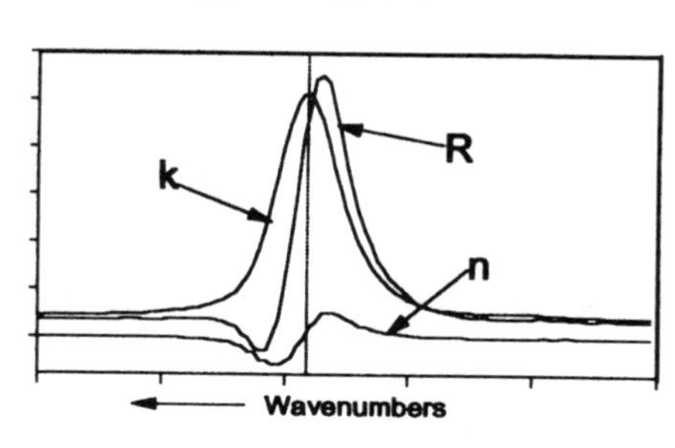

(b) Transflectance

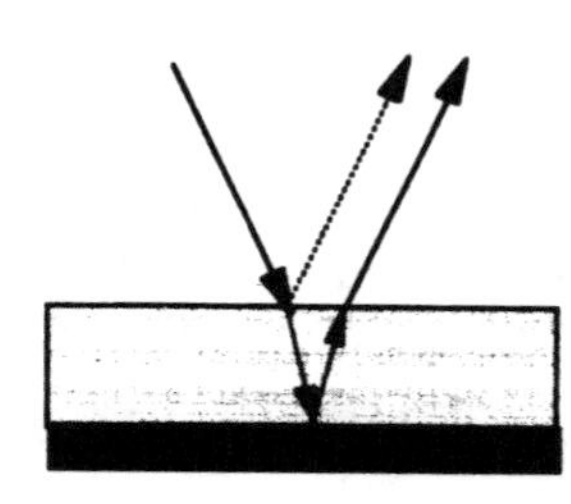

Grazing angle reflection-absorption spectroscopy.

(d) p polarisation

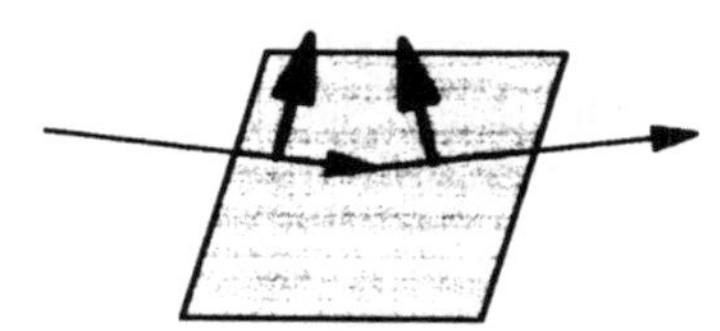

(e) s polarisation

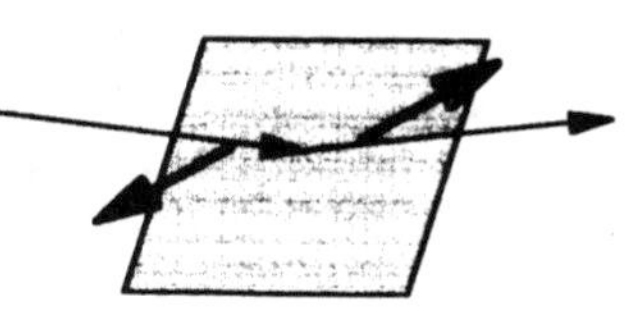

Figure 20 *Schematics of external reflection infrared sampling techniques:*
(a) *Specular reflectance;*
(b) *Transflectance;*
(c) *Schematic showing the relationship between the dispersion in a sample's refractive index, n, its absorption index, k, and R, its specular reflectance spectrum;*
(d) *p-polarisation grazing-incidence reflection–absorption; note how the electric field vectors (bold arrows) near the sample surface sum together;*
(e) *s-polarisation grazing-incidence reflection–absorption; note how the electric fields near the sample surface cancel*

tion algorithm, which yields both the refractive index dispersion spectrum and the much more analytically useful absorption index spectrum, see Figure 21(a). (K–K algorithms generally assume normal incidence radiation).

The technique is ideal for obtaining fingerprint spectra from highly reflective, optically thick polymers, with no sample preparation other than mounting in a

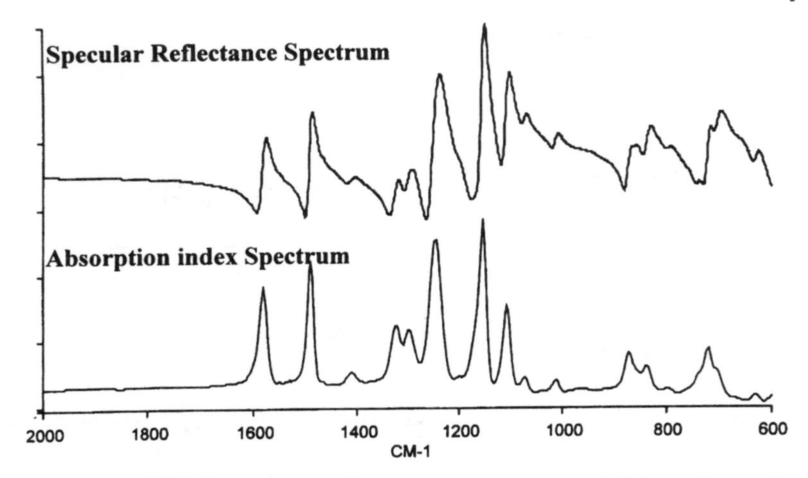

Figure 21(a) Top, *Infrared specular reflectance spectrum recorded from a* 3 mm *thickness plaque of a poly (aryl ether sulphone), PES;* Bottom, *Absorption index spectrum of PES generated from its specular reflectance spectrum by subjecting it to the Kramers–Kronig transform*

specular reflectance accessory, a DRIFT accessory, or under an FTIR-microscope, and ensuring that no reflection other than that from the front surface reaches the detector. The technique favours the more strongly absorbing polymers, such as aromatic polymers, since, remembering that refractive index and absorption go hand in hand, the more intense the absorption band the greater the refractive index dispersion. Consequently, the technique will be much less favourable for weaker absorbers, such as hydrocarbon polymers. Also, investigations are essentially limited to what can be discerned from within the signal-to-noise of the absorption index spectrum; the sample cannot be 'thickened' to increase signal intensity!

Reflection–Absorption Spectroscopy

Reflection–absorption (R–A) measurements are concerned with recording infrared spectra from thin films of materials supported on a reflective, non-absorbing substrate.

'Transflectance'

This is perhaps the most straightforward of the infrared external reflection spectroscopy measurements. It is carried out at near normal or low angles of incidence. As can be seen from the schematic for the measurement in Figure 20(b), it involves the incident radiation passing through the sample, being reflected from the reflective substrate, then again traversing the absorbing layer, before going on to the detector. This reflection–absorption approach has become commonly referred to as a *transflectance* measurement; it is *not* a

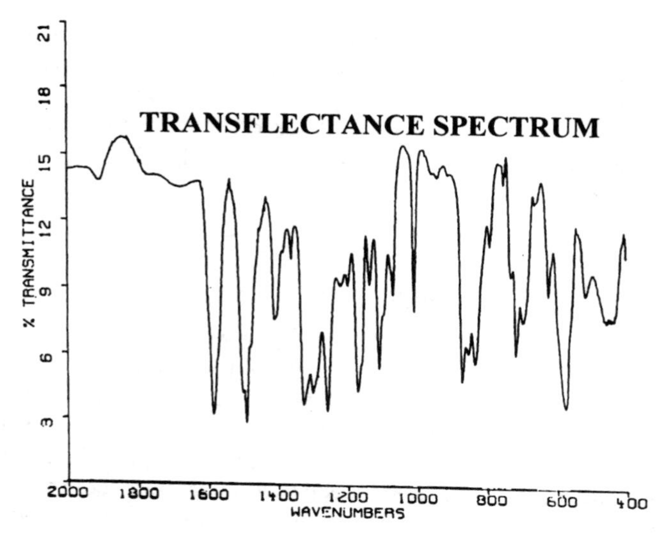

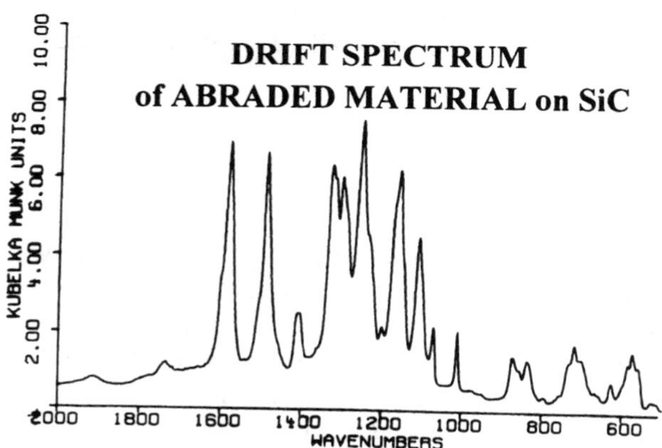

Figure 21(b) Top, *Transflectance spectrum recorded from a 'non-stick' PES coating applied to a baking tray;* Bottom, *DRIFT spectrum on SiC abrasive paper of material abraded from the surface of a PES moulding. Note: Close comparison of these spectra with those of Figure 21(a) will quickly reveal some effects (distortions, spectral contrast) associated with specular reflectance, which particularly affect the stronger absorption bands. (PES is essentially an amorphous polymer)*

specular reflectance spectrum, although superimposed on the 'double-transmission-like' reflection–absorption spectrum will be a weaker specular reflectance component, see figure and caption notes to Figure 21(b).

Highly reflecting surfaces with thin layers can be examined by the transflectance method, *e.g.* organic coatings/contamination on polished metal. Typical of these are the identification of coatings (~0.2 μm to 20 μm) on metallic substrates (*e.g.* beverage containers). They may also be used to test the effectiveness of anti-reflection coatings on infrared optics. A specialised application is the measurement of epitaxial layers on silicon wafers.

The same commercial accessory will serve for transflectance and specular reflectance measurements, although they are usually described in catalogues only as specular reflectance accessories.

Glancing-angle/Grazing-angle

For oblique incidence, the reflected intensity, R, depends on the polarisation of the incident beam. This is illustrated in Figures 20(d) and 20(e), where s (senskrecht) represents perpendicular polarisation (*i.e.* the electric field is perpendicular to the plane of incidence), and p polarisation means parallel polarisation (where the electric field is parallel to the plane of incidence).

Reflection–absorption infrared spectroscopy is here concerned with the change of reflectivity of the surface of a substrate introduced by a thin absorbing film. Absorption bands of the surface film are recorded by measuring the change of substrate reflectivity, ΔR, due to the surface film, which is approximated to:

$$\Delta R = 1 - \frac{R_0}{R} \tag{4}$$

where, R and R_0 are the reflectivity of the substrate with and without the film, respectively. The absorption band intensities will depend on the infrared standing-wave intensity at the surface as well as the absorption properties of the film. This type of R–A is most successfully applied to thin films (< 1000 Å) on metal surfaces, because under certain conditions the high reflectivity of the metal surface may be combined with an intense standing-wave at the metal surface. As can be intuitively deduced from Figure 20(e), for s-polarised light there is a 180° phase shift between the incident and reflected beams at the reflective surface. The two beams therefore superimpose to form a standing-wave with an electric field strength of zero at the surface. Consequently little or no interaction occurs between this infrared radiation and a thin surface film at the surface, for films thinner than $\lambda/4$ of the radiation. However, for p-polarised infrared radiation the phase shift varies with angle of incidence. At grazing incidence it is about 90°, resulting in large electric field values at the surface for high angles of incidence, maximising at 87°. Hence, optimum sensitivity is obtained at high incidence angles (usually about 80° is employed) with p-polarised radiation. The electric standing-wave at the surface is effectively polarised perpendicular to the surface, and will therefore only interact with dipole moments of species that are normal to the metal surface. This leads to the *Metal Surface Selection Rule*, which states that only vibrational modes with dipole components perpendicular to the metal surface may be excited, so not all the absorption bands of a sample film will be observed.

In addition to glancing-angle or grazing-angle, reflection–absorption spectroscopy, the approach also goes under the acronyms RAIRS (Reflection Absorption Infrared Spectroscopy) or IRRAS (Infrared Reflection Absorption Spectroscopy). The technique tends to be the preserve of more specialised applications, particularly adsorbed species on catalytic clean metal surfaces. It has also been used for contamination and corrosion studies, electrode/electrolyte interfaces, and similar studies relating to ultrathin organic films, monolayer or a few monolayers, on reflective metal surfaces.

7 Photoacoustic Spectroscopy (PAS)

Photoacoustic spectroscopy (PAS) has attracted considerable interest from infrared spectroscopists, since it was first coupled with FTIR instruments, because of its potential for examining samples with minimal sample preparation. In FTIR-PA spectroscopy, a sample is placed in a small-volume, sealed chamber containing an infrared transparent gas, usually helium. The modulated infrared beam from the spectrometer is directed onto the sample through a transparent window fitted within the PA-cell body, see Figure 22. (The modulation frequencies for continuous, rapid scan and step-scan operation of FTIR spectrometers are considered in more detail in Chapters 3 and 5.) Radiation absorbed by the sample is converted to heat, which generates thermal waves in the sample. These propagate to the sample surface–gas boundary, producing pressure fluctuations in the gas, which are then detected by a sensitive microphone, and subsequently processed as FTIR interferograms. (That is, the microphone *listens* to a sample becoming warm at its characteristic infrared absorbing wavelengths).

Since the incident infrared energy has to be absorbed by the sample and 're-emitted' as a thermal wave, the rate at which the signals can be detected depends on the thermal diffusion rate. The intensity of signal detected depends, not only on the optical modulation frequency and the sample thermal diffusivity, but also on both the optical absorption length and thermal diffusion length and the relationship between them, and possibly, also, on the sample thickness, see Figure 5 of Chapter 5 and pp. 187–188. These impose certain key experimental

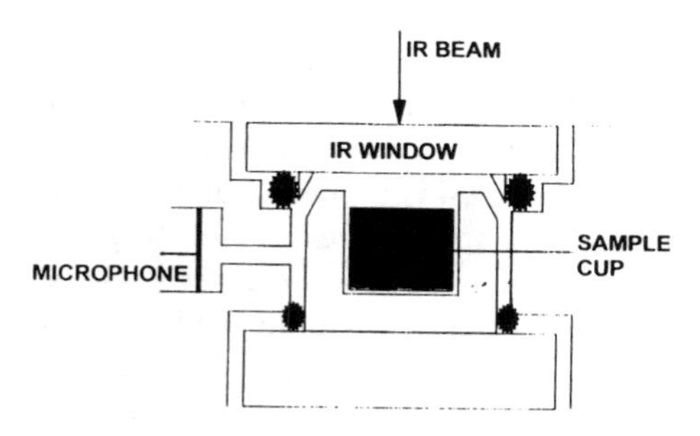

Figure 22 *Schematic of a photoacoustic cell*

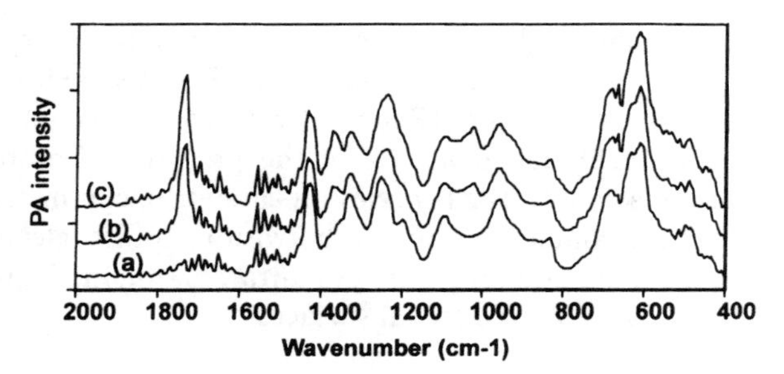

Figure 23 *FTIR-PA spectra of neat powders* (8 cm^{-1} *resolution,* 100 scans): (a) *Poly (vinyl chloride), PVC;* (b) *vinyl chloride/vinyl acetate (5%) (VC/VA) copolymer;* (c) *VC/VA(12%) copolymer. No desiccant was placed below the sample cell, so water vapour is very evident in these spectra. The single-beam FTIR spectra were ratioed against that of a C-black powder*

factors that need careful consideration, very particularly if quantitative analyses are to be undertaken. These will be briefly mentioned here, but the reader is recommended to pertinent texts listed in the bibliography at the end of this chapter. The sampling depth at any wavenumber, is dependent on the optical modulation frequency, and this, in continuous, rapid-scan mode, may be altered by changing the FTIR spectrometer mirror velocity. (Step-scan is discussed separately on pp. 112–117.) This is the basis of PAS depth-profiling. If the optical absorption length (μ_β) is less than the thermal diffusion length (μ_s) then photoacoustic signal saturation will occur, which will truncate strong bands. The best match with absorption spectra will therefore be obtained when $\mu_\beta > \mu_s$, although they will somewhat resemble ATR spectra in relative intensities, in that, for continuous, rapid-scan mode spectrum, the sampling depth is wavenumber dependent and increases with decreasing wavenumber, see p. 112. PA-FTIR spectra are displayed like absorbance spectra, see Figure 23. Typical sampling depths for polymer samples in PA-FTIR spectra are from a few μm to a hundred μm. To optimise the signal-to-noise ratio of PA-FTIR spectra it is common to reduce the continuous scan velocity of the interferometer to less than or about that normally used with a DTGS detector, see p. 95, in which case probably only a few scans will need to be co-added, although, at slow speeds the sampling depth will be greater, and the occurrences of photoacoustic saturation likely to increase.

Probably by far the majority of PAS cells in operation in industrial laboratories at this time have been manufactered by MTEC Photoacoustics, Inc., Ames, Iowa, USA. Figure 24 is a photograph of a MTEC model 300. To record a PA-FTIR spectrum, carbon black is used to generate the single-beam background reference spectrum. Originally carbon black powder was used, which often proved to be very messy to handle. MTEC now produce a polymer film which has a deposit of finely divided carbon on the surface. This is much easier to handle, and gives a very good background reference spectrum – in some cases, too good! Band intensities in PAS spectra of powders, like diffuse

Figure 24 *Photograph of MTEC 300 photoacoustic detector system* (Reproduced by kind permission of MTEC Photoacoustics, Inc., Iowa, USA)

reflectance spectra, are very dependent on particle size. Better comparison can be obtained when the particle size of the carbon matches the powder or carbon in a sample. This can be achieved by using different grades of carbon powder or placing one of the sample cups in a candle flame. (The effect is comparable to using an infrared microscope in reflection mode – a gold-coated mirror produces the best reference spectrum, but a more diffuse reflector which matches the properties of the sample, sometimes produces more aesthetically pleasing sample spectra!) The main sample cup of the MTEC cell will accommodate a sample 10 mm in diameter and 6 mm thick, although it is usual to use smaller samples placed in inner cups in combination with spacers, to displace surplus gas volume. The PA spectrum intensity is dependent on the gas and volume in the cell. Flushing and filling the cell with helium reduces the water vapour and CO_2 signals, and is important to spectrum quality. The effects of H_2O and CO_2 evolution from a sample into the cell can be minimised by pre-drying, or by placing desiccant in a second cup which is placed beneath the inner sample cup.

A wide variety of sample forms may be examined by PA-FTIR spectroscopy. Polymer chips, extruded film, fibres, or blown foam can be placed directly into a cup without preparation. Powders can also be placed directly into the cup without crushing, grinding, or diluting with alkali-halide powders, all of which might affect the polymorphic form. The technique therefore has considerable potential when dealing with polymorphs. In FTIR spectroscopy, the spectrum strength can be enhanced in transmission by introducing more sample into the beam, or by increasing the number of reflections in a reflection technique. In PAS, the signal strength is optimised by balancing the cell volume and sample surface area. Although a single polymer chip in a large cup will give a PAS spectrum, the same chip in a small cup will produce a more intense spectrum, as there is a smaller gas volume to compress. Ideal samples are loosely packed powders or open-cell polymer foams. Samples containing low levels of elemental carbon also give an increased signal, due to the absorption coefficient of carbon. Samples which are pigmented black with carbon can also sometimes yield strong spectra, but heavily carbon-filled samples suffer from too much absorption.

8 Emission Spectroscopy

The techniques described so far have been concerned essentially with measuring the absorption of infrared radiation. If a sample absorbs infrared radiation at characteristic wavenumbers, it is capable of emitting radiation at these wavenumbers. A thin sample of a material will emit radiation with a spectrum very similar to its absorption spectrum. An emission spectrum, which generally has the appearance of an inverted transmission spectrum, is a plot of emittance *vs.* wavenumber (or wavelength, or frequency). The emittance at a particular wavenumber ($\varepsilon_{\tilde{\nu}}$) is defined as:

$$\varepsilon_{\tilde{\nu}} = \frac{\text{Radiant energy emitted per unit of sample}}{\text{Radian energy emitted per unit area of a blackbody at the same temperature}} \tag{5}$$

If a is the absorptance of a sample, and t and r its transmittance and reflectance, respectively, then, at wavenumber $\tilde{\nu}$, by Kirchhoff's Law:

$$a_{\tilde{\nu}} = 1 - (t_{\tilde{\nu}} + r_{\tilde{\nu}}) = \varepsilon_{\tilde{\nu}} \tag{6}$$

so, since,

$$t_{\tilde{\nu}} + r_{\tilde{\nu}} + \varepsilon_{\tilde{\nu}} = 1 \tag{7}$$

it is relatively easy to deduce that in emission spectra the strong bands will exhibit similar refractive index dispersion effects to that described on p. 148 for certain IRS spectra. Also, that for very thick or opaque samples ($t_{\tilde{\nu}} = 0$), selective reflection will give rise to 'reduced emission' or inverted bands, in a manner somewhat analogous to that sometimes observed for the most intense bands of a transflectance spectrum.

Originally, samples had to be heated to above room temperature, typically to 40–100°C (to minimise sample degradation), and a blackbody source at the same temperature used as a reference. In such a case it is assumed that the detector is at room temperature. With FTIR instruments, which are often now fitted with cooled detectors, emission spectra can be recorded at room temperature, although care must be taken to ensure that under these circumstances it is the spectrum of the sample, and not those of surrounding materials or FTIR spectrometer components, that is being recorded; recommended procedures and means for accounting for these potential interferences can usually be found in the spectrometer manufacturer's manuals.

For the majority of analyses, there is little or no advantage to be gained in recording the emission rather than the absorbance spectrum of a condensed phase sample. In some circumstances, however, it may prove more convenient or practicable. Surface coatings on metal surfaces are frequently notoriously difficult to examine. Transflectance, specular reflection, glancing angle, and ATR can all be used with varying success, but they all largely depend on having a flat surface. Emission spectroscopy can be used on surfaces that are difficult to examine because of product geometry, and has, for example, been particularly useful for studying resins and inks on drinks cans, although, these may be more conveniently fingerprinted now by transflectance measurements, using a FTIR-microscope. Catalysts, where species are adsorbed/absorbed onto inorganic or metal surfaces, have also been studied successfully using emission spectroscopy techniques.

Important industrial applications are found outside the laboratory in remote sensing. A telescopic collection device mounted on an FTIR spectrometer can be used to examine chimney smoke and exhaust emissions. The Earth's atmosphere and interplanetary atmospheres have also been examined in this way.

9 Raman Sample Preparation and Handling

Raman spectroscopy, being a scattering phenomenon, is, of course, well known and justifiably publicised as a technique requiring a minimum of sample handling and preparation. However, not all samples can be readily examined directly and some means of containment or preparation may be necessary. Compared with IR spectroscopy, fewer specific Raman accessories are commercially available, since a number can be used for both techniques, or 'borrowed' or simply adapted from 'consumables' of other analytical techniques. Typical Raman accessories are powder sample holders, cuvette holders, small sample holders for glass tubes to facilitate liquid and powder analyses, and clamps for irregularly shaped objects. Commercial cell holders and the like designed specifically for FT-Raman use may be gold- or silver-coated, to provide the highest efficiency in the NIR region, where 180° back-scatter is usually the optimum sampling arrangement.

Many powders and liquids can be examined, directly, in the container in which they are supplied, *e.g.* glass bottles or vials; provided that the container is transparent to the laser and Raman radiation, and that the sample has a strong Raman scattering cross-section, *i.e.* gives a strong spectrum, then the shape and colour of the vial or bottle is often only of secondary importance. Samples have been examined in brown and plastic bottles as well as clear vials, (or even inside thin plastic packaging, see later and Figure 3(b) of Chapter 3). The only constraints are that the outside of the container is clean and free from fingerprints, which will most likely cause fluorescence, and that the labels don't obscure the sample. However not all samples may be examined directly; some may be prone to charring or burning, be excessively fluorescent, or, if a weak scatterer, be much more efficiently examined in a more considered manner. If the samples are weak Raman scatterers then the spectrum of the container will probably interfere with the sample spectrum. A number of techniques have been developed which reduce some deleterious effects and enhance spectra.

Gas cells are often simple cylindrical glass (fused silica) vessels, frequently adapted to enable the laser to multi-pass through the contained gas, which are aligned in the spectrometer for 90° collection of scattered radiation. Variable temperature, evacuable vapour, and pressure cells are also available. They tend to be similar to IR cells but simpler. They have the advantage that glass can also be used as a window material.

Liquids are typically analysed in glass capillary, melting-point, or NMR tubes, although spherical cuvettes covered with a reflective coating and with an aperture have been developed recently as optimum containers for FT-Raman sampling. For example, hydrocarbons are generally clear, and weak spectra can therefore be enhanced by placing the samples in silvered holders, which reflect signals back through the sample onto the detector. However, only small samples may be be examined in this way or the Raman radiation will probably be self absorbed. Liquids can suffer from fluorescence, especially brownish plant batches, but do not suffer from burning. This is due to their high mobility and hence high heat-sink capacity.

Neat powders may be examined in a glass tube of the type listed above for liquid samples, or may be compacted into a solids holder, which, typically, is a cavity in a metal block. The authors used this latter approach successfully, for example, with a crystalline, low density fungicide which was moved away from the bottle wall by the laser beam power, but gave a strong spectrum when 'fixed' in the holder. *However, one should not become over-confident with the simplicity of sampling that Raman offers, and a little forethought (or hindsight) may be well rewarded!* For instance, we found that mild physical stimulation such as gently pressing down terephthalic acid powder into a sample holder was sufficient pressure to convert the crystal form I to crystal form II; the same effect could be achieved by vigorously shaking a bottle of the powder, see paper referenced in Figure 7 of Chapter 5 for more details. Samples which are too small to fill the beam (usually only a limitation with FT-Raman investigations), or which burn in a bottle can be prepared in the form of an IR KBr disk. Intense spectra have been recorded at high laser powers (1400 mw), without burning, by this method. With samples that strongly absorb and burn in conventional dispersive Raman examinations, it is common practice to employ a sample rotating device in order to constantly refresh the sample in the beam. However, this can cause 'beats' in a FT spectrum. Very sensitive, highly coloured, or strongly absorbing samples may burn or char even at very low laser power levels. The preparation of samples as IR Nujol mulls, between salt flats, can give good strong Raman

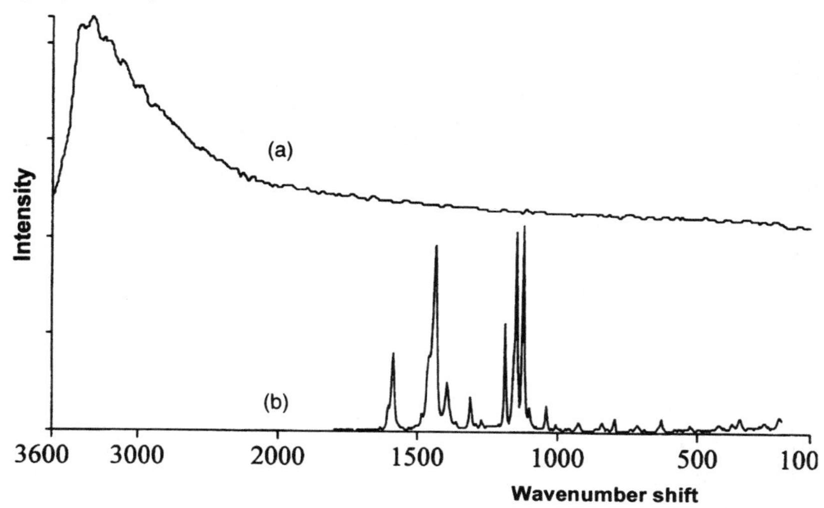

Figure 25 (a) *FT-Raman 'spectrum' recorded from laser 'burning/charring' of neat azo dye powder.* (b) *FT-Raman spectrum from azo dye powder prepared as a Nujol™ Mull. Spectra have been offset for clarity*

spectra, see Figure 25. Even some black samples have been examined successfully by one of us by NIR FT-Raman at a laser power of 1400 mw. This preparation technique, as for IR, also preserves physical form for polymorphism studies.

If fluorescence is encountered, and it is not an intrinsic property of the analyte molecule, a normal practice with conventional visible laser sources is to attempt to 'burn out' the fluorescence, simply by continuously irradiating the sample area to be analysed with the excitation laser. This may take a few seconds or several hours, or even days, and often not all the fluorescence is 'photobleached'. Fluorescence problems are frequently circumvented by using FT-Raman spectrometers with laser sources in the near-IR, most typically 1.064 μm. However, even at this wavelength some solid samples still exhibit fluorescence. We have found that this can sometimes be reduced by preparing the sample as a disk, using silver powder instead of KBr. Neat copper phthalocyanine (CuPc) gives a characteristic Raman spectrum, with a fluorescent background using visible lasers. At 1064 nm a characteristic fluorescent hump fills the spectrum: see Figure 26(a); by moving to 1300 nm the characteristic fluorescence is reduced. A cheaper method is to prepare a silver disk. At 1:1000 parts CuPc:Ag, although some background fluorescence still exists, a reasonably intense spectrum can be recorded which matches closely a 504 nm visible light excitation Raman spectrum. Red, yellow, some brown and even black samples can be examined by FT-Raman, but blues and greens based on CuPc still fluoresce. It appears that the problem is concerned with the transition metals. Metal-

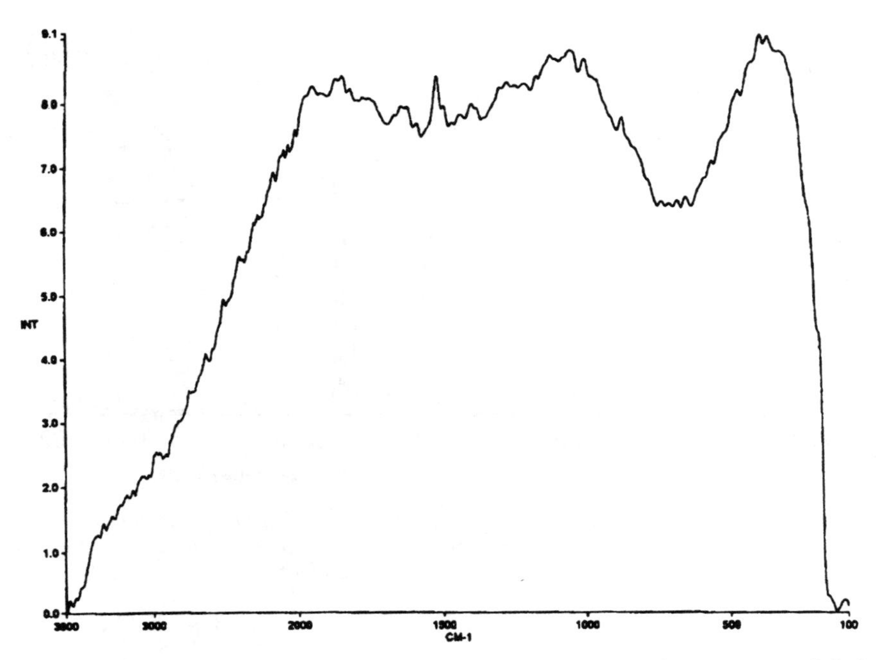

Figure 26(a) *NIR (1064* nm *excitation) FT-Raman spectrum of neat copper phthalocyanine (CuPc)*. For acknowledgement see legend to Figure 26(b)

free phthalocyanine gives a good intense NIR FT-Raman spectrum, as does TiOPc, see Figure 26(b). White or colourless samples still have to be examined on an empirical basis to ascertain how prone they may be to fluorescence in the NIR region.

Polymers in all shapes, forms, and sizes can be examined by Raman. Safety spectacles, rolls, thin films, bottles, car bumpers and moulded platens have all been examined readily by this technique. There is much published literature on the Raman spectra of polymers for identification, structural behaviour, and morphological properties. However, polymers are generally weak scatterers. This can either be a problem or used to advantage, particularly if one is interested in examining coloured synthetic materials. It was mentioned earlier that samples can be examined in plastic bottles. Sulphur can yield a very strong Raman spectrum with no evidence of bands from the plastic bottle wall material in the spectrum. On the other hand a 10% dye in a polymer fibre will probably show strong polymer bands. To examine a polymer film, one recommendation is to fold the film as many times as possible to create a 'thick' layer. Sometimes film samples are not big enough to fold. An enhanced strong spectrum can be recorded by placing a small single sheet flat across the mirrored back-face of a liquid holder. We have recorded spectra of a coloured polyester in this way, which had strong enough bands to see both the dye and the film. Coloured

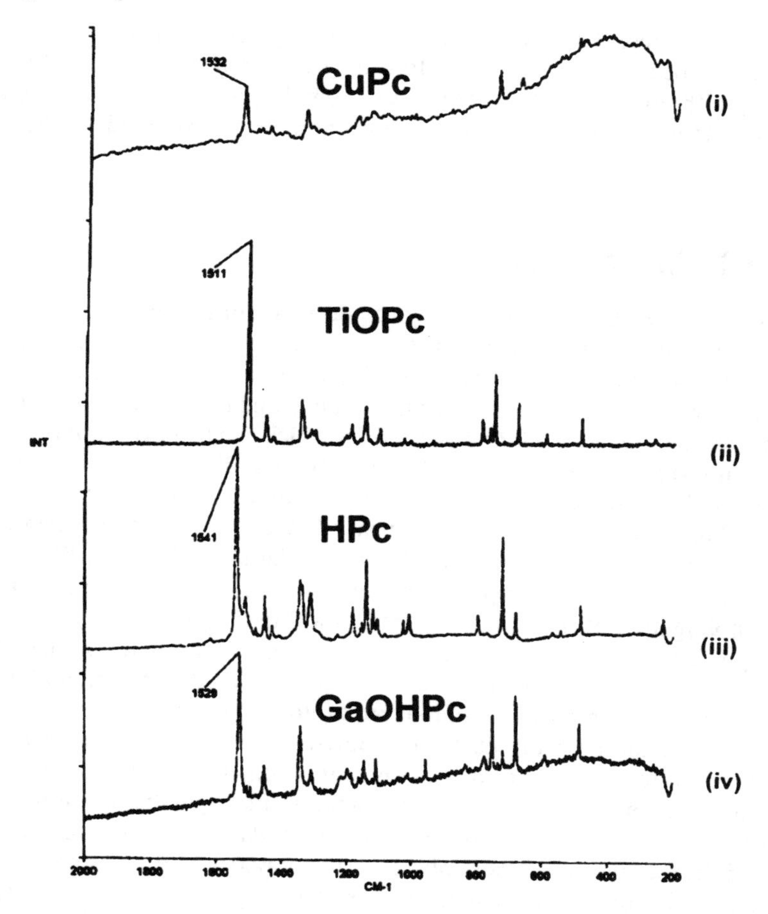

Figure 26(b) *NIR FT-Raman spectra recorded from:* (i) *CuPc prepared as a Ag disk;* (ii) *neat TiOPc;* (iii) *neat HPc; and* (iv) *neat GaOHPc. Spectra offset for clarity* (Full details are reported in *NIR FT-Raman Examination of Phthalocyanines at 1064 nm*, Dent G. and Farrell W.M., *Spectrochim. Acta*, 1997, **53A**, 1 21, © 1997, with kind permission of Elsevier Science–NL, Sara Burgerhartstraat 25, 1055 KV Amsterdam, The Netherlands)

cellulose film works even better in this respect; as cellulose is a very weak Raman scatterer the dye spectra show very clearly.

Initially, industrial Raman spectroscopy used visible laser sources and large instruments. Many of the sampling techniques were developed around these instruments. The advances of sources and detection systems, *etc.*, has widened the industrial applications into UV-Raman and NIR-Raman, and simplified visible Raman spectrometers, leading to many new developments in Raman sampling with microscopes and fibre optics. Other more specialised areas of the Raman technique such as resonance Raman, CARS, SERS, SERRS, and many other 'acronym' variations have also expanded the applications field. However, very few of these have yet made a major, general impact on industrial vibrational spectroscopy, and therefore they have not been dealt with in this chapter, although several of them are now beginning to be exploited in industrial laboratories, and some may eventually be exploited in the works environment.

10 Closing Remarks

In this chapter we have attempted to provide a flavour of the range of sampling accessories and techniques available to the vibrational spectroscopist. In one or two instances a basic method or principle has been set out in detail. The step change from dispersive to Fourier transform infrared instrumentation has led to the development of methods and accessories which can be fairly sophisticated in their own right *e.g.* DRIFT, PAS, IR-microscopes, fibre-optic probes. Others, such as the Disposable IR cards (3M Co., St. Paul, MN, USA) and the Screen Cell^{TM} (Janos Technology, Inc., Townshend, VT, USA), are designed as low cost, routine, convenience supports for non-volatile liquids, cast films, and a wide range of smearable sample types, such as many foodstuffs and cosmetics. The sampling area of the former consists of a thin film of a polymer substrate, either polyethylene or PTFE; the latter type has a sampling area designed around a glass fibre grid screen; both are in mounted card frames that fit directly into conventional sample beam slide-mounts, and which are readily archiveable. Together with the ease of spectrum 'improvement' via software techniques, these have led to an aura of 'magic' where it is sometimes forgotten that basic principles still apply. In most cases the method of sample preparation remains vitally important, and the old adage, *Simplest (traditional) is best*, remains true; *e.g.* for infrared, whenever possible, a correctly performed transmission measurement in the laboratory remains the preferred option; short cuts invariably lead to wasted time and 'impure' data.

Fibre-optic sampling, using a variety of probe-heads, is rapidly becoming established as a technique for remote sampling, particularly for NIR and Raman, where fibre lengths may be used effectively and routinely over several tens to a hundred metres or longer. Current technology however, limits commercial mid-IR cables, made from chalcogenide, to about 1.5 metres in length.

The range of sampling techniques available to the vibrational spectroscopist is almost as wide as the variety of sample types. Infrared in particular is very adaptable to many sample forms. With development of instrumentation and accessories, numerous sampling techniques have evolved whilst others have been forgotten or discarded. The correct method to use very much depends on the question being asked and the type of information being sought. Whichever technique or accessory is used, it must be remembered that the basic laws of optics still apply. The authors consider that this is arguably the simplest but most important factor in cost-effective and efficient industrial problem-solving. Prior thought must be given as to how a sample preparation or accessory may change or affect the spectrum that is eventually recorded. Poor interpretation, or reporting, can be avoided if this area is approached conscientiously.

11 Bibliography

This chapter contains a number of instructions and comments in sample preparation or use of accessories. These are not prescriptive, but based on experience, including getting it wrong. Industry-specific and local needs will determine optimal conditions and procedures. To reference each original or milestone publication, invention, or application would have been too daunting a task; many have been gradually developed and improved by numerous workers over many years. Occasional references are given in the text to point the reader to greater in-depth understanding of the principles behind some of the necessarily brief descriptions. In addition to the bibliography in Chapter 1 a short recommended bibliography relevant to this chapter is listed below:

Infrared Microspectrometry: Theory and Applications, eds R.G. Messerschmidt and M.A. Harthcock, Marcel Dekker Inc., New York, (1988).

Internal Reflection Spectroscopy, N.J. Harrick, Wiley Interscience, New York, (1967).

Practical Raman Spectroscopy, eds D.J. Gardiner and P.R. Graves, Springer-Verlag, Berlin Heidelberg (1989).

Analytical Raman Spectroscopy, eds J.G. Grasselli and B.J. Bulkin, John Wiley & Sons, Inc., New York (1991).

The Infrared Analysis of Solid Substances, G. Duyckaerts, *Analyst*, 1959, **84**, 201.

ASTM Designation: E 573–90, *Standard Practices for Internal Reflection Spectroscopy*, 1995 Annual Book of ASTM Standards, Volume 03.06, ASTM, Philadelphia, USA, © 1995, American Society for Testing and Materials.

Raman Spectrometry with Fibre-Optic Sampling, I.R. Lewis and P.R. Griffiths, *Appl. Spectrosc.*, 1996, **50**, 10, 12A.

CHAPTER 5

Quantitative Analysis

1 Introduction

Infrared and Raman spectroscopy are both used as quantitative techniques in the analysis of industrial materials, processes, and environments. To a first approximation, the intensity of the infrared absorbance or Raman scatter is linearly proportional to the number density of the vibrating species responsible for its occurrence.

Quantitative infrared spectroscopy has for many years played an important role in industrial analyses. In many situations it provides a more cost-effective, safer, simpler, and less tedious procedure than alternative analytical procedures. Applications cover a wide and diverse range, from support of research and development of novel products, through at-line product quality control, to in-line process monitoring. By comparison, uses of Raman spectroscopy for quantitative analysis were for a long time mostly confined to the specialist research laboratory; however, recent developments leading to FT and compact Raman systems now allow similar applications of this technique. The at-hand access to computerised advanced data processing mathematical treatments has considerably benefited both techniques, with marked increases in their applications to quantitative measurements. Potential human errors may be eliminated through more automated calculations, which may be particularly important for operations involving non-specialist personnel, for example at the process line; extended applicability has become feasible, and a proliferation of novel analyses have been realised. Notwithstanding, successful, reliable quantitative method development and application requires special and careful attention.

An experienced spectroscopist, examining a vibrational spectrum recorded from a well-prepared sample, may well be able to deduce species or recognise patterns simply from the mid-IR spectral region without being overly concerned with other than zero-order relative band intensities, even if there are minor 'imperfections' within the spectral detail. Such tolerance is not available to quantitative analysis, where reproducibility is paramount and which consequently needs a more diligent, procedural, and objective approach to both sample preparation and presentation, particularly in quality control applications.[1]

[1] Compton S.V. and Compton D.A., in *Practical Sampling Techniques for Infrared Analysis*, ed. Coleman P.B., CRC Press, Boca Raton, USA, 1993, ch. 8.

While we do not mean to imply that second-quality spectra will suffice for qualitative studies, we emphasise that the criteria set during method development *must be met* if quantitative results are to be meaningful. This applies whether the analysis is univariate or multivariate. Particle size can be a very important parameter in many solid sampling methods; the effects of sample morphology, *e.g.* crystallinity or molecular orientation, must be considered; intermolecular interactions, state of hydration, and many other physical and chemical properties might also need to be considered and defined, in addition to factors such as temperature and pressure or the presence of atmospheric absorbances. Also, representative reference samples should be used for the calibration set. Many laboratory or pilot-plant materials may not reflect exactly the spectral qualities of those from full-scale industrial operations. It is likely to be inappropriate to use *ultra-pure* materials if the process to be monitored, for instance, uses industrial-grade or recycled solvents, which may contain 'impurities' affecting the spectrum, whether, for example, clearly as additional absorption peaks or merely as a contribution to the overall absorbance background in a spectrum. For most quantitative analyses, in particular multivariate analyses, it is important that the spectral ranges analysed are free from artefacts such as room-light emission bands (Raman) and atmospheric absorptions (infrared: water vapour and CO_2; Raman: O_2 and N_2). Spectral artefacts that may be attributed directly to the sample preparation method, such as interference fringes and dispersion in infrared measurements, must also be minimised, see pp. 13 and 129–130.

2 Infrared Absorption

Beer's Law

Assuming no losses of radiation other than absorption by the sample, then for each thickness (path-length) element $\mathrm{d}t$ in a homogeneous sample the fractional decrease in radiation intensity $\mathrm{d}I/I$ will be a constant k', *i.e.*

$$-\frac{\mathrm{d}I}{I} = k'.\mathrm{d}t \tag{1}$$

where I is the radiation intensity incident on the thickness element, k' is a constant, characteristic of the sample and the radiation wavenumber.

Integration of Equation (1) defines the loss through absorption of incident radiation in a sample of thickness t, *i.e.*

$$\int_{I_0}^{I} -\frac{\mathrm{d}I}{I} = k't \tag{2}$$

where I_0 is the intensity of radiation incident on the sample of thickness t, and I is the emergent (transmitted) radiation intensity.

Equation (2) becomes:

$$I = I_0 e^{-k't} \tag{3}$$

The fundamental relationships (1)–(3) which describe the exponential decrease of transmitted radiation intensity with increasing sample thickness are attributed to Bouguer (1729) and Lambert (1760).

Equation (3) may be written as:

$$\ln I_0/I = k't \qquad \text{or} \qquad \log_{10} I_0/I = a't \tag{4}$$

k' has been defined as the absorption coefficient and a' as the extinction coefficient.

The amount of infrared radiation transmitted, T, by a sample is given by:

$$T = \frac{I}{I_0} \tag{5}$$

where, I_0 represents the intensity of infrared radiation incident on the sample, and I represents the amount of radiation transmitted (exiting from the sample). T is referred to as the transmittance, which varies from 1 to 0, and $100 \times T$ as the per cent transmission (%T).

The absorbance A is defined as $\log_{10} I_o/I$. Thus:

$$A = \log_{10} I_0/I = a't = \log_{10} 1/T = -\log_{10} T \tag{6}$$

In 1852, Beer observed that as the thickness of a sample changed its transmittance varied exponentially with the concentration of the absorbing species, giving:

$$-\frac{dI}{I} = kc.dt \tag{7}$$

so now, following the rationale on going from Equations (1) to (4), we obtain:

$$\ln I_0/I = kct \qquad \text{or} \qquad \log_{10} I_0/I = act \tag{8}$$

Combining equations (6) and (8) gives us:

$$A = \log_{10} I_0/I = act = \log_{10} 1/T \tag{9}$$

where, c is the concentration of the absorbing species, a is known as the absorptivity, and $a = A/ct$.

(a has been referred to in the past as the extinction coefficient, the specific extinction, or the absorbency index; the IUPAC term for a is 'specific absorption coefficient'. The product of a with the molecular weight of the absorbing substance is known as the molar absorptivity, ε, or by the IUPAC term 'molar absorption coefficient'.)

Equation (9) is commonly known as Beer's Law, although it has variously been labelled the Beer–Lambert Law, the Bouguer–Beer Law or the Bouguer–Lambert–Beer Law.

Beer's Law states that the absorptivity of a substance at a particular wavelength is a constant with respect to changes in concentration; ct represents the number of absorbing species in the sample thickness (path-length). The dimensions of a are ML^{-2}, and the units will be in accord with those used for c and t.

Equation (9) shows that the absorbance A is linearly related to c, and hence it is A that is usually measured in quantitative analyses. Beer's Law has the advantage that it is additive; that is, at a particular wavelength/wavenumber:

$$A = \sum_{0}^{i} a_i c_i t \tag{10}$$

This is an extremely important property for many multivariate analyses.

For gas analysis, Beer's Law is expressed as:

$$A = apt \tag{11}$$

where p in equation (11) replaces c in equation (9) and p is the pressure or partial pressure of the gas in a mixture of gases. The absorbance of a component in a gas mixture does not only depend on its partial pressure, but is also a function of the total pressure. The rotational fine-structure of gas-phase bands will broaden as a consequence of molecular collisions; a phenomenon known as *pressure-broadening*. So, although 50 mm Hg or less is a pressure at which many gases in a 10 cm path-length cell yield useful mid-IR spectra, it is common practice in quantitative analyses to keep the total pressure constant. In practice, this is achieved by adding a non-absorbing gas such as nitrogen up to a standard pressure, such as 760 mm Hg.

At the start of the derivations above leading to Beer's Law the requirements of a homogeneous sample and losses only by absorption were stated. Strictly, many other conditions are assumed to prevail.[2,3] The infrared radiation should be both collimated and monochromatic. The sample needs to be not only of a uniform distribution, at least on a scale comparable with the interrogating radiation, but also of a uniform thickness/path-length, *i.e.* it should present parallel surfaces perpendicular to the incident beam, and, for instance, not be wedged. In addition to there being implied no scatter or reflection loss, no stray light should be present, a condition that requires that all radiation reaching the detector has passed through the sample, and has done so only once. Deviations

[2] McClure G.L., in *Laboratory Methods in Vibrational Spectroscopy* 3rd Edn, ed. Willis H.A., van der Maas J.H., and Miller R.G.J., J. Wiley, Chichester, UK, 1987, ch. 7.

[3] Bauman R.B., *Absorption Spectroscopy*, J. Wiley, New York, USA, 1962.

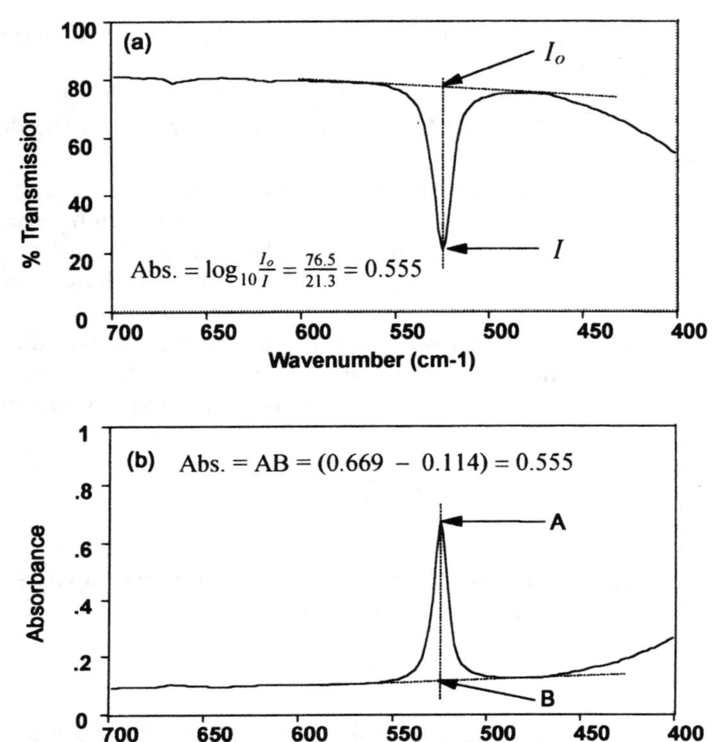

Figure 1 *Infrared spectrum from* 700 cm^{-1} *to* 400 cm^{-1} *of cyclohexane in a* 200 μm *path-length transmission cell with KBr windows:* (a) *shows the tangent base-line method used for determining the absorbance of the band at ca.* 525 cm^{-1} *from the recorded transmission infrared spectrum;* (b) *shows the tangent base-line method used for determining the absorbance of the band at ca.* 525 cm^{-1} *from the infrared spectrum plotted as an absorbance vs. wavenumber spectrum*

from these criteria will lead to non-linearities in the absorbance–concentration relationship. Notwithstanding, to be practical, compromises have to be tolerated or standardised (nulled) during the process of method calibration;[2,3] for instance, convergent beams are normally used and the spectrum intensity is measured with a finite resolution, usually a few wavenumbers, typically between 2 and 16 cm^{-1}, *i.e.* over a small spread of wavelengths. Reproducible reflection and systematic scatter losses in transmission measurements can sometimes be accounted for by the baseline method for determining I_0, see Figure 1.

Chemical and physical interactions will also lead to deviations from Beer's Law. For example, changes in hydrogen bonding or solvation will cause errors which may be substantial over a wide concentration range, and significantly limit the applicability of a quantitative procedure. The partial pressure of a gas needs to be considered, as might polar interactions in both solution and solid mixtures, each of which can result in divergences from Beer's Law. Ultimately, limitations in the linearity and precision of the detector and its associated electronics may restrict the range of a single procedure.

Absorbance Intensity

Equation (6) states that:

$$A = \log_{10} 1/T = \log_{10} I_0/I.$$

In normal infrared transmission spectroscopy, the spectrum recorded is a ratio of the radiation intensity emerging from the sample and reaching the detector to that incident on the detector in the absence of the sample *vs.* wavenumber (sometimes wavelength). This spectrum will almost certainly represent a summation of many effects, only one part of which is of interest in quantitative component analysis – the absorbance attributable to the concentration of absorbing species present. For instance, there will be reflection losses from sample surfaces, including those from any means of containment or support such as cell windows. Even when 'non-absorbing', these infrared transparent windows may contribute to general scatter loss, as may the sample itself. Scatter loss is particularly pertinent to dispersant methods of sample presentation such as alkali-halide discs. These losses contribute to a general backgound loss of radiation intensity, so while determining the transmittance value I at an absorbance peak maximum is straightforward, I_0 is rarely equivalent to 100%T.

Spectral background losses are sometimes minimised by ratioing against an appropriate 'blank sample' such as a clean support window(s) (beware of empty cells which may introduce interference fringes, see next section), rather than the intensity from an open-beam. Notwithstanding, the background is most commonly accounted for by employing the base-line method for determining I_0, see Figure 1.

Figure 1(a) depicts an example transmission spectrum which shows an absorption band superimposed on a background of other radiation intensity loss processes. This is a fairly typical situation met in single-component quantitative analysis, and Figure 1 also shows the most common compromise method of background compensation – the tangent base-line; I_0 is the ordinate value at the intersection of the tangent base-line with the analytical wavenumber.[4] The simplest form of analysis is where the absorption constituting a band is uniquely attributable to the analyte, the so-called 'key-band' situation. Nowadays, for quantitative purposes, it is more usual to print out the computerised analytical spectrum as a direct plot of absorbance *vs.* wavenumber; Figure 1(b) is the absorbance equivalent of Figure 1(a). The means of determining the analyte absorbance at the selected analytical wavenumber is also shown in the figure.

Figure 2 was generated from a solution of a phenolic antioxidant in

[4] *Annual Book of ASTM Standards*, Vol. 03.06, ASTM, Philadelphia, USA, 1995, Designation E168–92.

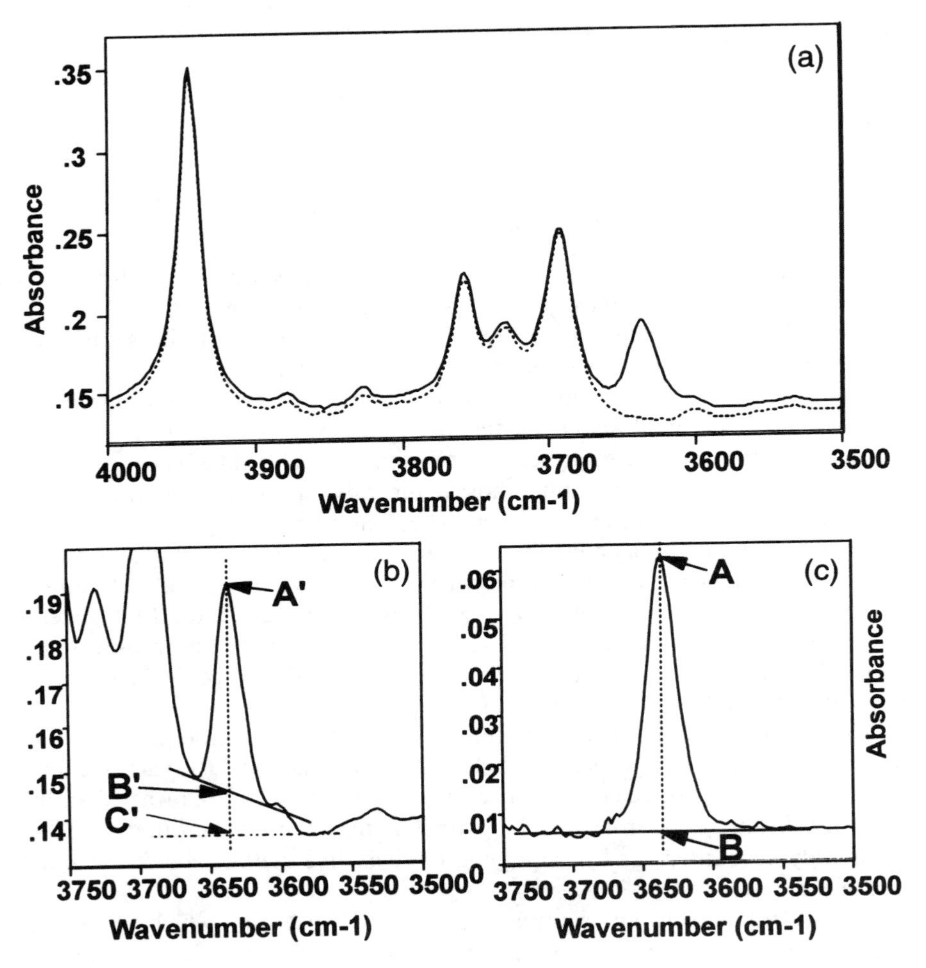

Figure 2 (a) *FTIR absorbance spectrum from* 4000 cm^{-1} to 3500 cm^{-1} *of a* ~0.5 % wt./vol. *solution of Irganox 1010 in 1,2 dichloromethane in a* 200 μm *path-length cell, KBr windows (solid trace); FTIR absorbance spectrum of the solvent, 1,2 dichloromethane, recorded in the same cell (dotted trace);* (b) *Scale-expanded absorbance spectrum from* 3750 cm^{-1} *to* 3500 cm^{-1} *of the solution of Irganox 1010 in 1,2 dichloromethane:* (c) *Absorbance spectrum of Irganox 1010 'analyte-band' at* ~3637 cm^{-1} *after compensation (absorbance-subtraction) of the absorbance contributions of the solvent 1,2 dichloromethane from the spectrum shown in* (b)
The absorbance AB in (c) *is measured as 0.0563. The absorbances A′B′ and A′C′ in* (b) *are measured as 0.0444 and 0.0543 respectively. (Note: the 'true* I_0*' value of the analyte band, see* (a), *would be below the point C′ shown in* (b). *The phenolic antioxidant, Irganox 1010, is pentaerythritol-tetrakis[3,5-di-tertbutyl-4-hydroxyphenyl propionate])*

dichloromethane (~0.5% wt/vol), and part of the losses contributing to the background at the analyte band wavenumber position are attributable to overlap by the wing of an absorption band of the solvent. Better manual spectral analysis (and greater sensitivity) prevails if this contribution is removed. This can be achieved by subtracting the absorbance spectrum of the pure solvent from that of the solution, see Figure 2(c). The tangent base-line is now much less of a compromise, and, hopefully, closely represents any uncompensated non-absorption losses. Note that the absorbance values calculated from Figure 2(b) are less than those from Figure 2(c), since in Figure 2(b) a truer base-line would have been curved (see also note in figure caption); this may give difficulties when measuring low concentrations of an analyte and in obtaining calibration plots which naturally pass through zero! A main advantage of the tangent base-line approach is that a calibration plot of path-length normalised absorbances *vs.* analyte concentration will be linear, and for the situation shown in Figure 2(c) will hopefully pass through, or very close to, the graph origin (*i.e.* co-ordinate (0,0)).

Providing the analysis is fit for the purpose, for example in industrial process monitoring, where the need for precision, reproducibility and simplicity may well outweigh requirements for exactness, then other backgrounds may be more appropriate,[2] particularly for at-line operation, see examples in Figure 3.

Measurement of Path-length for Transmission Measurements

The effective path-lengths of gas cells usually vary from a few centimetres to many metres and are therefore readily determined.

Quantitative analyses of liquid samples may typically involve use of cells from *ca.* 0.2 μm to a few mm thickness. The path-length of thicker cells may be inferred from their spacer thicknesses. For a cell which has flat, polished parallel windows, its path-length may be determined from a recording of the interference (channel) fringe pattern generated by the empty cell. Constructive and destructive interference between the primary transmitted beam and the beam which has been twice reflected inside the cell results in a sinusoidal variation of transmitted intensity, see Figure 4.

A cell's path-length (thickness) t is calculated from:

$$t = \frac{N}{2(\tilde{\nu}_2 - \tilde{\nu}_1)} \tag{12}$$

where N is the number of complete fringes between wavenumbers $\tilde{\nu}_2$ and $\tilde{\nu}_1$. (When the cell window surfaces which form the cavity are not parallel, the fringes become increasingly weaker to higher wavenumbers, and may even be absent at high wavenumbers, see spectrum (c') in Figure 4).

For solid, continuous film samples such as those typically used in polymer analyses, the thickness of the film may be determined using a dial-gauge micrometer. The film should be of a uniform thickness and its average thickness

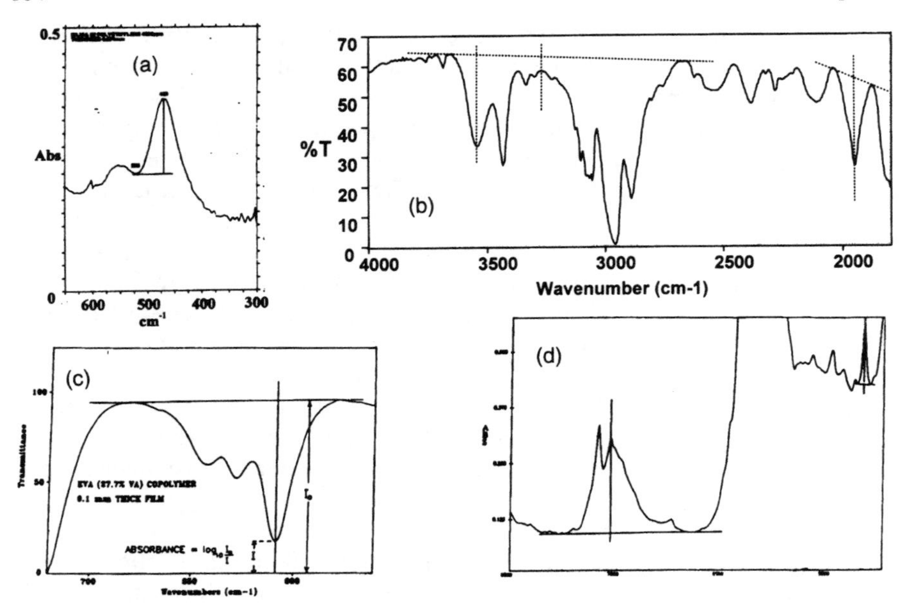

Figure 3 *Examples of various base-lines which have been/are used to measure absorbances of infrared spectral bands for quantitative analysis purposes:* (a) *absorbance spectrum of a* 0.574 mm *thickness polyethylene film containing* 4000 ppm *silica filler; this data was recorded on a dispersive IR spectrometer, and this baseline was chosen since the background data near* 700 cm^{-1} *was of poor S/N and irreproducible because of interferences from atmospheric CO_2 absorptions;* (b) *transmission spectrum of a dry, amorphous PET film of* ~0.1 mm *thickness used to determine the concentration of polymer end groups; the —OH end group occurs at* 3542 cm^{-1}; *the —COOH end group occurs at* 3256 cm^{-1}; *the* 1950 cm^{-1} *is used as a reference for thickness normalisation;* (c) *transmission spectrum of a* 0.1 mm *thickness ethylene/vinyl acetate (EVA) copolymer film containing* 27.7% *copolymerised VA;* (d) *absorbance spectrum of a* ~4 mm *thickness moulding EVA* (17.2%) *copolymer sample. In* (a) *and* (b), *since the base-line (I_0) values do not equate to zero analyte concentration, a correction factor must be applied to the absorbance determination*
[(a) reproduced from ref. 111, by kind permission of Perkin-Elmer Limited]

calculated over the area through which the infrared beam passes, by taking and averaging an appropriate series of micrometer readings.

If the spectrum of a solid film exhibits interference fringes then, using regions where there are no absorption bands, its thickness can be calculated in a similar manner to that for an empty cell, see above, but the equation must now include n, the average refractive index of the film material:

$$t = \frac{N}{2n(\tilde{\nu}_2 - \tilde{\nu}_1)} \tag{13}$$

However, it must be remembered that the presence of interference fringes will make backgrounds and true absorbance values difficult to determine and

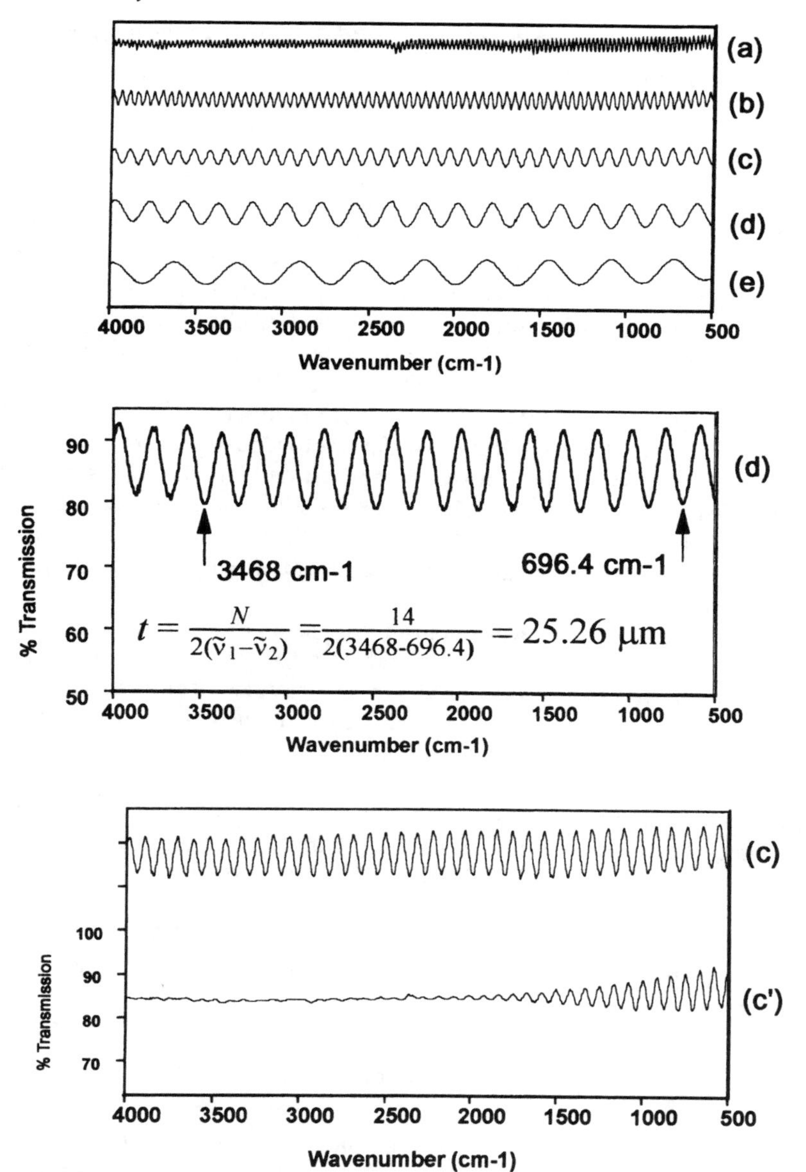

Figure 4 Top, *Infrared spectra* (100 scans, 4 cm^{-1} *resolution) of various pathlength (thickness) empty infrared transmission cells with KBr windows. Nominal spacer thicknesses:* (a) 200 μm; (b) 100 μm; (c) 50 μm; (d) 25 μm; (e) 12.5 μm. *Spectra offset for clarity, but each covers the approximate range* 100–80%T; Middle, *Example determination of the path-length (thickness, t) of the cell of nominal thickness* 25 μm *from the interference pattern generated by the empty cell;* Bottom, (c) *Transmission spectrum of empty* 50 μm *cell (Offset for clarity);* (c′) *Transmission spectrum of empty* 50 μm *cell with non-parallel cavity window faces; the non-parallelism was produced by loosening the retaining screws at one end of the cell assembly*

measure, and it is best to minimise/eliminate the causes of these fringes in solid samples, see pp. 128–131.

Normalised Absorbance Intensity

As indicated above and required by Beer's Law, measured absorbances need to be normalised against path-length. Path-length is relatively simple to control or measure in analyses using containment cells, for example, when using laboratory gas and solution transmission cells, and for solid, continuous free-standing films, such as those frequently formed for analytical purposes from polymer samples. However, there are numerous sample presentation methods for which the path-length may be neither readily controllable nor practically determinable. This includes such applications as those involving infrared techniques of internal reflection, diffuse reflectance, and photoacoustic spectroscopy on solid samples, and Raman applications. Standard 'mixes' may be employed in infrared dispersant approaches such as alkali-halide discs, but in many circumstances this proves to be less than satisfactory for other than semi-quantitative (low precision) measurements, highly uniform mixing and grinding being difficult to reproduce.

In circumstances where normalisation through a fixed or measurable path-length is impossible, extremely difficult to achieve, or imprecise, then internal standardisation is utilised. In its simplest form this requires a measurement of the ratio of the absorbance intensities of two appropriate bands within a spectrum. This is discussed in a general manner later in this chapter.

In *diffuse reflectance* measurements that are essentially free from dispersion effects and for weakly absorbing bands, it may prove effective to determine a band intensity ratio from the recorded data plotted in a form equivalent to that used in the measurement of absorbance from a transmission measurement, *i.e.* with an ordinate scale of $\log_{10} I/R$. The absorbance equivalent intensity A_R measured directly from a diffuse reflectance spectrum would be:

$$A_R = \log_{10} \frac{R_0}{R} \tag{14}$$

where, R is the amount of light reflected at the band peak and R_0 is the conventional base-line value. The equivalent Beer's law equation would be (*cf.* Equations (15) and (9)):

$$A_R = acl \tag{15}$$

where, l is regarded as a constant if the sample is 'infinitely thick', *i.e.* increasing its thickness does not alter its spectrum. This approach is widely used with NIR analyser spectra, but for quantitative purposes mid-infrared diffuse reflectance spectra are usually represented in terms of a relationship derived by Kubelka and Munk,[5–7] since relative band intensities within a spectrum more closely resemble those from an equivalent absorbance spectrum.[7] The Kubelka–Munk relationship may be stated as:

$$\left(\frac{1 - R_\infty}{2R_\infty}\right)^2 = \frac{2.303\varepsilon c}{S} \tag{16}$$

where, S is a scattering constant, ε is the molar absorptivity, c the molar concentration of the absorbing species, and R_∞ is the ratio of the diffuse reflectance of the sample at 'infinite depth' (*i.e.* at a depth beyond which the signal does not change, usually 2–3 mm) to that of a selected standard (*e.g.* finely ground powdered, dried KCl).[7]

While both univariate[8] and multivariate[9] data analyses of infrared *photo-acoustic* (PA) spectra can prove useful for quality assurance purposes, several relationships need to be considered. As pointed out in Chapter 4, pp. 165–168, in continuous, rapid-scan as opposed to step-scan operation, there is a dependency between the modulation frequency and infrared radiation wavelength, which leads to an effective increase in 'sampling-depth' with increasing wavelength and therefore (as with ATR measurements) bands at lower wavenumbers will be relatively more intense than those at higher wavenumbers in a PA spectrum compared to an absorbance spectrum.

For a Michelson interferometer, the modulation frequency $f_{\tilde{\nu}}$ is given by $f_{\tilde{\nu}} = 2V\tilde{\nu}$, where V is the velocity of the moving mirror (cm/s) and $\tilde{\nu}$ is in wavenumbers (cm^{-1}), (see p. 93).[8,9] (For a Genzel-type interferometer, see Figure 16(a) of Chapter 3, $f_{\tilde{\nu}} = 4V\tilde{\nu}$). In addition, one must consider the relationships between the optical absorption length (μ_a), the thermal diffusion depth (μ_s) and the sample thickness (t),[10,11] see Figure 5. (The optical absorption length is the reciprocal of the absorption coefficient). Depending on these relationships, samples may be classed as optically transparent or opaque and thermally thin or thick,[10,11] and it is easy to visualise from Figure 5 situations where PA saturation may occur (*i.e.* $\mu_a < \mu_s$) leading to relative intensity distortions. The amplitude dependence of the PA signal from a continuous homogeneous solid with respect to these effects have been described in a theory by Rosencwaig and Gersho.[10] In common with diffuse reflectance measurements, particle size, packing, *etc.* will also be important in powder analyses.[8]

[5] Kubelka P. and Munk F., *Z. Tech. Phys.*, 1931, **12**, 593.
[6] Kubelka P., *J. Opt. Soc. Am.*, 1948, **38**, 5, 448.
[7] Fuller M.P. and Griffiths P.R., *Anal. Chem.*, 1978, **50**, 13, 1906.
[8] Chalmers J.M. and Mackenzie M.W., in *Advances in Applied Fourier Transform Spectroscopy*, ed. Mackenzie M.W., J. Wiley, Chichester, UK, 1988, ch. 4.
[9] McClelland J.F., Jones R.W., Luo S., and Seaverson L.M., in *Practical Sampling Techniques for Infrared Analysis*, ed. Coleman P.B., CRC Press, Boca Raton, USA, 1993, ch. 5.
[10] (a) Rosencwaig A. and Gersho A., *J. Appl. Phys.*, 1976, **47**, 64; (b) Rosencwaig A. and Gersho A., *Science*, 1975, **190**, 556.
[11] Rosencwaig A., *Photo-acoustics and Photo-acoustic Spectroscopy*, J. Wiley, New York, USA, 1980.

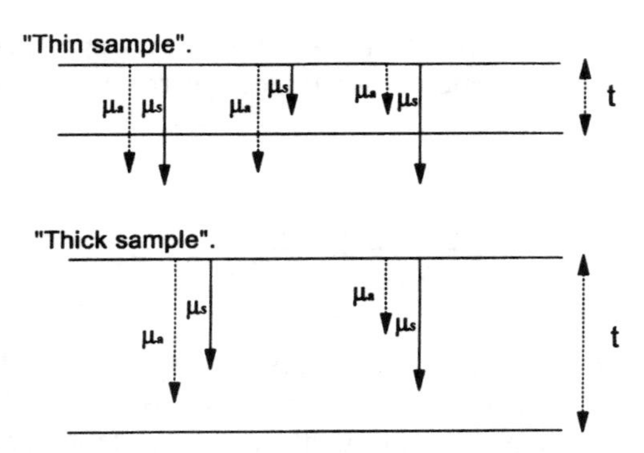

Figure 5 *Illustration of some relationships which might need to be considered in PA-FTIR spectra: μ_a represents the optical absorption length; μ_s represents the thermal diffusion length; t represents the sample thickness*

3 Raman Scatter Intensity

Since, compared to infrared absorption, Raman scatter is an inherently weak phenomenon and, in general spectra of poorer signal-to-noise ratio than transmission infrared are recorded, and because for many years fluorescence thwarted Raman analyses of many materials from industrial processes, the portfolio of industrial quantitative analyses by Raman is much smaller, particularly for routine quality control applications. However, the new generation of Raman instrumentation is doing much to dispel the widely held myth that Raman spectroscopy is unsuitable for robust quantitative methods, and their popular industrial acceptance is redressing this imbalance somewhat.

Although, the relative band intensities within a normal-Raman scatter spectrum will reflect direct proportionality to the concentrations of the exciting species present, the overall intensity of a spectrum will be dependent on the excitation source intensity and the Raman scattering cross-section within the sample. Since this latter parameter is very difficult to reproduce experimentally from sample to sample, by analogy with the infrared indeterminate path-length case, internal standardisation tends to be the norm with industrial quantitative analyses by Raman spectroscopy. Similarly to the infrared case, this may be achieved by a band-ratio determination, which may involve the use, for example, of a solvent or added 'marker' band in solution analyses, or normalisation against the band of the major constituent in component analyses of solids. General 'non-Raman' background emissions, such as low levels of fluorescence, are compensated for when measuring a peak intensity by use of a 'base-line' in a manner analogous to the infrared.

Ideally, as with infrared absorbance, one is looking for a linear response of signal intensity with increasing analyte concentration, and one must be alert to the same chemical and physical interactions which may cause deviations. While absorption of thermal infrared radiation can lead to a significant (a few degrees) increase in sample temperature in some infrared studies, and may need to be accounted for, it is much less a general problem than that of radiation absorption in a Raman experiment. Deleterious/destructive heating by absorption of the excitation radiation is usually easily observed and can frequently be overcome by attenuating (lowering) the source intensity, or circumvented by changing to another excitation wavelength. A particular difficulty arises when the absorption within a spectrum is wavelength dependent and different at each analytical wavenumber. In visible excitation Raman spectroscopy, self-absorption, which suppresses certain Raman bands within a spectrum relative to others, can occur when the Raman emissions are convoluted with electronic absorptions,[12–15] which are generally broad and have high extinction coefficients. These effects are clearly minimised in thin samples and reduced when back-scattering geometry is used.[12,13] (180° back-scatter is now perhaps the most widely used sample collection geometry, because it is the easiest to align for maximum signal, although 90° geometry would still be required for a full polarisation analysis measurement of molecular orientation.) Self-absorption is a particular complication associated with quantitative analysis of coloured samples under resonance Raman spectroscopy (RRS) conditions, where specific intensity enhancements occur.[15–18] Narrow-band selective self-absorption can be a particular problem for quantitative analysis with certain materials when using NIR FT-Raman spectroscopy.[19–21] This occurs when the absolute wavenumber of a Raman band coincides with a NIR band of the scattering material, *e.g.* for tetrahydrofuran (THF) excited by a Nd^{3+}:YAG laser emitting at 1.064 μm then the 917 Δcm^{-1} Raman band has an absolute wavenumber of *ca.* 8478 cm^{-1}, which is almost coincident with the peak of the NIR absorption band of the second overtone of the C—H stretching bands of THF.[19,21] Variations in relative band intensities within a FT-Raman spectrum have been observed and discussed in terms of sample alignment, analyte concentration, and collection optics design.[19–22] Awareness and attention to these experimental parameters may be vital to obtaining reproducible quantitative measurements from FT-Raman spectra where attenuation by self-absorption is a critical factor.

While it is expected that any calibrated Raman quantitative procedure will set criteria for source intensity and wavelength, it should be remembered there is a *ca.* ν^4 dependency on signal intensity with absolute wavenumber such that relative intensities within a Raman spectrum may differ slightly with different

[12] Turrell G., in *Practical Raman Spectroscopy*, eds. Gardiner D.J. and Graves P.R., Springer-Verlag, Berlin Heidelberg, Germany, 1989, ch. 2.
[13] Shriver D.F. and Dunn J.B.R., *Appl. Spectrosc.*, 1974, **28**, 4, 319.
[14] Strekas T.C., Adams D.H., Packer A., and Spiro T.G., *Appl. Spectrosc.*, 1974, **28**, 4, 324.
[15] Vickers T.J. and Mann C.K., in *Analytical Raman Spectroscopy*, ed. Grasselli J.G. and Bulkin B.J., J. Wiley, New York, USA, 1991, ch. 5.
[16] Ludwig M. and Asher S.A., *Appl. Spectrosc.*, 1988, **42**, 8, 1458.
[17] Rauch W. and Bettermann H., *Appl. Spectrosc.*, 1988, **42**, 3, 520.
[18] Ard J.S. and Susi H., *Appl. Spectrosc.*, 1978, **32**, 3, 321.
[19] Petty C.J., *Vibrational Spectroscopy*, 1991, **2**, 263.
[20] Schrader B., Hoffman A., and Keller S., *Spectrochim. Acta*, 1991, **47A**, 9/10, 1135.
[21] Everall N., *J. Raman Spectrosc.*, 1994, **25**, 813.
[22] Everall N. and Lumsdon J., *Vib. Spectrosc.*, 1991, **2**, 257.

excitation wavelength sources. Although the major differences in relative intensity will probably come from differing detector responses, which may be dramatic at the extremes of their response band envelope, they will be aggravated further close to filter cut-on/cut-off edges, where attenuation of band intensity by the filter characteristics may be severe.

4 Quantitative Analysis of Vibrational Spectroscopic Data

As normally practised, neither infrared nor Raman is an absolute method *per se*; calibration is essential if anything other than descriptive ('more-than', 'half-as-much', *etc.*) comparisons are to be made. Quantitative information can be extracted from single peak intensities after calibration against absolute standards *e.g.* 'known concentration solutions', but to allow for variations in sampling (*e.g.* path-length/scatter cross-section), instrument performance (*e.g.* detector non-linearity) and, perhaps, small but acceptable deviations from linearity, it is more usual to use a means of internal standardisation.

As intimated in the discussion so far, while there remain unique considerations in applying each technique, due attention must be given to a very similar range of sample properties if workable quantitative analytical methods are to be produced. In both cases, we shall usually be dealing with arrays of data, linear with respect to a wavenumber scale and with additive, probably base-line compensated, ordinate properties, whether infrared absorption or Raman intensities. So the data treatments, whether of a simple single-component two-band ratio manual measurement or of a complex multivariate regression, will be essentially identical. We have intermingled infrared and Raman quantitative considerations and applications in this chapter in order to emphasise the commonality in the arguments and considerations which prevail throughout the treatment of data suitable for quantitative analysis.

In addition to band intensity measurements, univariate analysis correlations between band-widths or positions may be particularly appropriate. (Band-widths are usually expressed in cm^{-1} as FWHH or FWHM, which stand for 'full-width at half-height' or 'full-width at half-maximum', respectively; care should be exercised if the measurement is made on a band superimposed on a sloping background, in which case a half-width (HWHH or HWHM) measurement might be more reliable). For instance, polymer spectra are sensitive to molecular conformation and order, and the carbonyl bandwidth in the Raman spectrum of poly(ethylene terephthalate) has been correlated with sample density (and by inference sample crystallinity),[23–25] see Figure 39(b) of Chapter 6 and associated text, while the position of the carbonyl band in the Raman spectrum of PEEK (poly(aryl ether ether ketone)) has been correlated directly

[23] Melveger A.J., *J. Polym. Sci. A2*, 1972, **10**, 317.
[24] Chalmers J.M., Croot L., Eaves J.G., Everall N., Gaskin W.F., Lumsdon J., and Moore N., *Spectros. Int. J.*, 1990, **8**, 14.
[25] Everall N., Tayler P., Chalmers J.M., MacKerron D., Ferwerda R., and van der Maas J.H., *Polymer*, 1994, **35**, 15, 3184.

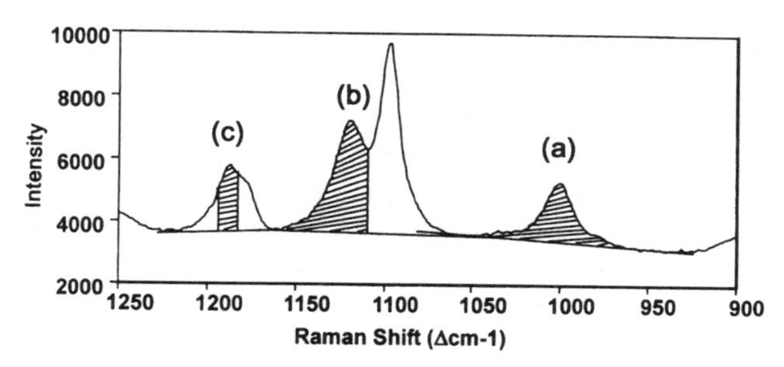

Figure 6 *Examples of measurements (shaded areas) of band area* (a) *or elements of area* (b) *and* (c). *The Raman spectrum is that of PET*

with sample X-ray crystallinity.[26–28] Band position has also been used to imply residual strain levels in carbon and other fibres, see pp. 314–323.

Molecular environment and interaction changes will most likely lead to a change in band shape and perhaps in peak position. In such circumstances, for minimal changes, it may be advantageous to determine an integrated band area or element of band area in preference to a peak intensity value,[2,4] since this may prove more reliable than a single-point intensity measurement, see Figure 6. Band area or element of band area may be calculated reasonably accurately by a process known as Simpson's method.[29]

Calibration of Quantitative Methods

For component quantification, measurements of infrared absorbance or Raman scatter intensity must be referenced to a calibration process involving standard samples, or be correlated with measurements made by an independent 'referee' method. Calibration may be followed by a validation procedure (for example, see p. 206). Although in principle only a single standard is necessary to convert measured intensities into concentrations, it is more advisable and usual to employ a set of standards, since this should 'average-out' systematic errors in the intensity measurement,[1,2,15,30] which may arise from spectral noise, sampling variations, spectral artefacts, and reference data errors, (see overdetermination analogy, p. 206).[31] Precision and reproducibility also need to be evaluated, by repeat measurements and multiple (*e.g.* triplicate)

[26] Louden J.D., *Polym. Commun.*, 1988, **27**, 3, 82.
[27] Everall N.J., Lumsdon J., and Chalmers J.M., *Spectrochim. Acta*, 1991, **47A**, 9/10, 1305.
[28] Everall N., Chalmers J.M., Ferwerda R., van der Maas J.H., and Hendra P.J., *J. Raman Spectrosc.*, 1994, **25**, 1, 43.
[29] Adams M.J., *Chemometrics in Analytical Spectroscopy*, Royal Society of Chemistry, Cambridge, UK, 1995.
[30] Hendra P., Jones C., and Warnes G., *Fourier Transform Raman Spectroscopy Instrumentation and Chemical Applications*, Ellis Horwood, Chichester, UK, 1991.
[31] Crocombe R.A., Olson M.L., and Hill S.L., in *Computerised Quantitative Infrared Analysis*, ed. McClure G.L., ASTM Special Technical Publication 934, ASTM, Philadelphia, USA, 1987, p. 95.

sampling respectively. A single standard may be recorded at the time of an analysis cycle with the prime purpose of validating instrument performance and method application. Typical of a set of reference samples might be: a prepared set of standard solutions covering the range of solute concentrations of interest; a gas cell containing the analyte gas at different defined partial pressures (see above); or homogeneous mixes of weighed amounts of two finely divided solids. 'Referee' methods need to be absolute or at least 'fit for purpose'; they may not be more precise! They may vary, for example, from direct correlations with NMR determinations, chemical titration, mass balance, elemental analysis, isotope exchange, X-ray results, other physical property measurements through to the utilisation of dopants in cast films, whose 'thickness' is determined subsequently, for example, a metal salt in combination with XRF analysis.

As stated above, quantitative analysis is essentially a two-stage process, although the first – calibration – may only need to be done at the outset of a new method.

For calibration:

Beer's Law may be expressed as: $a = A/ct$

Raman intensity as: $K' = I/c$

where K' is a normalised constant and I the measured Raman peak intensity.

For the determination, we revert to:

$$c = A/at \qquad \text{or} \qquad c = I/K' \tag{17}$$

as appropriate.

Band Ratio Measurements

Let us first consider the case for determining the concentration of an analyte in a two-component mixture, whose spectrum contains two suitable unique bands, one for each component, see Figure 7. A plot of the band intensity ratio I_1/I_2 will be linearly related to the concentration ratio of the components c_1/c_2 by, K, the slope of the graph, *i.e.*:

$$\frac{I_1}{I_2} = K\frac{c_1}{c_2} \tag{18}$$

(The derivation of Equation (10) for infrared from Beer's Law would follow: if A_1 is the absorbance of a concentration c_1 of component 1 at its peak wavenumber, and A_2 is the absorbance of a concentration c_2 of component 2 at its peak wavenumber, then, from Equation (9):

$$A_1 = a_1c_1t \tag{19a}$$

and

$$A_2 = a_2c_2t \tag{19b}$$

where t represents the sample path-length.

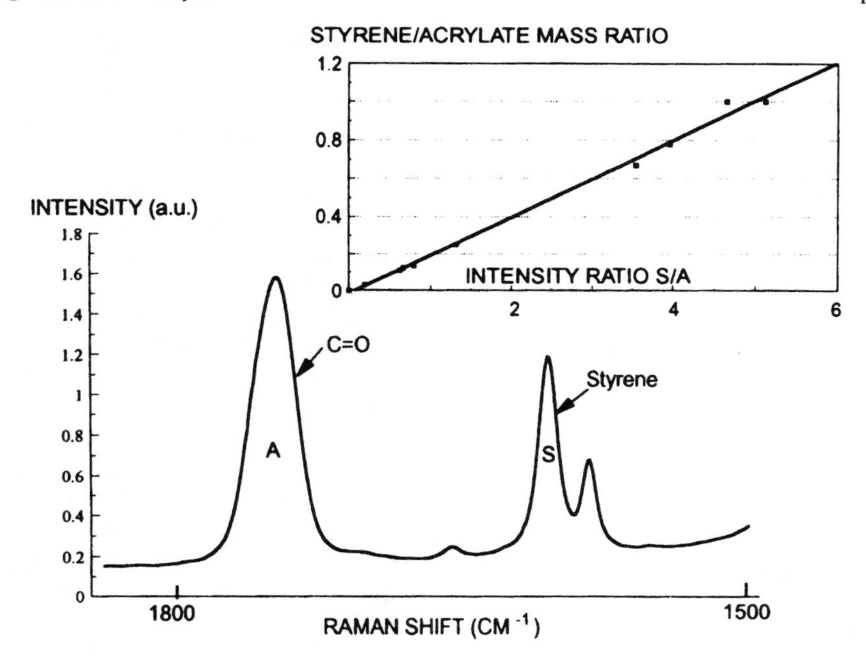

Figure 7 *Compositional analysis of co-monomer content of a cross-linked styrene-butyl acrylate (S-A) rubber. The ratio of the 'key' peak intensities S/A (aromatic ring stretch/C=O stretch) correlates linearly with the copolymerised styrene/acrylate concentration ratio. (a.u. = arbitrary units)*
(Reproduced with kind permission from Everall N.J. "Industrial Applications of Raman Spectroscopy", pp. 115–131, in *An Introduction to Laser Spectroscopy*, ed. Andrews D.L. and Demidov A., Plenum Press, New York, USA, 1995)

Dividing Equation (19a) by (19b), we can eliminate t, to obtain:

$$\frac{A_1}{A_2} = \frac{a_1 c_1 t}{a_2 c_2 t} = K\frac{c_1}{c_2} \tag{20}$$

cf. Equations (20) and (18).)

The requirements for Equation (18) to plot linearly and pass through the origin are that the intensity/concentration relationships of the individual components are linear and would pass through zero, *i.e.* 'base-lines' must have been selected such that at zero concentration of a component no intensity, positive or negative, would have been measurable at its analytical wavenumber. While linear intercept graphs of A against c would be readily workable for single-component analysis of each component, were t of Equations (19a) and (19b) determinable, if both exhibit an intercept their quotient will be a non-linear (curved) calibration plot with an intercept.[2]

In the case above, we were considering an example of a two-component mixture in which the concentrations of the components in a set of test samples were likely to vary over a wide range. There are numerous circumstances in industrial quality control applications where for practical purposes the concentration of one of the components may be considered as essentially invariant, and one may simply plot the measured intensity ratio against the changing concentration of the other, which is present as the minor or a trace component. The use of a non-overlapped solvent band to normalise for cell path-length in dilute solution analysis is a good example of this, where, assuming no solvent–solute interactions, a direct plot of the intensity ratio against level (*e.g.* %, wt.) of the minor component will probably give rise to a workable linear plot. Another situation frequently encountered occurs in polymer analyses, where the intensity of a band in the polymer spectrum is used to reflect a sample's path-length; this band should ideally be insensitive to changes in sample morphology (*e.g.* tacticity, conformation, configuration, chain packing) and molecular ordering (*e.g.* orientation). Alternatively, acceptable criteria for these need to be stipulated in the method procedure, see Figure 8, where the samples have to be 'as made', *i.e.* unsintered. In infrared measurements, ratio methods will almost certainly be required for univariate quantitative analyses using sample presentation methods other than transmission.

Univariate band intensity ratio methods may still give rise to workable linear correlation plots, albeit with an intercept, in instances where there is overlap of the analytical band of one component by spectral features of the other component, provided that the 'non-overlapped' band represents the numerator of Equations (18) and (20).[2]

So far we have confined our discussions to single-intensity measurements at analytical wavenumbers. The arguments developed apply equally to band area or elements of band area, which, as stated previously, can prove advantageous in certain circumstances.

The interested reader is recommended to ref. 2 for a fuller discussion of intensity ratio analyses.

Ratio of Spectra Method

In the previous section, the band ratio measurement was discussed in terms of univariate compositional analysis, for situations in which a band could be discerned within the spectrum of a mixture which could be uniquely attributed to the analyte. Calibration was effected simply either by the fact that the components in a mixture were essentially separable (*e.g.* solute and solvent) or by correlating measurements with those from an independent absolute method. A method has been proposed[32] and applied[32–35] which seeks to generate concentration coefficients without external calibration and also to isolate

[32] Hirschfeld T., *Anal. Chem.*, 1976, **48**, 721.
[33] Koenig J.L., D'Esposito L., and Antoon M.K., *Appl. Spectrosc.*, 1977, **31**, 4, 292.
[34] Koenig J.L. and Kormos D., *Appl. Spectrosc.*, 1979, **33**, 4, 349.
[35] Koenig J.L., *Pure & Appl. Chem.*, 1982, **54**, 2, 439.

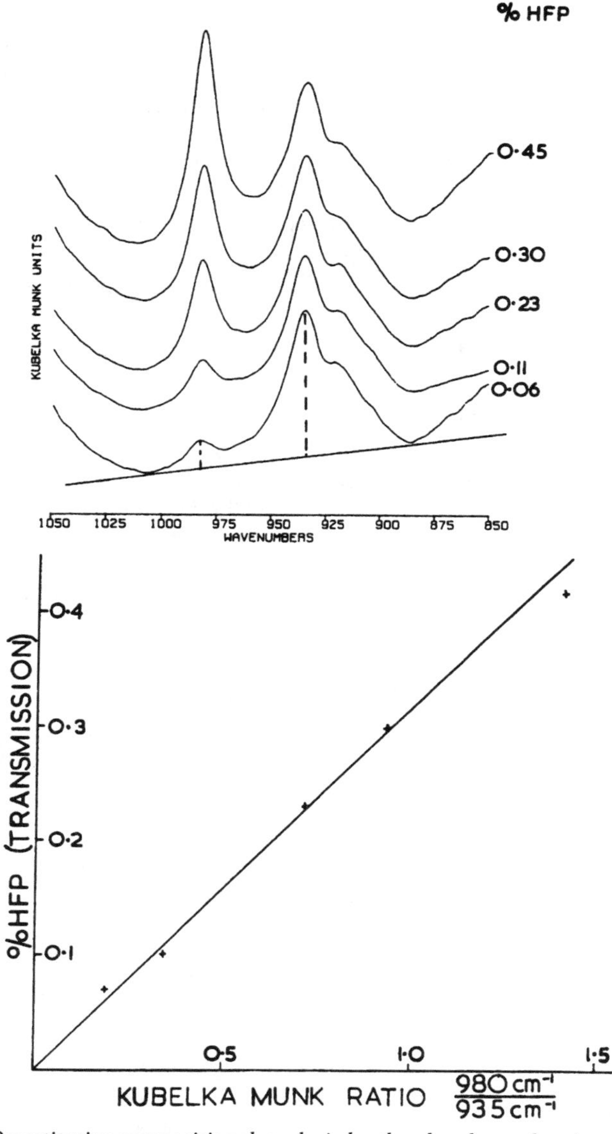

Figure 8 *Quantitative compositional analysis by the absorbance band-ratio method from the diffuse reflectance infrared spectra recorded from a series of tetrafluoroethylene/hexafluoropropylene (TFE/HFP) copolymer neat powders, using the relatively weak absorption bands at* 980 cm^{-1} *(HFP) and* 935 cm^{-1} *(TFE). The background used in the calculation of the peak heights is shown on the spectrum for the 0.06% HFP sample. The graph compares the Kubelka–Munk intensity ratio with the absorbance ratio results determined from a similar measurement undertaken using a transmission technique on solid room temperature compression formed films of nominal thickness* ~0.6 mm
(Figure reproduced from ref. 112, by permission of the American Society for Testing and Materials, © ASTM)

'pure' component spectra directly from mixtures. The method is based on ratioing point by point the spectra of mixtures in which the constituent components are present in different proportions.

For a binary mixture M_1 its spectrum might be represented as:

$$M_1(\tilde{\nu}) = a_1.f_1(\tilde{\nu}) + a_2.f_2(\tilde{\nu}) \tag{21a}$$

where, $f_1(\tilde{\nu})$ and $f_2(\tilde{\nu})$ are the spectra of the pure components 1 and 2, and a_1 and a_2 are the mole fractions of the two components respectively, $(a_1 + a_2 = 1)$. For a mixture M_2 of different component proportions of the same path-length, then:

$$M_2(\tilde{\nu}) = a'_1.f_1(\tilde{\nu}) + a'_2.f_2(\tilde{\nu}) \tag{21b}$$

and the ratio spectrum $R(\tilde{\nu})$ is given by:

$$R(\tilde{\nu}) = \frac{M_1(\tilde{\nu})}{M_2(\tilde{\nu})} = \frac{a_1.f_1(\tilde{\nu}) + a_2.f_2(\tilde{\nu})}{a'_1.f_1(\tilde{\nu}) + a'_2.f_2(\tilde{\nu})} \tag{22}$$

In spectral regions where the ratio is uniquely attributable to component 1 (*i.e.* $f_2(\tilde{\nu})$ has no absorptions above its baseline due to component 2), then:

$$\frac{M_1(\tilde{\nu}_1)}{M_2(\tilde{\nu}_1)} = \frac{a_1}{a'_1} = R(\tilde{\nu}_1) \tag{23a}$$

similarly,

$$\frac{M_1(\tilde{\nu}_2)}{M_2(\tilde{\nu}_2)} = \frac{a_2}{a'_2} = R(\tilde{\nu}_2) \tag{23b}$$

thus,

$$a_1 = a'_1.R(\tilde{\nu}_1) \quad \text{and} \quad a_2 = a'_2.R(\tilde{\nu}_2) \tag{24}$$

and

$$a_1 + a_2 = a'_1 + a'_2 = 1 \tag{25}$$

therefore,

$$a'_2 = \frac{1 - R(\tilde{\nu}_2)}{R(\tilde{\nu}_1) - R(\tilde{\nu}_2)} \quad \text{and} \quad a'_2 = \frac{1 - R(\tilde{\nu}_1)}{R(\tilde{\nu}_2) - R(\tilde{\nu}_1)} = \frac{R(\tilde{\nu}_1) - 1}{R(\tilde{\nu}_1) - R(\tilde{\nu}_2)} \tag{26}$$

In practice, prior to ratioing, the spectra are pre-processed to remove background and slope by baseline correction to help remove artefacts,[33] and the maximum and minimum values of the ratio spectrum are taken as the values for $R(\tilde{\nu}_1)$ and $R(\tilde{\nu}_2)$ respectively; increasing band overlap decreases the difference in

value between and $R(\tilde{\nu}_1)$ and $R(\tilde{\nu}_2)$, see Figure 9.[34] These values can be used to determine a_1' and a_2' using Equations (26) and then a_1 and a_2 using equations (24). The 'pure' component spectra $f_1(\tilde{\nu})$ and $f_2(\tilde{\nu})$ can then be generated using Equations (21a) and (21b) and the relationships of (26), *e.g.* for $f_1(\tilde{\nu})$:
from Equation (21b),

$$f_2(\tilde{\nu}) = \frac{M_2 - a_1'.f_1(\tilde{\nu})}{a_2'} \tag{27}$$

substituting for $f_2(\tilde{\nu})$ in Equation (21a), then

$$\begin{aligned} a_1.f_1(\tilde{\nu}) &= M_1 - a_2\left[\frac{M_2 - a_1'.f_1(\tilde{\nu})}{a_2'}\right] \\ &= M_1 - R(\tilde{\nu})[M_2 - a_1'.f_1(\tilde{\nu})] \end{aligned} \tag{28}$$

Substituting for a_1'$R(\tilde{\nu})$ for a_1 and using the relationship of equation (26) for a_1', then:

$$f_1(\tilde{\nu}) = \frac{M_1}{(1 - R(\tilde{\nu}_2))} - M_2\left(\frac{R(\tilde{\nu}_2)}{1 - R(\tilde{\nu}_2)}\right) \tag{29}$$

A similar argument may then be followed to 'generate' $f_2(\tilde{\nu})$.

Expansion of the procedure for ratioing of spectra to the analysis of more than two component mixtures has been discussed, but rapidly becomes impractical as the number of components increases.[35]

Least-squares Determinations and Multiple Linear Regression

Least-squares modelling of *n*-calibration values (usually in replicate) is commonly used in single component correlations to determine the best fit correlation (straight line fit) between the spectroscopically measured values and the reference values (*e.g.* mole fraction of a component), see Figure 10.[31–34] The form of the equation is:

$$\varepsilon = \sum_{i=1}^{n}(\hat{y}_i - y_i)^2 \tag{30a}$$

which becomes,

$$\varepsilon = \sum_{i=1}^{n}(\hat{y}_i - y_i)^2 = \sum_{i=1}^{n}(a + bx_i - y_i) \tag{30b}$$

where ε is the total error which is minimised, x_i is the measured value (intensity, absorbance), y_i is the dependent variable, $\hat{y}_i$ is the estimated linear model value

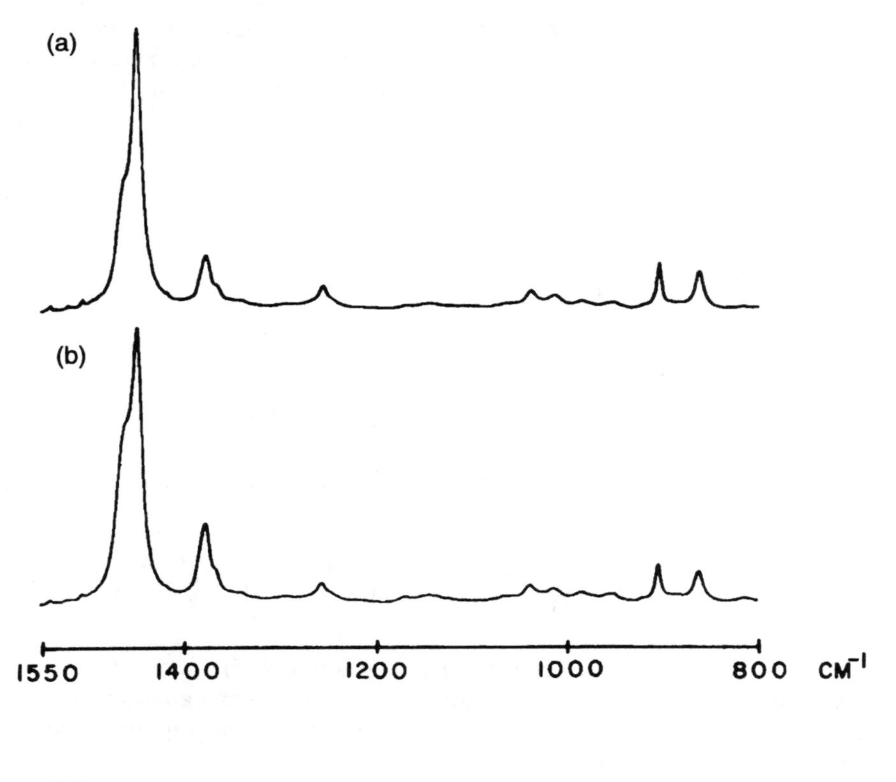

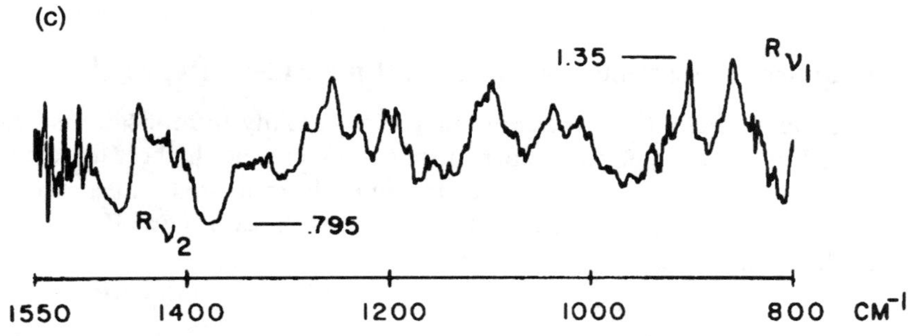

Figure 9 (a) *FTIR absorbance spectrum* (2 cm^{-1}, 64 scans) *of a hexane-cyclohexane mixture,* 50.9% *cyclohexane;* (b) *FTIR absorbance spectrum of a hexane-cyclohexane mixture,* 37.6% *cyclohexane;* (c) *Absorbance ratio spectrum of* (a) *and* (b). *The results calculated following the absorbance ratio method yielded cyclohexane values of* 49.9% *and* 36.9% *for mixtures* (a) *and* (b) *respectively* (Reproduced from ref. 34, Koenig J.L. and Kormos D., *Quantitative Infrared Spectroscopic Measurements of Mixtures without External Calibration*, Appl. Spectrosc., 1979, **33**, 4, 349, by kind permission of The Society for Applied Spectroscopy; © 1979)

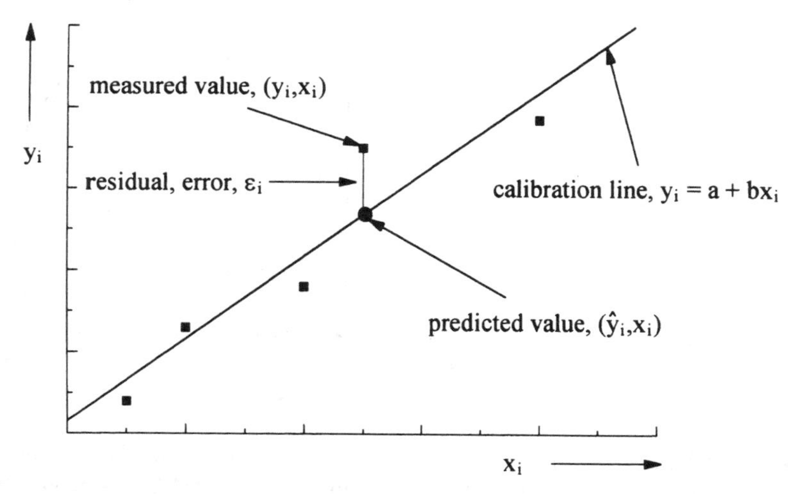

Figure 10 *Illustration of a least-squares modelling (straight line fit) between spectroscopically measured values (x_i) and six calibration (reference) values (y_i) for an analyte*
(Adapted from ref. 38)

of the dependent variable, *n* is the number of data points, and *a* and *b* are the intercept and slope for the best-fit line.[29,36–39]

The 'pure' component spectra of a multi-component mixture generated by a procedure such as that discussed above, or, as is more usually the case, recorded individually, can be curve-fitted (matched) by a least-squares criterion method,[2,35,40–42] see below, pp. 228–232 and Figure 24. In this instance, least-squares spectral multiple linear regression (MLR) routines seek to find the best match (proportional summation) of component reference spectra to the profile of the mixture spectrum, fitting in such a way that the sum of the squares of the residuals between the ratio of the model to the mixture spectrum are minimised ('best least-squares-fit'). This procedure of synthesising (matching) a mixture spectrum of unknown composition from a linear combination of the spectra of its components has been referred to as the *Q*-matrix approach.[2,43] Example applications have been discussed for xylene isomer

[36] Mark H. and Workman J., *Spectroscopy*, 1992, **7**, 1, 44.
[37] Workman J. and Mark H., *Spectroscopy*, 1992, **7**, 3, 20.
[38] Fearn T., *Spectroscopy World*, 1990, **2**, 3, 32.
[39] Osten D.W. and Kowalski B.R., in *Computerised Quantitative Infrared Analysis*, ed. McClure G.L., ASTM Special Technical Publication 934, ASTM, Philadelphia, USA, 1987, p. 6.
[40] Antoon M.K., Koenig J.H., and Koenig J.L., *Appl. Spectrosc.*, 1977, **31**, 6, 518.
[41] Haaland D.M. and Easterling R.G., *Appl. Spectrosc.*, 1980, **34**, 5, 539.
[42] Haaland D.M. and Easterling R.G., *Appl. Spectrosc.*, 1977, **31**, 6, 518.
[43] McClure G.L., Roush P.B., Williams J.L., and Lehmann C.A. in *Computerised Quantitative Infrared Analysis*, ed. McClure G.L., ASTM Special Technical Publication 934, ASTM, Philadelphia, USA, 1987, p. 131.

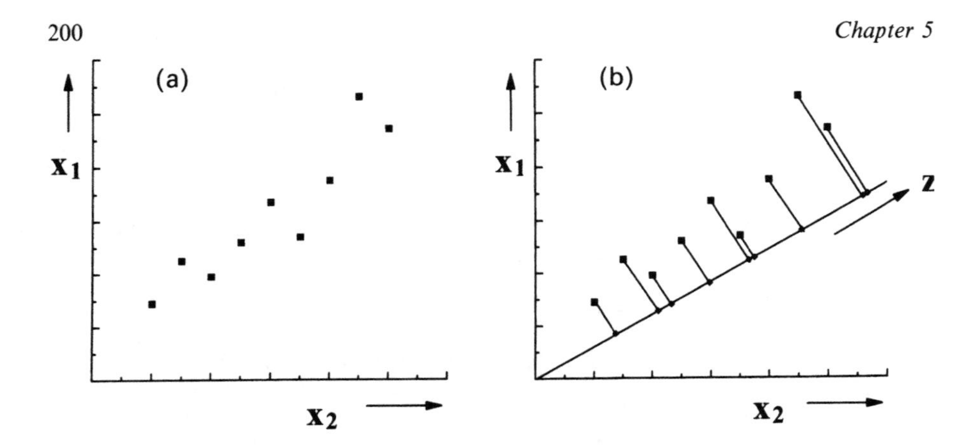

Figure 11 (a) *Two-dimensional plot of intensities, x_1 and x_2, at two spectral analytical data points from a series of mixture (multicomponent) spectra;* (b) *The projection from two dimensions (x_1 and x_2) of the points plotted in Figure 11(a) onto one dimension z*
(Adapted from ref. 38)

mixture analysis,[40,42,44] polymer blend analysis,[40] gas analyses,[41] and mineral matter in coal.[45]

In multivariate techniques, least-squares multiple linear regression lines are also generated to provide 'predicted' *vs.* 'actual' correlation plots. The procedure deals with the case for which there are more equations (*e.g.* S, standard spectra) than unknowns (*e.g.* n, sources of variation, components), and for which the responses due to each component at each wavenumber add linearly.[1,29,36–39,46] For example, simply consider the case for which to predict the concentration of one component in a multicomponent system it is necessary to use information from two spectral analytical data points x_1 and x_2, because of band overlap. For the calibration samples these data may be pictured as a two-dimensional plot,[38] see Figure 11(a). In Figure 11(b) the points shown on Figure 11(a) are shown projected onto the line z. The projection lines are orthogonal to z. Then for each calibration sample value of z, a new variable can be determined, which can be least squares plotted against the component concentration c, to give a prediction equation of the form $c = a_0 + a_1 z$. Since z is related to the original data values by an equation of the form $z = b_1 x_1 + b_2 x_2$, where b_1 and b_2 are constants which depend on the direction of z, then $c = a_0 + a_1 b_1 x_1 + a_1 b_2 x_2$. MLR seeks to choose projections for z such that the sum of the squares of the total prediction error is minimised (least squares fit) when plotted as a calibration against the component concentration, c.[38]

[44] Haaland D.M. and Easterling R.G., *Appl. Spectrosc.*, 1982, **36**, 6, 665.
[45] Painter P.C., Rimmer S.M., Snyder R.W., and Davis A., *Appl. Spectrosc.*, 1981, **35**, 1, 102.
[46] Workman J. and Mark H., *Spectroscopy*, 1993, **8**, 9, 16.

A brief outline of some of the more commonly quoted statistical expressions associated with univariate and more particularly multivariate analyses procedures may be found in Section 7.

5 Multicomponent Analyses and Multivariate Techniques

In the opening line of his abstract of reference 47, Tomas Hirschfeld, well-respected for his many profound and thought-provoking deliberations,[48] states: 'So much can be done in infrared (IR) spectroscopy with computers that we are beginning to lose the healthy paranoia that should accompany their use.' Uses of sophisticated computer processing techniques to sort, enhance, and analyse vibrational spectroscopic data have proliferated in recent years, and continue to expand and enable more and more complex analyses. Chemometric* software packages are almost considered a standard component of any new spectrometer system, and their application is commonplace in many industrial laboratories; yet 'belief in the computer output' leads to many abuses of their capabilities – if told to correlate, a computer programme most likely will! The good practitioner will have a healthy regard for their potential and capabilities, but be acutely aware of their limitations and requirements, and cautious of oversimplifying the analytical problem. Robust quantitative multivariate analytical methods need careful attention to the design of the standard set of samples, the experimental parameters, method validation and interrogation of the variances. So while this chapter cannot hope to deal with all aspects of these mathematical and statistical treatments, nor indeed are the authors experts in their intricacies, which are beyond the scope of this book, we have attempted to plot a course from the basics through to a simplistic picture of the advanced, since it is essential that appliers of the tools understand their inherent properties and become familiar with the jargon terms associated with these techniques, and, most importantly, their purpose. Discussion of much of the mechanics, constraints, and implementation of many of the chemometric or multivariate techniques now invoked to analyse or rationalise vibrational spectroscopic data can be found presented in more expert and thorough fashion in and through many of the references cited. To aid differing levels of interest we have included in the references several elementary, descriptive articles (no disrespect to the authors cited, indeed the opposite – it takes an expert to present a complex subject in a form understandable to the layman), as well as references to some more detailed mathematical texts.

* For our purposes, an appropriate definition of *Chemometrics* here might be, 'The application of mathematical and statistical methods to the solution of complex spectroscopic analyses and characterisations and to extract from vibrational spectroscopic data useful information which is often hidden or difficult to discriminate'.

[47] Hirschfeld T., in *Computerised Quantitative Infrared Analysis*, ed. McClure G.L., ASTM Special Technical Publication 934, ASTM, Philadelphia, USA, 1987, p. 169.
[48] *e.g.*, *Appl. Spectrosc.*, 1986, **40**, 6, 11A.

General Introduction to Quantitative Multicomponent Analysis

Much of the discussion so far has been concerned essentially with single-component analysis, in which for two-component systems there was an interdependence on concentration, or for systems in which the majority of the sample, as represented by the major component, could be considered as a spectrally invariant reference. However, there are many systems, both simple and complex, for which the industrial vibrational spectroscopist may wish or need to apply computerised multicomponent or multivariate analysis procedures.[49] It may simply be that a multivariate approach provides a more robust, faster, easier to operate at-line method for plant personnel, or it may be that there is no viable alternative since the problem is insoluble by more direct means. It may also be that the use of pure component spectra would be too extreme for generating robust calibrations, or pure component spectra are inappropriate to component mixture analysis.

Consider a multicomponent system in which the absorption or Raman bands overlap and in which the components have independently variable concentrations. None of the approaches considered so far are capable of generating a valid quantitative analysis, but analysis of digitised spectra by computer-based mathematical techniques will provide the solution in many instances.

The intensity I per unit path-length at wavenumber $\tilde{\nu}$ for a n component system may be represented by,

$$I_{\tilde{\nu}} = K_1c_1 + K_2c_2 + K_3c_3 + K_4c_4 + \ldots + K_nc_n \tag{31}$$

where c_1 to c_n represent the concentrations of component numbers 1 to n, and similarly K_1 to K_n their respective path-length normalised constants.

Equation (31) assumes that the contributions to the intensity $I_{\tilde{\nu}}$ of the components is additive; that is, each component follows a linear intensity/concentration relationship at $\tilde{\nu}$.

Any spectrum analysed from the multicomponent system can be considered to be composed of x such equations, where x equals the number of digitised data points over the wavenumber range to $\tilde{\nu}_1$ to $\tilde{\nu}_x$, providing the linear relationship holds throughout.

Matrix notation can be used to express the x equations:

$$\begin{bmatrix} I_{\tilde{\nu}_1} \\ I_{\tilde{\nu}_2} \\ I_{\tilde{\nu}_3} \\ I_{\tilde{\nu}_4} \\ \cdot \\ \cdot \\ I_{\tilde{\nu}_x} \end{bmatrix} = \begin{bmatrix} K_{\tilde{\nu}_1 1} & K_{\tilde{\nu}_1 2} & K_{\tilde{\nu}_1 3} & K_{\tilde{\nu}_1 4} & \ldots & K_{\tilde{\nu}_1 n} \\ K_{\tilde{\nu}_2 1} & K_{\tilde{\nu}_2 2} & K_{\tilde{\nu}_2 3} & K_{\tilde{\nu}_2 4} & \ldots & K_{\tilde{\nu}_2 n} \\ K_{\tilde{\nu}_3 1} & K_{\tilde{\nu}_3 2} & K_{\tilde{\nu}_3 3} & K_{\tilde{\nu}_3 4} & \ldots & K_{\tilde{\nu}_3 n} \\ K_{\tilde{\nu}_4 1} & K_{\tilde{\nu}_4 2} & K_{\tilde{\nu}_4 3} & K_{\tilde{\nu}_4 4} & \ldots & K_{\tilde{\nu}_4 n} \\ \cdot & \cdot & \cdot & \cdot & \ldots & \cdot \\ \cdot & \cdot & \cdot & \cdot & \ldots & \cdot \\ K_{\tilde{\nu}_x 1} & K_{\tilde{\nu}_x 2} & K_{\tilde{\nu}_x 3} & K_{\tilde{\nu}_x 4} & \ldots & K_{\tilde{\nu}_x n} \end{bmatrix} \begin{bmatrix} c_1 \\ c_2 \\ c_3 \\ c_4 \\ \cdot \\ \cdot \\ c_n \end{bmatrix} \tag{32}$$

[49] *Annual Book of ASTM Standards*, Vol. 03.06, ASTM, Philadelphia, USA, 1995, Designation E1655–94.

The matrix may be represented as:

$$\mathbf{I}_{\tilde{\nu}_x} = \mathbf{K}_{\tilde{\nu}_x n}\mathbf{c}_n \tag{33}$$

If there are S standard samples then there will be S sets of such matrices, which may be represented as:

$$\mathbf{I}_{\tilde{\nu}S} = \mathbf{K}_{\tilde{\nu}n}\mathbf{c}_{nS} \tag{34}$$

Criteria for successful development of a calibration using Equation (34) have been discussed by McClure in ref. 2, and in summary require that:

(i) n is equivalent to the number of all the sources of variation measured, and
(ii) $x \geqslant n$,
(iii) $S \geqslant n$, and
(iv) the matrix is non-singular, *i.e.*
  (a) no row or column is all zeros
  (b) no row or column is a multiple of another row or column
  (c) no row or column is a linear combination of two or more other rows or columns respectively.

Requirements (iv)(a) to (iv)(c) impose certain restraints on the calibration set of samples,[1,2,31,49] and a statistically well-designed set is important to the development of robust quantitative methods. Standards must be independent and not reproduce systematic errors. For example, in solution analyses standards *must not* be simple repetitive dilutions of a master standard solution, nor should the sum of the solute concentrations throughout the set add to a constant.[1,31] *It is also desirable that for each component there is at least one data point in a spectrum at which the component contributes more to the intensity than the rest of the components at that spectral location.*[2]

The *K*-matrix and *P*-matrix Methods

Equations (32–34) represent the basis for what is termed a ***K***-matrix or Classical Least Squares (CLS) manipulation method.[2,4,31,50–55] Since there is always some error associated with a measurement, a random error term, e, is added to these equations,[50–52] *e.g.* Equation (34) becomes,

$$\mathbf{I}_{\tilde{\nu}_x S} = \mathbf{K}_{\tilde{\nu}_x n}\mathbf{c}_{nS} + \mathbf{e}_{\tilde{\nu}_x S} \tag{35}$$

[50] PLSplus™ for GRAMS/386™. Add-On Applications. © Galactic Industries Corporation, 1991–1993.
[51] Haaland D.M. and Thomas E.V., *Anal. Chem.*, 1988, **60**, 11, 1193.
[52] Haaland D.M., Easterling R.G., and Vopicka D.A., *Appl. Spectrosc.*, 1985, **39**, 1, 73.
[53] Brown C.R., Lynch P.F., Obremski R.J., and Lavery D.S., *Anal. Chem.*, 1982, **54**, 1472.
[54] Kisner H.J., Brown C.R., and Kavarnos G.J., *Anal. Chem.*, 1983, **55**, 1703.
[55] Maris M.A., Brown C.R., and Lavery D.S., *Anal. Chem.*, 1983, **55**, 1694.

where, $\mathbf{I}_{\tilde{\nu}_x S}$ denotes a $\tilde{\nu}_x \times S$ matrix of calibration/training spectra, $\mathbf{K}_{\tilde{\nu}_x n}$ signifies a $\tilde{\nu}_x \times n$ matrix of path-length normalised constants (absorption coefficients), c_{nS} denotes a $n \times S$ matrix of component concentrations, and $\mathbf{e}_{\tilde{\nu}_x S}$ is a $\tilde{\nu}_x \times S$ matrix of random errors associated with each I.

Another commonly used approach is referred to as the ***P***-matrix method or Inverse Least Squares (ILS) method.[2,4,31,50–55] In our notation the basic equation would have the form:

$$\mathbf{c}_{nS} = \mathbf{P}_{n\tilde{\nu}_x}\mathbf{I}_{\tilde{\nu}_x S} \qquad (36)$$

or,

$$\mathbf{c}_{nS} = \mathbf{P}_{n\tilde{\nu}_x}\mathbf{I}_{\tilde{\nu}_x S} + \mathbf{e}_{nS} \qquad (37)$$

where, $\boldsymbol{P}$ is the matrix which relates intensities to concentration, and $\boldsymbol{e}_{nS}$ represents a matrix of random errors in the calibration concentrations.

For ease of perception we will now omit the subscripts of Equations (34) to (37) to give respectively:

$$\mathbf{I} = \mathbf{K.c} \qquad (38a) \qquad\qquad \mathbf{I} = \mathbf{K.c} + \mathbf{e} \qquad (38b)$$

$$\mathbf{c} = \mathbf{P.I} \qquad (39a) \qquad\qquad \mathbf{c} = \mathbf{P.I} + \mathbf{e} \qquad (39b)$$

In general analysis terms $\boldsymbol{I}$ is referred to as the response function or response matrix. In single-component, single analyte wavenumber analysis $\boldsymbol{K}$ would be the slope of the 'best-fit' (least-squares calibration) straight-line between the co-ordinate points of a plot of $\boldsymbol{I}$ against $\boldsymbol{c}$. The 'best-fit' prediction line from the calibration samples would be that in which the sum of the squares of the prediction errors is a minimum, *i.e.* the least-squares line generated by a MLR treatment, see pp. 197–201.

For Equation (38a) in calibration we know the component concentrations of the standards and have recorded and digitally stored their spectra, which contain the response (intensity) function information, so Equation (38a) might be represented as:

$$\mathbf{K} = \mathbf{I}\mathbf{c}^{-1} \qquad (38c)$$

However in matrix algebra, the conceptual equivalent of a reciprocal is the inverse of a matrix, which can only be performed on square matrices.[56,57] Therefore, to determine the sensitivity matrix $\boldsymbol{K}$, since the $\boldsymbol{c}$ matrix is not usually square, both sides of Equation (38a) [or (38c)] need to be multiplied by $\boldsymbol{c}^t$ the transpose of $\boldsymbol{c}$, to yield:

$$\mathbf{K} = \mathbf{I}\mathbf{c}^t.(\mathbf{c}\mathbf{c}^t)^{-1} \qquad (38d)$$

[56] Workman J. and Mark H., *Spectroscopy*, 1993, **8**, 7, 16.
[57] Workman J. and Mark H., *Spectroscopy*, 1993, **8**, 8, 16.

The transpose of $\begin{bmatrix} i_1 & j_1 & k_1 \\ i_2 & j_2 & k_2 \end{bmatrix}$ is $\begin{bmatrix} i_1 & i_2 \\ j_1 & j_2 \\ k_1 & k_2 \end{bmatrix}$.

For analysis, in order to determine the concentrations of components, a similar operation is undertaken, but this time involving the transpose of $\boldsymbol{K}$, *i.e.*

$$\mathbf{c} = \mathbf{IK}^t.(\mathbf{KK}^t)^{-1} \tag{38e}$$

For the $\boldsymbol{P}$-matrix case, the calibration steps may be represented by:

$$\mathbf{P} = \mathbf{c}.\mathbf{I}^{-1} \tag{39c}$$

$$\mathbf{P} = \mathbf{cI}^t.(\mathbf{II}^t)^{-1} \tag{39d}$$

but this time, a second matrix inversion is unnecessary, and the concentrations of components in test samples be found directly from:

$$\mathbf{c} = \mathbf{P}.\mathbf{I} \tag{39e}$$

Equations (39e) and (39a) are identical.

For the $\boldsymbol{K}$-matrix treatment of data the concentration of all potential components or sources of variation must be known and included in the calibration model. An advantage of the $\boldsymbol{P}$-matrix approach is that it is not a prerequisite to know the concentrations of all the components in a mixture, although, as with the $\boldsymbol{K}$-matrix method, *all potential sources of variation must still be present in the training set of spectra.* The $\boldsymbol{P}$-matrix approach seeks to determine correlations one at a time between intensities and known concentrations for the components under study. For the $\boldsymbol{P}$-matrix treatment if there are n components to be determined then at least n intensity values (data points) must be used. However, since the inverted matrix $[\boldsymbol{II}^t]$ has the dimensions $\tilde{\nu}_x$ then the number of selected wavenumbers used in the analysis is generally small,[4,31,50–52] also, it cannot exceed S, the number of training spectra.[31,51] Remembering that S, the number of standards, must be greater than n, then if a large number of data points were required to stipulate good correlations, as might be the case for severely overlapped bands, a large number of standards would be necessary! However, overfitting and collinearity can cause significant problems with ILS data analysis, and adding more standards to increase the number of analyte wavenumbers used may deteriorate the correlation. (*Overfitting* may lead to calibration errors or noise being modelled rather than real differences between training spectra;[4,31,51,52] highly correlated (linear or near-linear relationship) intensities at multiple wavenumbers are said to be *collinear*, they tend the matrix $[\boldsymbol{II}^t]$ towards singularity).[4,31,51,52] Appropriate wavenumber (feature) selection is also a particular concern with ILS analysis of data.[53–55,58] The $\boldsymbol{K}$-matrix has the advantage of offering some signal-averaging since it uses spectral regions rather

[58] Honigs D.E., Freelin J.M., Hieftje G.M., and Hirschfeld T.B., *Appl. Spectrosc.*, 1983, **37**, 6, 491.

than selected wavenumbers; the matrix [$\boldsymbol{KK}^t$], has the dimensions n the number of components, and is therefore independent of $\tilde{\nu}_x$, the number of wavenumbers used.[4,31,51] As stated in the previous section, S and $\tilde{\nu}_x$ must be equal to or greater than n.[2,4] *Overdetermination*,[1,26,45] addition of more useful spectral data and/or more calibration samples than there are components, is an important step to providing robust methods (subject to the analysis not being compromised by adding an unaccounted for source of intensity variation)[4] – better discrimination can be achieved between true component intensity and that due to noise, irreproducible sampling, *etc*., although the gains may not be immediately evident since, for a small number of calibration spectra, 'erroneous selections' may lead to apparently better correlations.[31] To test for calibration and analysis errors and ascertain if adequate overdetermination has been employed, a calibration is set up with $(n-1)$ standard samples and the 'missing' standard analysed against this calibration. The procedure is repeated for all or many of the spectra from the standard samples. This process is known as *cross-validation*.[31,59] If the analysis error exceeds the calibration error, it is likely that the method is unreliable and 'under-determined'. Multiple (repeat) sampling will reduce intrinsic errors.

The $\boldsymbol{Q}$-matrix (linear curve-fit of pure component spectra[40–42,44,45,52]), see p. 199, is often referred to as a CLS-type analysis,[51,52] since the $\boldsymbol{Q}$-matrix approach may be considered as a special case of the $\boldsymbol{K}$-matrix method in which the standard mixtures are pure components with unity concentration.[2,43] The subtle differences, advantages, and limitations of the $\boldsymbol{K}$- and $\boldsymbol{P}$-matrix methods are well described and exampled in references 2, 31, 39, 43, 49–52. References 53–55 contain examples of the use of ILS, including demonstrations of the value of overdetermination with this approach![54,55] ILS has proven more popular for the analysis of NIRA data than for mid-infrared or Raman spectra.[58,60,61]

Principal Component Analysis (PCA) and Factor Analysis (FA)

Principal Component Analysis (PCA) and Factor Analysis (FA) are closely related, but essentially distinct, multivariate data analysis methods.[29,59,62–67] A good introduction to the aims and workings of FA may be found in ref. 64 from which much of the qualitative description below is adapted. A good treatise on PCA containing many examples of uses of PCs in multivariate analyses in many

[59] Wold S., *Technometrics*, 1978, **20**, 4, 397.
[60] *Handbook of Near-Infrared Analysis*, ed. Burns D.A. and Ciurczak E.W., Marcel Dekker, New York, USA, 1992.
[61] Wetzel D.L., *Anal. Chem.*, 1983, **55**, 12, 1165A.
[62] Spragg R., *Spectroscopy World*, 1990, **2**, 2, 32.
[63] Sharaf M.A., Illman D.L., and Kowalski B.R., *Chemometrics*, J. Wiley, New York, USA, 1986.
[64] (a) Malinowski E.R. and Howery D.G., *Factor Analysis in Chemistry*, J. Wiley, New York, USA, 1980; (b) Malinowski E., *Factor Analysis in Chemistry*, 2nd Edn, J. Wiley, New York, USA, 1991.
[65] Fister J.C. and Harris J.M., in *Computer Assisted Analytical Spectroscopy*, ed. Brown S.D., J. Wiley, Chichester, UK, 1996, ch. 4.
[66] Kruskal J.B., in *International Encyclopedia of Statistics*, Vol. 1, ed. Kruskal W.H. and Tanur J.M., The Free Press, New York, USA, 1978, p. 307.
[67] Joliffe I.T., *Principal Component Analysis*, Springer-Verlag, New York, USA, 1986.

fields may be found in ref. 67, including discussions of the subtle differences between FA and PCA, about which, to the layman, there seems perhaps to be some confusion about the semantics and debate over precise definitions and differences![67,68]

Factor Analysis techniques are important in determining mathematically the key parameters in multicomponent, multivariate mixture characterisations. They have the potential of yielding the number of components contributing to a spectrum, the concentration of each component and its spectrum.[64,69] The initial objective of FA is to obtain a mathematical solution in which each data point in a data matrix (*e.g.* a set of mixture spectra) may be expressed as a linear sum of product terms.[64] The number of *factors* is the number of terms in the sum. The product terms constitute a *scores* matrix and a *loadings* matrix. Since no assumptions are made about the data matrix, the FA solution is mathematical, abstract, and therefore not implicitly directly related to any physical or chemical characteristic.[64] If from a set of mixture spectra we have a matrix representing the intensities I of S samples each recorded over the wavenumber range $\tilde{\nu}_x$ for which $\tilde{\nu}_x > S$ then, the matrix $I_{\tilde{\nu}_x S}$ can be decomposed as a row matrix and a column matrix:

$$\overset{\text{FACTOR}}{} \qquad \overset{\text{MIXTURE}}{}$$

$$\tilde{\nu}_x\overset{S}{[I]} = \text{WAVENUMBER}\begin{bmatrix} \tilde{\nu}_{11} & \tilde{\nu}_{12} & \cdots & \tilde{\nu}_{1f} \\ \tilde{\nu}_{21} & \tilde{\nu}_{22} & \cdots & \tilde{\nu}_{2f} \\ \cdot & \cdot & \cdots & \cdot \\ \cdot & \cdot & \cdots & \cdot \\ \cdot & \cdot & \cdots & \cdot \\ \tilde{\nu}_{x1} & \tilde{\nu}_{x2} & \cdots & \tilde{\nu}_{xf} \end{bmatrix} \overset{\begin{matrix} 1 & 2 & \ldots & S \end{matrix}}{\begin{bmatrix} I_{11} & I_{12} & \cdots & I_{1S} \\ I_{21} & I_{22} & \cdots & I_{2S} \\ \cdot & \cdot & \cdots & \cdot \\ \cdot & \cdot & \cdots & \cdot \\ I_{f1} & I_{f2} & \cdots & I_{fS} \end{bmatrix}} \text{FACTOR} \tag{40}$$

If f is the number of factors which account for the intensities within experimental error, then each intensity has the general solution:

$$I_{\tilde{\nu}_i k} = \sum_{j=1}^{f} w_{\tilde{\nu}_i j} m_{jk} \tag{41}$$

where, $w_{\tilde{\nu}_i j}$ is the jth abstract row cofactor associated with the $\tilde{\nu}_i$th wavenumber and represents the jth abstract column cofactor associated with the kth mixture, and f reveals the number of factors responsible for the intensities.

The abstract factor solution of the intensity data matrix may then be expressed as the product of wavenumber-factor and mixture-factor matrices:

$$[\mathbf{I}] = [\mathbf{W}]_{\text{abstract}}[\mathbf{M}]_{\text{abstract}} \tag{42}$$

[68] Wold S., Albano C., Dunn W.J., Esbensen K., Hellberg S., Johansson E., Lindberg W., and Sjostrom M., *Analusis*, 1984, **12**, 10, 477.

[69] Irish D.E. and Ozeki T., in *Analytical Raman Spectroscopy*, ed. Grasselli J.G. and Bulkin B.J., J. Wiley, New York, USA, 1991, ch. 4.

The next step is to convert this abstract solution into a solution that has physical or chemical significance by finding a transformation matrix which will yield:

$$[\mathbf{I}] = [\mathbf{W}]_{\text{real}}[\mathbf{M}]_{\text{real}} \tag{43}$$

Equation (34) might have been expressed as:

$$I_{\tilde{\nu}_i k} = \sum_{j=1}^{n} K_{\tilde{\nu}_i j} c_{jk} \tag{44}$$

So, since $f \equiv n =$ the number of components in the mixture spectra, a transformation matrix might be found which converts equation (42) into a meaningful chemical solution, to give:

$$[\mathbf{I}] = [\mathbf{K}]_{\text{real}}[\mathbf{c}]_{\text{real}} \tag{45}$$

(Since, in Equation (38b) the intensity matrix $\boldsymbol{I}$ is represented by two smaller matrices, $\boldsymbol{K}$ and $\boldsymbol{c}$, CLS might be considered a FA method, in which the columns of $\boldsymbol{K}$, the pure component spectra, are the factor loadings, while the chemical concentrations, $\boldsymbol{c}$, are the factor scores).[51,64]

PCA, also called Principal Factor Analysis (PFA), is a least-squares technique which calculates a mathematical solution to Abstract Factor Analysis (AFA),[64] and its solutions are by implication essentially devoid of any physical or chemical meaning[29,64,70] PCA is termed therefore an 'unsupervised' technique in that it makes no *a priori* assumptions about the training set. The mathematical method is the process of *eigenanalysis*, and yields a set (matrix) of abstract *eigenvectors* (abstract eigenfactors, loadings) and an associated set of abstract *eigenvalues* (factor matrix, scores). Mathematically, the PCs are vectors (eigenvectors) in multi-dimensional space, and each new variable (PC) must be orthogonal (mathematically independent, uncorrelated) to all others. PCA involves *diagonalisation* of the *covariance* matrix, which is the product of the data matrix premultiplied by its transpose.[29,64–67,70] Eigenvalues indicate the relative importance of the eigenvectors/factors, major factors being indicated by large eigenvalues, and unimportant factors by small eigenvalues.[64] So PCA is a data analysis method which seeks to find the maximum variations (and hence reduce the dimensionality) in a data set that has been normalised, to form a new set of variables known as *Principal Components* (PCs), which provide a simpler representation of the data.[29,62–67,71] In addition to being or having been normalised, the data set may commonly undergo other preprocessing treat-

[70] Malinowski E.R., in *Computerised Quantitative Infrared Analysis*, ed. McClure G.L., ASTM Special Technical Publication 934, ASTM, Philadelphia, USA, 1987, p. 155.

[71] (a). Davies A.M.C., *Spectroscopy World*, 1992, **4**, 1, 23; (b) Davies A.M.C., *Spectroscopy Europe*, 1992, **4**, 2, 38.

ments. These might include centring and maybe scaling.[49–51,63,64] A common centring pretreatment is known as *mean-centring*.[49–51,64,72]

Descriptively, PCs are established on a 'learning/training' set of spectra by a process known as spectral decomposition.[50,51,64,72] For example, the first factor (eigenvector displayed as a function of wavenumber), PC1, of a data set comprising spectra from a series of mixtures might most closely resemble the average spectrum, see Figure 12, since this probably accounts for as much of the variance in the original data set as possible.[62] Successive factors, PC2 *etc*., are established, similarly, by accounting for as much of the data remaining as possible after subtracting the previous PC, continuing until there is no distinction between residual variation and noise-level. If we re-represent Equation (34) as:

$$\tilde{\nu}_x\overset{S}{\begin{bmatrix}\text{Mixture}\\ \text{Spectra}\end{bmatrix}} = \tilde{\nu}_x\overset{n}{\begin{bmatrix}\text{Normalised}\\ \text{Constants}\end{bmatrix}}\overset{S}{\begin{bmatrix}\text{Component}\\ \text{Concentrations}\end{bmatrix}}n \quad (46)$$

then, its PCA counterpart could be represented as:

$$\tilde{\nu}_x\overset{S}{\begin{bmatrix}\text{Mixture}\\ \text{Spectra}\end{bmatrix}} = \tilde{\nu}_x\overset{f}{\begin{bmatrix}\text{Principal}\\ \text{Component}\\ \text{Loadings}\end{bmatrix}}\overset{S}{\begin{bmatrix}\text{Principal}\\ \text{Component}\\ \text{Scores}\end{bmatrix}}f \quad (47)$$

(Equations (46) and (47) adapted from ref. 62). Thus PCA is a tool for predicting the number of sources of variation within a data set. In Equation (47) if we assume there are no other sources of variation then the number of PCs, f, should equal the number of components n. Examination of the PC loadings as a function of $\tilde{\nu}_x$, *i.e.* the 'loadings spectrum' (elements of the eigenvectors), highlights the weights given to each spectral point over the spectral range.[29,64,73] A common practice is to factor analyse a normalised data set that has been mean-centred preprocessed, since this overcomes any requirement to fit a non-zero intercept in any subsequent regression analysis.[29,50,51] The process proceeds by calculating from the training data set its average spectrum. Each spectrum in the training set is then compared to the average spectrum and a new spectrum created (the first loading vector) which reflects the variances between them at each spectral data point.[50] It then calculates the amount (the scores) of the loading vector in each spectrum in the training set. The contribution of the loading (its vector × the score) to each spectrum in the training set is then calculated and subtracted from each spectrum in the training set. The resultant training set is then similarly used to determine the next factor, and so on until the required number of factors have been established.

[72] Malinowski E.R., *Anal. Chem.*, 1977, **49**, 4, 606.
[73] Cowe I.A. and McNicol J.W., *Appl. Spectrosc.*, 1985, **39**, 2, 257.

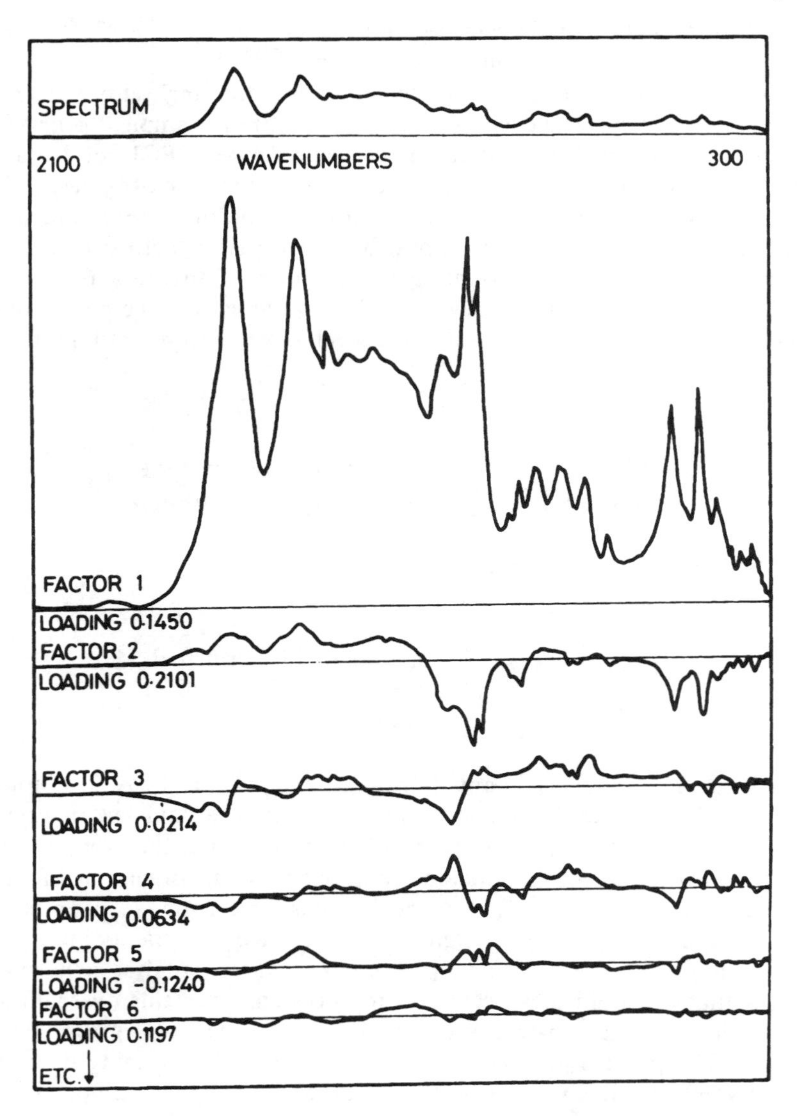

Figure 12 *Factor analysis of the* 2100–300 cm^{-1} *region of a set of coal FTIR absorbance spectra. One spectrum from the set is reproduced at the top. (Note: the labelling convention adopted in this diagram is different from that cited in the text, where the factors are described as the Loadings spectra and their weighting by the Scores, see refs. 64 and 65–68 for discussions of FA conventions. For the convention used in this diagram, the absorbance spectrum is reproduced by: (0.1450* × Factor 1) + (0.2101 × Factor 2) + (0.0214 × Factor 3) *etc.)*
(Reproduced from ref. 76, Fredericks P.M., Lee J.B., Osborn P.L., and Swinkels D.A.J., *Materials Characterisation Using Factor Analysis of FT-IR Spectra. Part I: Results*, Appl. Spectrosc., 1985, **39**, 2, 303, by kind permission of the Society for Applied Spectroscopy; © 1985)

Linear combinations of the PCs, or linear combinations of the PCs plus the mean spectrum, will reproduce the original spectra, the appropriate combinations being assembled by a set of weighted coefficients (PC scores, eigenvalues) for each 'PC loadings' spectrum. Having calculated the full PCA solution, the factor model may then be 'compressed' to include only those factors which are important, eliminating those associated essentially with experimental error and noise.[49,59,63,64,67] Finding the true/significant 'rank of the matrix' is accomplished by mathematical techniques based on estimating the errors associated with reproducing/predicting the original data matrix within experimental error;[64,74] one approach for selecting the optimum number of factors is based on cross-validation determining when the PRESS (Prediction Residual Error Sum of Squares) statistic reaches a minimum.[50,59]

FA and PCA may be thought of as key data preprocessing tools for further data analysis treatments.[29,64,70] The next important step may be to derive more meaningful or recognisable solutions, from which one can gain insights into the chemical or physical significance of the solutions.[29,64,70] Transformation processes by which this may be achieved are termed *rotation* and *target-testing*.[29,64,67,75] 'Matrix rotation' tends to yield more easily aligned/interpretable PC solutions than those generated by PCA,[64] whereas Target Factor Analysis (TFA) aims to identify typical or basic real factors.[64,69,75] In target-transformation, each factor may be tested individually against the PCA solution in a least-squares operation, and new models built up from sets of real or typical factors.[64,70]

Examples of the application of FA and PCA may be found in refs. 65, 76–87, and 89. Sets (CsI discs, DRIFT, solutions in transmission cells) of infrared spectra from well-characterised samples of various mineral materials have been factor analysed to determine the minimum set of factors and associated eigenvectors required to reproduce the spectra.[76,77] Linear regression of the eigenvalues against a variety of measured properties has then been used to establish correlations, which in turn have been used to predict/estimate these property values in similarly constituted unknown samples.[76] For example, Figure 13 shows plots of infrared predicted values *vs*. measured values of ash content determined by combustion and specific energy measured by calorimetry

74 Malinowski E.R., *Anal. Chem.*, 1977, **49**, 4, 612.
75 (a) Hopke P.K., *Chemometrics and Intelligent Laboratory Systems*, 1989, **6**, 7; (b) Hopke P.K., in *Chemometrics Tutorials II*, Elsevier Science, Amsterdam, Holland, 1992, ch. 11.
76 Fredericks P.M., Lee J.B., Osborn P.R., and Swinkels D.A., *Appl. Spectrosc.*, 1985, **39**, 2, 303.
77 Fredericks P.M., Lee J.B., Osborn P.R., and Swinkels D.A., *Appl. Spectrosc.*, 1985, **39**, 2, 311.
78 Brown C.W., Obremski R.J., and Anderson P., *Appl. Spectrosc.*, 1986, **40**, 6, 734.
79 Antoon M.K., D'Esposito L., and Koenig J.L., *Appl. Spectrosc.*, 1979, **33**, 4, 351.
80 Gillette P.C. and Koenig J.L., *Appl. Spectrosc.*, 1982, **36**, 5, 535.
81 Gillette P.C., Lando J.B., and Koenig J.L., *Appl. Spectrosc.*, 1982, **36**, 6, 661.
82 Gillette P.C., Lando J.B., and Koenig J.L., *Anal. Chem.*, 1983, **55**, 4, 630.
83 Malinowski E.R., Cox R.A., and Haldna U.L., *Anal. Chem.*, 1984, **56**, 4, 778.
84 Guzonas D.A. and Irish D.E., *Can. J. Chem.*, 1988, **66**, 1249.
85 Starkweather H.W., Ferguson R.C., Chase D.B., and Minor J.M., *Macromolecules*, 1985, **18**, 9, 1684.
86 Koenig J.L. and Rodriquez M.J.M.T., *Appl. Spectrosc.*, 1981, **35**, 6, 543.
87 Lin S-B. and Koenig J.L., *J. Polym. Sci.: Polym. Phys. Ed.*, 1982, **20**, 2277.

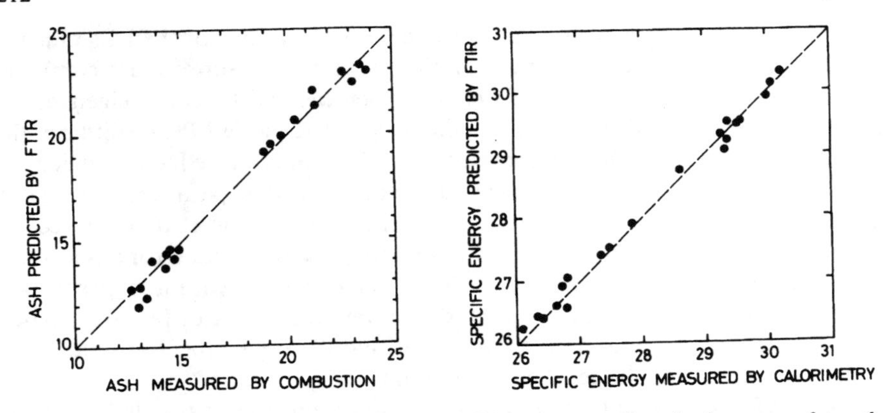

Figure 13 Left, *Plot of measured ash* (wt%, *db) against predicted ash content from the FTIR CsI pellet spectra of 20 'unknown' coal samples from a single mine using factor analysis, with 60 coals used in the calibration set. RMS (root mean square) difference* = 0.50%; Right, *Plot of measured specific energy* (MJ/kg) *against predicted specific energy from the CsI pellet spectra of 20 'unknown' coal samples from a single mine using factor analysis, with 60 coals used in the calibration set. RMS difference* = 0.11 MJ/kg
(Reproduced from ref. 76, Fredericks P.M., Lee J.B., Osborn P.L., and Swinkels D.A.J., *Materials Characterisation Using Factor Analysis of FT-IR Spectra. Part I: Results*, Appl. Spectrosc., 1985, **39**, 2, 303 by kind permission of the Society for Applied Spectroscopy; © 1985)

for coal samples from a single mine.[76] PFA has been utilised in a comparative study for the quantitative analysis of infrared spectra of mixtures represented in either the spectral domain or the Fourier domain,[78] the latter yielding the optimum results (lower SEE, see Section 7). FA has been applied to different spectral regions of a model system comprising transmission infrared spectra of solutions of mixed xylene isomers,[79] and also used to determine the number of independently absorbing species in some polymer systems.[79] It has been used to reconstruct better quality signal-to-noise ratio infrared spectra from noisy spectra of binary mixtures, by reconstructing the data matrix using only the two eigenvectors corresponding to the two largest eigenvalues.[80] Its potential has been explored as a tool in conjunction with a reference infrared spectral library for the qualitative analysis of unknown mixtures.[81] FA has been used to extract pure component spectra from a series of hexane/chloroform mixture infrared spectra.[82] It has also been employed to isolate the Raman spectra of aqueous sulphuric acid components,[83] followed by application of TFA to generate concentration profiles of these components over the 16 mixture sample set extending over the acid mole range 0.0087 to 0.9723. Similarly spectral isolation factor analysis has produced distinct Raman spectra of 1,4-diazabicyclo[2.2.2]octane and its protonated and diprotonated forms.[69,84] Other reported applications of FA include polymer studies on: separating crystalline and amorphous phase component infrared spectra and assigning their bands for poly(tetrafluoroethylene);[85] determining the number of components in the FTIR spectra of compatible blends of polyphenylene oxides and polystyrene;[86]

and determining the number of detectable conformational sequences in the infrared spectra of poly(ethylene terephthalate) (PET).[87] This last realised two components, the *trans* and *gauche* isomers, the 'pure' component spectra of which were calculated by the ratio of spectra method, see pp. 194–197, which were in turn then least-squares curve-fitted (spectra matched), see pp. 197–201, to analyse spectra recorded from PET prepared under various annealing conditions.[87,88] PCA has also been used in a study to determine rate of cure and internal temperature in an epoxy system in real time via a fibre-optic probe coupled with Raman spectroscopy,[89] while FA has been undertaken of transient Raman scattering data to resolve spectra of ground- and excited-state species.[65]

Principal Component Regression (PCR) and Partial Least Squares (PLS) Modelling

PCR and PLS are two commonly applied chemometric methods from a family of biased regression methods.[29,39,49–51,63,68,90–92] At the time of writing this monograph, Principal Component Regression (PCR) and Partial Least-Squares (PLS) modelling are probably the two most popular chemometric techniques used in industry for analysing multivariate vibrational spectroscopic data sets, for which other more traditional multivariate methods are inappropriate. The procedures seek to encompass the advantages of CLS, signal-averaging by utilising many responses, and ILS, single-component calibration in complex mixtures, at the expense of their disadvantages: CLS, all component concentrations in the model must be known; ILS, wavenumber selection. They are termed biased methods since some of the input data is discarded, and both methods involve an inverse model.[45,93]

As was discussed in the previous section, PCA decomposes the response matrix (spectra) into factors, by a process known as *singular value decomposition*,[91] which can be used to reduce the dimensionality of the data matrix, by defining each measured variable as a linear combination of latent variables (scores), which are related by corresponding loading vectors (PCs). PCA, leading to a 'reduced rank matrix', followed by a separate stepwise regression analysis of the scores matrix against each reference value of interest (*e.g.* each known component concentration/sample property) constitutes modelling by Principal Component Regression (PCR).[29,49–51,90–92] An example of PCR modelling of coal spectra was quoted in the previous section.[76,77]

The process for Partial Least Squares (PLS) Regression (partial least squares path modelling with latent variables) is very similar to PCR, except that latent variables are extracted which both model the response matrix (spectrum) and correlate with the component concentration (or equivalent) matrix. This has the

[88] Lin S-B. and Koenig J.L., *J. Polym. Sci.: Polym. Phys. Ed.*, 1983, **21**, 2365.
[89] Aust J.F., Booksh K.S., and Myrick M.L., *Appl. Spectrosc.*, 1996, **50**, 3, 382.
[90] Kowalski B.R. and Seasholtz M.B., *J. Chemometrics* , 1991, **5**, 129.
[91] Lorber A., Wangen L.E., and Kowalski B.R., *J. Chemometrics*, 1987, **1**, 19.
[92] Geladi P. and Kowalski B.R., *Anal. Chim. Acta*, 1986, **185**, 1.
[93] Halaand D.M. and Thomas E.V., *Anal. Chem.*, 1988, **60**, 11, 1202.

potential advantages over PCR in that the models can be more easily interpreted and should be better at prediction, since in PCR the loadings vectors derived to represent the spectral data may not be the optimum for the desired concentration/property prediction.[39,50,51,90–94] In PLS concentration/property information is used in the decomposition process, such that the first factor describes the vector of highest variance that also correlates with the desired reference value, *etc.*[49,51,90–95] (The more common and better approach is to use the PLS1 algorithm (PLS regression) which performs a calibration/prediction analysis one component/property at a time,[51,90] since the PLS2 algorithm, in which two or more components/properties are simultaneously calibrated/analysed, seems better suited to classification or pattern recognition applications).[51]

The advantages and limitations of PLS, PCR, CLS, and ILS as quantitative calibration methods and as means to extracting qualitative information have been compared and discussed in ref. 51. PLS, PCR, and CLS have also been compared as methods for the quantitative analysis of infrared spectra of silicate-based glasses,[93] and for the analysis of B, P, and film thickness for infrared spectra collected as transmission spectra and as 60° incident angle reflection spectra from borophosphosilicate films deposited on Si wafers,[96] in which PLS and PCR both proved to be more precise than CLS. PCR and PLS have also been used to calibrate and validate idealised models for the compositional analysis of binary and ternary mixtures of xylene isomers from their FTIR spectra.[97]

A PLS2 implementation for the quantitative compositional analysis of commercial surfactants used as laundry detergents has been reported.[98,99] The FTIR spectra were recorded using a trough-ATR cell, because of its reproducible path-length and ease of cleaning. For a series of 14 calibration and 7 validation samples, high component correlation coefficients were determined for both the calibration set (>0.998) and validation set (>0.991) for the components base detergent, 1,2-propanediol, glycerol, polypropylene glycol, sodium benzoate, and 2-propanol.[98] The effects of spectral ranges and zero- and first-derivative modes have been compared for the PLS2 simultaneous determination of vulcanised rubber additives in solution; first-derivative spectra generally yield improved quantitation by reducing band overlap and suppressing baseline drift, although at the expense of sensitivity for some of the more weakly absorbing components.[100] Mixtures of natural, butadiene, and styrene-butadiene rubbers are commonly used in the tyre production process.[101] Semi-quantitative (precision ~10%) has been demonstrated for the compositional

[94] Martens H., *Spectroscopy World*, 1991, **3/4**, 26.
[95] Geladi P. and Kowalski B.R., *Anal. Chim. Acta*, 1986, **185**, 19.
[96] Halaand D.M., *Anal. Chem.*, 1988, **60**, 11, 1208.
[97] Cahn F. and Compton S., *Appl. Spectrosc.*, 1988, **42**, 5, 865.
[98] Fuller M.P., Ritter G.L., and Draper C.S., *Appl. Spectrosc.*, 1988, **42**, 2, 228.
[99] Fuller M.P., Ritter G.L., and Draper C.S., *Appl. Spectrosc.*, 1988, **42**, 2, 217.
[100] Blanco M., Coello J., Iturriaga H., Maspoch S., and Bertran E., *Appl. Spectrosc.*, 1995, **49**, 6, 747.
[101] Lutz E.T.G., Luinge H.J., van der Maas J.H., and van Agen R., *Appl. Spectrosc.*, 1994, **48**, 8, 1021.

analysis of these in high carbon-black filled, up to 35 wt%, rubber materials through application of PLS regression to first-derivative derived spectra from external reflection FTIR measurements made on freshly cut surfaces (in order to avoid surface ageing effects).[101]

The application of the PLS (PLS1) algorithm has become particularly popular for calibrating polymer spectra data sets.[102] The low cost and ease of operation and maintenance of vibrational spectrometers, compared to techniques such as MS, NMR, and X-ray, makes them more cost-effective, rapid, and appropriate for routine quality assurance applications.[102] Good agreement has been reported between the mean results for wt% EO determined from a NMR analysis correlated by PLS analysis with FTIR spectra of solid films evaporated onto a horizontal ATR element of polyoxyethylene/polyoxypropylene (EO/PO) condensates in the range 70 wt% to 85 wt% EO, with the FTIR values showing a higher precision![102] High precision univariate analysis of EO/PO spectra in this composition range would be very difficult, perhaps impossible, to achieve, see spectra of Figure 14. PLS modelling of FTIR spectra from a mixed set of unsulphonated, fully sulphonated, and partially sulphonated aryl-ether-sulphone/aryl-ether-ether-sulphone (ES/EES) copolymer solvent cast films has been demonstrated to be potentially very effective in providing a single analytical method for their compositional analysis and indicating the degree of sulphonation,[102] see Figures 15(a) and 15(b). PLS compositional correlations have also been reported for the FTIR spectra of unsulphonated ES/EES copolymers recorded by transmission, DRIFT and ATR techniques.[103] More rigorous and robust methods, compared to univariate treatments, for determining the % crystallinity of poly(aryl-ether-ether-ketone) (PEEK) samples from their Raman spectra have been derived from PLS regression analysis of FT-Raman spectra,[28,102] albeit that in this instance separate correlations were established for isotropic and uniaxially drawn samples, because of a less-than-appropriate sample set.[28] For a similar but more detailed study, FT-Raman spectroscopy and multivariate data analysis have been combined successfully to provide a single calibration for predicting the density (and by implication crystallinity) of poly(ethylene terephthalate) (PET) for both isotropic and anisotropic (uniaxially and biaxially oriented) samples.[25] (For industrial purposes, density may often be an acceptable alternative to crystallinity for characterising the influence of process conditions on final product properties, although this does not yield a detailed insight into polymer microstructure.)[25] A cross-validated SEP (standard error of prediction) of 0.0024 g cm^{-1} was established for the combined data sets recorded from the isotropic chip and oriented film samples, see Figure 16. This was only marginally less precise than that from the individual models developed for the two types of sample sets. However, neither of these individual models was capable of satisfactorily predicting the densities of samples in the other set. The key

[102] Chalmers J.M. and Everall N.J., *TrAC*, 1996, **15**, 1, 18.
[103] Luinge H.J., de Koeijer J.A., van der Maas J.H., Chalmers J.M., and Tayler P.J., *Vib. Spectrosc.*, 1993, **4**, 3, 301.

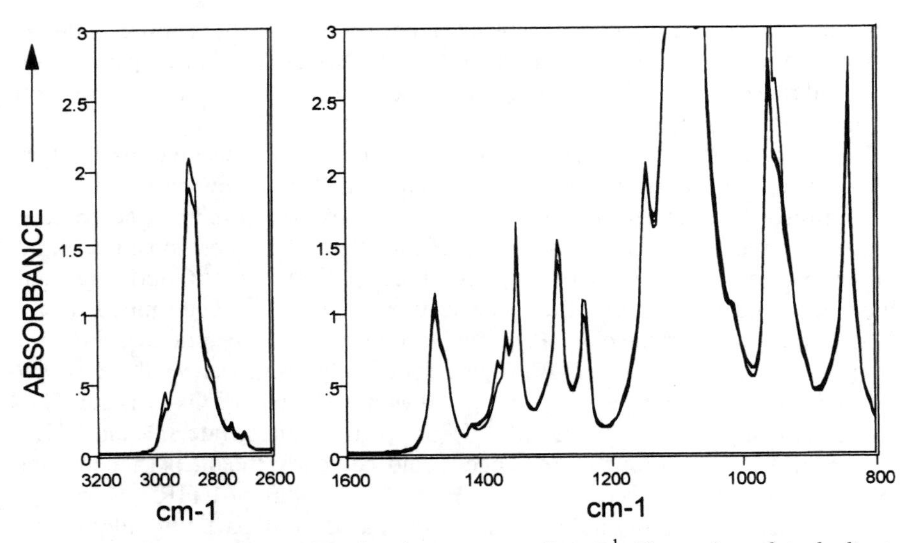

Figure 14 *Overlaid FTIR-ATR absorbance spectra* (8 cm^{-1}, 50 scans), *each in duplicate, of two different EO/PO copolymers, containing ca.* 72 wt% *and ca.* 84 wt% *EO. A precise compositional analysis procedure was established by PLS1 modelling of such spectra, which were recorded from a series of EO/PO copolymer samples within the range 70% to 85% EO, prepared as solid films evaporated from standard methanol solutions onto the trough sampling surface of a horizontal ZnSe ATR element*
(Reprinted from ref. 102, *Trends in Analytical Chemistry*, **15**, 1, J.M. Chalmers and N.J. Everall, *FTIR, FT-Raman and chemometrics: applications to the analysis and characterisation of polymers*, 18–25, 1996, with kind permission of Elsevier Science – NL, Sara Burgerhartstraat 25, 1055 KV Amsterdam, The Netherlands)

success of the particular PLS model developed was that it was able to decouple the effects of orientation and crystallinity, since distinguishing fully between the effects on a vibrational spectrum of molecular ordering and molecular packing can often be very difficult. In this study the potential of HCA (hierarchical cluster analysis) to classify samples according to their physico-chemical properties was also explored,[25] see Figure 17. In a subsequent study on the oriented film samples set, matched performance (precision) was obtained using a process Raman (holographic grating) analyser coupled to a 100 m fibre-optic probe and non-contacting head, with the advantage of using much lower laser power and shorter data accumulation times.[104] Good cross-validated correlations for polymer density using PLS have also been obtained between micro-Raman spectra recorded from a series of polyethylene samples[105] and from a series of oriented PET films.[106]

[104] Everall N., Owen H., and Slater J., *Appl. Spectrosc.*, 1995, **49**, 5, 610.
[105] Williams K.P.J. and Everall N.J., *J. Raman Spectrosc.*, 1995, **26**, 427.
[106] Everall N., Davis K., Owen H., Pelletier M.J., and Slater J., *Appl. Spectrosc.*, 1996, **50**, 3, 388.

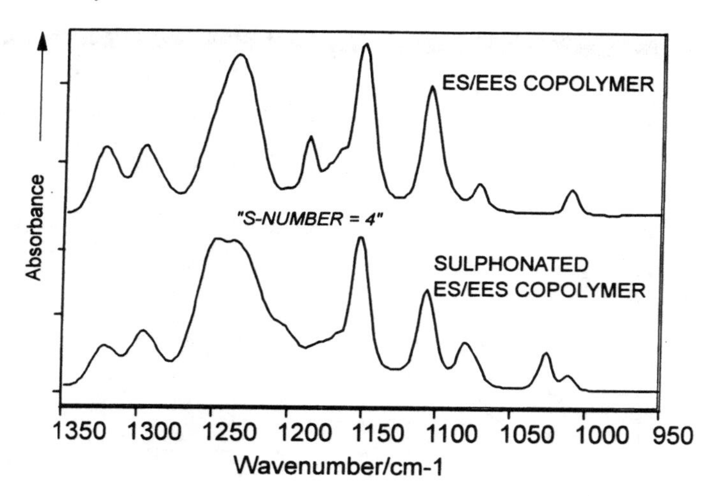

"S-number":

The "***s-number***" is defined as the ratio:

$-SO_2-C_6H_4-O- \; / \; -O-C_6H_4-O-$

Thus: PEES is $s = 2$ and PES is $s = \infty$.

$s = 5$ is :

3x $-O-C_6H_4-O-C_6H_4-SO_2-C_6H_4-$

+

2x $-O-C_6H_4-SO_2-C_6H_4-$

i.e. 3 units of EES and 2 units of ES.

Figure 15(a) *FTIR absorbance spectra in the range* 1350–950 cm^{-1} *of solvent cast films of an ES/EES 's = 4' copolymer:* Upper trace, *unsulphonated;* lower trace, *fully monosulphonated (salt form). Also shown is a definition of the 's-number' invoked to facilitate quantitative compositional analysis, because of the similarity in the comonomer sub-structural units. PES stands for the homopolymer poly(aryl ether sulphone); PEES is used to denote the homopolymer poly(aryl ether ether sulphone)*
(Reprinted from ref. 102, *Trends in Analytical Chemistry*, **15**, 1, J.M. Chalmers and N.J. Everall, *FTIR, FT-Raman and chemometrics: applications to the analysis and characterisation of polymers*, 18–25, 1996, with kind permission of Elsevier Science – NL, Sara Burgerhart straat 25, 1055 KV Amsterdam, The Netherlands)

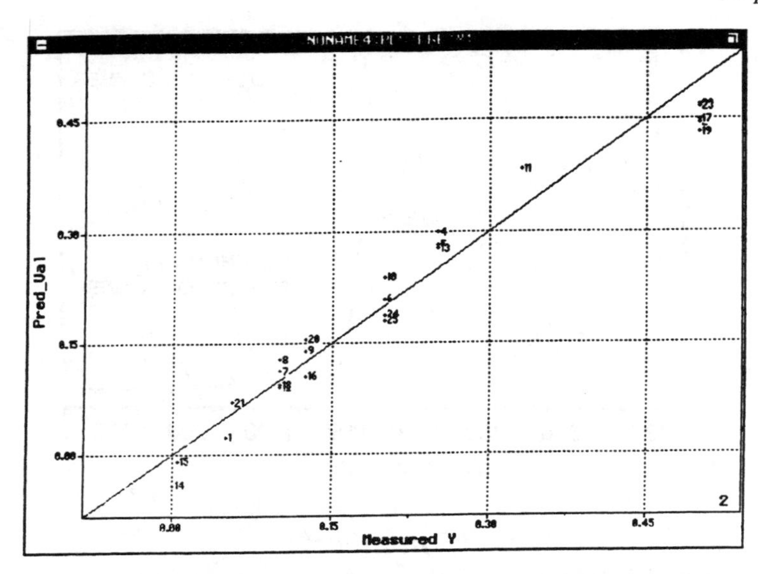

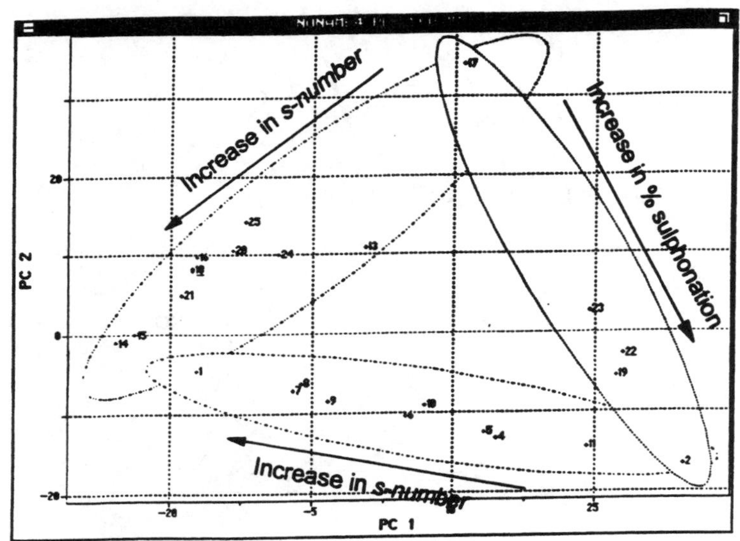

Figure 15(b) Upper graph, *cross-validated PLS calibration model for predicting 1/s-number* (SEP = 0.034) *of unsulphonated and sulphonated ES/EES copolymers.* Lower plot, *Factor scores plot for the cross-validated PLS model. Ellipses have been overlaid to highlight trends. The lower ellipse contains the fully monosulphonated copolymers. The ellipse on the right contains unsulphonated and partially and fully monosulphonated homopolymers. The other ellipse contains the unsulphonated polymer samples. The samples cover the range s = 2 (PEES and SPEES) to s = ∞ (PES), (SPEES denotes fully sulphonated PEES)*

(Reprinted from ref. 102, *Trends in Analytical Chemistry*, **15**, 1, J.M. Chalmers and N.J. Everall, *FTIR, FT-Raman and chemometrics: applications to the analysis and characterisation of polymers*, 18–25, 1996, with kind permission of Elsevier Science – NL, Sara Burgerhart straat 25, 1055 KV Amsterdam, The Netherlands)

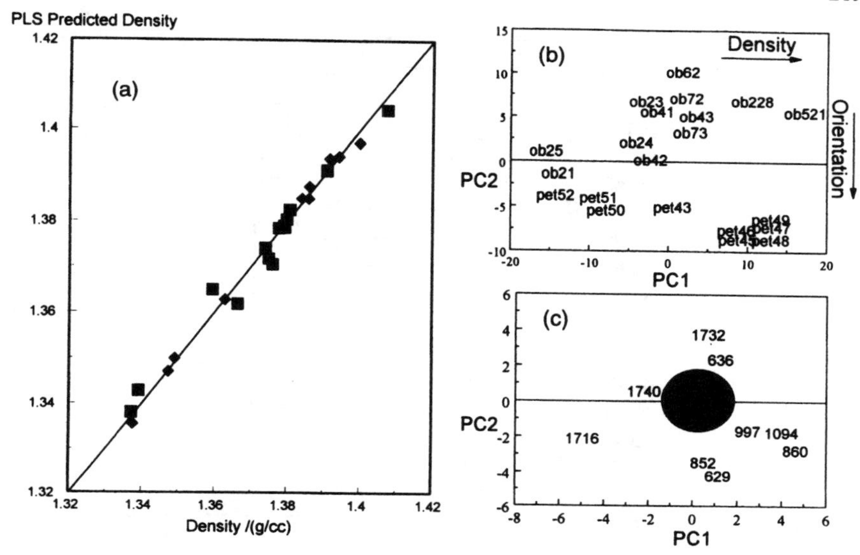

Figure 16 *PLS modelling of PET density from normalised, mean-centered FT-Raman spectra* (1800–1650 Δcm^{-1} *and* 1500–600 cm^{-1}): (a) *Cross-validated prediction plot for PLS calibration (SEP* = 0.0024 g cm^{-3}) *using four factors and a combined PET sample set of isotropic annealed chips (■, ob set) and uniaxially and biaxially oriented films (◆, pet set);* (b) *Scores plot, showing that the first factor (PC1) essentially models variation in density, while factor 2 (PC2) differentiates samples primarily on the basis of orientation. Factors 1 and 2 accounted for 47% and 22% respectively of the variance in this model;* (c) *Loadings plot revealing which bands most strongly contribute to the factors, e.g.* νC=O to density and ~630 Δcm^{-1} *to orientation*
(Reproduced from ref. 25, *Polymer*, **35**, 15, Everall N., Tayler P., Chalmers J.M., MacKerron D., Ferwerda R., and van der Maas J.H., *Study of density and orientation in poly(ethylene terephthalate) using Fourier transform Raman spectroscopy and multivariate data analysis*, 3184–3192, © 1994, with kind permission from Elsevier Science Ltd., The Boulevard, Langford Lane, Kidlington OX5 1GB, UK)

6 Spectral Enhancement and Band Resolution Techniques

In the previous section the use of first-derivative spectra to aid quantitation was alluded to in two of the examples referenced.[100,101] Several methods are available for enhancing the apparent contrast within a spectrum that cannot be satisfactorily improved further by experimental conditions. Bands separated by less than their half-widths may be severely overlapped, or band recognition may be seriously impaired by poor signal-to-noise spectral characteristics. This section will discuss various methods which have been used to enhance spectral contrast or resolve higher level information.

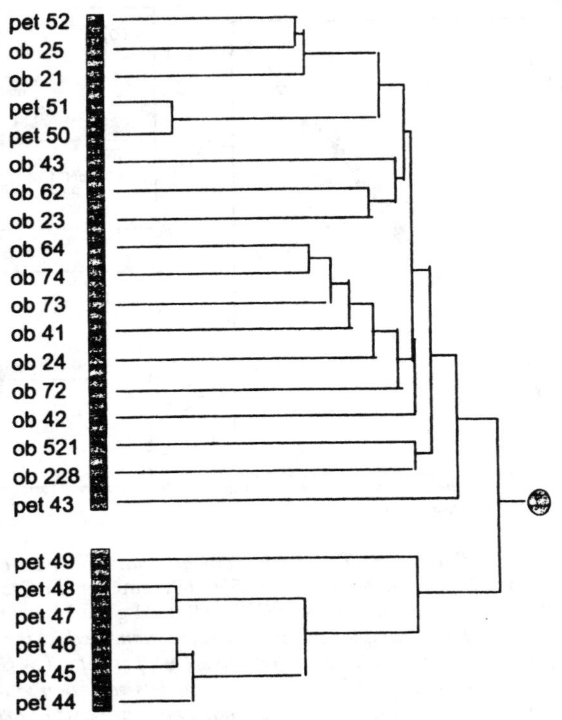

Figure 17 *Hierarchical Cluster Analysis (HCA) of the sample data set used to generate Figure 18. HCA classifies spectra on the basis of their similarity, for example, the 'grouping' (pet52–pet50) contains the low density materials, while the cluster (pet 49–pet 44) contains dense oriented samples, (see ref. 25 for a more detailed description)*
(Reproduced from ref. 25, *Polymer*, **35**, 15, Everall N., Tayler P., Chalmers J.M., MacKerron D., Ferwerda R., and van der Maas J.H., *Study of density and orientation in poly(ethylene terephthalate) using Fourier transform Raman spectroscopy and multivariate data analysis*, 3184–3192, © 1994, with kind permission from Elsevier Science Ltd., The Boulevard, Langford Lane, Kidlington OX5 1GB, UK)

Derivative Spectroscopy

Mathematical differentiation of a spectrum is a sometimes useful data presentation procedure, which has found favour in certain circumstances, since first-, second- and higher-order derivative spectra, by emphasising changes of slope, can yield spectra which have apparently higher band resolution than their zeroth-order parent spectrum. Even-order differentiated spectra do, however, yield peaks with negative side lobes, see Figure 18. The effects of various treatments for calculating derivative spectra and their effects on different lines/band-shapes with differing background noise

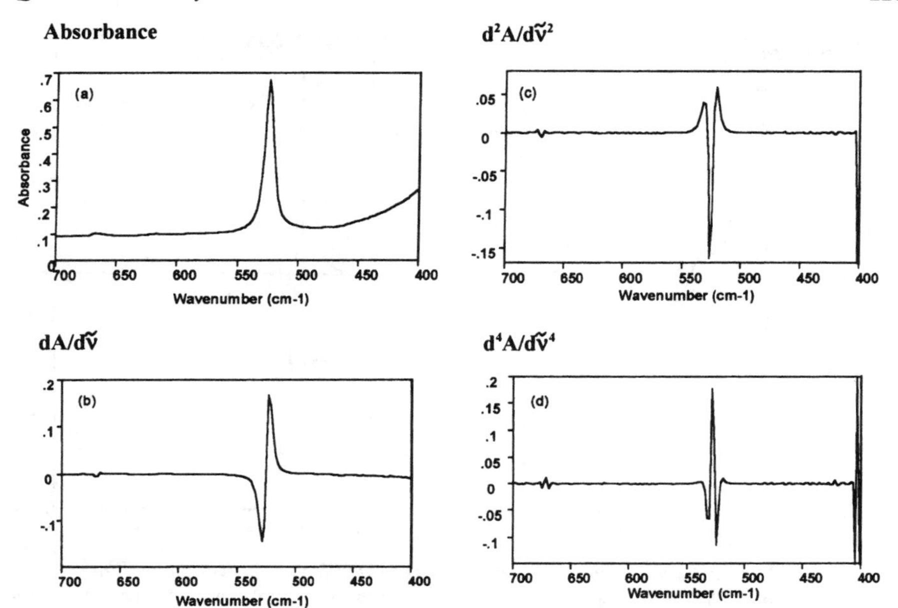

Figure 18 (a) *Zero-order (absorbance);* (b) *first-order* ($dA/d\tilde{v}$)*;* (c) *second-order* ($d^2A/d\tilde{v}^2$)*; and* (d) *fourth-order* ($d^4A/d\tilde{v}^4$) *derivative spectra of the cyclohexane spectrum shown in Figure 1(b). Note the degradation in signal-to-noise level with increasing differentiation, cf. the regions* 500–450 cm^{-1}. *The increase in 'signal' near* 400 cm^{-1} *is caused by the discontinuity (cut-off) of the data array*

characteristics have been studied on both synthetic and experimental infrared spectra.[107–110]

Derivative spectroscopy has been traditionally used for contrast enhancement of UV/visible spectra and has become a popular pre-processing routine used with the quantitative analysis of NIR absorbance spectra,[60] since in these regions the bands are generally much broader than those encountered in mid- and far-infrared and Raman spectra. Also, UV/visible and NIR spectra frequently have a very high signal-to-noise ratio (S/N), which can be important, particularly for univariate analysis treatments, since the S/N is degraded with each successive derivative calculation. However, very high S/N is not always a prerequisite for successful application of derivative spectroscopy.

Compositional concentrations calculated from differentiated mid-infrared overlapping absorption bands have compared favourably with other univariate data analyses approaches. For example, in the determination of the methyl

[107] Maddams W.F. and Mead W.L., *Spectrochim. Acta*, 1982, **38A**, 4, 437.
[108] Hawkes S., Maddams W.F., Mead W.L., and Southon M.J., *Spectrochim. Acta*, 1982, **38A**, 4, 445.
[109] Maddams W.F. and Southon M.J., *Spectrochim. Acta*, 1982, **38A**, 4, 459.
[110] Wiliams A.D. and Spragg R.A., *Perkin-Elmer Infrared Bulletin 96*, Perkin-Elmer, Beaconsfield, UK, 1982.

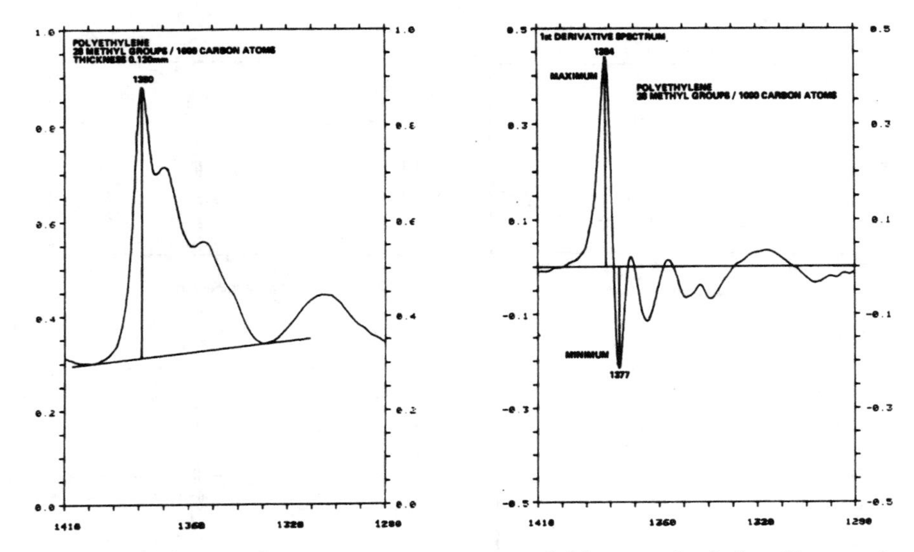

Figure 19 Left, *infrared absorbance spectrum recorded from a polyethylene film containing 28 methyl groups per 1000 C atoms;* Right, *first-derivative of the polyethylene absorbance spectrum*
(Reproduced from ref. 111, by kind permission of Perkin-Elmer Limited)

group concentration in poly(ethylenes) using the 1378 cm^{-1} band,[111,112] see Figure 19, and measuring the concentration of a phenolic antioxidant in mineral oil via its band at 3643 cm^{-1}.[2] However, nowadays these analyses are most likely to be undertaken following computer subtraction of the absorptions of the interfering (overlapping) species. With the advent of computerised vibrational spectroscopy and the arrival of other data processing and enhancement techniques, uses of derivative spectroscopy in the quantitative analysis of mid-infrared data declined. However, derivative spectra can be particularly useful when no suitable or reproducible background/baseline can be found. In pre-processing of NIR analyser data it is widely employed to minimise the effects of sample scatter and other causes of differing background slopes and levels prior to multivariate analyses treatments.[60] This attribute is now being increasingly used with mid-infrared and Raman multivariate data analysis sets.

In a method developed for the density mapping of poly(ethylene terephthalate) films using a fibre-coupled Raman microprobe,[106] prior to developing the PLS calibration the spectra were pre-processed by a second-order derivative

[111] Chalmers J.M., Willis H.A., Cowell G.M., and Spragg R.A., *Perkin-Elmer Infrared Bulletin 99*, Perkin-Elmer, Beaconsfield, UK, 1982.

[112] Willis H.A., Chalmers J.M., Mackenzie M.W., and Barnes D.J., in *Computerised Quantitative Infrared Analysis*, ASTM Special Technical Publication STP 934, ASTM, Philadelphia, USA, 1987, p. 58.

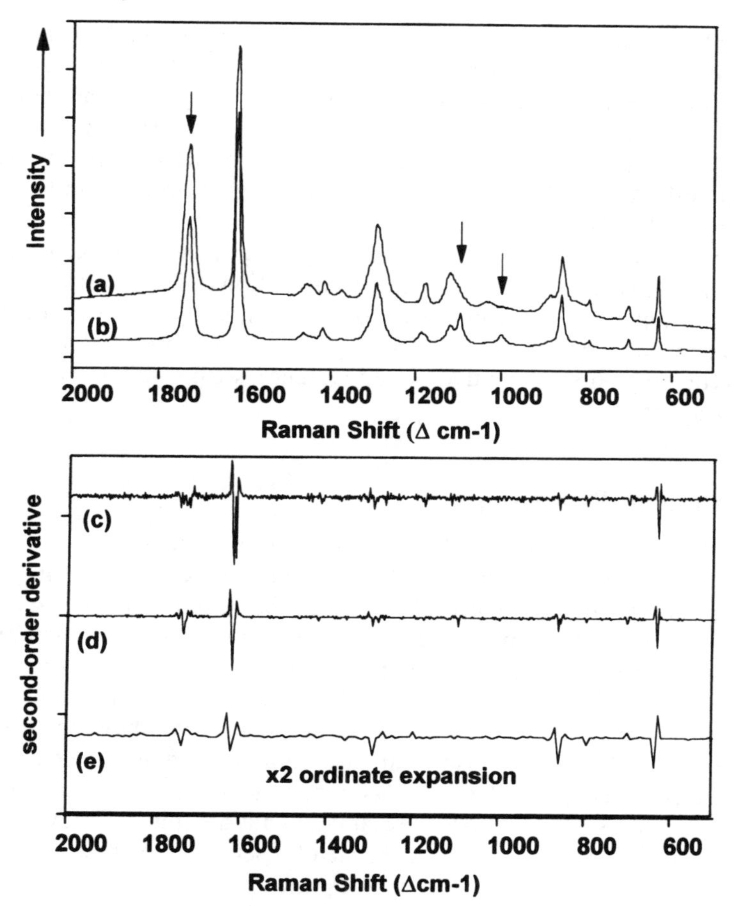

Figure 20 *Comparison of PET film fibre-coupled Raman microprobe spectra and some preprocessing steps used prior to developing a PLS model calibration for polymer density:* (a) *amorphous film;* (b) *crystalline film;* (c) *second derivative spectrum of* (a); (d) *second derivative spectrum of* (b); (e) × *2 ordinate scale-expanded, 8-point block-averaged spectrum of* (d). *Note flat baselines of* (c), (d), *and* (e). *The arrows in* (a) *and* (b) *indicate spectral features that are most sensitive to crystallinity changes*
(Figure produced from data used in ref. 106 with the corresponding author's permission)

algorithm to remove the very broad weak signals arising from sample fluorescence, see Figure 20. This method was preferred to subtracting a baseline curve since it was readily automated and entirely non-subjective. Optimal calibration models (lowest SEP) were, however, realised only when these data were then block-averaged (8-point) or smoothed, which degraded the spectral resolution but improved the S/N, suggesting that it may be possible to maintain precise, robust predictive models with much poorer spectral resolution data than those typically recorded for visual inspection (*i.e.* 2–8 cm^{-1})![106] In the infrared study of PLS calibration models for aromatic copolymers of aryl ether sulphone/aryl ether ether sulphone based on transmission, diffuse reflectance, and ATR spectra, the best (minimum error) regression was reported from normalised second-order derivative, 9-point quadratic polynomial smoothed, multiplicative scatter corrected diffuse reflectance spectra.[103]

First derivatives of specular reflectance spectra have also been used as alternatives to Kramers–Kronig generated absorption index spectra generated from specular reflectance measurements for the quantitative study of molecular orientation in polymer films and mouldings.[113–115] (NB pure specular reflectance spectra have a first-derivative-like appearance, see pp. 160–162 and Figure 21(a) of Chapter 4.)

Fourier Domain Processing

Once a vibrational spectrum has been recorded, processing of the spectrum in the Fourier (or interferogram) domain has many potential advantages for improving spectral contrast and probing the structure of band contours.[116–122] For example, the inverse Fourier transform may be convoluted with a parabolic function to generate, when re-Fourier transformed, a derivative spectrum.[116,118–121] Potential improvements of fitting spectral data to unknown mixtures via FA of Fourier rather than spectral domain data have already been noted.[78] Fourier domain processing can also be used to reduce the effects of a large overlapping background, such as a fluorescence emission overlaid on a set of Raman scatter peaks,[15,123] since the Raman bands are much narrower. In the Fourier domain, the first term may be rejected (or set to some other value) since it describes a signal with zero frequency, which represents the average DC (direct current) level, to which fluorescence is usually the main contributor.[15]

Another method of computationally resolving (sharpening) features within overlapping band contours, which cannot be separated instrumentally because of their intrinsic linewidth, is to use the process of Fourier self-deconvolution

[113] Everall N.J., Chalmers J.M., Local A., and Allen S., *Vib. Spectrosc.*, 1996, **10**, 253.
[114] Bensaad S., Jasse B., and Noel C., *Polymer*, 1993, **34**, 8, 1602.
[115] Jansen J.A.J., Paridaans F.N., and Heynderickx I.E.J., *Polymer*, 1994, **35**, 14, 2970.
[116] Cameron D.G. and Moffat D.J., *J. Test. Eval.*, 1984, **12**, 78.
[117] (a) Kauppinen J.K., Moffat D.J., Mantsch H.H., and Cameron D.G., *Appl. Spectrosc.*, 1981, **35**, 3, 271; (b) Kauppinen J.K., Moffat D.J., Cameron D.G., and Mantsch H.H., *Appl. Optics*, 1981, **20**, 10, 1866.
[118] Gilbert A.S., in *Computing Applications in Molecular Spectroscopy*, ed. George W.O. and Steele D., Royal Society of Chemistry, Cambridge, UK, 1995, ch. 2.
[119] Griffiths P.R., in *Laboratory Methods in Vibrational Spectroscopy*, 3rd Edn, ed. Willis H.A., van der Maas J.H. and Miller R.J.G., J. Wiley, Chichester, UK, 1987, ch. 6.
[120] Griffiths P.R. and de Haseth J.A., *Fourier Transform Infrared Spectrometry*, J. Wiley, New York, USA, 1986.
[121] Kauppinen J.K., Moffat D.J., Mantsch H.H., and Cameron D.G., *Anal. Chem.*, 1981, **53**, 1454.
[122] Griffiths P.R. and Pariente G.L., *TrAC*, 1986, **5**, 8, 209.
[123] Mann C.K. and Vickers T.J., *Appl. Spectrosc.*, 1987, **41**, 3, 427.

(FSD).[116–120,122,124–130] This is usually applied only to the spectral region encompassing the overlapping bands of interest, for which the first step is to take its Fourier transform. This inverse FT will be the summation of the cosines representing each band in the Fourier (or displacement) domain, which has an exponential decay profile determined by the line width and shape of the original overlapping bands; the broader the line, the more rapidly its Fourier transform cosine will decay. If this is then multiplied point-by-point (convoluted) by an exponential rise function, then the forward Fourier transform of this convoluted data will yield a band contour for which the bands have been narrowed (resolution enhanced). However, in practice there are several limitations imposed on the extent to which this process can be carried out.

Row A of Figure 21[117] shows the Fourier transform of a Lorentzian line, which is, of course, truncated; the maximum displacement (L) of this inverse FT is determined by the spectral resolution used to record the original spectrum, and which *cannot* be exceeded through application of FSD.[117] If, in the process of FSD, the exponential decay profile of the Fourier transform were to be convoluted with a matched exponential rise profile, the inverse Fourier transform of these data will yield a much narrower (sinc function) lineshape at the same position, but having significant side-lobes, see row B of Figure 21. These side-lobes may be reduced by additional apodisation,[117,122,126,129–131] at the expense of band width. Rows C and D in Figure 21 show other apodisation functions. In addition to line shape and width, another important property is the S/N of the data.[117,118,122,126,130] Noise will be distributed essentially evenly throughout both the spectral and Fourier domain data, and so the maximum useful improvement in band enhancement will be limited to convoluting only those data in the Fourier domain for which the exponentially decaying cosine intensity is greater than the noise level. Another key consideration to any practical use of FSD to optimising the enhancement of the spectral contrast for an overlapping band contour, is that all bands within the profile are unlikely to have the same shape and width, so deconvolution is usually set such that overdeconvolution (side-lobes) are avoided for the narrowest band.[117]

FSD is a subjective tool, for which some prior knowledge of the underlying components (especially their relative bandwidths) giving rise to an overlapping band contour is very desirable in real-world analytical applications. In application-customised FSD software, particularly FSD routines embedded within proprietary software for data instrument control and manipulation, the default

[124] McClure W.F., *Spectroscopy World*, 1991, **3**, 1, 28.
[125] Gilbert A.S., in *Analytical Applications of Spectroscopy II*, ed. Davies A.M.C. and Creaser C.S., Royal Society of Chemistry, Cambridge, UK, 1991, p. 275.
[126] James D.I., Maddams W.F., and Tooke P.B., *Appl. Spectrosc.*, 1987, **41**, 8, 1362.
[127] Compton D.A.C., Young J.R., Kollar R.G., Mooney J.R., and Grasselli J.G., in *Computerised Quantitative Infrared Analysis*, ed. McClure G.L., ASTM Special Technical Publication 934, ASTM, Philadelphia, USA, 1987, p. 36.
[128] Parker S.F., in *Computing Applications in Molecular Spectroscopy*, ed. George W.O. and Steele D, Royal Society of Chemistry, Cambridge, UK, 1995, p. 181.
[129] Tooke P.B., *TrAC*, 1988, **7**, 4, 130.
[130] Smeller L., Goossens K., and Heremans K., *Appl. Spectrosc.*, 1995, **49**, 10, 1538.
[131] Yang W-J. and Griffiths P.R., *Computer Enhanced Spectroscopy*, 1983, **1**, 3, 157.

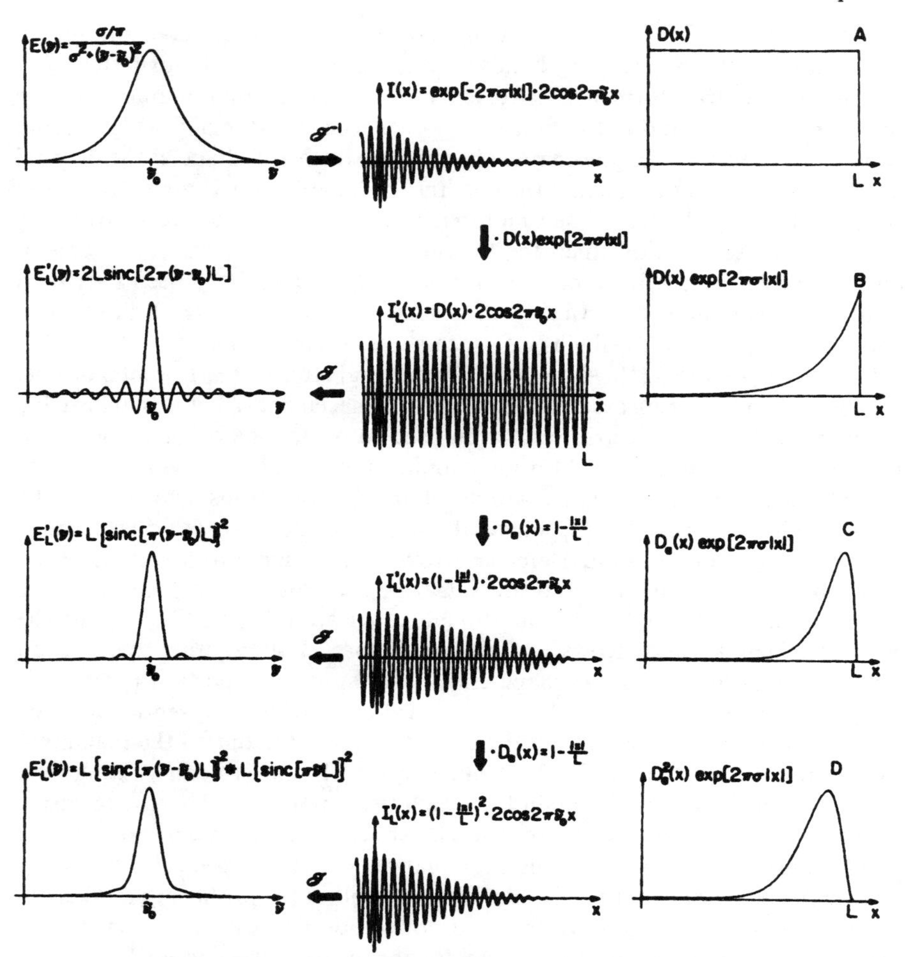

Figure 21 *Illustration of the process of Fourier self-deconvolution (FSD) on a Lorentzian line. The left-hand column shows the initial Lorentzian line (top), below which, in rows B, C, and D, are the lineshapes produced by the Fourier transform of the interferograms in the middle column, which resulted from applying the apodisation functions shown in the right-hand column on the interferogram of the Lorentzian line (middle, top row)*
(Figure reproduced from ref. 117a, by permission of the Society for Applied Spectroscopy; © 1981)

lineshape function is usually set as Lorentzian and the operator is required typically to input and iterate empirically between two values until satisfied – *i.e.* an estimated optimum half-bandwidth (or bandwidth) of the peaks constituting the overlapping band profile and an enhancement factor.[129] There may also be an option to choose the type of apodisation function implemented, triangular-sinc,[117,126,128,130] triangular-squared,[130,132] Gaussian,[133,134] and Bessel[135] being among those commonly used as default or standard, the balance being between maximum resolution enhancement and optimum suppression of the noise,[117,128–130] (see also Figure 12 of Chapter 3).

FSD has been used in the investigation and analysis of both infrared and Raman spectra, both as a peak-picking tool to aid band assignments and as a pre-processing procedure to improve band separation for quantitation. The νCH region of the diffuse reflectance spectra of bituminous coal have been deconvolved to yield more detailed information on the structure of whole powdered coals,[119,136] see Figure 22. Both the infrared[134,137] and Raman[133,138–139] Amide I band profiles of proteins have been Fourier deconvolved to peak-pick band positions and assign these to various conformations, which have then been curve-fitted to estimate their relative concentrations semi-quantitatively.[133,134] In these studies, more detailed results were obtained when a Gaussian function was used, rather than the more usual Lorentzian approximation.[133,134] The FSD enhancement routine has also been applied to several polymer systems. For poly(vinyl chloride)s, PVCs, of differing syndiotacticities, it has been employed in the C—Cl stretching region of the infrared spectrum[135] to aid assignment of bands to different conformational sequences. Potential Raman evaluations have also been undertaken.[132,140] FSD has been applied to several regions of the infrared spectra of ethylene/acrylic and methacrylic acid copolymers.[141] Quantitative ethylene sequencing and composition has been achieved in propylene/ethylene copolymers by FSD band separation enhancement of FTIR spectra followed by band area measurements.[142,143] The νC=O band profiles in the FTIR spectra of oxidised polyethylenes have been resolution enhanced through application of FSD in order to resolve the absorption envelope into its components,[128,144] revealing a new weak feature at 1697 cm^{-1}, the intensity of which was very dependent on polyethylene type (density). In another FTIR polymer study, FSD has been applied to the spectra of a series of linear poly(dimethyl siloxane) oligomers, in order to quantitatively relate

[132] Bowley H.J., Collin S.M.H., Gerrard D.L., James D.I., Maddams W.F., Tooke P.B., and Wyatt I.D., *Appl. Spectrosc.*, 1985, **39**, 6, 1004.
[133] Susi H. and Byler D.M., *Appl. Spectrosc.*, 1988, **42**, 5, 819.
[134] Byler D.M. and Susi H., *Biopolymers*, 1986, **25**, 469.
[135] Compton D.A.C. and Maddams W.F., *Appl. Spectrosc.*, 1986, **40**, 2, 239.
[136] Wang S-H. and Griffiths P.R., *Fuel*, 1985, **64**, 2, 229.
[137] Yang W-J, Griffiths P.R., Byler D.M., and Susi H., *Appl. Spectrosc.*, 1985, **39**, 2, 282.
[138] Incardona N.L., Prescott B., Sargent D., Lamba O.P., and Thomas G.J., *Biochemistry*, 1987, **26**, 1532.
[139] Thomas G.J., *Spectrochim. Acta*, 1985, **41A**, 1/2, 217.
[140] King J. and Bower D.I., *Proc. VII Int. Conf. Raman Spectrosc.*, 1980, p. 242.
[141] Harthcock M.A., *Proc. Int. Conf. Four. Comp. Infrared Spectrosc.*, ed. Grasselli J.G. and Cameron D.G., SPIE, 1985, **553**, 245.
[142] Chalmers J.M., Bunn A., Willis H.A., Thorne C., and Spragg R., *Proc. 6th Int. Conf. Four. Trans. Spectrosc.*, August 24–28, Vienna, 1987, *Mikrochim. Acta*, I, 1988, 287.
[143] Willis H.A., Chalmers J.M., Bunn A., Thorne C., and Spragg R., in *Analytical Applications of Spectroscopy*, ed. Creaser C.S. and Davies A.M.C., Royal Society of Chemistry, London, UK, 1988, p. 188.
[144] Maddams W.F. and Parker S.F., *J. Polym. Sci.: B: Polym. Phys*, 1989, **27**, 1691.

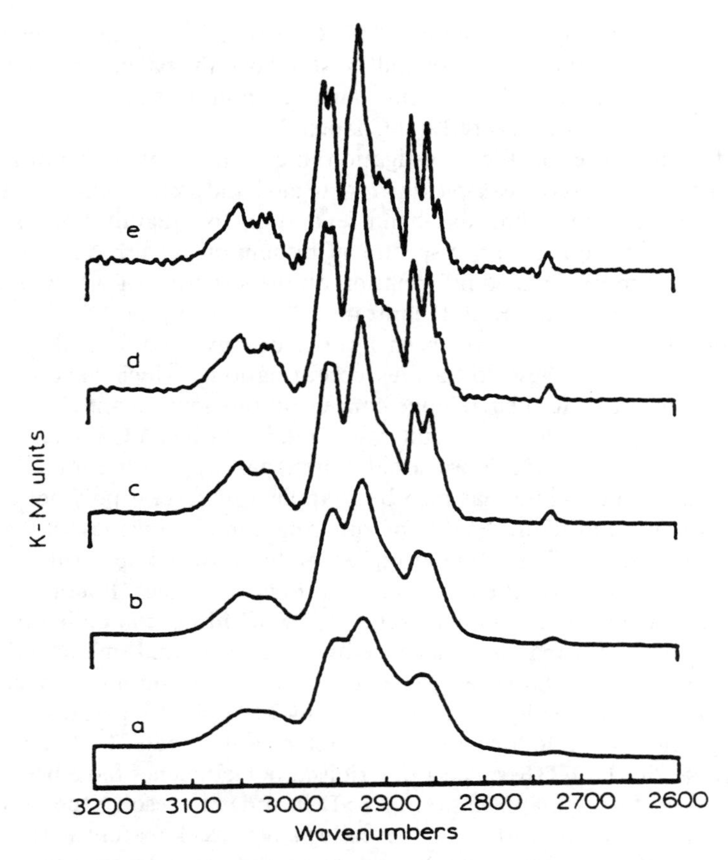

Figure 22 *Series of progressively deconvolved (FSD) spectra of medium-volatile bituminous coals:* (a) *original spectrum;* (b)–(e) *calculated by multiplying the FT of* (a) *with different weighting functions and then computing the reverse FT, see ref. 136 for more details*
(Reproduced from ref. 136, *Fuel*, **64**, S-H Wang and P.R. Griffiths, *Resolution enhancement of diffuse reflectance i.r. spectra of coals by Fourier self deconvolution. 1. C—H stretching and bending modes*, pp 229–236, © 1985, with kind permission from Elsevier Science Ltd., The Boulevard, Langford Lane, Kidlington OX5 1GB, UK)

spectral features to chain lengths of oligomers up to a molecular weight of 15 000,[145] see Figure 23.

Curve-fitting

On pp. 199–200 we briefly discussed a spectrum-matching procedure (sometimes referred to as curve-fitting) in which a least-squares residual optimisation

[145] Lipp E.D., *Appl. Spectrosc.*, 1986, **40**, 7, 1009.

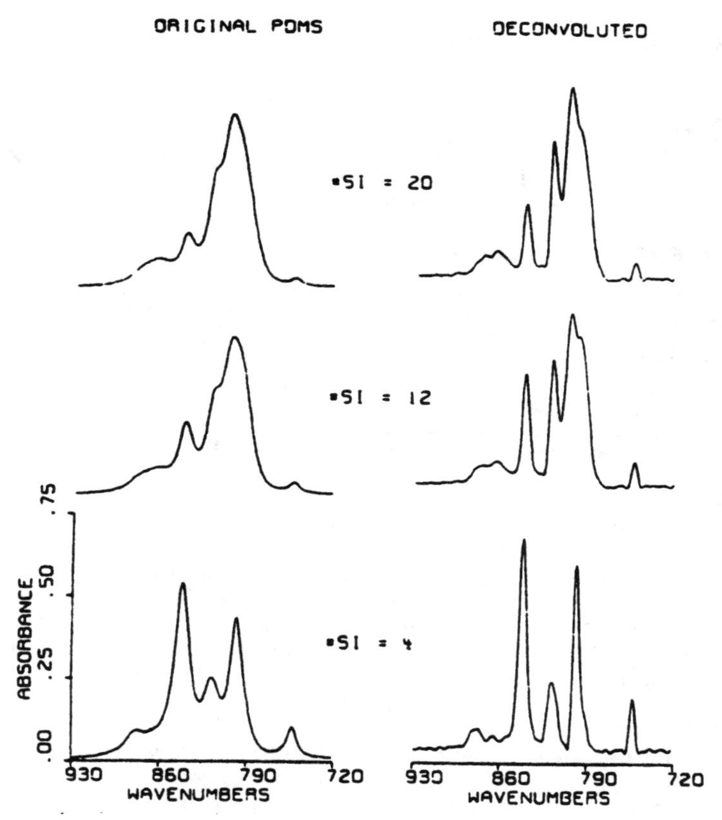

Figure 23 *The original absorbance and Fourier self-deconvoluted spectra of the three polydimethyl siloxane (PDMS) oligomers. #SI equals the number of Si atoms (x + 2) in the structure* $(CH_3)_3Si—(OSi(CH)_2)_x—OSi(CH_3)_3$. *A plot of the* $Si(CH)_2/Si(CH)_3$ (*ca.* 830–775 cm^{-1}/*ca.* 850–830 cm^{-1}) *peak areas vs. the number of Si atoms per oligomer gave a straight line, with a correlation coefficient of 0.99976*

(Figure reproduced from ref. 145, Lipp E.D., *Application of Fourier Self-Deconvolution to the FT-IR Spectra of Polydimethylsiloxane Oligomers for Determining Chain Length, Appl. Spectrosc.*, 1986, **40**, 7, 1009, by permission of the Society for Applied Spectroscopy; © 1986)

procedure may be used for the summation of the full path-length-normalised individual spectra or spectral regions of the components of a mixture, as an aid to the quantification of the mixture spectrum,[2,31,40–45] provided that essentially there are no molecular interactions of the components when present in the mixture. An example is given in Figure 24.

Curve-fitting (as opposed to component-matching a spectrum) *per se* does not attempt to improve spectral contrast by resolution enhancement, but rather by finding the best fit to an overlapping band contour of a prescribed set of

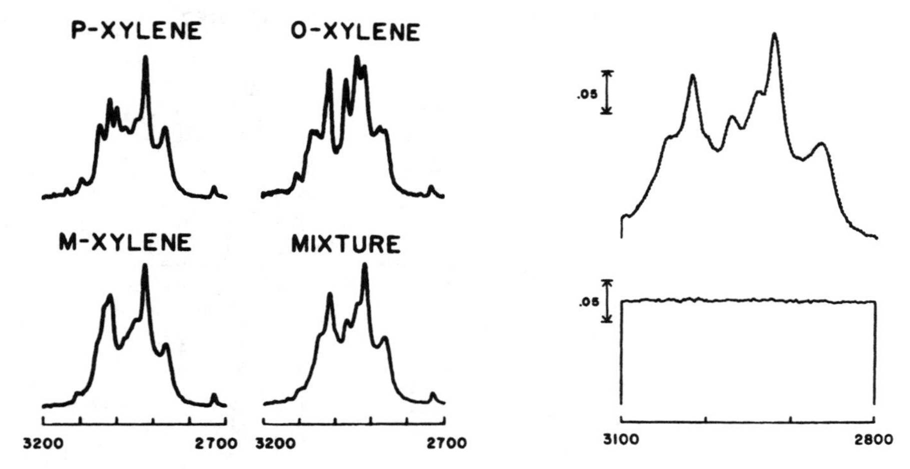

Figure 24 *Curve-fitting: least-squares spectra matching of component spectra to a mixture spectrum.* Left *and* middle columns, *FTIR* (2 cm^{-1}) *absorbance spectra of equal volumes of three pure xylene isomers and a mixture consisting of approximately equal volumes of the three isomers.* Right column: top, *overlaid spectra of experimental xylene mixture (solid line) and curve-fitted* (3100–2800 cm^{-1}) *xylene mixture spectrum (dotted line) generated by the linear addition of the three pure component spectra using least-squares coefficients;* bottom, *random noise error residual remaining after subtracting the two spectra above*
(Figure reproduced from ref. 40, Antoon M.K., Koenig J.H., and Koenig J.L., *Least-squares Curve-fitting of Fourier Transform Infrared Spectra with Applications to Polymer Systems, Appl. Spectrosc.*, 1977, **31**, 6, 518, by permission of the Society for Applied Spectroscopy; © 1977)

individual peaks.[119,146–150] The optimised fit is based on an iterative minimisation of the least-squares residuals between the band contour and a summation of the individual peak profiles. Pre-selection of the number of bands, their width, shape (*e.g.* Cauchy/Lorentzian, Gaussian, or Gaussian perturbed Lorentzian), and position may often be largely subjective, and therefore requires significant prior knowledge of the origins (chemical species) and parameters of the component bands, since, clearly, the solutions are not necessarily unique. Iterative software procedures are available which allow for constrained shifts/changes in the initial parameter set, while seeking optimisation. Figure 25 is an example of a curve-fitting optimisation applied to a Raman study of acrylonitrile partition between aqueous and polymer phases in a polybutadiene rubber latex.[151] A curve (peak) fit routine was used to resolve the overlapping peaks

[146] Pitha J. and Jones R.N., *Can. J. Chem.*, 1966, **44**, 3031.
[147] Fraser R.B.D. and Suzuki E., *Anal. Chem.*, 1969, **41**, 1, 37.
[148] Young R.P. and Jones R.N., *Chem. Rev.*, 1971, **71**, 2, 219.
[149] Schwartz L.M., *Anal. Chem.*, 1971, **43**, 10, 1336.
[150] Gold H.S., Rechsteiner C.E., and Buck R.P., *Anal. Chem.*, 1976, **48**, 11, 1540.
[151] Hergerth W-D. and Codella P.J., *Appl. Spectrosc.*, 1994, **48**, 7, 900.

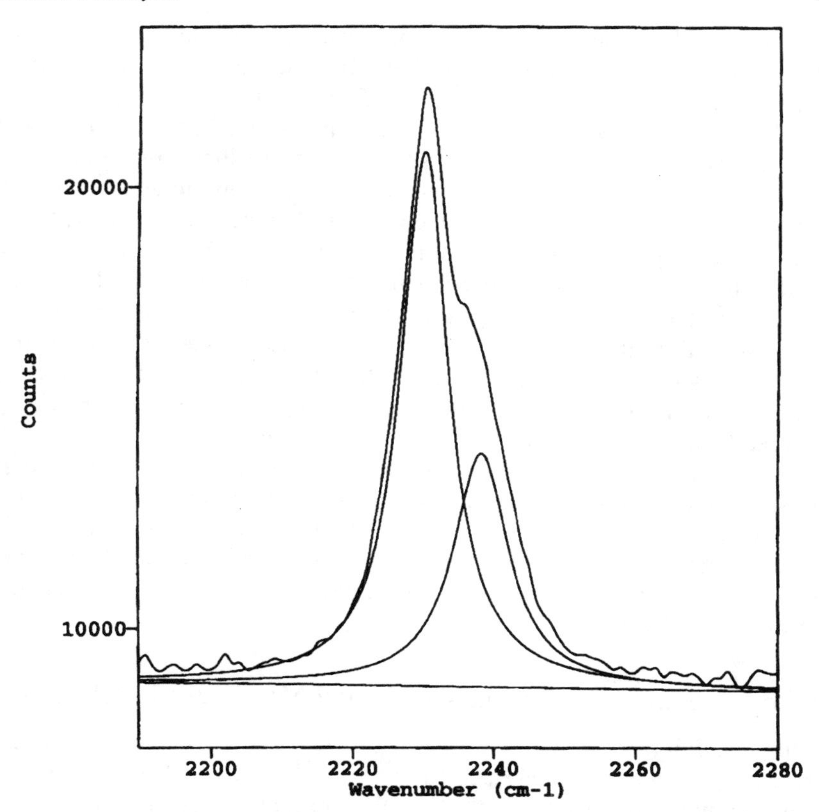

Figure 25 *Raman spectrum of the —C≡N stretching region of a polybutadiene rubber latex saturated with acrylonitrile (ACN), curve-fitted to separate the contributions due to ACN at* 2230 Δcm^{-1} *within the polymer particles from that at* 2237 Δcm^{-1} *dissolved in the continuous water phase*
(Reproduced from ref. 151, Hergeth W.-D. and Codella P.J., *Monomers in Polymer Dispersions. Part IV: Partition of Acrylonitrile in Rubber Latex as Studied by Raman Spectroscopy, Appl. Spectrosc.*, 1994, **48**, 7, 900, by kind permission of the Society for Applied Spectroscopy; © 1994)

corresponding to acrylonitrile monomer in the two phases. The quantitative analysis was based on the solvent-induced wavenumber shift of the νC≡N band, which was used to yield the partition values. As stated in the previous section, FSD using a Gaussian apodisation function of protein spectra followed by curve-fitting the deconvolved spectra by Gaussian line-shape bands has been used to analyse the Amide I band envelope in both FTIR[134] and Raman[133] spectra, in order to estimate the concentrations of secondary structures. In a study of Raman band profiles of aqueous indium(III) chloride solutions, the spectra were qualitatively assigned to four components using both PCA and Gaussian–Lorentzian band-fitting.[152] Further examples of the application of

[152] Jarv T., Bulmer T.J., and Irish D.E., *J. Phys. Chem.*, 1977, **81**, 7, 649.

PCA, to establish the number of components in a complex band profile, accompanied by curve the band envelope, include both infrared and Raman studies of equilibria in solution[153] and Raman studies of aqueous Zn(II) chloride solutions.[154] In a similar study, FA and a matrix-rotated FA method have been used to isolate component spectra from the low-wavenumber region of the Raman spectra of zinc bromide complexes in dimethyl sulphoxide solution.[155] The methyl (1050–850 cm^{-1}) and methylene (850–650 cm^{-1}) rocking mode region of propylene/ethylene copolymers have been curve-fitted to elucidate contiguous monomer sequence distributions from overlapped band contours.[156] The wavenumber positions of peaks, shoulders, and inflection points were first determined from first- and second-derivative spectra; a Lorentzian band-shape function was employed in the curve-fitting procedure; the contributions of structural units were calculated from the band areas measured from the computer-resolved band parameters.

Curve-fitting procedures have also been applied extensively to the detailed analysis of polarised infrared spectra recorded from uniaxially drawn PET film subjected to constant strain,[157] and to elucidate FTIR spectra obtained from coals varying in rank from peat to anthracite, to gain additional information on coal structures and the primary structural changes that take place during the process of coalification.[158]

Maximum Entropy Method (MEM) and Maximum Likelihood

Worthwhile differentiation and FSD are both very limited by the S/N (signal-to-noise ratio) of the recorded spectrum;[118,128,159,160] also, they both produce negative values in the result.[118,128,159,160] Conceptually, they may be thought of as 'backwards' approaches in that they attempt to line-narrow spectral features by mathematical filtering processes. Maximum Entropy Methods (MEMs)[118,128,159,161–163] and Maximum Likelihood[164,165] techniques may be considered essentially as 'forwards' or 'restoration' approaches. They are 'probabilistic' methods, which use Bayesian statistics, to generate ('reconstruct/restore') the 'most-probable/most-likely' solution. They are signal (data) processing methods which can lead to substantial (astonishing!) improvements

153 Shurvell H.F. and Bulmer J.T., in *Vibrational Spectra and Structure*, Vol. 6, ed. Durig J.R., Elsevier Scientific, Amsterdam, The Netherlands, 1977, ch. 2.
154 Shurvell H.F. and Dunham A., *Can. J. Spectrosc.*, 1978, **23**, 5, 160.
155 van Heuman J., Ozeki T., and Irish D.E., *Can. J. Chem.*, 1989, **67**, 2030.
156 Drushel H.V., Ellerbe J.S., Cox R.C., and Lane L.H., *Anal. Chem.*, 1968, **40**, 2, 370.
157 Hutchinson I.J., Ward I.M., Willis H.A., and Zichy V., *Polymer*, 1980, **21**, 55.
158 Munoz E., Moliner R., and Ibarra J.V., *Coal Sci. Technol. (Coal Sci. Vol 1)*, 1995, **24**, 115.
159 Ni F. and Scheraga H.A., *J. Raman Spectrosc.*, 1985, **16**, 5, 337.
160 Kauppinen J.K., Moffat D.J., Hollberg M.R., and Mantsch H.H., *Appl. Spectrosc.*, 1991, **45**, 3, 411.
161 *Maximum Entropy in Action*, ed. Buck B. and Macaulay V.A., Clarendon, Oxford, UK, 1991.
162 *MaxEnt. A Practical Guide for Spectroscopy and Chromatography*, MaxEnt Solutions, Isleham, Ely, UK, 1992.
163 Wright K.M., *Spectroscopy World*, 1990, **2**, 6, 37.
164 DeNoyer L.K. and Dodd J.G., *Am. Lab.*, 1990, **22**, 5, 21.
165 DeNoyer L.K. and Dodd J.G., *Am. Lab.*, 1991, **23**, 12, 24D.

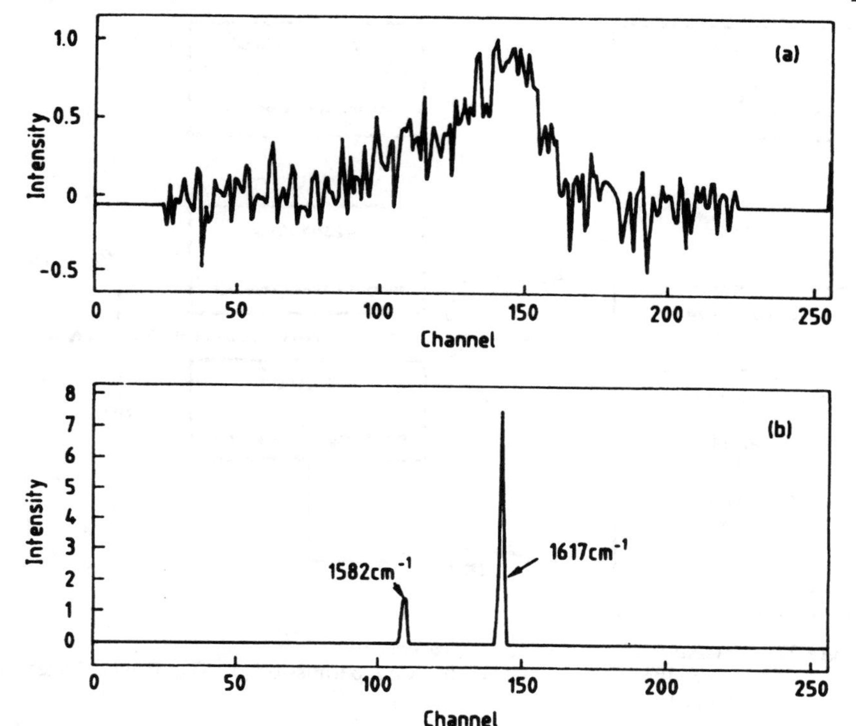

Figure 26 (a) *Raman spectrum of a graphite/$FeCl_3$ intercalate;* (b) *MEM reconstruction using a Lorentzian blurring function with FWHM of* 25 cm^{-1}
(Reproduced from ref. 166, *Maximum Entropy in Action*, ed. B. Buck and V.A. Macaulay, by kind permission of Oxford University Press)

in S/N and spectral resolution/contrast in vibrational spectra, see Figure 26, but they are *not* substitutes for good experimental practice![166] For instance, the 'preferred solution' of implementing a line-narrowing algorithm will not necessarily be the 'correct/real/true' spectrum, but will depend on the choice of the deconvolution function.[159,160,166,167] They seek to provide the solution which maximises the information content (entropy/likelihood) of the output spectrum, for which there is evidence in the raw spectrum. Although they make no assumptions about the number of components, like FSD, they are usually applied to spectral regions rather than full spectra and are best suited to situations for which all the peaks have essentially the same shape and width, as is frequently the case with NMR and MS data.[128,161] The FSD and MEM approaches have been compared and contrasted,[159,160] and new methodologies proposed which take advantage of the efficacies of both.[159,160] Both are more demanding of computational time than FSD.

[166] Davies S., Packer K.J., Baruya A., and Grant A.I., in *Maximum Entropy in Action*, ed. Buck B. and Macaulay V.A., Clarendon, Oxford, UK, 1991, ch. 4.
[167] Kauppinen J.K. and Saario E.J., *Appl. Spectrosc.*, 1993, **47**, 8, 1123.

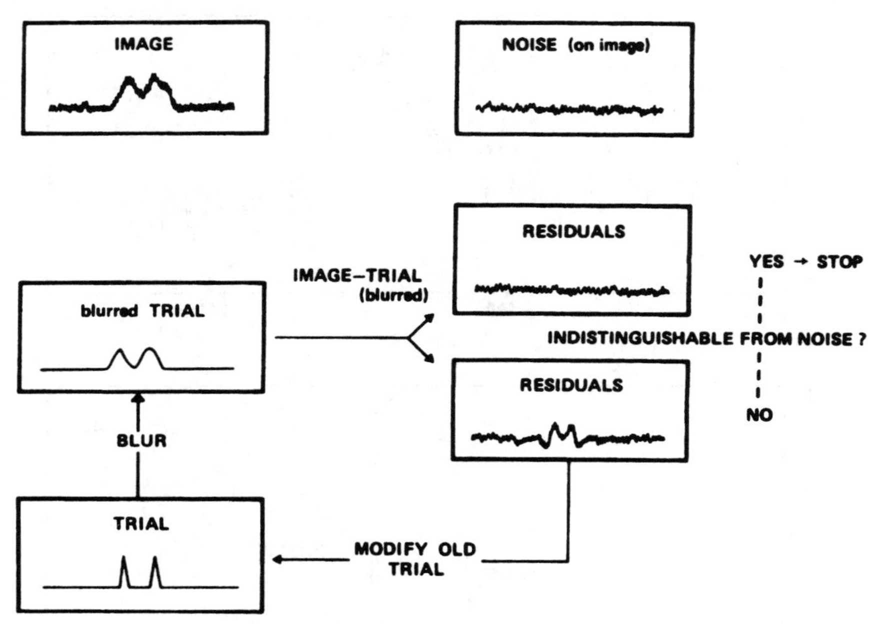

Figure 27 *Schematic of iterative MEM procedure*
(Reproduced from ref. 118 by kind permission of The Royal Society of Chemistry)

The principle underlying linear and non-linear MEM is that it seeks to find (reconstruct) the 'true object', or, in our case, the 'true or real (or image) spectrum'. A recorded spectrum may be thought of as the 'true' spectrum which is 'blurred' by extrinsic (*e.g.* instrumental resolution) and intrinsic factors, and which is then degraded by noise.[118,128,162,166] The MEM process may be described as follows: a candidate, 'mock' or 'trial' spectrum is defined and compared to the recorded spectrum; if the residuals are distinguishable from the overall noise characteristics, then the residuals are utilised to modify the 'trial spectrum' with the proviso that no values are allowed to go negative; the process is then repeated with the new 'trial spectrum' and the iteration continued until the residuals and the noise characteristics are indistinguishable to yield the best possible estimate for the image spectrum,[118,128,161,162] see schematic in Figure 27. The 'trial' spectrum might be represented as:

$$S_i^t = \sum_{k=1}^{n} O_{ik} S_k^r + \sigma_i \tag{48}$$

where, S_i^t represents the 'trial' spectrum, S_k^r is the 'real' or most-probable estimated spectrum, O_{ik} its associated 'blurring' function, n is the number of digitised data points, and σ_i represents the noise on each data point, which is assumed independent of S_k^r. The final 'mock/trial' spectrum should look like the raw data spectrum without the noise, that is smoothed (enhanced S/N) but without peak broadening. Known information, *e.g.* line shape and width (as a point spread function (PSF)), although not essential, can and should be used to influence the 'blurring'/convolution function O, which governs the line-shape narrowing (resolution enhancement).[162] Since, there may be many different images consistent with a noisy data set, the 'logical choice' criterion is determined as that 'spectrum' which has the maximum configurational entropy E (principle of maximum entropy, minimum information content, negation of information) associated with a probability distribution p_i, according to an equation (Shannon's expression)[118,161,163] of the form of Equation (49), subject to the constraint that the FT of that spectrum is a good statistical fit to the raw spectrum:

$$E = -K\sum_{i=1}^{n} p_i \log_n p_i \tag{49}$$

where, K is a constant, n is the number of intensity values, and p_i is the normalised intensity at point i, defined as:

$$p_i = \frac{S_i^r}{\sum_{j=1}^{n} S_i^r} \tag{50}$$

MEM processing has been applied to the in-plane stretching region of Raman spectra of a graphite/$FeCl_3$ intercalate,[166] to enhance a series of Raman spectra recorded from a study of an acrylate copolymerisation reaction,[166] and to improve signal recovery of low wavenumber Raman spectra of TiO_2.[168] MEM and FSD have been applied to enhance the inherently low resolution of the Raman spectra of peptides and proteins.[159]

Maximum Likelihood reconstruction is a non-linear technique which is conceptually similar to MEM, and can be used to generate both smoothed[164,169] and deconvoluted[165,169] spectra, which are the most probable underlying an experimentally recorded spectrum. Figure 28 shows a Gaussian Maximum Likelihood smoothed micro-Raman spectrum of carbon.

7 Statistical Expressions

The statistical tests and terms associated with regression and chemometric analyses can appear a minefield of bewilderment to the non-expert (such as us, the authors!), with different but very similar terms and terminology seemingly almost invoked at will to evaluate, optimise, and validate calibration models. Mathematical expressions for many, sometimes with purpose descriptions, can

[168] Craggs C., Galloway K.P., and Gardiner D.J., *Appl. Spectrosc.*, 1996, **50**, 1, 43.
[169] Square Tools®, Spectrum Square Associates, Ithaca, NY, USA.

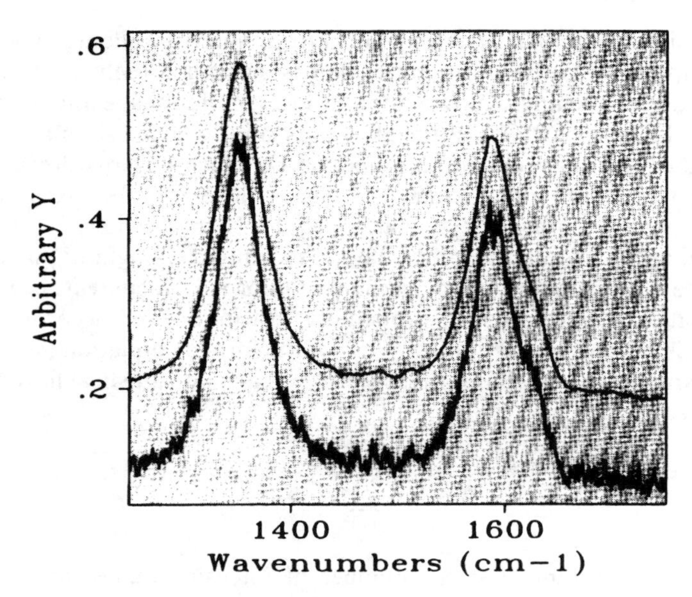

Figure 28 Lower spectrum, *a Raman microprobe spectrum recorded from a carbon;* Upper spectrum, *a Gaussian Maximum Likelihood generated smoothed spectrum of the recorded carbon spectrum; note that the process retains definition of the high wavenumber shoulder on the* 1600 Δcm^{-1} *band* (Reproduced from ref. 164, by kind permission of American Laboratory, International Scientific Communications, Inc)

be found in several places, particularly in refs. 49 and 170–172, from which much of the information below is reproduced or adapted to aid those wishing for a quick, simplistic, convenient glossary.

In Equation (30a) the error ε to be minimised in a least-squares modelling of n calibration values was given as:

$$\varepsilon = \sum_{i=1}^{n} (\hat{y}_i - y_i)^2 \tag{30a}$$

where, $\hat{y}_i$ is the estimated/predicted linear model value of the dependent variable, y_i is the reference value of the dependent variable, and n the number of calibration values.

In multivariate analysis ε is the calibration error vector, usually between the output estimated/predicted value and the input value for the reference dependent variable. This error, the *sum of the squares of the residuals* associated with

[170] Workman J. and Mark H., *Spectroscopy*, 1992, **7**, 8, 14.
[171] Workman J., in *Handbook of Near-Infrared Analysis*, ed. Burns D.A. and Ciurczak E.W., Marcel Dekker, New York, USA, 1992, ch. 10.
[172] Mark H. and Workman J., *Statistics in Spectroscopy*, Academic, San Diego, USA, 1991.

the best possible fit line is sometimes denoted by a term such as SS_{RES}[49,170–172] (in ref. 49, SS_D is used instead), *i.e.*

$$SS_{RES} = \varepsilon = \sum_{i=1}^{n} (\hat{y}_i - y_i)^2 \tag{51}$$

The *variance* (distribution) associated with these deviations is given by Equation (51) divided by the number of degrees of freedom, given as $(n - 2)$, *i.e.* the sample size minus the number of parameters estimated for the slope and intercept of the line,[29] so the variance,

$$\sigma^2_{RES} = \frac{SS_{RES}}{(n-2)} = \frac{\sum_{i=1}^{n} (\hat{y}_i - y_i)^2}{(n-2)} \tag{52}$$

The goodness of fit of a straight line (or calibration model) to the experimental data can be assessed by R^2, where R is the *correlation coefficient*. For a 'perfect' fit $R^2 = 1$, where

$$R^2 = \frac{SS_{REG}}{SS_{TOT}} \tag{53}$$

SS_{REG} denotes the *sum of the squares for regression*, and SS_{TOT} denotes the *total sum of the squares*,[49] where

$$SS_{REG} = \sum_{i=1}^{n} (\hat{y}_i - \bar{y})^2 \tag{54}$$

where, $\bar{y}$ is the mean y (known) value for all the samples, and,

$$SS_{TOT} = \sum_{i=1}^{n} (y_i - \bar{y})^2 \tag{55}$$

$(SS_{REG} = SS_{TOT} - SS_{RES})$.[170,172]

The *root mean squared deviation*, RMSD, is expressed as:

$$RMSD = \sqrt{\frac{\sum_{i=1}^{n} (y_i \hat{y}_i)^2}{n}} \tag{56}$$

The *mean square for regression*, MS_{REG}, is given as:

$$MS_{REG} = \frac{SS_{REG}}{df_{REG}} = \frac{SS_{REG}}{k-1} \tag{57}$$

where, df_{REG} represents the degrees of freedom for regression, that is the number of wavenumbers, k.

Similarly, MS_{RES} the *mean square for residuals*, is given by:

$$MS_{RES} = \frac{SS_{RES}}{df_{RES}} = \frac{SS_{RES}}{n-k} \tag{58a}$$

or, for mean-centered data,

$$MS_{RES} = \frac{SS_{RES}}{n-k-1} \tag{58b}$$

The Prediction Residual Error Sum of Squares, PRESS,[49,50,93,171] see p. 211, is of the form:

$$PRESS = \sum_{i=1}^{n}(y_i - \hat{y}_i)^2 \tag{59}$$

A PRESS plot against the number of factors in PCR and PLS modelling can be used to determine the optimum number of factors to be used in the calibration model, via the process of *cross-validation*. For example conceptually, for a set of k calibration samples, the first sample would be left out and a decomposition for one factor of the data matrix corresponding to the remaining $(k-1)$ samples used to calculate a calibration matrix, which would then be used to predict a particular dependent variable value for the first sample. The 'residual', *i.e.* the difference between the predicted and reference value for the first sample, is then squared to yield a single PRESS value for the first factor for this sample. This process is then repeated, returning the first sample to the training set but this time leaving out the second sample, and so on through the k samples, summing the singular PRESS values as the process progresses. The whole process is then repeated using two factors, and so on until a satisfactorily diagnostic PRESS *vs.* Factor Number plot is reached. The plot should pass through a minimum at the optimum number of factors to be included in the calibration model for that particular dependent variable. A lesser number of factors indicates *underfitting*, and a greater number *overfitting*.[49–51,171]

The standard error of estimate, SEE, or standard error of calibration, SEC, for samples within a calibration set is given by:

$$SEE = SEC = \sqrt{MS_{RES}} \tag{60}$$

The standard error of prediction, SEP,[50,93,171] or standard error of cross validation, SECV,[49,171] has an expression of the form:

$$SEP = SECV = \sqrt{\frac{\sum_{i=1}^{n}(y_i - \hat{y}_i)^2}{(n-1)}} = \sqrt{\frac{PRESS}{(n-1)}} \tag{61}$$

The F-test statistic (F-ratio, F for regression, t-squared)[49,51,171] may be expressed as:

$$F = \frac{MS_{REG}}{MS_{RES}} = \frac{R^2(n - k - 1)}{(1 - R^2)k} \tag{62}$$

F will increase during regression analysis as the equation models more of the variation in the training set. It can be seen from equation (62), that if R^2 kept constant, F should increase with increasing n, and decrease with increasing k. If data associated with an unimportant wavenumber is removed from the training set data matrix, then F will tend to increase. If a sample is deleted from the training set and F increases or if a sample is added to the training set and F decreases, then the sample should be suspected as an *outlier*, *i.e.* a rogue sample that does not conform with others in the training set (*e.g.* it may have an extreme composition which exercises undue leverage on the regression model, the reference value measured/given for the dependent variable was incorrect, or the sample was contaminated). Conversely, if a sample is deleted from the training set and F decreases, then that sample is not considered an outlier, and if a sample is added to the training set and F increases, then that sample may lead to a better model of the variance. Other outlier statistics are also employed during development of a calibration model, such as the *Mahalanobis* distance (D^2),[29,49,171,173] a scalar measure in multivariate parameter space of the 'similarity' between spectra.

Monitoring changes in F, alongside other values for statistical tests, such as PRESS and R^2, are therefore diagnostically important to establishing robust calibration models.

8 Closing Remarks

In this chapter, we have sought to steer a course from the simple 'pencil and ruler' quantitative analysis approaches applied routinely to vibrational spectroscopic data, through to those procedures necessitating sophisticated software algorithms and/or computer manipulations. Several of the latter are now well-established, standard practices in many industrial laboratories; others have been tried, but sparsely and remain peripheral approaches to most industrial spectroscopists. The use of these advanced data treatments has undoubtedly broadened considerably the scope, application, and effectiveness of vibrational spectroscopy quantitative analyses in industry, with some now proving invaluable tools, in particular linear PLS and PCA. However, if used with minimal knowledge of their workings and limitations, and too high expectations (imparted perhaps by over-zealous, non-expert salespersons!), they can be extremely dangerous, and prove very disappointing. Notwithstanding, such tools are now a fact of life in many industrial

[173] Buco S.M., *Spectroscopy World*, 1990, **3**, 3, 28.

laboratories and QC/QA installations, and others will surely follow, supplement, or supersede them. Waiting in the wings are, certainly, non-linear multivariate regression techniques, including PLS and PCR, and feed-forward neural networks,[29,174] possibly, genetic algorithms[175] and, maybe, even the likes of *Massive Inference*![176]

[174] Brown S.D. and Blank T.B., in *Computer Assisted Analytical Spectroscopy*, ed. Brown S.D., J. Wiley, Chichester, UK, 1996, ch. 5.
[175] Leardi R., *J. Chemometrics*, 1994, **8**, 65.
[176] Skilling J., *Massive Inference*, MaxEnt Solutions, Cambridge, UK, presented at Bruker Spectrospin, Coventry, UK, MaxEnt Information and User Meeting, March 1996.

CHAPTER 6

Vibrational Spectroscopy – Microsampling and Microscopy

1 Introduction

Many industrial analytical problems require that vibrational spectroscopic data is obtained either from very small samples or at high spatial resolution, whether as a consequence of limited amounts of the analyte or of a need to probe or map the inhomogeneity or anisotropy of a sample. The Raman technique has an obvious application to microsampling, with its narrow monochromatic laser source that can be readily focused to a sub-millimetre focal cylinder within a sample,[1] while miniaturisations of many of the infrared transmission sample presentation procedures and accessories covered in Chapter 4 are well established, and readily applicable to micro and ultra-micro investigations;[2–4] the amount distinctions for a solid sample have been proposed as 1–50 μg and 50 ng–1 μg, respectively.[2]

Although sensitivity may be a major requirement for many investigations, sample robustness, handling, and manipulation may, perhaps be even more important. Also, for many applications there will be a prime need to be both clinical and meticulous in approach. In all these respects, coupled systems with optical microscopes serve the industrial spectroscopist extremely well. They have led to major advances in and simplification of applications of micro-analyses of solid samples using vibrational spectroscopy instrumentation, and decreased the required sample amount to sub-nanogram levels.

In this chapter we will begin by briefly considering the micro-sampling approaches, many of them traditional, which are usually undertaken without direct utilisation of a microscope to facilitate analysis, before moving on to a

1 Louden J.D., *Laboratory Methods in Vibrational Spectroscopy*. 3rd Edn, ed. Willis H.A., van der Maas J.H., and Miller R.G.J., J. Wiley, Chichester, UK, 1987, ch. 22.
2 Coates J.P., *Am. Lab.*, 1976, **8**, 67, or, *Int. Lab.*, 1977, January/February, 39.
3 Whitehouse M.J. and Curry C.J., in *Laboratory Methods in Vibrational Spectroscopy*. 3rd Edn, ed. Willis H.A., van der Maas J.H., and Miller R.G.J., J. Wiley, Chichester, UK, 1987, ch. 11.
4 *Annual Book of ASTM Standards*, Vol. 3.06, ASTM, Philadelphia, USA, 1993, Designation E 334-90.

more detailed discussion of the vibrational spectroscopy–microscopy coupled systems, in particular their application advantages.

2 Micro-sampling by Raman Spectroscopy

Two characteristics of Raman spectroscopy clearly point to it being an ideal tool for fingerprinting and analysing small samples. Firstly, the source is a laser, which typically has a beam diameter of about 1–3 mm, which is reduced by focusing to about 20–100 μm at the sample position. Secondly, the technique requires a minimum, if any, of sample preparation. A third advantage is the 'transparency' of glass; samples may be examined mounted on microscope slides or contained in thin-walled capillary tubes, a favourite being a glass melting-point or capillary tube. (A 3 mm continuous liquid column in a melting-point tube is approximately 2 μl).[5] However, the micro-sample must be able to withstand the source irradiance (power per area cross-section). All too often a micro-sample has been 'lost' through inattention to laser power; thermal damage may change or destroy the analytical evidence of relevance to the problem, or sample degradation and decomposition may even result in a catastrophic loss of the sample (*e.g.* 'charring' of polymers to carbon). Of course, fluorescence emission has precluded the ready analysis of many micro-samples by Raman spectroscopy, since in an industrial context these are frequently discoloured contaminants. Notwithstanding the drawbacks, and with due attention to the problem in-hand and the properties of the sample, examination of micro-samples by Raman spectroscopy has proven extremely fruitful.

The simplicity of sampling and advantages of Raman are perhaps nicely illustrated in Figure 1, taken from a contaminant investigation undertaken by us in about 1975.[6] The 'defect' was isolated from the surrounding matrix film by a skilled optical microscopist and passed on to the vibrational spectroscopists for further characterisation. An additional advantage of the Raman technique here was its high functional-group sensitivity to unsaturation, which served to highlight the polymer degradation. The 'defect' area was characterised by an additional band at 1660 cm^{-1}, see Figure 1(c), which was attributable to the trans chain unsaturated group —CH=CH— formed in the ethylene/vinyl (EVA) acetate copolymer following from the evolution of acetic acid as a consequence of thermal decomposition. The spectra of low density and high density polyethylene (LDPE and HDPE respectively) are included to show the similarity between the four spectra – they all resemble polyethylene, since the bands due to oxygen containing groups are weak in the Raman spectrum of the low VA content copolymer. The LDPE and HDPE are distinguished mainly by the band at 1640 cm^{-1} in the spectrum of the HDPE sample, arising from the terminal vinyl (—CH=CH_2) group, *cf.* Figures 1(e) and 1(d).

[5] Freeman S. K., *Applications of Laser Raman Spectroscopy*, J. Wiley, New York, USA, 1974.
[6] Willis H.A., in *Molecular Spectroscopy*, ed. West A.R., Heyden, London, UK, 1977, ch. 27.

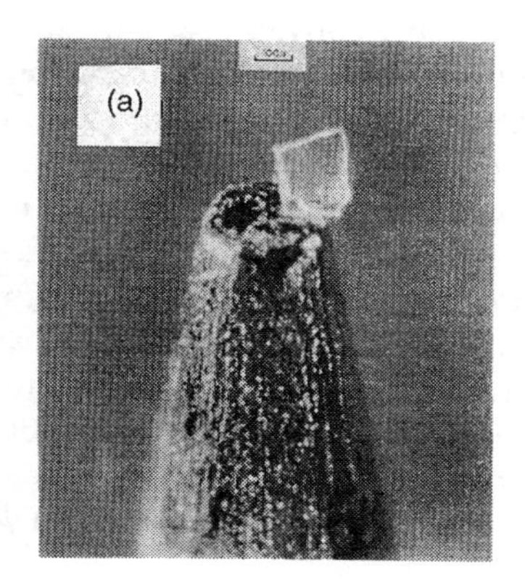

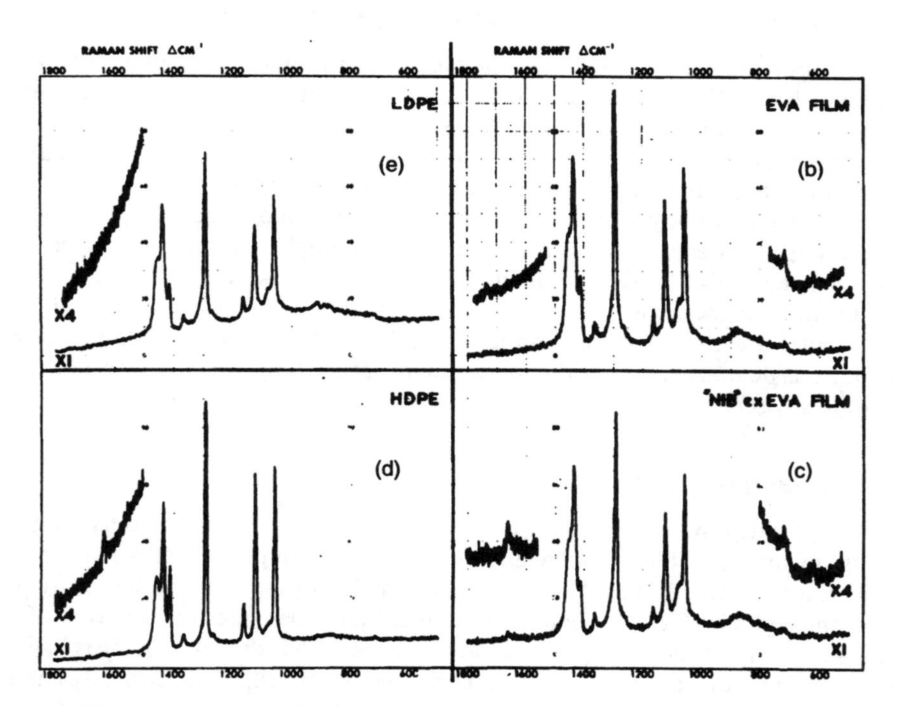

Figure 1 *Photograph of 'defect' ('nib') area cut from EVA copolymer film mounted on a pin for measurement in a Raman spectrometer (the scale bar represents* 100 μm*), and Raman spectrum from 'defect' area of polymer film and spectra of associated materials, see text for explanatory details*
(Reproduced from ref. 6, pp. 566 and 567, by kind permission of The Institute of Petroleum, London, UK)

References 7–10 are sources for other examples of conventional Raman spectroscopy to analysing micro-samples. They are taken from early in the history of laser Raman spectroscopy, since, almost from the moment lasers became the preferred source, it was realised that the higher intensity focused laser beam made it eminently suitable for micro work,[7] and it was not too long before the diffraction-limited focus of a laser beam was used for obtaining microprobe spectra from femtogram amounts of material,[11] with Raman microprobe and microscopes quickly following on.[12] The examples in refs. 7–10 include the spectrum of a single strand (50 μm diameter) nylon-66 fibre,[7] a low wavenumber spectrum of a single cystine crystal weighing less than 10 μg[7] and the spectrum of a single crystal (about 3 μg) of chlordiazepoxide.HCl,[8] the spectrum of 2 nl cysteine.HCl (20% aqueous solution) in a 0.1 mm i.d. capillary,[8] and the spectrum of 0.1 mg GLC trapped benzoic acid in a 0.5 mm bore capillary.[9] (For micro-sampling of gases see below).

3 Micro-sampling by Infrared Spectroscopy

As mentioned in the introduction the 'key' here for analysing liquid and solid samples by traditional presentation techniques is 'miniaturisation', frequently used in conjuction with a set of beam-condensing optics, which better match the infrared radiation beam size to that of the sample.[3] Reference 4 covers many techniques that are of general use in securing and analysing micro-samples by infrared spectroscopy, for both general micro-sampling and analyses using microscope systems. The throughput and multiplex advantages accrued through the use of FTIR instrumentation are major assets to micro-sampling by traditional methods. Diffuse reflectance and photoacoustic measurement techniques are also both well adaptable to the analysis of small amounts of a sample, as demonstrated for the former in its use for 'fingerprinting' chromatographically separated fractions, see Chapter 7.

Micro-sampling of Gases

Although single-pass micro-path-length flow-through transmission cells for gas sampling are available commercially, for example, with path-lengths as narrow as 8 mm (capacity 150 μl),[13] (and low-volume light-pipe cells are the heart of flow-through GC-FTIR systems, see Chapter 7), in one sense trace analysis of a gas is contrary to the concept of miniaturisation, in that the cell volume is likely to be considerably larger than a 'standard' laboratory cell, which typically has a path-length of 5 or 10 cm. The aim is to enable interaction to occur many times

[7] Sloane H.J., *Appl. Spectrosc.*, 1971, **25**, 4, 430.
[8] Freeman S.K., Reed P.R., and Landon D.O., *Mikrochim Acta*, 1972, 288.
[9] Freeman S.K. and Landon D.O., *Anal. Chem.*, 1969, **41**, 2, 398.
[10] Bailey G.F., Kint S., and Scherer J.R., *Anal. Chem.*, 1967, **39**, 8, 1040.
[11] Hirschfeld T., *J. Opt. Soc. Am.*, 1973, **63**, 476.
[12] Rosasco G.J., in *Advances in Infrared and Raman Spectroscopy*, Vol. 7, ed. Clark R.J.H. and Hester R.E., Heyden, London, UK, 1980, ch. 4.
[13] 'Micro Gas Cell' and 'Micro Liquid Cell', Graseby Specac Ltd., Orpington, Kent, UK.

between the few molecules of the gas and the infrared radiation so as to increase absorption intensity. This is achieved through use of a multi-pass gas cell in which the radiation beam can be folded back and forth many times within the cell by reflecting mirrors, thereby increasing the effective path-length (up to as much as 10 m, or more) for interaction, but minimising the length dimension of the cell. Similar cells may also be used to enhance the intensity of Raman scatter from trace gas samples.[14,15] (The interested reader is reminded of the pressure-broadening effect associated with gas sampling, see p. 179.)

A selection of multi-pass and micro-gas cells is shown in Figure 2.

Micro-sampling of Liquids

For volatile liquids and solutions, minimum volume short path-length cells are commercially available. For example, micro-cavity cells fabricated from single crystal blocks of NaCl or KBr are manufactured with a capacity as low as 0.2 μl and a path-length of 50 μm.[16] Demountable micro-liquid cells with interchangeable spacers are also available, which have, for example, a minimum volume at 25 μm path-length of 0.28 μl.[13] (Micro-cells are also an important component of flow-through liquid chromatography (LC) and supercritical fluid chromatography (SFC) combinations with infrared spectroscopy, see Chapter 7.) Routine success in satisfactorily filling short path-length cells will be dependent on sample volatility and viscosity, see pp. 132–133.

An infrared spectrum from a micro-amount of a non-volatile liquid sample may sometimes be recorded successfully by other approaches: use of a micro-ATR accessory may prove a simple option for a rapid identification; low vapour pressure samples may be amenable to a micro-KBr disc preparation, although viscous samples will probably need to be first dissolved in a minimal amount of a volatile solvent to allow transfer, deposition, and dispersion through the powdered support; similarly, a deposition onto the surface of powdered alkali halide in a micro-sample cup followed by measurement of a diffuse reflectance spectrum may prove an easier alternative; a sandwich between alkali-halide polished plates may work, but the problem of masking (excluding stray-light) often presents considerable difficulty.

Micro-sampling of Solids

The main difficulties with analysing micro-amounts of solids lie in sample handling and manipulation. Ingenuity and patience are disciplines that are well rewarded, as are learning skills from experienced microscopists and utilising appropriate 'tools-of-their-trade', and ensuring a clean, distraction-free work-

[14] Ferraro J.R., in *Raman Spectroscopy. Theory and Practice*, ed. Szymanski H.A., Plenum Press, New York, USA, 1967, ch. 2.

[15] Baranska H., Labudzinska A. and Terpinski J. (Translation ed. Majer J.R.), *Laser Raman Spectrometry Analytical Applications*, Polish Scientific, Warsaw, Poland and Ellis Horwood, Chichester, UK, 1987.

[16] Spectra Tech Inc., Stamford, USA.

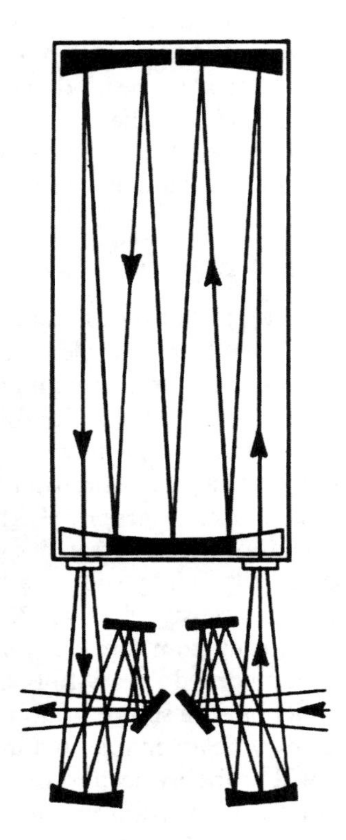

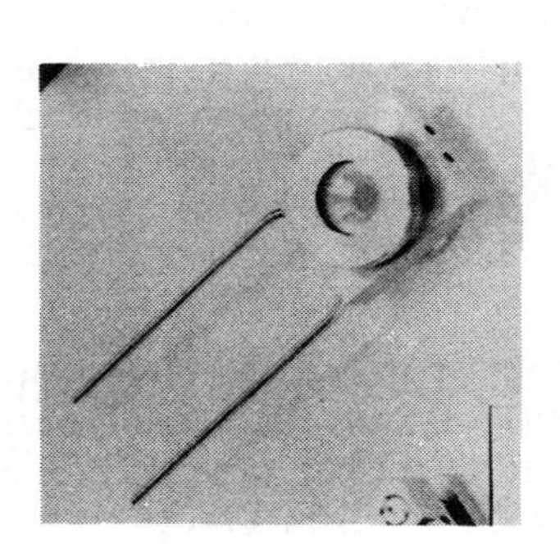

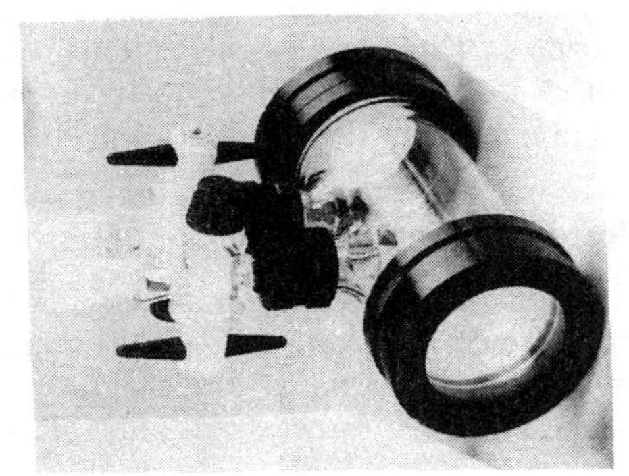

place. The choice of the sample presentation method which is most appropriate will be dependent on the form and physical properties of the material.

Micro-sampling of Solids: Transmission Methods

For brittle or finely-powdered solids micro-KBr disc type preparations are a sound approach. Ancillary equipment such as masks (*e.g.* metal discs with central apertures) are available to help concentrate the analyte into a small volume. It is imperative that the apparatus is clean and the operator is clinical in applying the procedure – 'impurities' will also be concentrated. Examples of some micro-disc making equipment are shown in Figure 3. An adaptation of this technique is the use of the Wick-Stick™[17] approach[3,18] (see Figure 3), in which a dissolved sample is eluted to the tip of a piece of wedged, compacted, powdered KBr. The solvent is evaporated, leaving a concentrate of analyte at the Wick-Stick™ tip, which is then broken off and converted into a micro-KBr disc; alternatively it may be suitable for a micro-diffuse reflectance measurement. In both cases a comparison of the analyte sample spectrum with that from a 'blank' sample is recommended; the 'blank' sample is prepared by the procedure in an identical manner, but using a similar volume of the solvent alone. Reasonable quality spectra should be obtained from 10–50 μg of analyte.[3]

Micro-, miniaturised, mull cells for supporting liquid paraffin mulls and the like are also supplied by most accessory manufacturers.

Micro-samples from malleable solids may cold flow enough to be flattened in a laboratory hydraulic press to a thickness suitable for a direct micro-transmission examination, after appropriate mounting over the aperture of a mask. We have used a KBr press for this, sandwiching, for example, an isolated polymeric contaminant between the die faces prior to applying the pressure. Of course, microtomed sections which contain an 'area' to be investigated may be prepared from continuous materials, and selectively analysed by masking – the problem outlined in Figure 4 was handled in this way,[19] having been analysed prior to the availability of FTIR-microscope systems. Note that here the preferred

[17] Wick-Stick™, Harshaw Chemical Co., Solon, Ohio, USA.
[18] Garner H.R. and Packer H., *Appl. Spectrosc.*, 1968, **22**, 2, 122.
[19] Curry C.J., Whitehouse M.J., and Chalmers J.M., *Appl. Spectrosc.*, 1985, **39**, 1, 174.

Figure 2 *A selection of gas cells for infrared transmission spectroscopy.* Top left, *a fixed pathlength (*8 mm*), low volume (*150 μl*) micro gas cell*
(Reproduced by kind permission of Graseby Specac Limited, Orpington, UK)
Bottom left, 10 cm *pathlength gas cell*
(Reproduced by kind permission of Graseby Specac Limited, Orpington, UK)
Top right, *multiple reflection variable pathlength (*1.2 m, 8 passes *to* 7.2 m, 48 passes*)* 530 ml *volume gas cell*
(Reproduced by kind permission of Infrared Analysis, Inc., Anaheim, CA, USA)
Bottom right, *schematic of multi-pass gas cell with 6-mirror transfer optics for use in a centre-focus sample compartment*
(Reproduced by kind permission of Infrared Analysis, Inc., Anaheim, CA, USA)

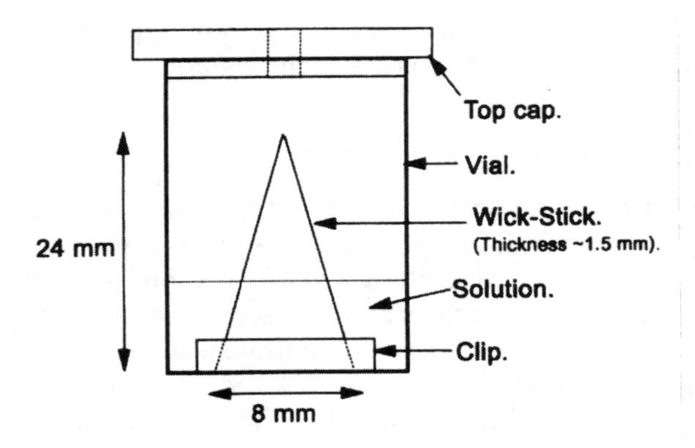

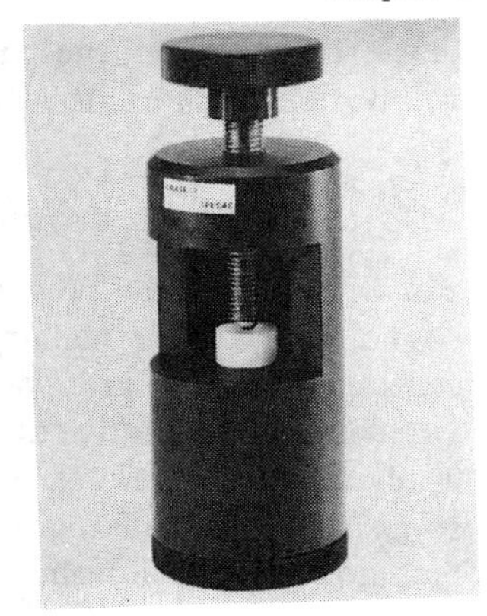

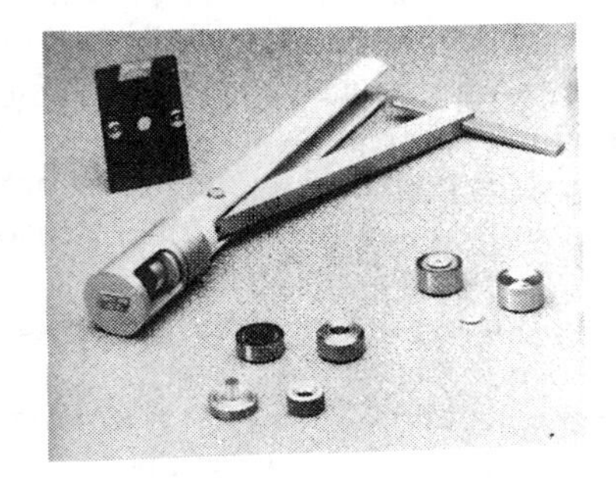

Figure 3 *A selection of some equipment associated with micro alkali-halide disc examinations.* Top left, *schematic of 'Wick-Stick™' approach – the solute (analyte) elutes and is concentrated at the tip of the KBr wedge;* Bottom left, *a hand-held press for producing alkali-halide discs with diameters* 1, 3 *or* 7 mm
(Reproduced with kind permission of Spectra-Tech, Inc.)
Top right, *a hand-held micro-pellet die for producing alkali-halide discs with diameters* 1, 2 *or* 3 mm
(Reproduced by kind permission of Graseby Specac Limited, Orpington, UK)
Bottom right, *a 4 × beam condenser accessory*
(Reproduced with kind permission of Spectra-Tech, Inc.)

approach was infrared since copolymerised ethylene is not as distinctly observed in the Raman spectrum, *cf.* the choice for the problem described for Figure 1. Examinations[19–24] involving aperture or pin-hole mounting and masking of contaminants and similar served to highlight and prove the potential of FTIR for micro-analyses, prior to the introduction of modern FTIR-microscope systems in the early to mid-1980s.

[20] Anderson D.H. and Wilson T.E., *Anal. Chem.*, 1975, **47**, 14, 2482.
[21] Humecki H.J. and Muggli R.Z., *Microbeam Anal.*, 1982, **17**, 243.
[22] Humecki H.J., *Solid State Technol.*, 1985, **28**, 4, 309.
[23] Cournoyer R., Shearer J.C., and Anderson D.H., *Anal. Chem.*, 1977, **49**, 14, 2275.
[24] Lacy M.E., *Proc. Inst. Environ. Sci.*, 1982, 185.

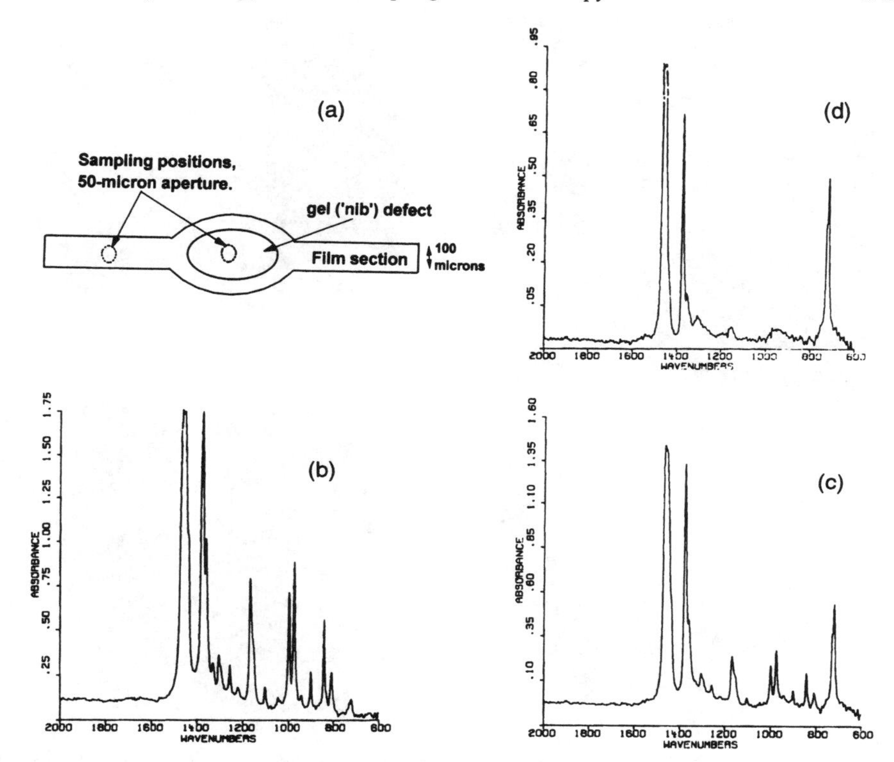

Figure 4 (a) *Schematic of a microtomed section from a* 100 μm *thickness film made from a blend of polypropylene (PP) and an ethylene/propylene (E/P) copolymer rubber. The film had poor profile, exhibiting prominences (defects) termed 'nibs'. The* 50 μm *aperture-masked areas selected for examination, which produced the spectra of* (b) *and* (c), *are indicated;* (b) *FTIR absorbance spectrum of* 50 μm *pinhole-masked area of base film, recorded using an 8 × beam-condenser;* (c) *FTIR absorbance spectrum of* 50 μm *pinhole-masked area of 'nib', recorded using an 8 × beam-condenser;* (d) *Absorbance difference spectrum,* (c) *minus* (b). *The difference spectrum clearly shows that the 'nibs' were caused by poor dispersion of the rubber phase*
((b), (c) and (d) reproduced from ref. 19, Curry et al., *Appl. Spectrosc.*, 1985, **39**, 1, 174–180, by kind permission of the Society for Applied Spectroscopy; © 1985)

Another means of 'thinning' a sample is to make use of a compression cell. Diamond anvil cells were developed originally for the purpose of examining solids under high pressure,[25,26] up to 50 000 atmospheres (70 kbar). A solid is compressed and held between opposing faces of two diamond windows. Such cells, see examples in Figure 5, are ideal for squashing samples, particularly

[25] Weir C.E., Lippincott E.R., Van Valkenburg A., and Bunting E.N., *J. Res. Nat. Bur. Standards* 1959, **63A**, 1, 55.

[26] Lippincott E.R., Weir C.E., Van Valkenburg A., and Bunting E.N., *Spectrochim. Acta*, 1960, **16**, 58.

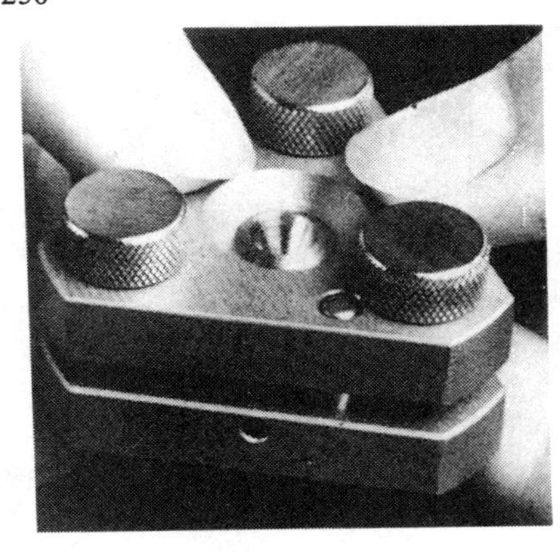

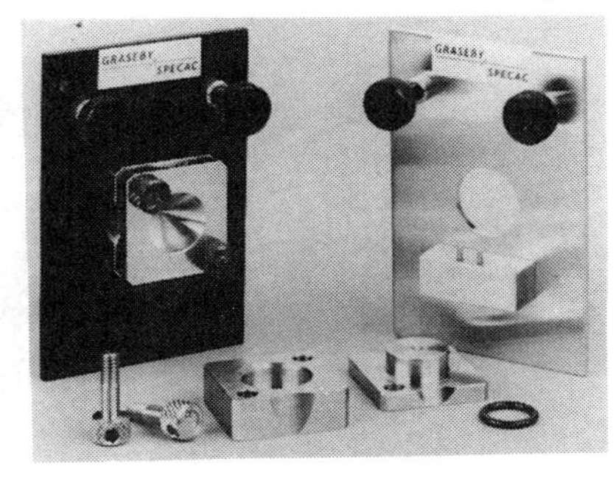

Figure 5 *Compression cells for FTIR-microscopy:* Top left, *'The Miniature Diamond Anvil Cell', High Pressure Diamond Optics, Inc., Tucson, USA, depicting flattening of a sample with finger pressure;* Top right, *'The Diasqueeze' Diamond Compression Cell, Graseby Specac Limited, Orpington, UK;* Bottom, *'Micro Compression' Cells, Spectra-Tech, Inc., Stamford, USA* (Figures reproduced by kind permission of the manufacturers/suppliers)

those that are intractable, into a thickness suitable for infrared transmission studies. The sample may be examined *in situ* in the cell or removed and aperture-masked. Industrial and forensic science applications have included studies on paints,[27–30] fibres,[31] explosives,[32–34] and hair.[35,36] In the studies of refs. 31

[27] Rodgers P.G., Cameron R., Cartwright N.S., Clark W.H., Deak J.S., and Norman E.W.W., *Can. Soc. Forensic Sci. J.*, 1976, **9**, 1, 1.
[28] Rodgers P.G., Cameron R., Cartwright N.S., Clark W.H., Deak J.S., and Norman E.W.W., *Can. Soc. Forensic Sci. J.*, 1976, **9**, 2, 49.
[29] Rodgers P.G., Cameron R., Cartwright N.S., Clark W.H., Deak J.S., and Norman E.W.W., *Can. Soc. Forensic Sci. J.*, 1976, **9**, 3, 103.
[30] Tweed F.T., Cameron R., and Rodgers P.G., *Forensic Sci.*, 1974, **4**, 211.
[31] Read L.K. and Kopec R.J., *J. Assoc. Off. Anal. Chem.*, 1978, **61**, 3, 526.
[32] Midkiff C.R. and Washington W.D., *J. Assoc. Off. Anal. Chem.*, 1976, **59**, 6, 1357.
[33] Kopec R.J., Washington W.D., and Midkiff C.R., *J. Forensic Sci.*, 1978, **23**, 57.
[34] Miller P.J., Piermarini G.J., and Block S., *Appl. Spectrosc.*, 1984, **38**, 5, 680.
[35] Strassburger J. and Breuer M.M., *J. Soc. Cosmet. Chem.*, 1985, **36**, 61.
[36] Brenner L., Squires P.L., Garry M., and Tumosa C.S., *J. Forensic Sci.*, 1985, **30**, 420.

and 33, the workers also used a sapphire window compression cell as a companion to their diamond window cell studies. Type II diamond windows are partially opaque over the region 4000–1800 cm^{-1}, while sapphire windows have good transmission characteristics over the range 4000–1600 cm^{-1}, thus together the cells provide good transmission properties over the range 4000–200 cm^{-1}.[31,33,37] The miniature diamond anvil cell accessory,[38] with the smallest diamond face having a diameter of 0.6 mm, is designed for use with a beam-condenser in the sample compartment of a FTIR spectrometer, where, for example, it has been used to examine automotive paint chips,[39] or directly on the stage of a FTIR-microscope[40,41] for polymer investigations. Compression cells have increased considerably in popularity through their utilisation in FTIR-microscopy examinations, see microscopy applications later.

Micro-sampling of Solids: Reflectance Methods

Internal Reflection Spectroscopy. Micro-sampling is strictly limited to miniaturised versions of the single-reflection ATR technique. It is particularly useful for analysing small samples of deformable materials such as rubbers or foams, flakes of paint *etc.*, and adhesives, remembering that the spectrum recorded will be representative of the surface layer (to a depth of about 1–3 μm) in contact with the ATR element and may not be representative of the bulk! In some circumstances, touching the micro-ATR crystal onto, say, a contaminated surface may result in transfer of enough contaminant material for subsequent 'isolated' analysis. This may prove effective, for example, in adhesive or 'additive-bleed' investigations. The minimum surface area for successful analysis is likely to be of the order of 1 mm^2.

Relatively recently a new accessory has been introduced which has been developed specifically for the analysis by FTIR internal reflection spectroscopy of physically small samples or small areas of large samples.[42,43] Known as the 'SplitPea™'[44], see Figure 6, it incorporates a 3 mm diameter hemispherical internal reflection element (IRE) into a set of beam condensing optics, which provide a six-times linear reduction of the source image onto the sample surface, with an option of a viewing eyepiece. The short pathlength element allows silicon to be a suitable choice for the crystal material, which has advantages in both its hardness and inertness.[42,43] The flat surface of the hemisphere is bevelled on the edge to provide a sampling area on the element which is slightly

[37] Lippincott E.R., Welsh F.F., and Weir C.E., *Anal. Chem.*, 1961, **33**, 1, 137.
[38] High Pressure Diamond Optics Inc., Tucson, USA.
[39] Schiering D.W., *Appl. Spectrosc.*, 1988, **42**, 5, 903.
[40] Lang P.L., Katon J.E., Schiering D.W., and O'Keefe J.F., *Polym. Mater. Sci. Eng.*, 1986, **54**, 381.
[41] Katon J.E., Lang P.L., Schiering D.W., and O'Keefe J.F., in *The Design, Sample Handling and Applications of Infrared Microscopes*, ed. Roush P.B., ASTM Special Technical Publication 949, ASTM, Philadelphia, USA, 1987, p. 49.
[42] Harrick N.J., Milosevic M., and Berets S.L., *Appl. Spectrosc.*, 1991, **45**, 6, 944.
[43] Harrick N.J., Milosevic M., and Berets S.L, *Am. Lab.* 1992, February, 50MM.
[44] Harrick Scientific Corporation, Ossining, USA.

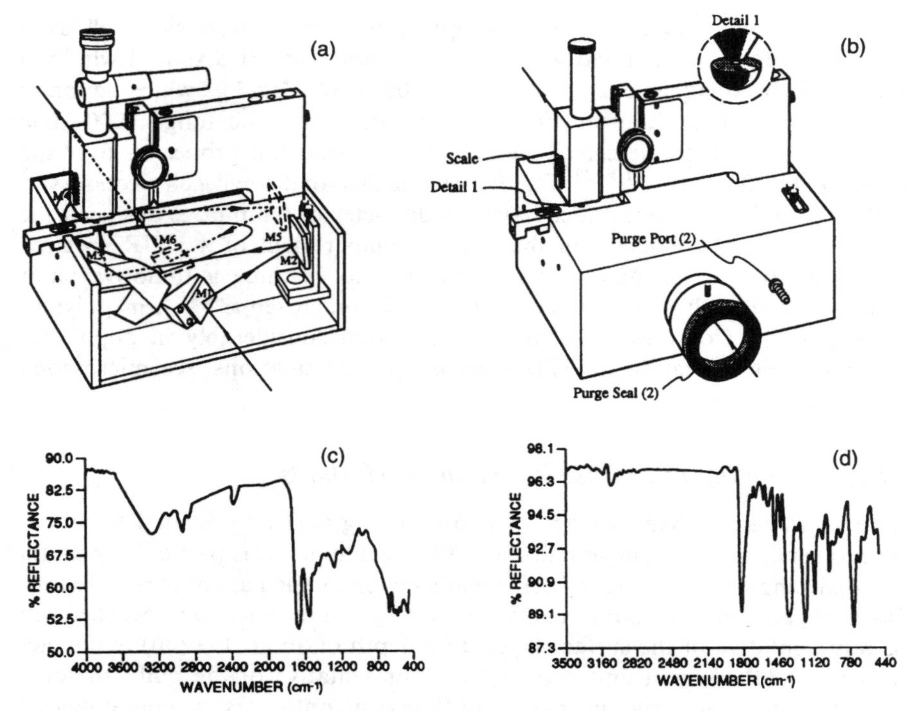

Figure 6 (a) *Schematic of the optics of the SplitPea*™, *with the optional viewing microscope. The mirror configuration provides a* 6 × *linear reduction of the source image on the sampling surface;* (b) *Schematic of the SplitPea*™ *enclosed in its purgeable box. Detail 1 illustrates the* 3 mm *diameter hemispherical internal reflection element (IRE);* (c) *FTIR spectrum from a single strand of a human hair recorded using a Si hemisphere IRE;* (d) *FTIR spectrum from a* 20 mm *diameter polyethylene terephthalate fibre recorded using a ZnSe hemisphere IRE*
(Figures reproduced by kind permission of Harrick Scientific Corporation, Ossining, NY, USA)

larger than 200 μm.[44] This small contact area allows very high localised contact clamping pressures to be generated, such that ATR absorbance spectra from hard and intractable materials like human hair and paint chips may be recorded with ease.

Other recent innovations use small contact area diamond ATR elements.[45, 46] In one accessory[45] the contact surface of the element is 'upward facing'. Samples are placed directly on this surface, and a calibrated plunger produces reproducible pressure on the sample. The ATR element is a composite sensor compris-

[45] Milosevic M., Sting D., and Rein A., *Spectroscopy*, 1995, **10**, 4, 44.
[46] 'The Golden Gate' Single Reflection Diamond ATR, Graseby Specac Ltd., Orpington, UK.

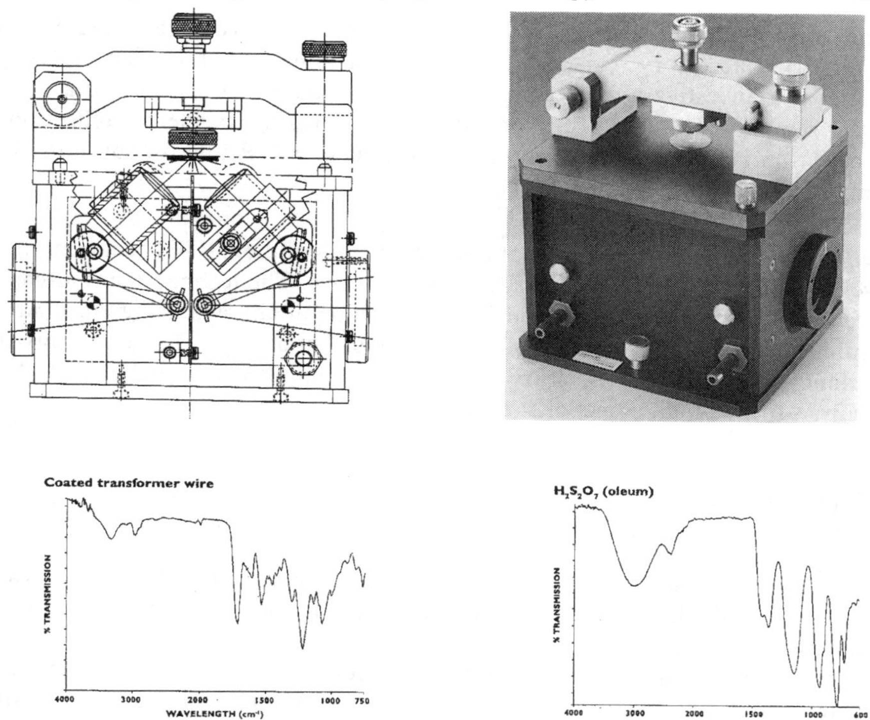

Figure 7 Top, *schematic and photograph of the 'Golden Gate' single reflection ATR accessory, Graseby Specac Ltd., Orpington, UK.* Bottom, left, *spectrum recorded from a coated transformer wire;* right, *spectrum recorded from* $H_2S_2O_7$ *(oleum)*
(Reproduced by kind permission of Graseby Specac Limited, Orpington, UK)

ing a <4 mm diameter diamond sensor area supported on an infrared focusing crystal, typically of ZnSe. In another design, the 'Golden Gate™', a 2 mm single reflection type IIa diamond is brazed in tungsten carbide.[46] It has a focused sample aperture of 0.6 mm, and the beam condensing coupling optics is fitted with either ZnSe (standard) or KRS-5 (for extended working range) lenses. Diamond has excellent mechanical strength and chemical stability, and its optical surface will be scratch and abrasion resistant. Pressures of up to 3 kbar may be applied reproducibly to the sample in contact with the diamond element by using the sapphire anvil and a torque wrench. Example spectra recorded with this device are shown in Figure 7.

External Reflection. With appropriate masking or mounting, small areas (~1 mm × 1 mm) of a sample surface may be subjected to infrared radiation and a specular reflectance or a 'transflectance' spectrum recorded in an appropriate accessory. This may prove the most convenient approach, certainly initially, to investigating defects in coatings on metallic substrates. For a small area sample of a very thin coating, where grazing incidence reflection–absorption is appropriate, then grazing-incidence FTIR-microscopy must be employed, see pp. 273–276.

Diffuse Reflectance. The micro-sampling attributes of the DRIFT technique are well established[47] and were discussed in Chapter 4. The technique has already been mentioned twice in Section 3, and its value in characterising chromatographically separated eluates alluded to, see Chapter 7. The SiC abrasive technique (see p. 159) is clearly a way of obtaining a minimal sample for analysis from a bulk object or sample.

Micro-sampling of Solids: Other Methods

Photoacoustic. The potential and limitations for recording FTIR fingerprint spectra directly from small samples using a photoacoustic cell have been exampled.[48–50] They include single fibres,[48,49] coal particles,[48] and surface contamination of alkyd painted panels.[50] Figure 8 shows the micro-sampling holder for a photoacoustic cell and some examples of single fibre nylon spectra recorded using it.[48] Spectral contrast is clearly dependent on fibre diameter and geometry, and in some cases poor since the strong absorbance bands of the nylon fibres are truncated due to photoacoustic signal saturation. Nevertheless, generic polymer typing is relatively easy, and in many favourable instances second-order information may be readily observed.

Emission. Emission accessories are also available which facilitate the sampling of small areas.[51,52] Test results have been reported on the examination of samples of thin silicate coatings on electrochemical steel surfaces and hydrated silica gel surfaces[51] and a single 12 μm diameter filament. In the surface studies,[51] black-body masks with different hole sizes were placed over a sample in order to control the area from which the emission spectrum was recorded, although the size required needed to be 0.5 mm diameter or greater; a magnifying eyepiece viewer is fitted to the accessory.

[47] Fuller M. P. and Griffiths P.R., *Appl. Spectrosc.*, 1980, **34**, 5, 533.
[48] McClelland J.F., Jones R.W., Luo S., and Seaverson L.M., in *Practical Sampling Techniques for Infrared Analysis*, ed. Coleman P.B., CRC Press, Boca Raton, USA, 1993, ch. 5.
[49] McClelland J.F, Luo S., Jones R.W., and Seaverson L.M., *Proc. 8th Int. Conf. Four. Trans. Spectrosc., SPIE,* 1991, **1575**, 226.
[50] Davis D.M. and Hoffland L. D., *Proc. Four. Comp. Infrared Spectrosc.*, SPIE, 1985, **553**, 146.
[51] Handke M. and Harrick N.J., *Appl. Spectrosc.*, 1986, **40**, 3, 401.
[52] DeBlase F.J. and Compton S., *Appl. Spectrosc.*, 1991, **45**, 4, 611.

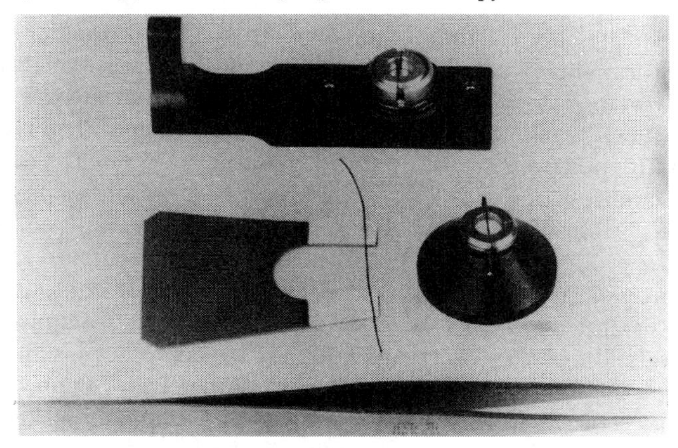

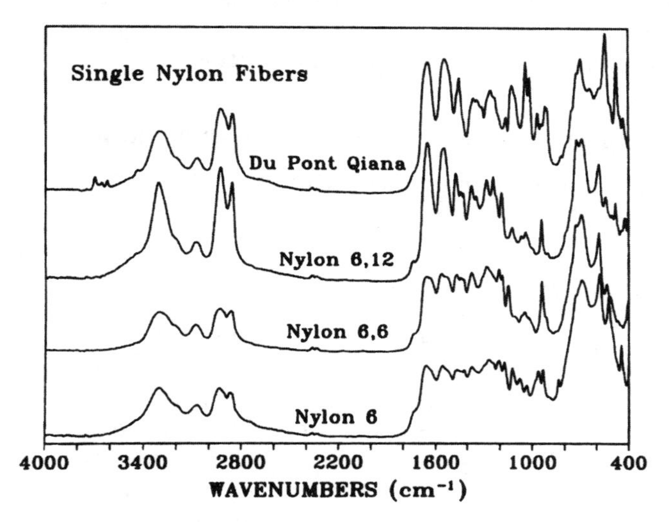

Figure 8 Top, *photograph of a black thread depicting a fibre being mounted and placed in the photoacoustic cell micro-sampling holder over a covered desiccant (magnesium perchlorate) cup; the PA accessory was filled with dry helium gas atmosphere purge, (full mounting and manipulation details may be found in ref. 48)* Bottom, *PA-FTIR spectra of four types of nylon single fibres. Some of the spectra are truncated due to PA signal saturation*
(Reprinted with permission from ref. 48, *Practical Sampling Techniques for Infrared Analysis*, ed. Coleman P.B. (1993). Copyright CRC Press, Boca Raton, Florida, USA, © 1993)

4 Vibrational Spectroscopy–Microscopy

We are in little doubt that the coupling of optical microscopes to FTIR spectrometers, initiated in the 1980s, was a major step in industrial problem-solving by infrared spectroscopy, and opened up many opportunities for novel spatially-resolved applications. Together with their Raman counterparts, they provide industrial analytical and research spectroscopists with a most powerful combination of complementary tools, not only in the respective chemical and physical information they yield, but also in their appropriateness to various problems; a synergy and complementarity we hope to demonstrate later in this chapter. Firstly, however, we will give a brief description of the development and instrumentation of each vibrational spectroscopy–microscopy combination, followed by specific sections on sample presentation. Both combinations, in addition to producing spatially resolved spectra for structure analysis, are capable of providing spatially resolved intensity images or contour or axonometric plots, for multi-dimensional visualisation of the spectroscopic properties. Contour and axonometric plots are built up by computer techniques from selected wavenumber regions extracted from successively scanned point-by-point spectra recorded from consecutive areas of a sample, *i.e.* the sample is 'analyte (functional-group) intensity mapped'. Images are grey-scale or colour maps of intensity within specific or narrow wavenumber regions recorded directly by line-scanning or global illumination methods.

Vibrational Spectroscopy–Microscopy Instrumentation

Outlines of the principles and modes of operation of each combination are considered in the next two sections.

Instrumentation: Raman-microscopy

Raman-microscopy has been used almost since its inception both to identify components at the microscopic level within a heterogeneous sample and to provide a picture of their spatial resolution.[12,53,54] In analyses using the Raman spectrometer–optical microscope combination, the radiation source can be a tightly focused laser beam, which may have a focal cylinder waist-diameter of only about 1–2 μm . This can therefore be readily positioned (focused) at a small area on or within a sample. As a consequence, the term microprobe analysis is frequently used to describe this mode of operation, in which a single-point fingerprint wavenumber-shift *vs.* intensity spectrum is recorded. This sometimes (although the two terms, Raman-microprobe analysis and Raman-microscopy, are frequently interchanged) distinguishes it somewhat from the mode in which it is set to show a Raman functional-group magnified image over a large area, *e.g.* of 100–200 μm or greater diameter, for visual perception.[55,56] Selected Raman images are obtained by passing the scattered radiation through essentially a narrow tuned band-pass filter or monochromator, in order to

[53] Louden, J.D., *Practical Raman Spectroscopy*, ed. Gardiner D.J. and Graves P.R., Springer-Verlag, Berlin Heidelberg, 1989, ch. 6.
[54] Delhaye M. and Dhamelincourt P., *Microbeam Anal.*, 1990, **25**, 220.
[55] Delhaye M. and Dhamelincourt P., *J. Raman Spectrosc.*, 1975, **3**, 33.
[56] Dhamelincourt P., Wallart F., Leclercq M., N'Guyen A.T., and Landon D.O., *Anal. Chem.*, 1979, **51**, 3, 414A.

isolate characteristic spectral features.[54–69] Some of the filtering mechanisms employed have used: a triple monochromator,[55] a concave grating double monochromator,[56] a stigmatic monochromator using holographic gratings,[55] a multilayer dielectric,[60] a dielectric with tuneable wavelength excitation,[64] and a dual holographic grating tuneable filter.[65] Other designs have used AOTFs (acousto-optic tuneable filters) and LCTFs (liquid crystal tuneable filters),[67] and a Haddamard transform Raman microprobe.[68] Raman intensity maps may be constructed by direct imaging or reconstructed from data collected by scanning techniques.[12,53–69] (The focusing of the laser to a near diffraction-limited spot has repercussions on the laser power that should be used in order to avoid sample damage; source irradiance will be high at the sample, and the laser power at the head may well need to be reduced to 10 mW or less.)

A schematic of the laser focusing, sample viewing, and scattered light collection geometry of a Raman microprobe system is given in Figure 9.[57,69] The objective lens is used to both focus the laser beam onto the sample and collect the back-scattered radiation. For single point investigations at the '1–5 μm size range', a typical measurement objective would have a magnification of 100 × and a Numerical Aperture (NA) of 0.9,[53,56] lower magnification objectives being used to aid sample location and manipulation. Illumination of the sample by a white light source may be used to provide a visual image. A wide range of optical-microscopy illumination techniques, (such as transmitted, reflected, phase contrast, interference contrast, *etc.*), may be utilised to improve this image contrast or highlight phase differences within a sample.[53] A good, concise account of practical Raman-microscopy and many examples of industrial applications was published in 1989.[53]

Raman microprobes were introduced in the mid to late 1970s.[55,56,70–74] One of these systems, which was sold commercially, the MOLE (Molecular Optical Laser Examiner) used an optical microscope to provide both single-point analysis (monochannel detection) and an imaging capability (multichannel

[57] Barbillat J., Dhamelincourt P., Delhaye M., and da Silva, E., *J. Raman Spectrosc.*, 1994, **25**, 1, 3.
[58] Bowden M., Gardiner D.J., Rice G., and Gerrard D.L., *J. Raman Spectrosc.*, 1990, **21**, 37.
[59] Bowden M., Dickson G.D., Gardiner D.J., and Wood D.J., *Appl. Spectrosc.*, 1990, **44**, 10, 1679.
[60] Batchelder D.N., Cheng C., Muller W., and Smith B.J.E., *Makromol. Chem., Macromol. Symp.*, 1991, **46**, 171.
[61] Batchelder D.N., Cheng C., and Pitt G.D., *Adv. Mater.*, 1991, **3**, 11, 566.
[62] Williams K.P.J. and Batchelder D.N., *Spectroscopy Europe*, 1994, **6**, 1, 19.
[63] Meier R.J. and Kip B.J., *Microbeam Anal.*, 1994, **3**, 2, 61.
[64] Puppels G.J., Grond M., and Greve J., *Appl. Spectrosc.*, 1993, **47**, 8, 1256.
[65] Pallister D.M., Govil A., Morris M.D., and Colburn W.S., *Appl. Spectrosc.*, 1994, **48**, 8, 1015.
[66] Gardiner D.J. and Bowden M., *Microsc. Anal.*, 1990, **20**, 27.
[67] (a) Morris H.R., Hoyt C.C., and Treado P.J., *Appl. Spectrosc.*, 1994, **48**, 7, 857.
(b) Morris H.R., Hoyt C.C., Miller P., and Treado P.J., *Appl. Spectrosc.*, 1996, **50**, 6, 805.
[68] Treado P.J. and Morris M.D., *Appl. Spectrosc.*, 1989, **43**, 2, 190.
[69] Dhamelincourt P., Barbillat J., and Delhaye M., *Spectroscopy Europe*, 1993, **5**, 2, 16.
[70] Rosasco G.J., Etz E.S., and Cassatt W.A., *Appl. Spectrosc.*, 1975, **29**, 5, 396.
[71] Rosasco G.J. and Etz, E.S., *Res. Development*, 1977, **28**, 20.
[72] Blaha J.J. and Etz E.S., *Anal. Chem.*, 1978, **50**, 7, 892.
[73] Cook B. W. and Louden J.D., *J. Raman Spectrosc.*, 1979, **8**, 5, 249.
[74] Corset J., Dhamelincourt P., and Barbillat J., *Chem. Brit.*, 1989, **25**, 6, 612.

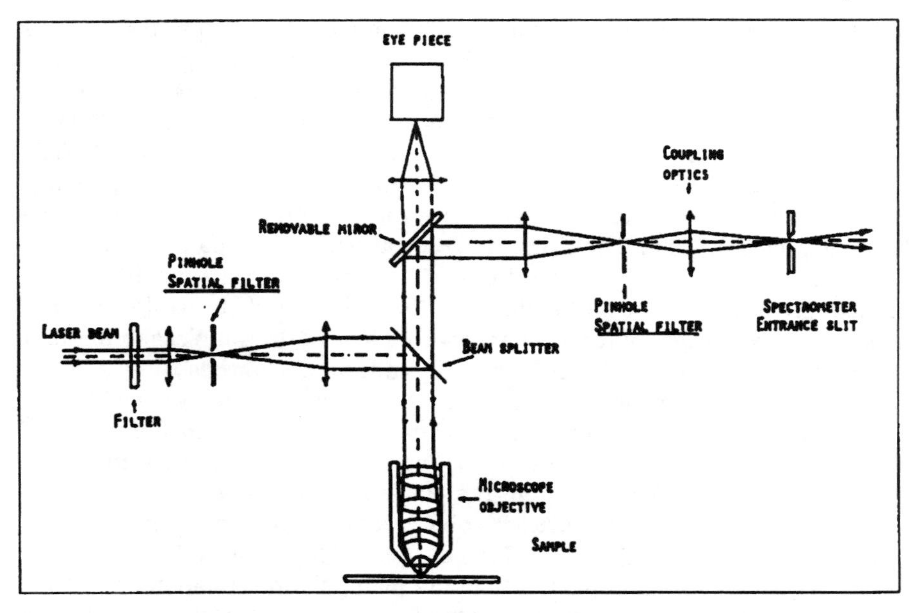

Figure 9 *Optical scheme of the illumination and collection optics of a micro-Raman spectrometer*
(Reprinted with permission from ref. 69, Dhamelincourt P., Barbillat J., and Delhaye M., *Spectroscopy Europe*, **5**, 2, 16–26 (1993))

detection).[53–56,75] Global imaging of the sample is achieved by illuminating the whole field of view of the microscope with the laser beam, and then, in the original design, selecting with the spectrometer monochromator the wavenumber range to be focused onto an intensified closed circuit television (CCTV) camera. This arrangement has now been superseded; a holographic notch filter is used to reject the exciting radiation more efficiently and the intensified TV camera has been replaced by a CCD array detector.[57,75] The properties of holographic notch filters (HNFs) and wavelength specific, fast (*e.g.* *f*/1.8) minimal-aberration optics, mean that for many applications a single grating spectrograph can replace the high dispersion monochromators traditionally used to filter out the Rayleigh scatter.[76] At the laser wavelength, optical densities in excess of 4 ($<0.01\%$ transmission) are achieved, with better than 80% transmission at wavelengths outside the rejection band. Current technology restricts observations to wavenumber shifts of greater than about 50 Δcm^{-1}, and bands at below $\sim 150\ \Delta cm^{-1}$ close to the filter cut-off, may be strongly attenuated.[76] HNFs also permit the use of very low laser powers. The principle of Raman imaging is shown in Figure 10;[57,69] here a small amount of the elastically scattered Rayleigh radiation is used to develop the full sample

[75] Dilor S.A., Lille, France.
[76] Everall N., *Appl. Spectrosc.*, 1992, **46**, 5, 746.

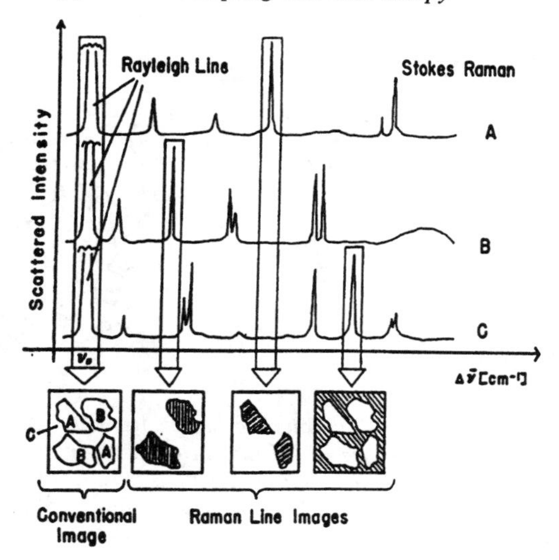

Figure 10 *Schematic showing the principle of Raman imaging*
(Reprinted with permission from ref. 69, Dhamelincourt P., Barbillat J., and Delhaye M., *Spectroscopy Europe*, **5**, 2, 16–26 (1993))

image, while selected band-passes are imaged on the detector to highlight component concentrations.

An alternative method of constructing an image is to scan the sample surface. This has been achieved by using fast-moving optics to line scan the sample, which is sequentially passed under the field of view by, for example, a 0.5 μm stepper-motor-driven *x-y* microscope stage.[58,59] More recently, cylindrical optics have been employed to provide line (typical length 100 μm) focus illumination.[77,78] The sample is scanned in a linear step-wise manner across the focal plane, and the scattered Raman intensity collected by a two-dimensional CCD array detector. One axis of the array detector represents the spatial discrimination; the other axis contains the spectral information. Computer images are then reconstructed by retrieving the sequential requisite wavenumber region intensity information. This indirect procedure affords improved stray-light rejection and better spectral resolution than the direct global illumination method.[57] A schematic of a Raman microline focus spectrometer is given in Figure 11.[63,77] In another design for a surface-scanning micro-Raman system, the relative movement between the laser light and the sample is achieved by applying appropriate electric fields to a pair of computer-controlled piezo-crystal translators which are fixed to the microscope objective.[79,80] Originally, a

[77] Bowden M., Donaldson P., Gardiner D.J., Birnie J., and Gerrard D.L., *Anal. Chem.*, 1991, **63**, 24, 2915.
[78] Gardiner D.J. and Bowden M., *Microsc. Anal.*, 1990, **20**, 27.
[79] Lankers M., Gottges D., Materny A., Schaschek K., and Kiefer W., *Appl. Spectrosc.*, 1992, **46**, 9, 1331.
[80] Lankers M. and Kiefer W., *Fresenius J. Anal. Chem.*, 1994, **349**, 224.

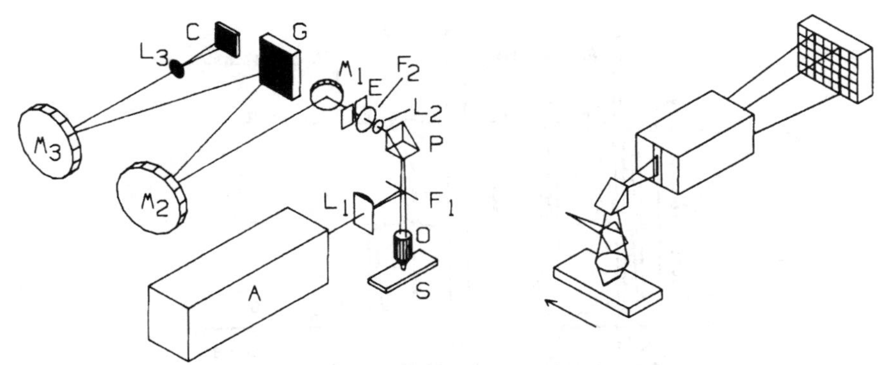

Figure 11 *Schematics of a Raman microline focus spectrometer:* A, *air-cooled laser;* C, *CCD camera;* E, *entrance slit;* F1, *dichroic beam-splitter;* F2, *edge filter;* G, *grating;* L1, *cylindrical lens;* L2, *matching lens;* L3, *correcting lens;* M1, M2, M3, *spectrographic mirrors;* O, *objective;* P, *prism;* S, *sample; Enlargement (right) depicts in more detail the geometry of the line-focus and the image on the CCD*
(Left, reprinted with permission from ref. 77, Bowden *et al., Anal. Chem.*, **63**, 24, 2915–2918 (1991). Copyright (1991) American Chemical Society. Right, reproduced by kind permission of Professor D.J. Gardiner)

prime objective of this device was to minimise sample heating and decomposition of highly absorbing samples.[79]

In the early 1990s many advances for Raman-microscopy applications were concerned with highlighting the use of confocal microscopes to minimise spectral contamination from 'surrounding material' caused by the depth of focus.[57,69,81–84] High-sensitivity low-noise CCD (charge coupled device) array detectors also became generally available which greatly enhanced both sensitivity and speed of analysis. Used in conjunction with holographic optical elements (HOEs), rapid, of the order of a few minutes, high signal-to-noise spectra can now be obtained routinely.[82] The spectral dispersion (Raman scatter) of a single-point measurement exiting from the spectrograph may be line-focused onto a row of pixels of the array detector for a spectrum display, or the system may be set to record the Raman intensity emitted over a narrow wavenumber region of interest and a line-focus successively scanned across a sample to build up a Raman image on the CCD array. A schematic of such a system is shown as Figure 12.

The principles of confocal laser microscopy are shown schematically in Figure 13, while a demonstration of its advantages may be seen in Figure 14.[82]

[81] Tabaksblat R., Meier R.J., and Kip B.J., *Appl. Spectrosc.*, 1992, **46**, 1, 60.
[82] Williams K.P.J., Pitt G.D., Batchelder D.N., and Kip B.J., *Appl. Spectrosc.*, 1994, **48**, 2, 232.
[83] Puppels G.J., Colier W., Olminkhof J.H.F., Otto C., de Mul F.F.M., and Greve J., *J. Raman Spectrosc.*, 1991, **22**, 217.
[84] Sharanov S., Chourpa I., Valisa P., Fleury F., Feofanov A., and Manfait M., *Microsc. Anal.*, 1994, **44**, 9.

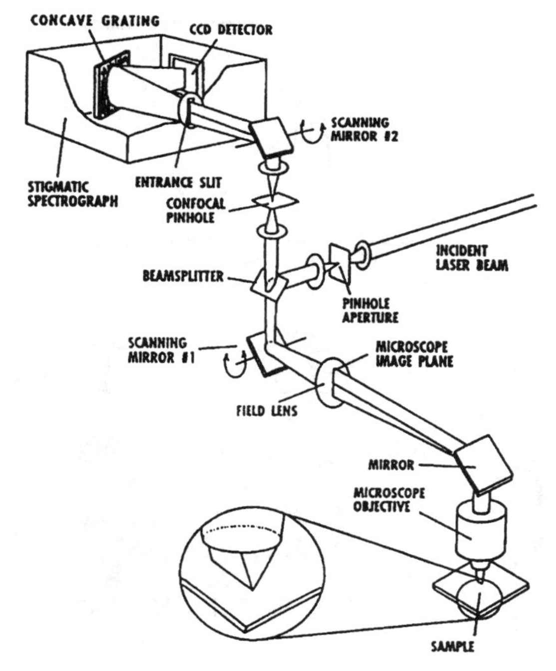

Figure 12 *Schematic of the optical layout of a confocal line illumination Raman microprobe system*
(Reprinted with permission from ref. 69, Dhamelincourt P., Barbillat J., and Delhaye M., *Spectroscopy Europe*, **5**, 2,16–26 (1993))

Confocal Raman microscopy provides an efficient means to obtaining 'interference-free' spectra and 2- or 3-dimensional selective images of small specimens embedded within a matrix.[57] While it offers some improvement in lateral resolution, the main advantage of confocal microscopy lies in its potential for 'depth profiling'. 'Optical sectioning' (layer discrimination) in transparent samples is achieved through the use of optically conjugate pin-hole diaphragms, which spatially filter the Raman light emanating from the sample. All light emitted from the focal plane within the sample reaches the detector; light emitted from planes above and below the focal plane are partially attenuated.[57,69,81–84]

FT-Raman microprobe systems have also been developed.[85–90] Commercial systems are available from a few manufacturers of FTIR/FT-Raman spectro-

[85] Messerschmidt R. and Chase D.B., *Appl. Spectrosc.*, 1989, **43**, 1, 11.
[86] Bergin F.J. and Shurvell H.F., *Appl. Spectrosc.*, 1989, **43**, 3, 516.
[87] Bergin F.J., *Spectrochim. Acta*,1990, **46A**, 2, 153.
[88] Sommer A.J. and Katon J.E., *Appl. Spectrosc.*, 1991, **45**, 4, 527.
[89] Sommer A.J. and Katon J.E., *Spectrochim. Acta*, 1993, **49A**, 5/6, 611.
[90] Brenan C.J.H. and Hunter I.W., *Appl. Spectrosc.*, 1995, **49**, 7, 971.

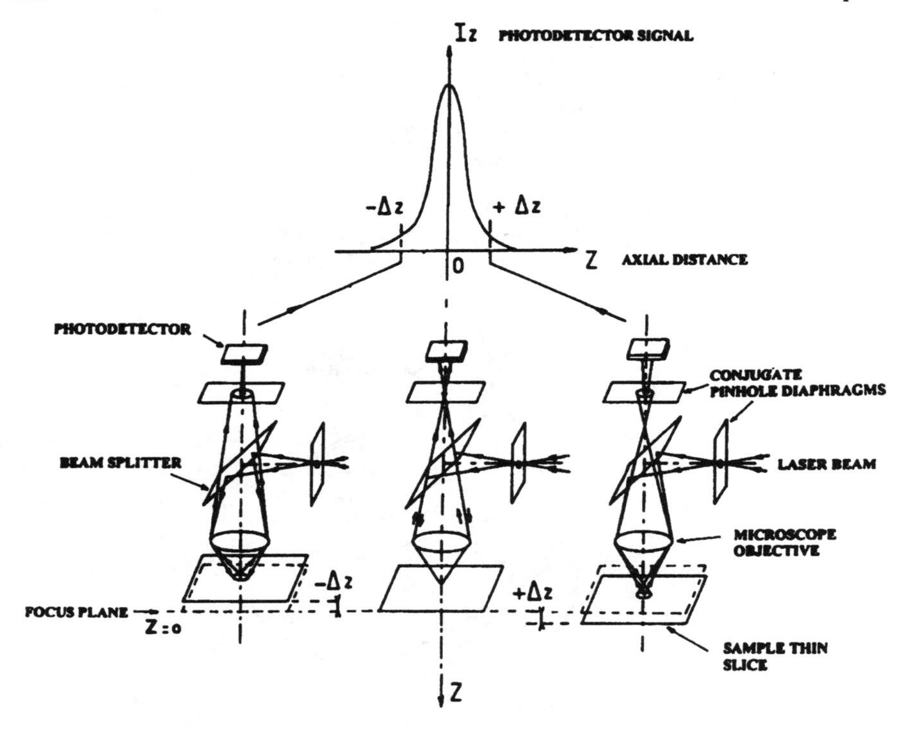

Figure 13 *Schematic of the principle of confocal microscopy and depth discrimination. The optically conjugate pinhole diaphragm placed before the detector 'isolates' the light originating from the small region of the sample coincident with the illuminated spot, and effectively 'eliminates' the contributions from the out-of-focus zones ($\pm\Delta z$) above and below the focus plane ($z = 0$)*
(Reproduced from ref. 57, *Raman Confocal Microprobing, Imaging and Fibre-Optic Remote Sensing: a Further Step in Molecular Analysis*, Barbillat J., Dhamelincourt P., Delhaye M., and da Silva E., *J. Raman Spectrosc.*, **25**, 1, 3–11 (1994). Copyright (1994). Reprinted by permission of John Wiley & Sons, Ltd)

meters. However, the use in these of the longer wavelength radiation at 1.064 μm of a Nd^{3+}:YAG laser detracts from the spatial resolution achievable with a visible excitation system, with a practical limit of ~5 μm spatial resolution being achievable, in optimum circumstances. Their sustained value needs to be considered alongside developments towards sensitive near-infrared CCD and diode array detectors, which in many circumstances may well provide a much better option when used in conjuction with an appropriate laser for circumventing fluorescence emissions, while still recording high signal-to-noise ratio (SNR) data from small samples. A dispersive system has been reported based around a tunable Ti-sapphire laser (640–900 nm radiation), an optical microscope, a red optimised beam splitter, a holographic grating spectrograph, and a CCD

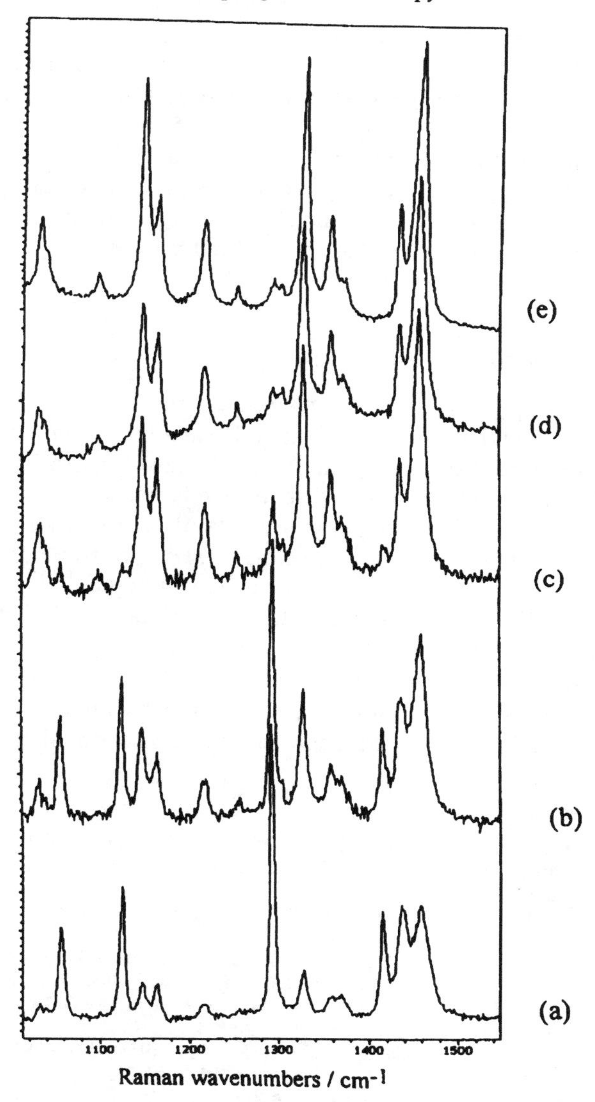

Figure 14 *Confocal Raman spectra recorded using a stigmatic spectrograph equipped with a CCD detector. The Raman data were obtained by focusing through a thin polyethylene (PE) film (*2 μm *thickness) positioned on top of a thick polypropylene (PP) layer. The spectra were recorded with the use of* 6 mW, 632.8 nm *radiation at the sample, 50 × objective, and accumulation times between* 3 *and* 5 *min:* (a) *focusing on the PE film;* (b)–(d) *focusing through the PE and into the PP;* (e) *PP reference spectrum.* (d) *shows only PP;* (a) *shows some weak PP features, since the thickness of the PE film is below the depth resolution of the set-up used*

(Reproduced from ref. 82, Williams *et al.*, *Appl. Spectrosc.*, **48**, 2, 232–235 (1994), by kind permission of the Society for Applied Spectroscopy; © 1994)

detector.[91,92] Comparable SNR data were recorded from a 20 μm diameter polyethylene fibre, albeit that the FT-Raman spectrum was recorded using 1W 1064 nm excitation over a spectrum accumulation time of 45 min, whereas for the dispersive experiment using 300 mW at 752 nm excitation a spectrum acquisition time of 10 s was all that was necessary![92] The authors also show good quality spectra from 2 and 20 μm diameter particles of TiO_2, using 760 nm excitation,[91,92] and from a 15 μm particle of the dye Rhodamine 580, using 820 nm excitation.[92] Multichannel dispersive Raman microanalysis systems using 1064 nm excitation have also been reported.[93,94]

Instrumentation: Infrared-microscopy

Infrared-microscopy, sometimes perceived as the newer cousin of Raman-microscopy, in fact has its roots in developments around 1950,[95–97] which included a commercial system by the Perkin-Elmer Corporation,[97] and while a low spectral resolution commercial dispersive microspectrometer system (based around a circular variable filter) was introduced in the 1980s,[98–101] it was the introduction in 1983 of the first commercial FTIR spectrometer–optical microscope combination which really launched the technique.[102,103] Since then, the advance of the technique has been rapid,[104,105] with one manufacturer offering a fully integrated system.[106] FTIR–microscope systems are now commonplace in industrial laboratories. They may be operated in both transmission and reflection modes, while, in addition, some systems offer the capabilities of ATR-microscopy[107–110] and grazing-incidence reflection–absorption spectro-

[91] Williams K.P.J., Dixon N.M., and Mason S.M., *Proc. 13th Int. Conf. Raman Spectrosc.*, ed. Kiefer W., Cardonna M., Schaack G., Schneider F.W., and Schrotter H.W., J. Wiley, Chichester, UK, 1992, 1070.

[92] Mason S.M., Conroy N., Dixon N.M., and Williams K.P.J., *Spectrochim. Acta*, 1993, **49A**, 5/6, 633.

[93] Barbillat J., da Silva E., Roussel B., and Howard S.G., *Microbeam Anal.*, 1991, 88.

[94] Barbillat J., da Silva E., and Roussel B., *J. Raman Spectrosc.*, 1995, **22,** 7, 383.

[95] Barer R., Cole A.R.H., and Thompson H.W., *Nature (London)*, 1949, **163**, 4136, 198.

[96] Cole A.R.H. and Jones R.N., *J. Opt. Soc. Am.*, 1952, **42**, 5, 348.

[97] Coates V.J., Offner A., and Siegler E.H., *J. Opt. Soc. Am.*, 1953, **43**, 11, 984.

[98] Hausdorff H.H. and Coates V.J., *Microbeam Anal.*, 1982, **17**, 233.

[99] Ramsey J.N. and Hausdorff H.H., *Microbeam Anal.*, 1981, 91.

[100] Scott R.M. and Ramsey J.N., *Microbeam Anal.*, 1982, **17**, 239.

[101] NanoSpec^{TM}/20IR, Nanometrics Inc., Sunnyvale, USA.

[102] BioRad, Digilab Division at the 1983 Pittsburgh Conference.

[103] Witek H., Krishnan K., and Kuehl D., *Abstracts 1983 Int. Conf. Four. Trans. Spec*, Durham, 5–9 September, 1983, Poster paper 1.12.

[104] Katon J.E. and Sommer A.J., *App. Spectrosc. Rev.*, 1990, **25**, 3/4, 173.

[105] Katon J.E. and Sommer A.J., *Anal. Chem.*, 1992, **64**, 19, 931A.

[106] (a) IRμs^{TM}, Spectra-Tech Inc., Stamford, USA.
(b) Reffner J.A. and Wihlborg W.T., *Microbeam Anal.*, 1991, 240.

[107] Reffner J.A., Wihlborg W.T., and Strand S.W., *Am. Lab.*, 1991, **23**, 6, 46.

[108] Reffner J.A., Alexay C.C., and Hornlein R.W., *Proc. 8th Int. Conf. Four. Trans. Spectrosc*, SPIE, 1991, **1575**, 301.

[109] Reffner J.A. and Wihlborg W.T., *Microbeam Anal.*, 1991, 95.

[110] Sweeney M., Gaillard F., Linossier I., Boyer N., and Stevenson I., *Spectra Analyse* 1994, **176**, 33.

scopy–microscopy[107–112] for spatially resolved surface and surface-layer characterisations.

A schematic of a commercial FTIR-microscope set-up is shown in Figure 15(a). A schematic of another commercial FTIR-microscope set-up is shown in Figure 15(b), alongside a photograph of such a system. The condenser and objective are both on-axis, all-reflecting Schwarzchild-type Cassegrainian lenses, and the infrared and visible optical trains are collinear and parfocal. The sample is supported on the microscope sample stage. For transmission studies the support for small samples is most usually some form of infrared transparent window; a 13 mm diameter KBr disc is commonly used. The objective lens focuses the radiation from the interferometer onto the sample; the radiation transmitted by the sample is collected by the condenser. This produces a magnified image at its conjugate image plane, which may then be masked so that essentially only radiation from the area of interest is received by the detector.[104] For reflection studies, the upper lens serves to both focus the radiation onto the sample and collect the radiation reflected from it. Typical objectives used for the infrared are 15 ×, with a NA of 0.58 or, for higher magnification, 32 ×, with a NA of 0.65. Working distances of up to about 15 mm are available typically for a 15 × objective or 24 mm with a 6 × objective, with the full field of view for the visual image being of the order of 1.3 mm diameter; typical diameters for the focused infrared beam at the sampling stage are between 500 and 1000 μm.

Since the diameter of the infrared beam at the sample focal plane is usually significantly greater than the dimensions of the sample or sample area being interrogated, then delineating (aperturing, masking-off of the surround) of the 'analyte' must take place to preclude unwanted radiation reaching the detector, whether it be from stray-light or from radiation that has interacted with material adjacent to the sample area of interest. In the microscope this is easily practicable through the use of a variable aperture sited in a remote image plane of the optical train. A variable aperture based on four independent high contrast blackened 'knife-edge' blades is probably the most convenient and widely used, enabling square and rectangular shaped apertures of differing dimensions to be readily formed. View through (glass blade, infrared opaque) apertures are also available from some manufacturers of infrared microscopes. These allow full detail of the magnified sample image to be viewed after the apertures have been set.

Unlike Raman-microscopy, source irradiance is rarely a problem with infrared-microscopy, localised heating/melting only becoming a consideration when the sample has a thermal or phase transition near ambient. In fact, the coherence and higher brilliance of mid-infrared radiation available from a

[111] Reffner J.A., SPIE Vol. 1437 *App. Spectrosc. Mater. Sci.*, 1991, 89.
[112] Reffner J.A. and Wihlborg W.T., *Proc. 8th Int. Conf. Four. Trans. Spectrosc.*, SPIE, 1991, **1575**, 298.

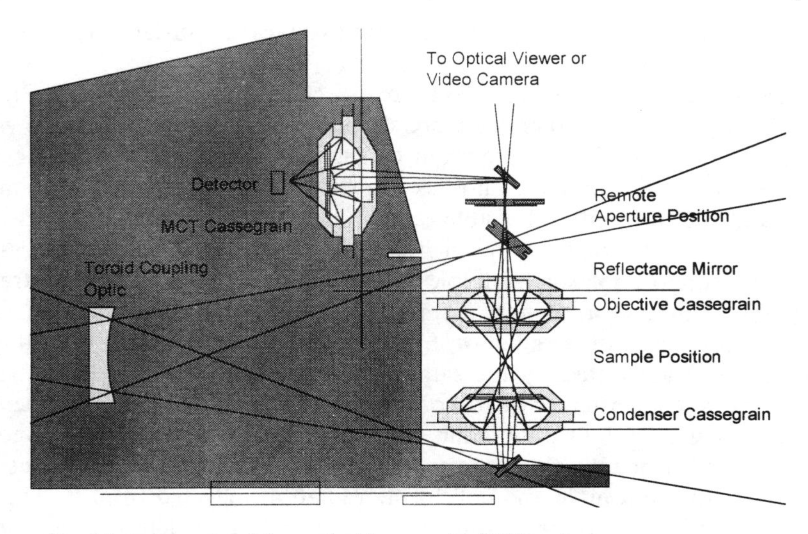

Figure 15 (a) *Schematic of the optical layout of a FTIR-microscope*
(Picture by courtesy of Perkin-Elmer Ltd., Beaconsfield, UK)

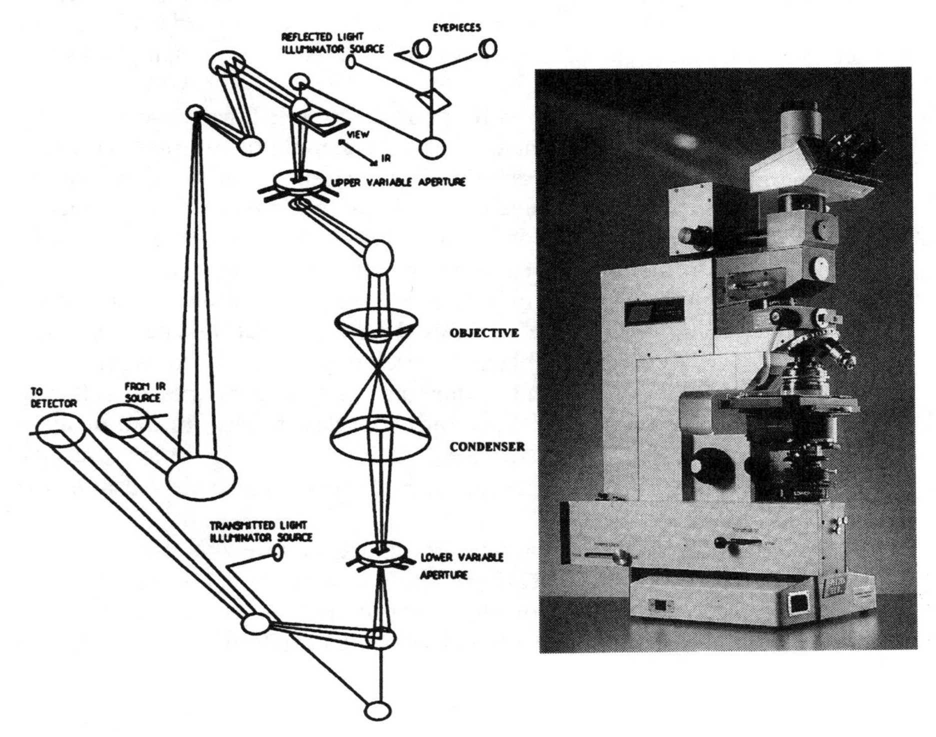

Figure 15 (b) *Schematic of a FTIR-microscope and a photograph of the Spectra-Tech IR-Plan® Research FTIR-microscope*
(Schematic reprinted from ref. 123, p.5 by courtesy of Marcel Dekker, Inc. Photograph reproduced by kind permission of Spectra-Tech, Inc.)

synchrotron beam opens up the possibilities for working closer to the diffraction limit.[113,114]

For very small samples, particularly those embedded within a matrix, which will invoke the use of very small apertures, it is particularly important to minimise diffraction-induced spectral impurity. Diffraction causes spill-over of radiation into regions outside the defined optical path.[115–117] The collection of this must be minimised if optimum spectral contrast is to be achieved, or optimum discrimination of the spectrum of the area of interest is to result, *i.e.* any stray-light component, which may contain spurious intensity contributions from the matrix or surrounding material, must be reduced. In transmission studies a widely used way of achieving this is to have a second aperture in another remote (confocal) image plane on the other side of the sample, see Figure 16. (This has been curiously trade-marked as 'Redundant Aperturing®',[118] since the purpose of the first aperture is to define (mask) the sample, while the function of the second (confocal) aperture is solely to minimise the recording of diffraction induced stray-light.)

For optimum performance, small area (typically, 250 μm × 250 μm) infrared detectors, usually a MCT, are fitted directly to the microscope, which itself is usually coupled to an external port of the spectrometer. Compromise, lower cost accessories which fit directly into the sample compartment are also available, see Figure 17. These have no independent infrared detector, and afford minimal sample manipulation, yet for routine contaminant work can be extremely rewarding. The spectra of Figure 4 were recorded with an early version of such an accessory, the SpectraScope™ marketed by Spectra-Tech, Inc., who also marketed for a limited time the Surface-Scope™, a grazing-incidence accessory. More recently other in-sample compartment accessories have become available. The SplitPea™[42–44] ATR accessory referred to earlier might be considered as another example, see Figure 6. These types of in-sample compartment microsampling accessories might be thought of as sophisticated beam-condensers, providing a capability of visual inspection.

Contour and axonometric plots and functional group maps may be generated from consecutive point-by-point absorbance spectra synchronised with step

[113] (a) Reffner J.A. and Williams G.P., *First Workshop on Applications of Synchrotron Radiation to Infrared Microspectroscopy*, February 3, 1994, National Synchrotron Light Source, Brookhaven National Laboratory, United States Department of Energy, 1994;
(b) Reffner J.A., Carr G.L., and Williams G.P., *Proc. 29th Ann. Conf. Microbeam Anal. Soc.*, Breckenridge, Colorado, August 6–11, 1995, p. 113.

[114] Ugawa A., Ishii H., Yakushi K., Okamoto H., Mitani T., Watanabe M., Sakai K., Suzui K., and Kato S., *Rev. Sci. Instrum.*, 1992, **63**, 1, 1551.

[115] Messerschmidt R.G., in *The Design, Sample Handling and Applications of Infrared Microscopes*, ed. Roush P.B., ASTM Special Technical Publication 949, ASTM, Philadelphia, USA, 1987, p. 12.

[116] Messerschmidt R.G., in *Infrared Microspectroscopy. Theory and Applications*, eds Messerschmidt R.G. and Harthcock M.A., Marcel Dekker, New York, USA, 1988, ch. 1.

[117] Sommer A.J. and Katon J.E., *Appl. Spectrosc.*, 1991, **45**, 10, 1633.

[118] Redundant Aperturing™ is a registered trademark of Spectra-Tech Inc.

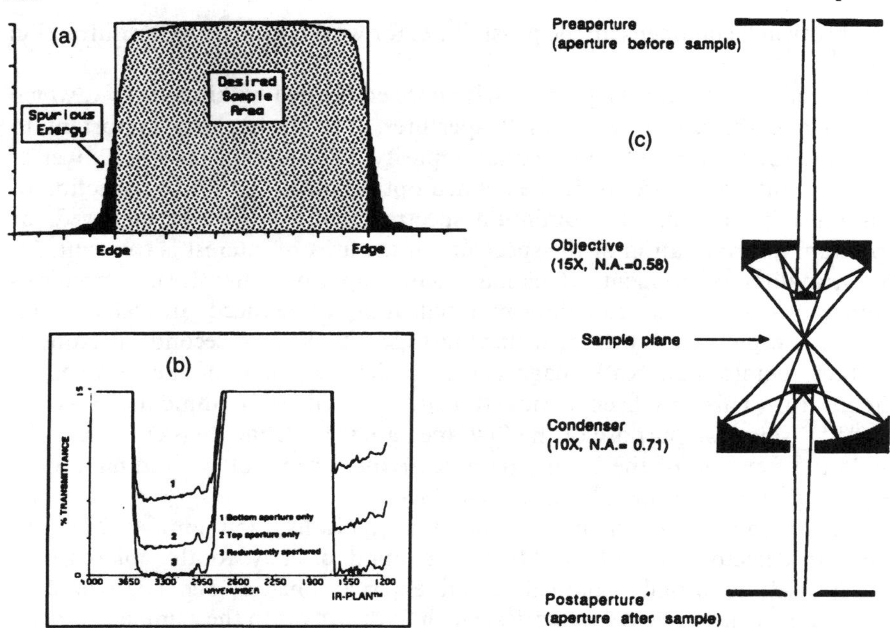

Figure 16 (a) *The imaging of two parallel high contrast edges, as would be used to delineate a specimen, in the specimen plane of a FTIR microscope. The spurious energy imaged outside the desired area manifests itself as stray-light*
(Schematic reprinted from ref. 123, p.32 by courtesy of Marcel Dekker, Inc.)
(b) *Overlaid transmission spectra of a single filament hair sample, which has a thickness such that major parts of the spectrum should be totally absorbed. Spectra 1 and 2 were produced with remote aperturing below and above the sample respectively. Spectrum 3 was produced with both remote apertures in place (Redundant Aperturing™), and demonstrates the virtual elimination of the stray-light component*
(Reproduced with kind permission from Spectra-Tech Inc., *IR-Plan Application Notes* Vol. 1, No. 3)
(c) *Schematic of the optical configuration of the Spectra-Tech Inc., IR-Plan® microscope, showing the pre- and post-sample plane remote aperturing*
(Reproduced from ref. 117, Sommer A.J. and Katon J.E., *Appl. Spectrosc.*, 1991, **45**, 10, 1633–1640, by kind permission of the Society for Applied Spectroscopy; © 1991)

displacements of a motorised microscope *x–y* sample-stage.[119,120] Rapid high spatial resolution imaging of absorption spectra in the visible and near-infrared regions 400–1900 nm (0.4–1.9 μm) have been demonstrated using an AOTF and CCD detector with an infinity-corrected microscope.[121]

[119] Harthcock M.A. and Atkin S.C., in *Infrared Microspectroscopy. Theory and Applications*, ed. Messerschmidt R.G. and Harthcock M.A., Marcel Dekker, New York, USA, 1988, ch. 2.
[120] Harthcock M.A. and Atkin S.C., *Appl. Spectrosc.*, 1988, **42**, 3, 449.
[121] Treado P.J., Levin I.W., and Lewis N., *Appl. Spectrosc.*, 1992, **46**, 4, 553.

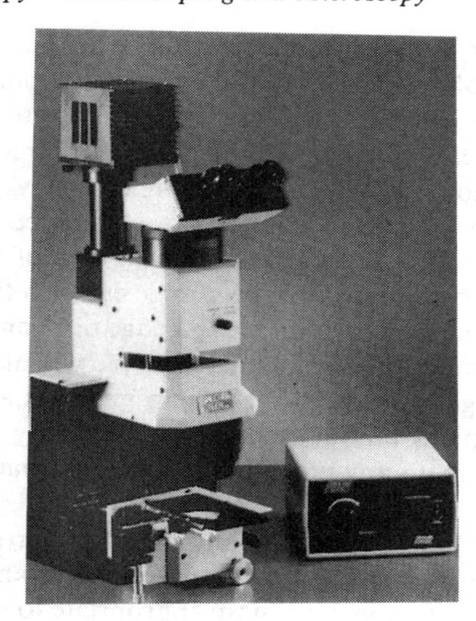

Figure 17 *Photograph of the EZ-Scope*^{TM}*, in-sample compartment FTIR-microscope* (Reproduced by kind permission of Spectra-Tech, Inc.)

Vibrational Spectroscopy–Microscopy Sample Presentation Methods

In this section common sample presentation methods for each vibrational technique are briefly discussed.

Collecting, transferring, mounting, sectioning, and thinning of samples require many skills and tools borrowed from the optical microscopists' armoury. A stereoscan microscope is an almost essential aid to sample visualisation and manipulation. Tweezers, scalpels, fine needles (*e.g.* tungsten needles) and pointed probes, razor blades, micropipettes, microtome, grid supports, compression and roller devices, *etc.* may all be necessary tools for preparing or mounting the optimum sample for investigation. Thermomicroscopy and polarisation studies will require appropriate hot/cryogenic cells and polarisers for the microscope, respectively, while mapping experiments will almost certainly benefit from a motorised *x–y* stage.

Sample Presentation: Raman-microscopy

Particle-like samples (*e.g.* powders, crystal mixtures) may simply be spread onto the surface of a glass microscope slide. Discrete particles may then be identified by visual examination, and their individual Raman spectra recorded. Isolated contaminants may also be characterised in this manner, although in many circumstances they may be identified *in situ*, utilising the 'microprobe' advantage of Raman-microscopy. Individual solid, liquid, and gaseous phases of fluid inclusions trapped within minerals have been studied directly in many geochemical, mineralogical, or precious stone related investigations. These are normally made on sections of about 0.5–1 mm thickness with a polished face(s). Micro-Raman spectroscopy has also been used successfully to analyse the gaseous constituents of bubbles in glass, by focusing the laser directly into the bubble.

The depth resolution of confocal Raman-microscopy greatly facilitates the direct analysis of an embedded analyte or sub-surface layers. However for depths $> ca.$ 6 μm, a section which contains an area of the solid analyte will probably need to be microtomed, or the sample cut such that the component(s) of interest is exposed at or near a surface. An edge or clean cut thickness surface of a suitably clamped multi-layer laminate sample will also be appropriate for microprobe layer fingerprinting. If the 'clamp' includes a support matrix (*e.g.* wax, ice, epoxy resin) to hold a flexible sample steady during microtoming, attention must be given to any potential transgress of matrix material into the sample specimen.

Fibre samples may be examined laterally either supported on a platform such as a microscope slide or free-standing mounted across an aperture. An end or cross-sectional examination may be more appropriate to some studies.

Fluid-like samples of low volatility may be examined as a drop or smear on a microscope slide, or liquids may be contained within a thin-walled glass tube.

Sample Presentation: Infrared-microscopy

By far the majority of infrared-microscopy applications have been undertaken using transmission measurements; although, since its introduction, the ATR objective has rapidly become a popular tool in industrial laboratories that need to study microscopic samples. The external reflection techniques of specular reflectance, 'transflectance' and grazing-incidence also have their place in the armoury of problem-solving tools. Many sample preparation and presentation methods are outlined in ref. 4, and discussed with reference to the forensic examination of paint samples in ref. 122. Numerous useful tips and examples may also be found in ref. 123.

Transmission methods. A high contrast absorbance spectrum that is essentially free from aberrations will only be recorded from a thin, uniform thickness specimen, for which the other dimensions are greater than the interrogation wavelengths and the delineating aperture, although interference fringes may still be a problem from very flat, transparent specimens. As the sample size or aperture reduces to the order of that of the infrared radiation wavelengths or less, diffraction increasingly becomes an important consideration. At an aperture or sample edge, diffraction will redistribute some of the radiation beyond the boundary and the detector will see stray-light,[115–117,123] see Figure 16. If the

[122] Allen T.J., *Vib. Spectrosc.*, 1992, **3**, 217.
[123] *Practical Guide to Infrared Microspectroscopy*, ed. Humecki H.J., Marcel Dekker, New York, USA, 1995.

sample is significantly larger than the delineating aperture then this spill-over will essentially not be perceived. An essentially flat specimen is necessary since other geometries such as a fibre will refocus the radiation beam and introduce spectral distortions. However, while not advocating second-quality spectra, generic typing may still be possible with less than perfect spectra. This is illustrated in the examination of fibres, (see also pp. 293–297). Fibres may be examined, after appropriate aperturing, supported on or in infrared transparent windows (*e.g.* 13 mm KBr discs), or as free-standing specimens mounted over an aperture. Their geometry does lead to spectral distortions, but these may be minimised or eliminated by 'flattening' the fibre, see Figure 18. For some fibres this may be possible by using a tool such as a roller-knife, see Figure 19, while for others it will be necessary to press them in a compression cell. However, such treatments may alter certain characteristics of the fibre, such as its morphology, that are important to the investigation, so caution in their use must be exercised. The diamond anvil or compression cell is an extremely effective means of containing, flattening, and reducing the thickness of a wide range of samples. However the high degree of parallelism between the window faces frequently gives rise to the appearance of interference fringes in the recorded spectrum.[124] For non-retractable samples this may often be avoided by opening the cell after the compression, since it is usual for the compressed sample to remain attached to one of the cell windows. It is then only necessary to present this window to the micro-spectrometer, thus eliminating the interference fringes caused by the full cell configuration.[124] The single beam background mid-infrared spectrum for the FTIR measurement is usually recorded from a single particle of an infrared transparent alkali-halide crystal such as KBr, which has been compressed in the cell at the same time alongside the sample.

For investigating the layer structure of multi-layer or laminated materials, it is common practice to microtome a thin section through the thickness of the sample such that when this direction is presented perpendicular to the micro-focused infrared beam, each layer may be successively interrogated with an appropriate rectangular (slit-shaped) aperture.

To facilitate sampling, several new preparation tools have been introduced. These include the roller-knife[16] mentioned above, which can also double as a sample collector and support for transflectance measurements; for example, it may be used to gather analyte specimens such as micro amounts of liquids from surfaces by rolling the wheel across the surface. This action will transfer some of the analyte onto the wheel's surface, where it may be examined *in situ*. Other tools,[16] see Figure 20, include diamond cleaving knives, diamond-edge or carbide-steel blade micro planes, and gold-coated filters. Cleaving knives with either a straight or 60° edge are available for scraping or shaving off thin sections of a sample; the micro planes may be used to plane or scrape a uniform thickness slice from the surface or thickness direction of a polymer or other solid material. Gold coated filters, diameter 13 mm, pore size of 0.8 μm, may be used to separate traces of suspended solids from liquids. They are used in a syringe

[124] Lin-Vien D., Bland B.J., and Spence V.J., *Appl. Spectrosc.*, 1990, **44**, 7, 1227.

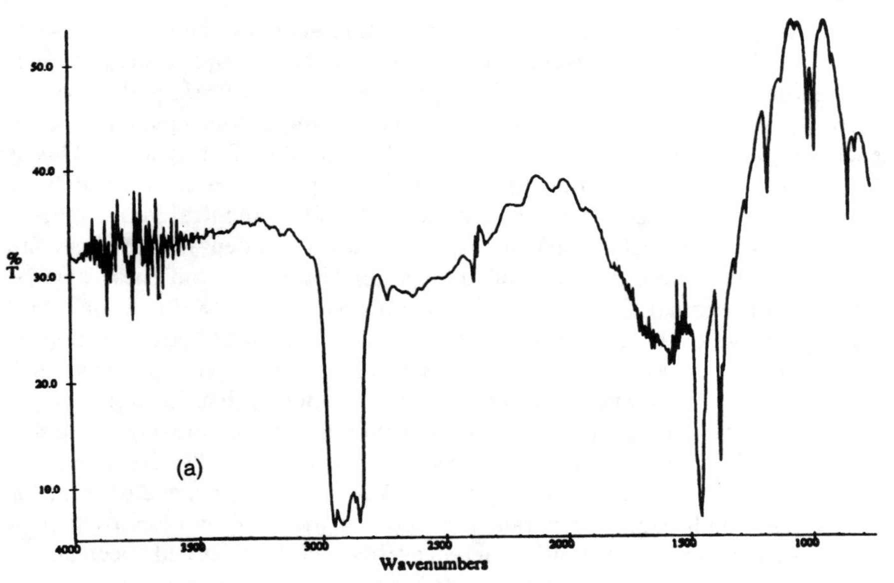

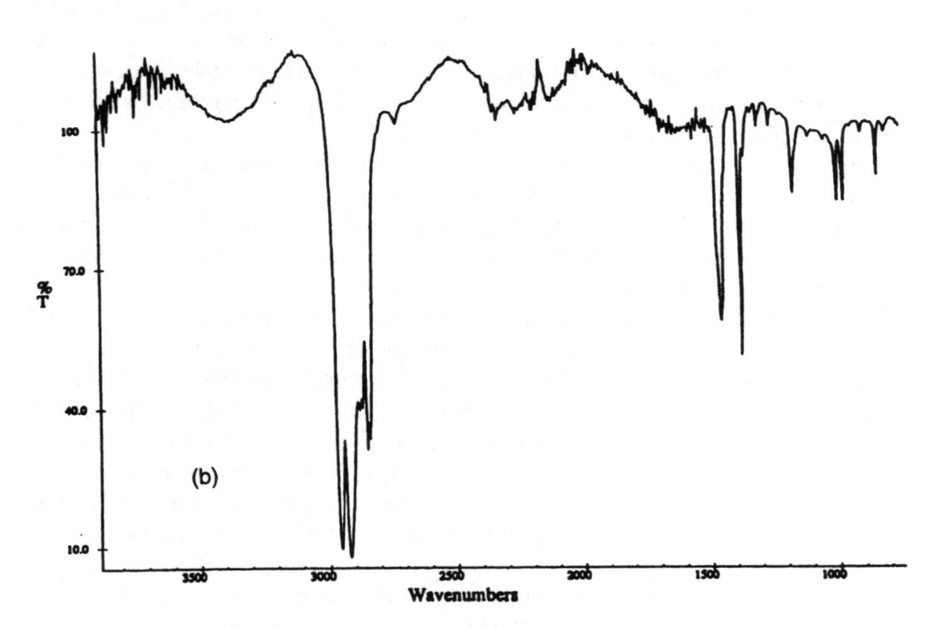

Figure 18 *FTIR-microscopy spectra recorded from a 16* μm *diameter polypropylene fibre:* (a) *original fibre;* (b) *fibre after flattening. Note: although* (b) *shows interference fringes superimposed on the fibre spectrum, the spectral distortions are much less severe than in* (a), *and the relative band intensities much truer in* (b)

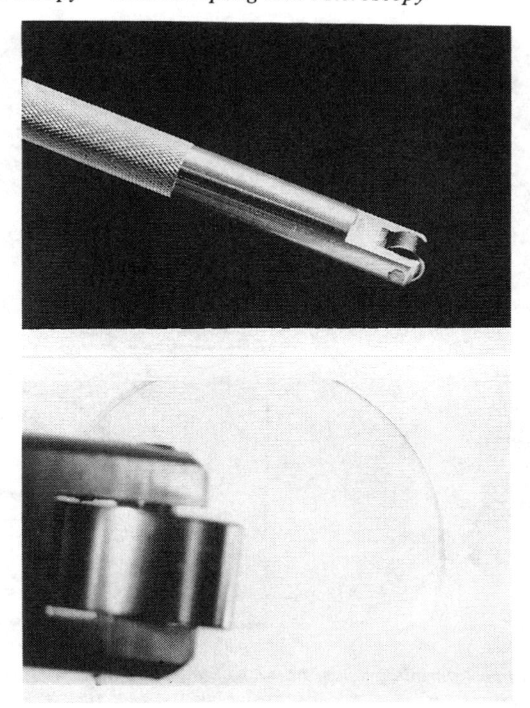

Figure 19 *Photograph of a Roller Knife, and expanded view of the roller wheel* (Reproduced by kind permission of Spectra-Tech, Inc.)

holder to isolate trace contaminants in samples such as injectable drugs; the isolated contaminant may be examined *in situ* on the filter by the reflection–absorption technique, see below.

Reflection Methods. The reflection–absorption or 'transflectance' approach is one of the most commonly used reflection techniques for examining microsamples. It is particularly appropriate to thin samples on non-absorbing reflective substrates. The 'standard' support and background reference for these measurements is a gold-coated microscope slide or metal disc. It must be remembered that the reflection–absorption measurement is essentially a superimposition of a specular reflectance component from the uppermost sample surface onto an essentially double-thickness transmission spectrum, the latter component being derived from that radiation which has passed through the sample twice having been reflected from the reflective substrate in contact with the sample back surface. The specular reflectance component is normally relatively weak, but in the case of strong absorption features, where there is a concomitant large change in refractive index, apparent band inversions may occur. In some circumstances, such as thick specimens or strongly absorbing or highly filled samples, useful qualitative absorbance spectra may be extracted from specular reflectance measurements by apply-

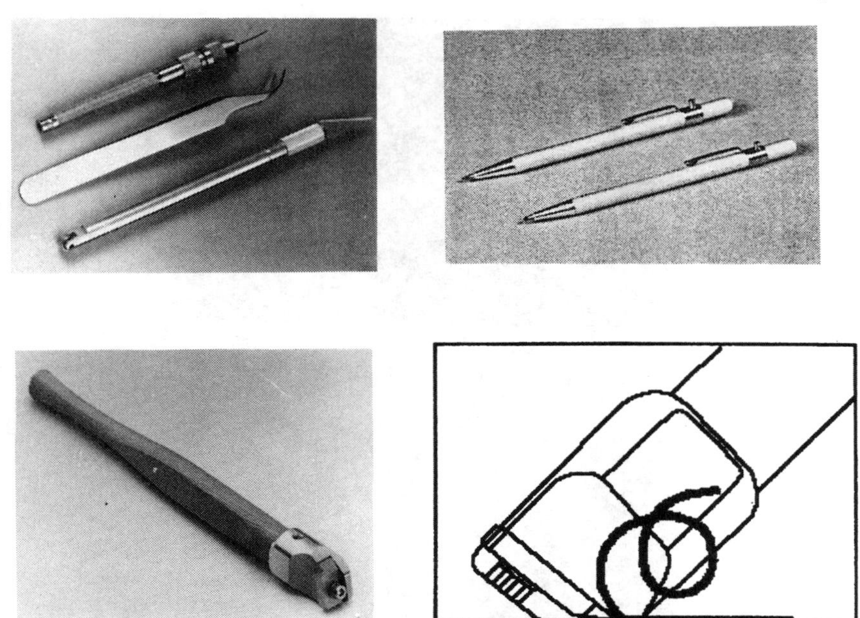

Figure 20 *Some micro-sample preparation and handling tools:* Top left, *probe, stainless steel tweezer and roller knife (showing its dual functionality, with the razor knife incorporated at the other end;* Top right, *Diamond Knives;* Bottom left, *a Micro Plane tool;* Bottom right, *schematic of the Micro Plane being used to produce a uniform thickness slice from a surface*
(Reproduced by kind permission of Spectra-Tech, Inc.)

ing the Kramers–Kronig algorithm.[125,126] In these cases the observed radiation must be that which is solely reflected from the topmost surface of the sample. A diffuse reflectance accessory has been adapted into a 'low-cost' reflectance FTIR-microscope for transflectance and specular reflectance measurements.[127,128]

For very thin films on reflective substrates then grazing-incidence reflection–absorption needs to be employed.[108–112] A schematic of a grazing-incidence objective is shown in Figure 21; the objective has a NA of 0.996 with a maximum angle of incidence of 85°.[111] The applicability of the technique to tribological studies such as friction, lubrication, and wear mechanisms has been demonstrated.[111,112] Figure 22 shows the spectrum of a fluorocarbon lubricant recorded from a 100 μm diameter spot on the surface of a 2 mm diameter ball-bearing intended for use in a gyrocompass for a satellite.[112]

[125] Chalmers J.M., Croot L., Eaves J.G., Everall N., Gaskin W.F., Lumsdon J., and Moore N., *Spectroscopy*, 1990, **8**, 13.
[126] Reffner J.A. and Wihlborg W.T., *Int. Lab.*, 1990, July/August, 19.
[127] Jansen J.A.J., van der Maas J.H., and Posthuma de Boer A., *Appl. Spectrosc.*, 1991, **45**, 7, 1149.
[128] Jansen J.A.J., Paridaans F.N., and Heynderickx *i.e.*J., *Polymer*, 1994, **35**, 14, 2970.

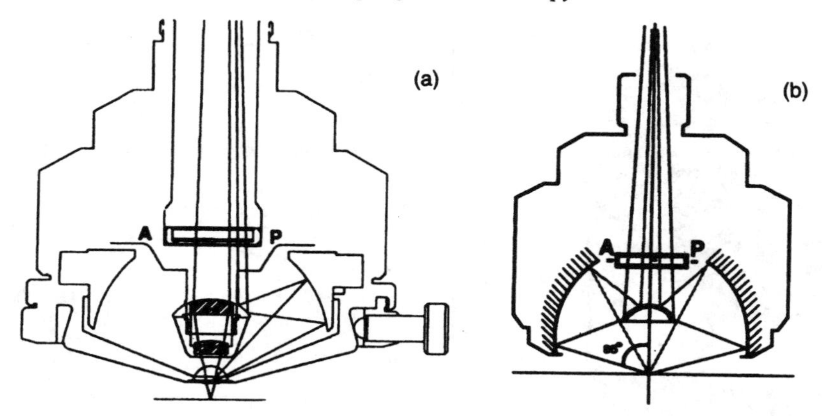

Figure 21 (a) *Schematic diagram of an ATR objective lens for a FTIR microscope;* (b) *Schematic diagram of a grazing-angle incidence objective lens for a FTIR-microscope.* AP *is the position for inserting annuli that define angular ranges for viewing and spectral analysis*
(Reproduced from Spectra-Tech *Product Data Sheets, PD-4* and *PD-8*, by kind permission of Spectra-Tech, Inc.)

For surface or surface-layer related studies then FTIR ATR-microscopy is probably the optimal technique.[108,110] A schematic of an ATR objective is shown in Figure 21. The internal reflection element is a modified hemisphere, with a bevelled edge to give a contact surface area of about 300 μm diameter.

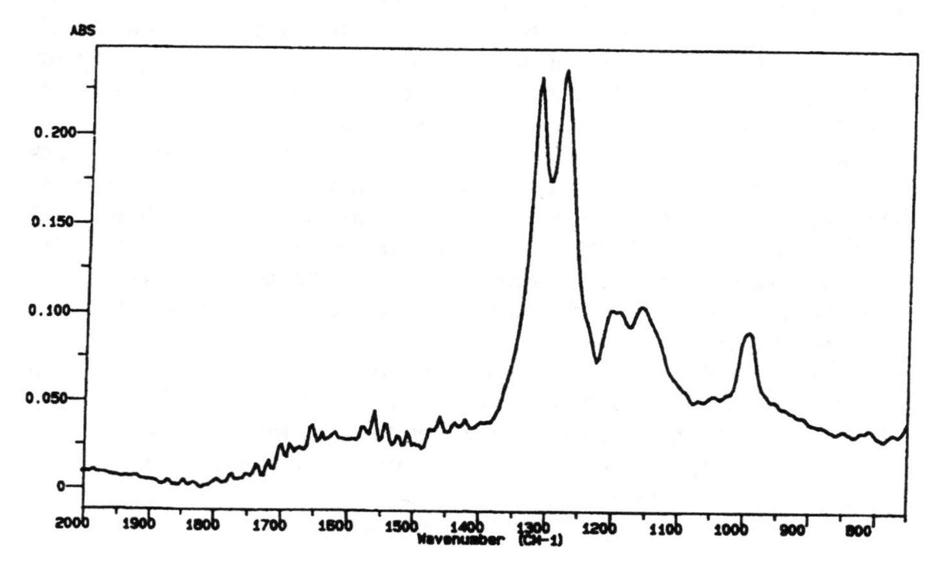

Figure 22 *FTIR spectrum of a fluorocarbon lubricant on a* 2 mm *ball bearing acquired using a grazing angle objective (*1000 scans, 8 cm^{-1} *resolution)*
(Reproduced from Spectra-Tech *Product Data Sheet PD-4*, by kind permission of Spectra-Tech, Inc.)

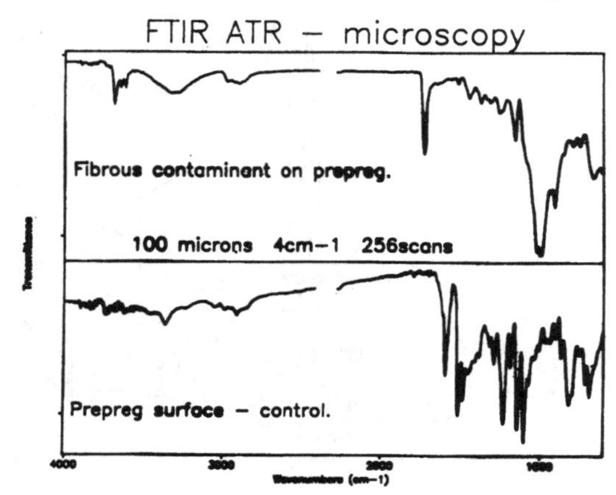

Figure 23 Left, *Photo-micrograph of a 'fibrous' contaminant in the surface of a C-fibre polyaromatic/epoxy prepreg. (The prepreg is about a 50:50 volume mix of* 7 μm *diameter uniaxially oriented C-fibres in the polyaromatic/epoxy resin.)*; Right, *FTIR-ATR (ZnSe prism) microscopy spectra of 'fibrous' contaminant (top) and prepreg surface (bottom). The top spectrum has features characteristic of the brown paper which serves as a backing to the prepreg, viz. a cellulosic material and kaolin filler, suggesting that the contaminant was 'debris' from the prepreg backing paper*

The interrogated surface area may be apertured down in size by a mask placed in a conjugate image plane. Another type of ATR objective uses a spring-loaded crystal, which enables automated point-by-point mapping of surfaces.[288] *Remember* the higher refractive index ATR element will focus (refract) the incident radiation beam to a reduced sampling diameter spot by a factor equivalent to 1/refractive index of the ATR element.

ATR objectives are available with a ZnSe crystal, which has a refractive index (RI) of 2.4, or for harder samples with a diamond crystal (RI = 2.4). For the examination of higher refractive index samples or shallower sampling depths then a Ge (RI = 4.0) or Si (RI = 3.4) crystal may be used. Unfortunately, these last two crystal materials are opaque to visible radiation, so contact of the crystal with the sample surface (wetting-out) cannot be observed, and a contact (pressure) alert system should be employed.[16] Figure 23 illustrates an application of the FTIR ATR-microscopy technique. In ATR mode the system shown in Figure 21(a) has a NA of 0.87, and the angle of incidence is 45°.[16]

References 104, 123, 129, and 130 are useful sources for further information on FTIR-microscopy instrumentation and sample handling, as well as giving many application examples.

[129] *Infrared Microspectroscopy. Theory and Applications*, ed. Messerschmidt R.G. and Harthcock M.A., Marcel Dekker, New York, USA, 1988.

[130] *The Design, Sample Handling and Applications of Infrared Microscopes*, ed. Roush, P.B., ASTM Special Technical Publication 949, ASTM, Philadelphia, USA, 1987.

5 Vibrational Spectroscopy–Microscopy Applications

Infrared and Raman microscopy/microprobe spectroscopy techniques have proven to be invaluable tools for the spatially-resolved identification, study, and characterisation of industrial materials and products. Applications are diverse and cover a wide range of manufacturing, fabrication, and processing industries. They extend from the more simple fingerprinting of bulk or surface contaminants or inhomogeneities, through to detailed studies of the chemical composition or physical characteristics of a material or product in order to gain insight into its constitution, properties, or performance characteristics. A review of applications of vibrational microspectroscopy to chemistry, published in 1994, contains nearly 400 references covering general chemistry, polymer chemistry, applied chemistry, biochemistry, geochemistry, and art and archaeometry.[131] (It does not include forensic applications.) Many applications of vibrational spectroscopy–microscopy in polymer science and characterisation have also been considered in separate publications;[125,132,133] while, many examples of Raman applications may be found in refs. 134 and 135.

The two vibrational spectroscopy techniques provide both complementary information and complementary attributes to microscopic investigations. These will be exampled in this section. Also, in an industrial problem-solving environment, a natural synergism exists between light microscopy, in particular polarised light microscopy, and vibrational spectroscopy–microscopy.[136] The optical properties, thermal behaviour (*e.g.* melting-point), solubility, colour and morphology, and phase discrimination are key features, particularly in particle analysis, product contaminations or defect/failure analysis (see below). The information derived from energy-dispersive X-ray (EDX) analysis is also vital in that it provides elemental data, which, when combined with the evidence from optical examinations, are signposts for the vibrational spectroscopist, indicating, for example, whether the analyte is homogeneous or heterogeneous, an inorganic, a crystalline solid, a natural product, a thermoplastic, or a cross-linked resin.

To illustrate the scope of the applications they are conveniently categorised into two types: those involving mainly chemical structure characterisations, and those involving essentially physical property studies. Each of these may be further sub-divided into some of the more common application areas. While perhaps not all may be classed as strictly industrial applications, they have been chosen to illustrate the breadth, scope, and importance of these types of study.

[131] Katon J.E., *Vib. Spectrosc.*, 1994, **7**, 3, 201.
[132] Pastor J.M., *Makromol. Chem., Macromol. Symp.*, 1991, **52**, 57.
[133] Chalmers J.M. and Everall N.J., *Macromol. Symp.*, 1995, **94**, 33.
[134] Adar F., *Microchemical J.*, 1988, **38**, 50.
[135] Huong P.V., *Spectra Analyse*, 1995, **182**, 35.
[136] Reffner J.A., Coates J.P., and Messerschmidt R.G., *Int. Lab.*, 1987, April, 38.

Vibrational Spectroscopy–Microscopy Applications: Chemical Studies

This is probably the most widespread type of application, finding extensive use in, for example, contaminant and defect studies of materials, hair analysis, fibre and laminate structure characterisations, probing inclusions in geological samples, semiconductor inhomogeneity studies, and biological tissue mapping. In addition, FTIR-microscopy enjoys a popular position within industry as a detector of chromatographically separated fractions, both in-line and trapped; this attribute is discussed separately in Chapter 7.

Particle Analysis

A main interest in the identification of discrete fine particles by microspectroscopy techniques stems from their environmental effects, such as air pollution and associated health concerns.[137] One of the first Raman microprobe systems, which was a modified conventional laser Raman spectrometer, was developed to obtain useful spectra from discrete solid particles as small as 0.7 μm in linear dimension,[70] and subsequently used for the analysis of individual urban airborne particulates and carbonaceous material associated with them.[138] Although most ambient airborne particles are too small for discrete analysis by FTIR-microscopy, conventional FTIR transmission spectroscopy on fine-particulate matter collected on Teflon® filters has been shown to be an effective method for the sulfate speciation and quantitation of ammonium bisulfate in ambient aerosol samples.[139,140] At the other end of the pollution cycle, Raman microprobe analysis has been employed to identify inhaled particles in the size range 3–4 μm present as inclusions in sections of lung and lymph-node tissues.[141] The particulate species identified included calcite and magnesite in the lung tissue and silica in the lymph-node tissue, from patients medically diagnosed to have died from silicosis.

A Raman-microscopy and FTIR-microscopy assessment of the nature of particles in settled office and laboratory dust has been carried out,[142] to provide an awareness of prevalent extraneous contaminants, in order to ensure that evidence relevant to a product failure or defect investigation may be correctly established. The most prevalent constituents in the dust samples were shown to be cellulosic, mainly wood cellulose from paper. The spectra also showed evidence from many common paper additives such as kaolin, $CaCO_3$, poly(methyl methacrylate), epoxy resin and photocopying toner. Polyamides were the next most common from sources such as skin, hair, and

[137] Van Grieken R. and Xhoffer C., *J. Anal. At. Spectrom.*, 1992, **7**, 81.
[138] Blaha J.J., Rosasco G.J., and Etz E.S., *Appl. Spectrosc.*, 1978, **32**, 3, 292.
[139] Krost K.J. and McClenny W.A., *Appl. Spectrosc.*, 1994, **48**, 6, 702.
[140] McClenny W.A., Krost K.J., Daughtrey E.H., Williams D.D., and Allen G.A., *Appl. Spectrosc.*, 1994, **48**, 6, 706.
[141] Buiteveld H., De Mul F.F.M., Mud J., and Greve J., *Appl. Spectrosc.*, 1984, **38**, 3, 304.
[142] Lang P.L., Sommer A.J., and Katon J.E., in *Particles on Surfaces 2*, ed. Mittal K.L., Plenum Press, New York, USA, 1989, p. 143.

clothing. Others constituents of the dust included amorphous carbon, talc, amorphous silica, glass, wool, cotton, synthetic polymer fibres, and cosmetic residues.[142] Micro-FTIR has also been used to characterise individual interplanetary dust particles,[143] for which the sample mounts were made by pressing crushed fragments (1–5 μm thickness) of a single particle into a freshly cleaved KBr crystal.

Contaminant Analysis

The need to readily obtain molecular information on microscopic 'foreign-material' in or on industrial products may be key to many process investigations or commercial support activities. As a consequence, one of the primary applications of vibrational microspectroscopy techniques has been their use in contaminant analysis. An extraneous particle or similar contaminant species may often be essentially cleanly removed from its surroundings with, for instance, a fine pointed probe such as a tungsten needle. It may then be mounted on an appropriate support for a spectroscopic examination: a glass microscope slide for a Raman study; a KBr disc for an infrared transmission measurement. In some case studies it may also be possible, using either technique, to obtain adequate selective molecular structure information on a 'buried' contaminant by microtoming a section through it from the host material. The probe beam advantage of Raman may also offer the opportunity for an *in situ* examination. Surface contaminants may frequently be examined directly on their host substrate by either Raman microscopy or FTIR ATR-microscopy.

In our experience, time spent on assembling a reference spectral library of prevalent or common sources of contaminants is time well spent, see p. 278. This should include cellulosic sources, such as wood, paper and cotton fibres, carbon types (for Raman), silicone rubber, silica, skin, natural fibres (*e.g.* hair), synthetic polyamides, *etc.*

Figures 24(a)–24(d) illustrate some examples of contaminant characterisations in which we or our colleagues have been involved. They have been chosen to illustrate the advantages of differing spectroscopic approaches. Figure 24(a) shows the FTIR ATR (ZnSe element)-microscopy spectra recorded from minute droplets contaminating the surface of a polyethylene cast tube. The 'surface defect' in this case is clearly identified as poly(dimethyl siloxane), (PDMS), probably from a source of silicone oil or grease. The characteristic absorption bands of PDMS are 1260 cm^{-1} (sharp), near 1100 cm^{-1} (broad doublet) and at 800 cm^{-1}. In this instance, attempts to transfer enough contaminant onto the surface of a roller knife for a 'transflectance' examination failed; the droplets also proved too fluorescent for fingerprinting by a micro-Raman examination. The example of Figure 24(b) concerned identifying *in situ* the contaminant at the base of a pit defect in a paint test panel. This was inaccessible to FTIR techniques, but readily probed by micro-Raman, and

[143] Wopenka B. and Swan P.D., *Mikrochim. Acta*, 1988, **I**, 183.

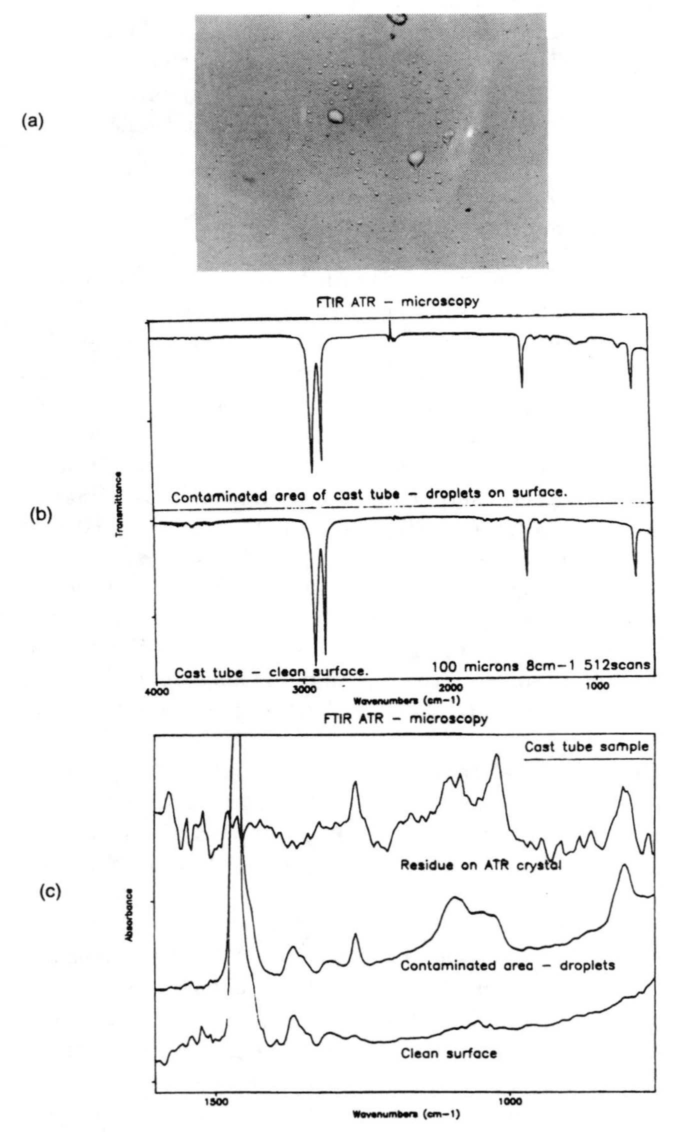

Figure 24(a) (a) *Photo-micrograph (* × 800*) of 'liquid' droplets contaminating the surface of a polyethylene cast tube;* (b) *FTIR-ATR (ZnSe prism) spectra (*100 μm *diameter aperture) recorded from the contaminated area of the cast tube (upper) and a clean area of the surface (lower). Weak bands can be observed in the upper spectrum in the region* 1300 cm^{-1} *and* 750 cm^{-1} *which are characteristic of poly(dimethyl siloxane) (PDMS), suggesting contamination by silicone oil/grease;* (c) *FTIR-ATR absorbance spectra over the range* 1600 cm^{-1} *to* 750 cm^{-1} *of the clean surface (lower), the contaminated area (middle) and some 'residual' PDMS transferred onto the ATR prism after it was withdrawn from contact with the cast tube surface*

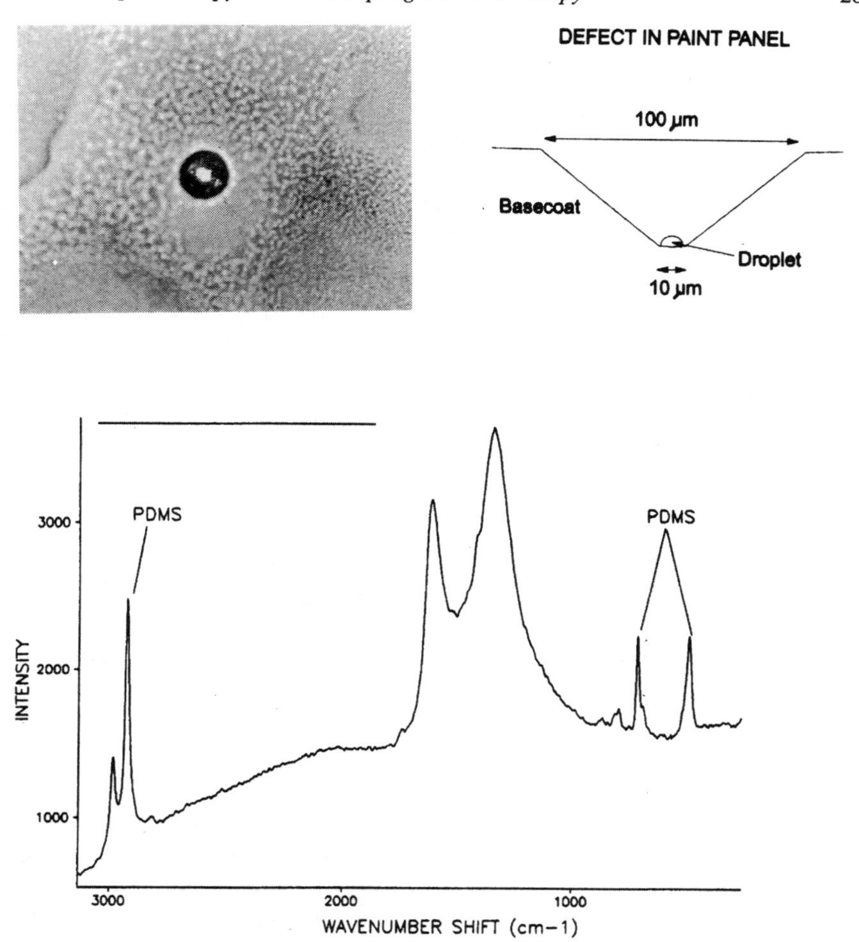

Figure 24(b) *In situ identification by micro-Raman spectroscopy of the contaminant at the base of a pit defect in a paint test panel.* Top left, *photo-micrograph (× 351) of the defect area;* Top right, *schematic of ca.* 10 μm *diameter 'droplet' contaminant at base of ca.* 100 μm *defect in paint basecoat;* Bottom, *Micro-Raman spectrum recorded in situ from 'droplet' contaminant showing the presence of both carbon and poly(dimethyl siloxane) (PDMS)*
(Data kindly supplied by Dr. Michael Claybourn)

shown to contain both carbon and PDMS. The spectra of contaminants shown in Figure 24(c) were recorded by FTIR transmission from the contaminants isolated from an acrylic moulding, which were then supported on a KBr disc.[125] The example of Figure 24(d), reproduced from ref. 53, uses a Raman microprobe to identify SiO_2 as the particulate species contaminating a polyethylene film.

Contaminants are, of course, a major source of many complaints concerning industrial products and materials. They may adversely affect a product's

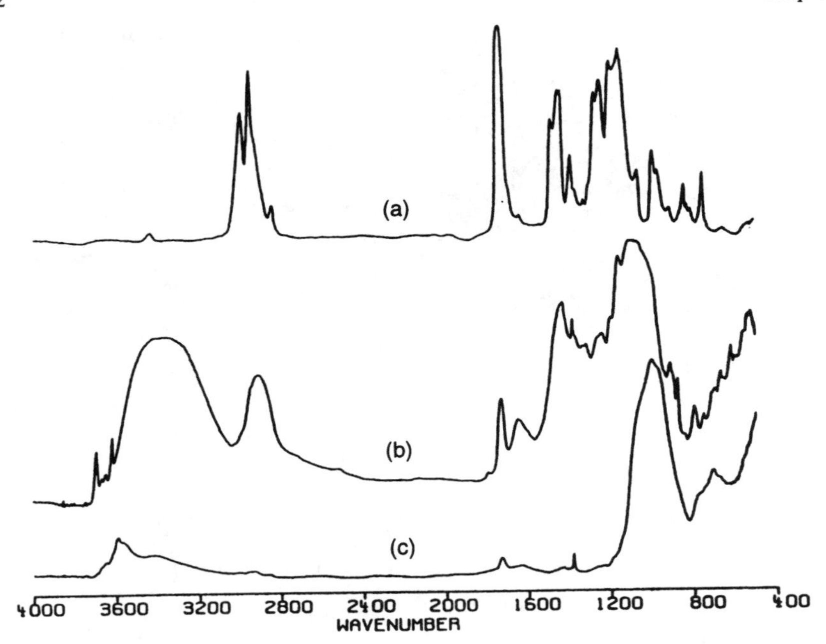

Figure 24(c) *FTIR-microscopy absorbance spectra of contaminants isolated from a glass-filled acrylic moulding:* (a) *a spectrum of the acrylic, shown for comparison;* (b) *cellulose;* (c) *silica. Each particle was ca.* 100 μm × 100 μm *and supported on a KBr window*
(Reproduced from ref. 125, by kind permission of IOS Press, Amsterdam)

performance or merely detract from its aesthetic appearance, although either may have high commercial consequences. FTIR reflectance-microspectroscopy techniques have been used to identify small contaminants that result in magnetic signal losses (dropouts) in magnetic media[144] – they originated from magnetic coating agglomerates, the abrasion limiting lubricant, and the polyester base film; while, for the tobacco industry, FTIR-microspectroscopy techniques have been used to characterise blemishes on cigarette papers.[145] Many other examples of characterising contaminants in or on materials may be found for micro-Raman in refs. 53 and 134, for FTIR in refs. 129 and 130.

Inclusion Analysis

Raman microprobe offers a distinct advantage to the analysis of the constituents of bubbles in glass, since it does not require the fracture and destruction of the glass samples.[146] Detailed study of the gases entrapped in bubbles is key to

[144] Webb R.E. and Young P.H., *Spectrosc. Lett.*, 1990, **23**, 5, 679.
[145] Thompson M.M., *Vib. Spectrosc.*, 1991, **2**, 1, 15.
[146] Lee S.W., Condrate R.A., Malani N., Tammaro D., and Woolley F.E., *Spectrosc. Lett.*, 1990, **23**, 7, 945.

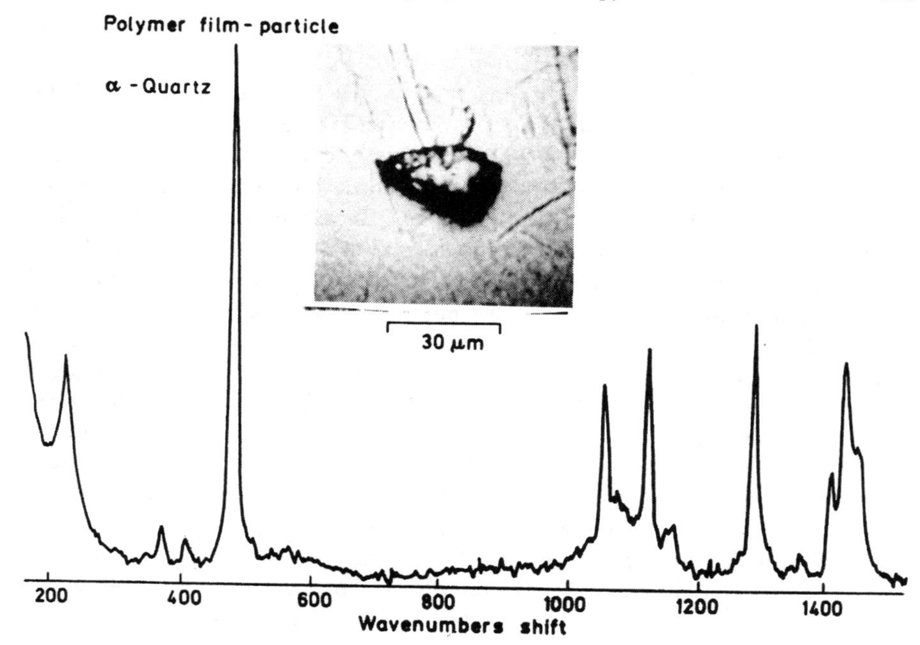

Figure 24(d) *Micro-Raman examination of particulate contamination in polyethylene film. The low wavenumber peak at* 465 cm^{-1} *is a characteristic of* α-SiO_2*; the bands between* 1000 cm^{-1} *and* 1500 cm^{-1} *are largely due to the polyethylene*
(Reproduced from ref. 53, by kind permission of Springer-Verlag GmbH & Co. KG, Germany)

elucidating the mechanisms for bubble formation in stages of glass manufacture such as melting and refining.[146,147] Although the vibrational bands may be observed and are less complex, they are much weaker than the rotational lines, which are usually used for diagnosis. Bubbles with diameters down to 15 μm have been successfully analysed; Figure 25(b) shows the Raman microprobe spectrum of a 24 μm diameter bubble containing oxygen in a silicate glass.[146] Raman microscope analysis has also been reported for the characterisation of gaseous and solid inclusions in fluoride glass optical fibres and preforms.[148] In this study the vibration band of O_2 at 1556 cm^{-1} was detected from illumination of a 4 μm diameter bubble. The solid inclusions identified by Raman included various allotropes of carbon as well as a LiF_2 crystal, which under white light had a hexagonal shape about 20 μm across.

Extensive use has also been made of Raman microprobe for both the qualitative and quantitative analysis of individual fluid inclusions in geological

[147] Janssen R.K. and Krol D.M., *Appl. Optics*, 1985, **24**, 2, 275.
[148] Bowden M., Dixon N.M., Gardiner D.J., and Carter S.F., *J. Mat. Sci.: Mat. Electron.*, 1990, **1**, 34.

samples.[149–157] These are intracrystalline cavities in minerals, which may be filled with a gas or liquid, or sometimes a solid, with a size range from a few microns to a few hundreds of microns.[156,158] FTIR microscopy has also been used to examine fluid inclusions[157–159] and a comparative study made with Raman microprobe analysis.[157] While in certain circumstances a FTIR microscopy investigation can be of value, the technique imposes a lower limit for inclusions to be of the order of 20 μm diameter. Raman however can be used to identify and quantify much smaller inclusions ($\geqslant 3$ μm).[157] Raman will also detect homonuclear diatomic species such as H_2, O_2, and N_2 which are infrared inactive; ready access to the low wavenumber region also aids the ready identification of many mineral inclusions.[157] A comparative study of NIR excitation FT-Raman, visible excitation Raman and FTIR microspectroscopies has been reported for the analysis of natural and synthetically produced hydrocarbon inclusions in host crystals.[160] For species that were fluorescent, no molecular detail was observed using the visible-Raman system, whereas vibrational bands were readily observed with the FT-Raman system, albeit that they were superimposed on much weaker fluorescent backgrounds.

Another form of inclusion analysis for which the Raman microprobe has proved to be well suited is the examination of inclusions in cut and uncut precious stones.[161,162] Figure 25(a) shows the unambiguous identification of a mixed pyrope–olivine inclusion in a diamond.

Mineral Analysis

The attributes of the Raman microprobe technique, as for inclusion analyses, are particularly appropriate to other forms of minerals microanalysis. Examples in petrology include distinguishing phase differences in feldspars and providing information on structural states[163] and characterising micron-sized minerals in eclogite high-pressure metamorphic rocks.[164] It has also been used to analyse individual crystals as small as 5×6 μm of calcium silicates in Portland

[149] Dhamelincourt P., Beny J-M., Dubessy J., and Poty B., *Bull. Mineral.*, 1979, **102**, 600.
[150] Pasteris J.D., Wopenka B., and Seitz J.C., *Geochim. Cosmochim. Acta*, 1988, **52**, 979.
[151] Wopenka B. and Pasteris J.D., *Appl. Spectrosc.*, 1986, **40**, 2, 144.
[152] Higgins K.L. and Stein C.L., *Microbeam Anal.*, 1986, 31.
[153] Rosasco G.J. and Roedder E., *Geochim. Cosmochim. Acta*, 1979, **43**, 1907.
[154] Dubessy J., Poty B., and Ramboz C., *Eur. J. Mineral.*, 1989, **1**, 517.
[155] Pasteris J.D., Seitz J.C., and Wopenka B., *Microbeam Anal.*, 1985, 25.
[156] Touray J.C., Beny-Bassez C., Dubessy J., and Guilhaumou N., *Scanning Electron Microsc.*, 1985, **1**, 103.
[157] Wopenka B., Pasteris J.D., and Freeman J.J., *Geochim. Cosmochim. Acta*, 1990, **54**, 519.
[158] Barres O., Burneau A., Dubessy J., and Pagel M., *Appl. Spectrosc.*, 1987, **41**, 6, 1000.
[159] O'Grady M.R., Conroy C.M., Taylor L.T., Knight C.L., and Bodnar R.J., *Proc. 7th Int. Conf. Four. Trans. Spectrosc.*, SPIE, 1989, **1145**, 613.
[160] Pironon J., Sawatzki J., and Dubessy J., *Geochim. Cosmochim. Acta*, 1991, **55**, 3885.
[161] Dhamelincourt P. and Schnubel H-J., *Rev. Gemnol.*, 1977, **52**, 11.
[162] Dele-Dubois M.L., Dhamelincourt P., and Schnubel H.J., *L'actualité chimique*, 1980, Avril, 39.
[163] Purcell F.J. and White W.B., *Microbeam Anal.*, 1983, 289.
[164] Boyer H. and Smith D.C., *Microbeam Anal.*, 1984, 107.

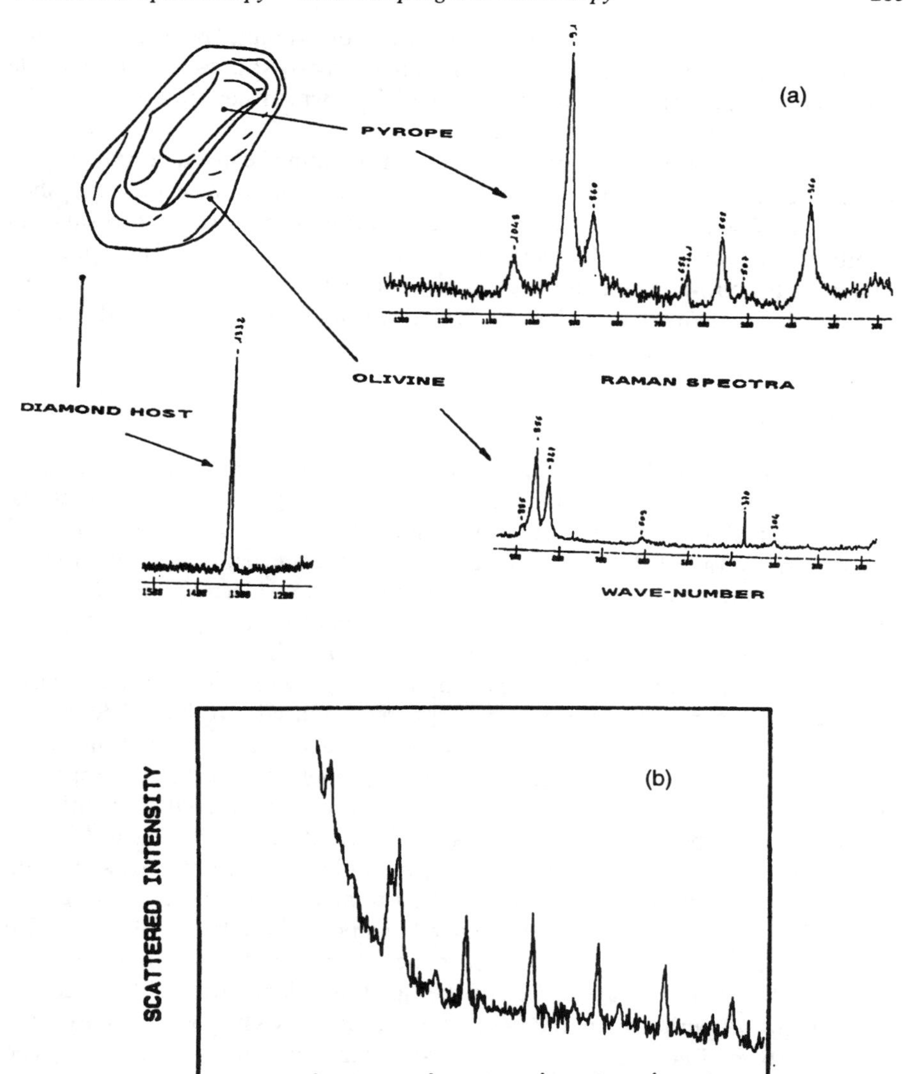

Figure 25 (a) *Micro-Raman identification of a mixed pyrope–olivine inclusion in un diamond taille brillant*
(Reproduced from *Utilisation de la spectrometrie Raman en gemmologie*, Lasnier B., *Analusis 23*, **1**, M16–M18 (1995))
(b) *Raman microprobe spectrum of a bubble (*24 μm *diameter) containing oxygen in a silicate glass*
(Reprinted from ref. 146, p. 958 by courtesy of Marcel Dekker, Inc.)

cements,[165] highlighting important differences, for example, between the white and grey clinkers; whereas polarised FTIR-microscopy studies have been made on single crystals of kaolinite and dickite,[166] in which new insights into the location of —OH groups were obtained.

In another infrared study, the feasibility of mapping semi-quantitatively the mineral distribution within a 1×1 cm grid on the surface of a polished specimen of sedimentary rock by using a scanning motorised *x–y* stage in combination with a FTIR-microscope was demonstrated.[167] The specimen prepared for SEM examination gave an excellent surface for specular reflectance measurements, from which the relative spatial abundances of dolomite, quartz, and illite could be ascertained.

Product 'Defect' Analysis

Information on the character and structure of process- or fabrication-induced product defects or imperfections has importance in product failure analysis, quality control and perhaps cause ownership. As with most 'problem-solving' situations, the fullest definition of cause and effect is often only realised when the vibrational spectroscopy–microscopy data is considered alongside other evidence, in particular, that deduced from examinations by optical microscopy techniques and energy dispersive X-ray spectroscopy (EDXS).

As with contaminant analysis, the application of vibrational spectroscopy–microscopy techniques to investigate process-induced product defects has widespread use in the plastics industry. Many such defects (visual imperfections) arise from included material which has different rheological properties from those of the bulk. This may arise because the defect material has a different copolymer or blend composition, is of a different molecular weight, has different end groups, or is oxidised, degraded, or cross-linked. Figure 26 shows infrared absorbance spectra recorded from a microtomed section taken through an observable flow-defect ('nib', 'gel-particle') in a poly(aryl ether sulphone) (PES) development film.[133] The reduced absorbance at 760 cm^{-1} shows that the defect area is deficient in the aryl—Cl end group; this might imply either another substituent or perhaps an increase in molecular weight, which causes the material to exhibit a different rheology at film-forming temperatures. The more likely cause was the latter, since optical microscopy examination had revealed that the nucleus of the defect area contained microcrystals of KCl; the polycondensation of PES proceeds via the reaction of —Cl and —OH(K) functional groups which is catalysed by K_2CO_3. The combined evidence therefore suggests that the flow defect arose from further reaction taking place on a catalyst residue particle(s) during polymer fabrication. Raman would have been inappropriate in this case, since PES generally exhibits a high fluorescence background, particularly with visible excitation; also, the high sensitivity of

[165] Conjeaud M. and Jaeschke-Boyer H., *L'actualité chimique*, 1980, Avril, 56.
[166] Johnston C.T., Agnew S.F., and Bish D.L., *Clays and Clay Minerals*, 1990, **38**, 6, 573.
[167] Liang K.K., *Microbeam Anal.*, 1991, 81.

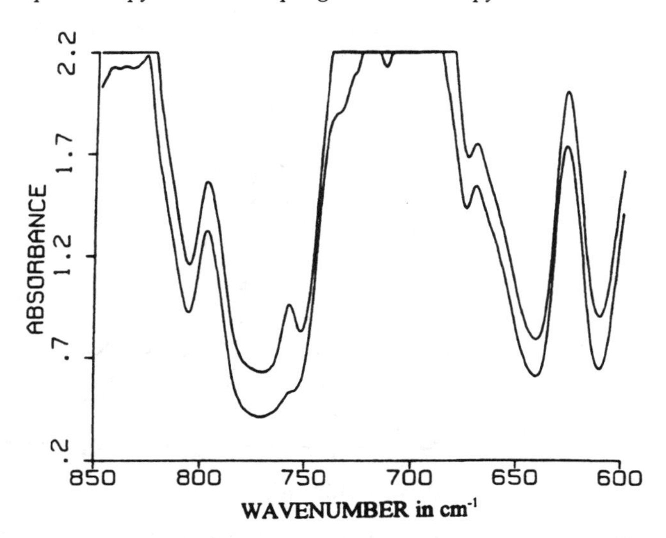

Figure 26 *FTIR-microscopy spectra recorded from a microtomed section of PES film, aperture* 100 μm: Top, *base film;* Bottom, *defect area. The spectra have been offset for clarity*
(Reproduced from ref. 133, by kind permission of Huthig & Wepf Publishers, Zug, Switzerland)

infrared to polar functional groups clearly acts as an advantage. Neither vibrational technique would have been capable of 'spotting' the KCl micro-crystals!

The product 'defect' case study featured in Figure 1 highlights the sensitivity of Raman to olefinic unsaturation; a similar benefit applied in the problem case study depicted in Figure 27. Here the two microprobe spectra recorded directly from the acrylic moulding show distinct differences associated with the amount of residual monomer present,[133] as evidenced, for example, by the relative intensities of the band near 1650 cm^{-1}. The clear areas show a much higher concentration, while in the hazy areas much of the monomer has been evolved, leaving behind voids which give rise to the visible scatter (defect). The Raman examination implies that optimum processing conditions had probably not been employed, but clearly shows that the fault was not due to an inorganic contaminant, as had been originally anticipated. An additional benefit Raman offered to this investigation was speed, since there was no need of any sample preparation.

(New/Advanced) Materials Analysis

The attributes of laser Raman micro-analysis have also been successfully applied to characterising compositional heterogeneities and identifying impurity phases in ceramic and thin film high-T_c superconductors.[168] For highly

[168] Etz E.S, Schroeder T.D., and Wong-Ng W., *Microbeam Anal.*, 1991, 113.

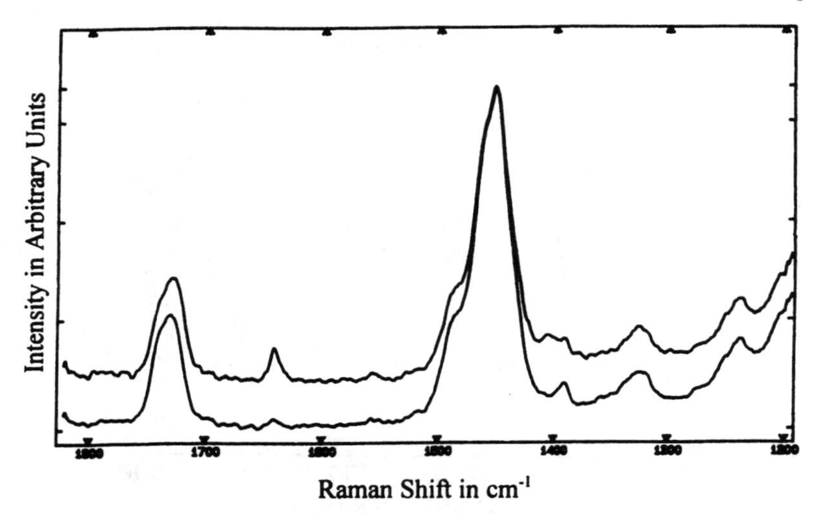

Figure 27 *Raman-microprobe spectra recorded from an acrylic moulding:* Top, *clear area;* bottom, *hazy area*
(Reproduced from ref. 133, by kind permission of Huthig & Wepf Publishers, Zug, Switzerland)

absorbing dark phases, low power density of the laser beam was found to be essential at the sample, equivalent to about 2–5 mW in a 2–3 μm diameter focused spot. Deviations from stoichiometry and composition lead to significant variability in spectral detail and band positions and intensity, see Figure 28. The oxygen content of non-stoichiometric solids such as the Y–Ba–Cu–O superconductors may be inferred from the position-sensitive bands, since the 'oxygen vacancies' may be correlated with the Raman line wavenumber positions of certain bands.[169] Other examples of chemical application micro-Raman studies to new (advanced) materials may be found in refs. 169 and 170; they include homopolar diatomic oxygen speciation (O_2^+, 1865 Δcm^{-1}; O_2, 1580 Δcm^{-1}; O_2^-, 1097 Δcm^{-1}; O_2^{2-}, ~766 Δcm^{-1})[170] in dopants in semiconductor inclusions or partners, impurity identifications in silicon, heteroatom bond observations in alloys, and interface chemical bonding in bilayers.

Laminated Structure Identifications

Extensive use has been made of vibrational spectroscopy–microscopy techniques in molecular structure fingerprinting of the layers in multilayer or laminated products, particularly polymeric films such as those used for packaging. Food packaging films are often complex structures consisting of

[169] Huong P.V., Verma A.L., Chaminade J.-P., Nganga L., and Frison J.-C., *Mater. Sci. Eng.*, 1990, **B5**, 2, 255.
[170] Huong P.V., *J. De Physique IV (Colloque C6)*, 1991, **1**, C6, 151.

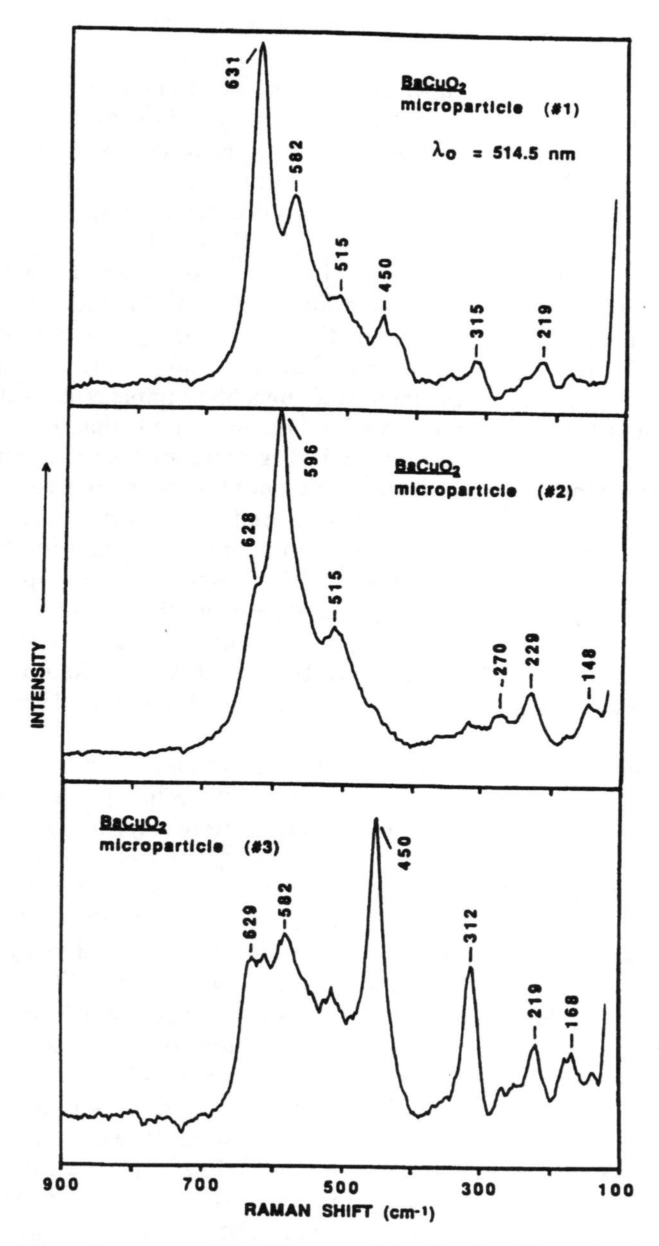

Figure 28 *Micro-Raman spectra of three particles (each* ~10 μm *in size) of ceramic powder of nominal composition $BaCuO_2$*
(Reproduced from ref. 168, by kind permission of San Francisco Press, Inc., CA, USA)

many layers which may have been assembled by combinations of ways such as co-extrusion, adhesion, coating, or lacquering. The laminate may also contain thin metallic layers (*e.g.* one side of a film layer may be aluminised), or be laid down on a substrate such as paper or cardboard. Each layer imparts a certain property to the finished article, whether it be strength, prevention of air or moisture ingress, to serve as a hot melt adhesive, to act as a key for over-printing, or externally merely for aesthetic reasons, *etc.*

In principle, Raman-microscopy offers two distinct advantages over FTIR-microscopy in these types of analyses.[125,171] Firstly, there is a significant increase in spatial resolution. In Raman by visible excitation the laser may be tightly focused to about a 1–2 μm focal cylinder, whereas diffraction effects limit the application of the longer wavelength infrared to applications involving characterisations on regions with dimensions 8–10 μm or greater. Secondly, although for convenience of specimen handling a film sample cross-section slice is often prepared,[171,172] Raman examination requires simply that one clean edge of the laminate is exposed, while for a FTIR-microscopy transmission study a thin ($\sim$ 10–20 μm thickness) transverse section needs to be microtomed from the sample.[171,173,174] However, these advantages are frequently not fully realisable because some layers may exhibit high levels of fluorescence, and very thin layers often prove to be highly photosensitive.[125,171] FT-Raman microscopy has been used to successfully characterise 20 μm layers in the lid of a packaged microwave meal.[175] For thin polymer samples, confocal Raman microspectroscopy has recently been shown to provide good depth resolution ($\sim$2 μm) spectra, thereby providing directly both good spatial and good lateral resolutions.[81,82]

The case studies illustrated in Figures 29–32 serve to show both the potential and some of the limitations of the microspectroscopy techniques, as well as their complementarity. The three spectra of Figure 29 were recorded from consecutive layers within a six-layer laminated structure.[125] The spectrum of the middle of these layers (8 μm wide) indicates the presence of a polyurethane adhesive. However it is overlapped by stray-light absorptions from the adjacent materials. The interferences come from the 50 μm wide polypropylene and 10 μm wide polyester layers either side of it. The spectrum shown in Figure 30 was recorded from an adhesive which was sandwiched between two poly(ethylene terephthalate) films, but in this instance the sample microtomed transverse section was squashed in a compression cell fitted with a pair of diamond windows in order to specifically increase the width of the softer EVA layer, so that an essentially 'clean' spectrum of the adhesive could be recorded. The spatial resolution advantage available with Raman could not be realised since the adhesive layer was highly fluorescent to visible excitation. (A FT-Raman microscope has been used to record the spectrum of a 50 μm spot size from a polyurethane

[171] Jawhari T. and Pastor J.M., *J. Mol. Struct.*, 1992, **266**, 205.
[172] Andersen M.E., *ACS Adv. Chem. Ser.*, 1983, **203**, 383.
[173] Shaw T.I., Karl F.S., Krishen A., and Porter L.E., *Spectroscopy*, 1993, **8**, 8, 45.
[174] Harthcock M.A., Lentz L.A., Davis B.L., and Krishnan K., *Appl. Spectrosc.*, 1986, **40**, 2, 210.
[175] Turner P.H., *Bruker Report*, 1994, **140/94**, 36.

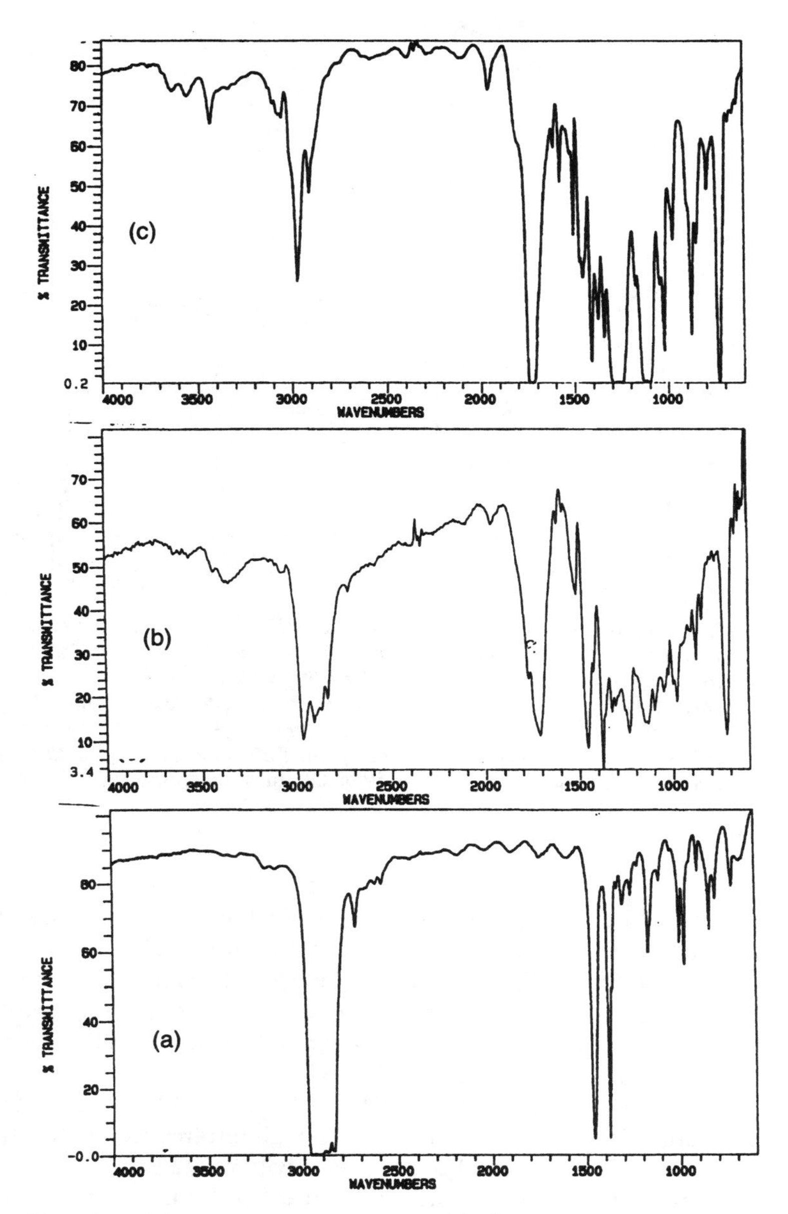

Figure 29 *Micro-FTIR transmission spectra of three consecutive layers of a thin transverse section cut from a multilayer polymer laminate:* (a) 50 μm *wide polypropylene layer;* (b) ~8 μm *wide layer showing presence of a polyurethane (adhesive) layer and interference from adjacent layers;* (c) 10 μm *wide polyester layer*
(Reproduced from ref. 125, by kind permission of IOS Press, Amsterdam)

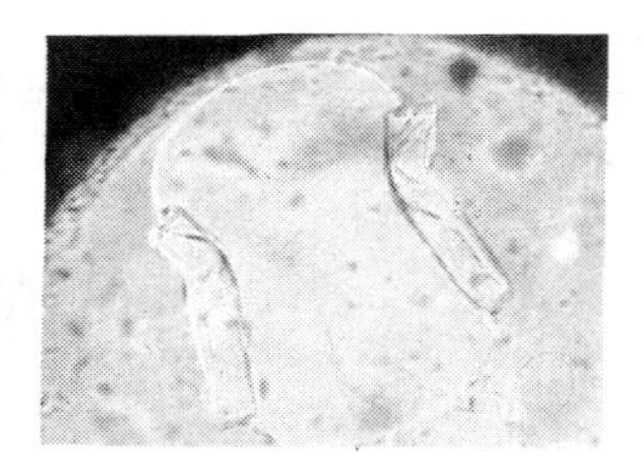

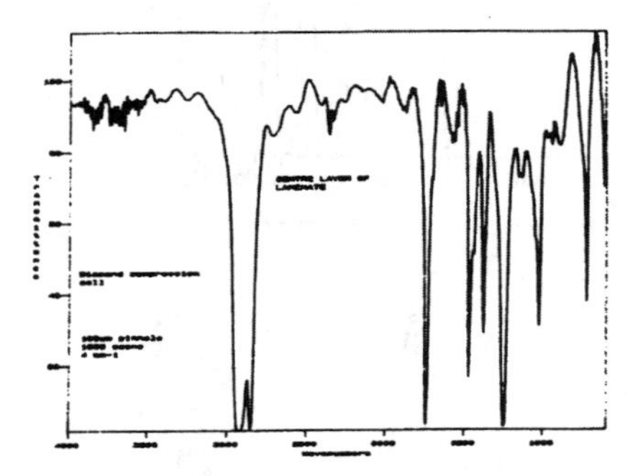

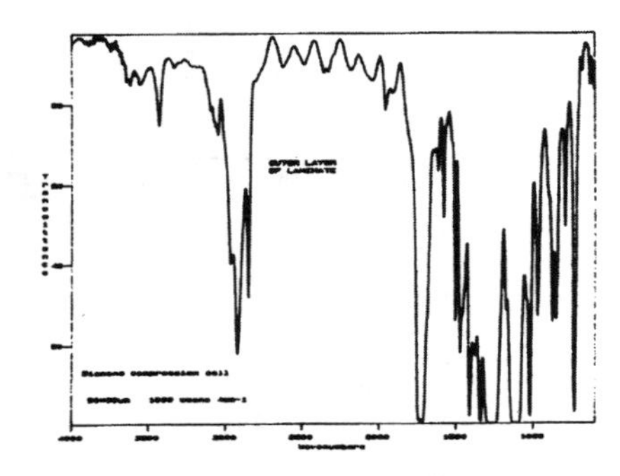

Figure 30 Top, *photograph of PET/EVA 'adhesive'/PET film laminate section squashed in a diamond window micro-compression cell, showing the spread of the EVA layer sandwiched between the two PET films, which remain relatively much less spread;* Bottom left, *FTIR-microscopy spectrum recorded from squashed EVA 'adhesive', aperture* 100 μm *pin-hole,* 1000 scans, 4 cm^{-1}; Bottom right, *FTIR-microscopy spectrum recorded from PET outer layer of laminate while held in compression cell, aperture* 50 × 60 μm, 1000 scans, 4 cm^{-1}

elastomer[87]). The FT-Raman spectra shown in Figure 31 were recorded from three consecutive 20 μm layers in a multilayer lid of a packaged microwave meal.[175] In the example of Figure 32, the higher spatial resolution advantage of Raman is put to good use in characterising the middle layer of a three-layer film laminate.[133] For this sample, mixed component information could be gleaned from a transmission infrared spectrum through the film structure. Raman and multiple internal reflection infrared (MIR) methods could be used to identify the composition of the two surface layers, although infrared would perhaps be the better option, since, as seen here, the sensitivity of Raman to oxygenated species is comparatively much weaker. The pattern of Raman bands in Figure 32 that are attributable to organic species is almost identical to that obtained from a low density polyethylene (remember the case study of Figures 1)! Transmission FTIR-microscopy through a transverse microtomed section could be used to fingerprint the thicker ethylene/vinyl acetate (EVA) copolymer outer layer, but would not be capable of distinctly resolving the two filled EVA layers. However, this is readily achievable using Raman-microscopy with visible excitation. The Raman spectrum of the sandwiched layer is clearly shown to

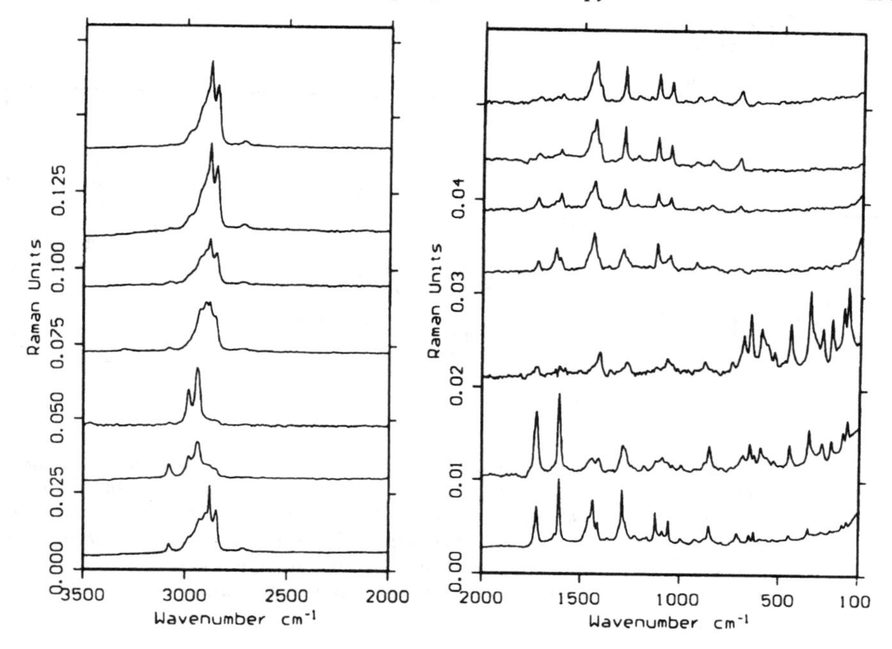

Figure 31 *FT-Raman microscopy spectra of a laminated polymer film lid of a packaged microwave meal. The bottom spectrum is that of the bulk film laminate (*120 μm *thick) recorded in macro mode, with the intensity scale reduced by* ×20. *The other traces have been taken across the film edge; the spectra from* 20 μm *wide apertured elements show the second trace from the bottom to be essentially PET, the third trace to be essentially that of poly(vinylidene chloride), the next trace is essentially that of the Nylon 6 layer, while the remaining spectra are essentially that of the remaining layer (*~60 μm *thick) comprising component contributions attributable to a polyethylene, a polyisobutylene and an ester. The spectra between* 2000 *and* 100 cm^{-1} *are intensity scale expanded compared with those over the range* 3500–2000 cm^{-1}.
(Reproduced from ref. 175, by kind permission of Bruker Spectrospin Limited)

contain titanium dioxide, which was not present in either of the surround layers; additionally, it could be characterised as substantially the rutile crystalline form, information a mid-infrared examination would be incapable of providing. This analysis succeeded since the EVA laminated film was neither highly photosensitive nor significantly fluorescent.

Fibre Identification

An investigation into the complementary attributes of infrared and Raman microspectroscopies to fibre identifications has demonstrated that compared to other methods, both have the advantages of being non-subjective, selective,

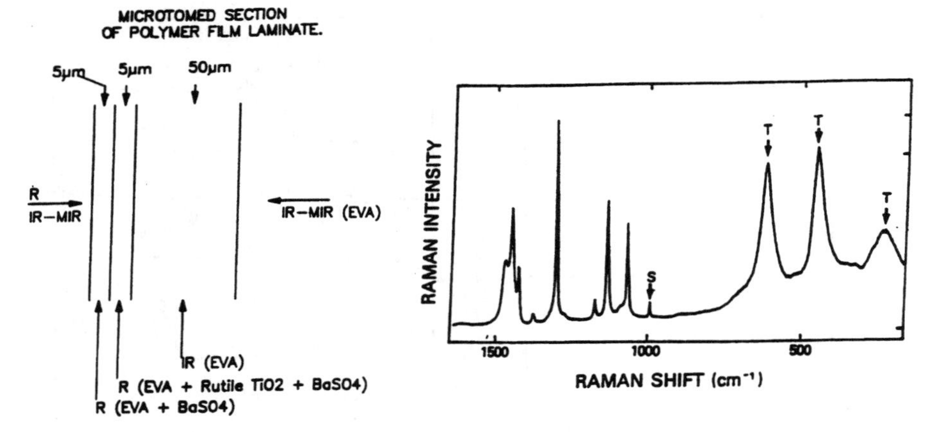

Figure 32 *Raman-microscopy examination of a three-layer polymer film laminate section:* Left, *laminate section schematic;* Right, *Raman spectrum of middle layer. (*S *denotes $BaSO_4$;* T *denotes TiO_2;* EVA *denotes ethylene/vinyl acetate copolymer;* IR *denotes infrared radiation;* IR-MIR *denotes infrared multiple internal reflection technique;* R *denotes Raman measurement)*
(Reproduced from ref. 133, by kind permission of Huthig & Wepf Publishers, Zug, Switzerland)

involving no sample degradation and requiring only a single fibre as small as 5–10 μm length.[176]

Provided that the specimen is neither highly fluorescent nor photosensitive, single fibre samples of a few microns diameter present little difficulty to examination by Raman-microscopy. A sample is simply appropriately oriented and held at the focus of the laser beam, *e.g.* taped at the ends onto a microscope slide. Indeed extensive use has been made of this technique for detailed studies of the physical characteristics of both polymeric and carbon fibres, see pp. 314–325. In many instances generic typing may be easily achieved from the infrared spectrum recorded from a single fibre held taut (straight) across an aperture mount or an infrared transparent window. However, high contrast spectra that exhibit minimal aberrations will only be recorded from specimens that have been appropriately flattened, particularly in the case of a narrow diameter fibre.[41,171,176–179]

Short sub-millimetre lengths of fibres, like particulates, are common types of contaminants that have been identified in many products. They also often form key evidence in many forensic investigations.[177–180] There are many sources for these, and they may be either natural (*e.g.* wool, hair, cotton, cellulose),

[176] Lang P.L., Katon J.E., O'Keefe J.F., and Schiering D.W., *Microchemical J.*, 1986, **34**, 319.
[177] Tungol M.W., Bartick E.G., and Montaser A., *Appl. Spectrosc.*, 1993, **47**, 10, 1655.
[178] Tungol M.W., Bartick E.G., and Montaser A., *Appl. Spectrosc.*, 1990, **44**, 4, 543.
[179] Bartick E.G., in *The Design, Sample Handling and Applications of Infrared Microscopes*, ed. Roush P.B., ASTM Special Technical Publication 949, ASTM, Philadelphia, USA, 1987, p. 64.
[180] Shearer J.C., Peters D.C., and Kubic T.A, *Trends Anal. Chem.*, 1985, **4**, 10, 246.

synthetic (*e.g.* polymeric, clothing, or packaging fibres, such as acrylic, nylon, rayon, and polyester), or inorganic (*e.g.* asbestos, glass). Since many fibre contaminants may be coloured or fluorescent, then, despite the more stringent requirements for anomaly-free absorbance spectra, FTIR-microscopy has probably been the most widely used for identifying fibres present as contaminants in industrial products and consumer articles, *e.g.* pharmaceuticals[181] and cigarette pack wrapping.[145] However, the potential of FT-Raman-microscopy to record good analytical spectra directly from single fibres with diameters as narrow as 5 μm has been demonstrated.[182] While both molecular microspectroscopy techniques have been shown to be effective methods for identifying and quantifying additives to polypropylene single fibres,[183] where fluorescence and sample heating were not problematic Raman analysis proved particularly effective over a range of pigments at both identification and quantification (0.1% to 10%).

Characterising the components of a heterofil (core/sheath) fibre is clearly much more readily accomplished by a microprobe Raman analysis.[125,184] The separate layers may be examined individually from a fibre-end or cross-section. Alternatively, from a lateral observation near-discrete spectra of the two components may be obtained by appropriate focusing of the laser beam, see Figure 33, and these will be optimally resolved if a confocal micro-Raman configuration is used. Figure 33 also shows that FTIR-ATR microspectroscopy has the potential to discriminate such a fibre as a 'two-layer' sample. With minimal contact pressure of the ATR element objective, then the fingerprint spectrum recorded is essentially that of the polyethylene sheath; with increasing pressure 'breakthrough' of absorption bands due to the polypropylene core is clearly observed. However one must be careful. For this fibre sample too much pressure led to its 'destruction' – delamination occurred!

FTIR-microscopy has been variously applied to the analysis of human hair.[185–187] For example, oxidative damage to single hair fibres treated with hydrogen peroxide or metabisulphite solutions has been assessed, and comparisons made between cysteic acid levels from hair root to hair tip, between naturally weathered and bleached hair.[185] Different regions of human hairs in differing stages of their ageing cycle (anagen, catagen and telogen) have also been analysed using FTIR-microscopy, in combination with a diamond window compression cell.[186] In both the above studies second-derivative absorbance data were employed to enhance spectral contrast and differentiation (see pp. 220–224). The potential of determining drug abuse from cross-sectionally or

[181] Clark D.A. and Nichols G., *Anal. Proc.*, 1990, **27**, 19.
[182] Sawatzki J., *Fresenius J. Anal. Chem.*, 1991, **339**, 267.
[183] Bouffard S.P., Sommer A.J., Katon J.E., and Godber S., *Appl. Spectrosc.*, 1994, **48**, 11, 1387.
[184] Everall N.J., in *An Introduction to Laser Spectroscopy*, ed. Andrews D.L. and Demidov A.A., Plenum Press, New York, USA, 1995, p. 115.
[185] Joy M. and Lewis D.M., *Int. J. Cosmet. Sci.*, 1991, **13**, 249.
[186] Bramanti E., Ronca F., Teodori L., Trinca M.L., Papineschi F., Benedetti E., Spremolla G., Vergamini P., and Benedetti E., *J. Soc. Cosmet. Chem.*, 1992, **43**, 285.

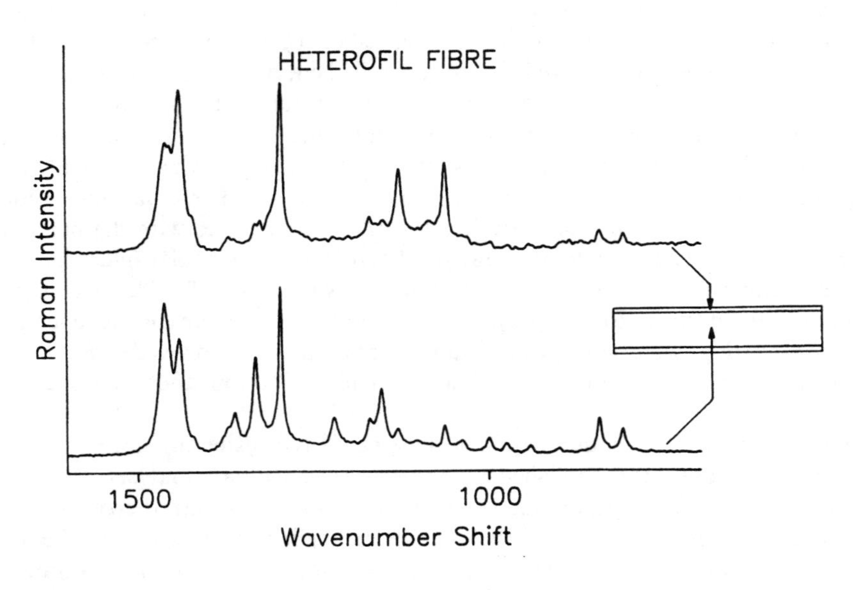

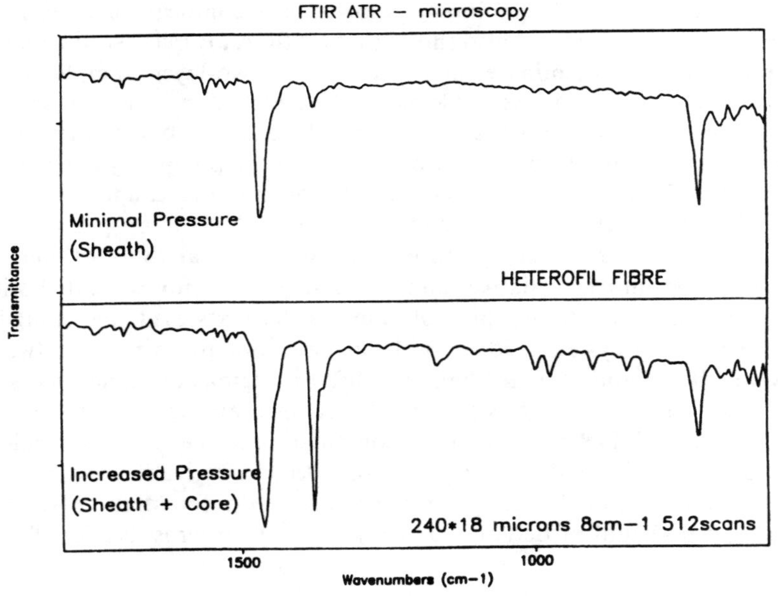

Figure 33 Top, *Micro-Raman (non-confocal) spectra recorded from a heterofil-fibre comprising a* ~50 μm *diameter polypropylene core surrounded by a* ~3 μm *thick polyethylene sheath. The arrows indicate the points at which the laser was focused;* Bottom, *FTIR-ATR (ZnSe prism)*, 240 × 18 μm *aperture, spectra recorded from the heterofil-fibre with minimal contact pressure (upper) and increased pressure (lower)*

laterally microtomed hair has also been reported.[187] Infrared spectroscopic examination of the interior of the hair can differentiate between passive contamination and drugs absorbed into the hair from ingestion.

Domain Structure Characterisation

The value of Raman-microprobe point-by-point mapping of phase-contrasted surfaces is well established.[53] Raman spectra are recorded from selected regions or features which have been highlighted by different optical illumination techniques, such as transmitted and reflected polarised light and incident and transmitted differential interference contrast. These are used to locate and expose strain effects and birefringence in samples, and differences in surface topography of materials, which are not readily discernible by transmitted and incident (reflected) bright-field illumination.[53]

With the increased sensitivity now available through advances in filter technology and CCD detectors, direct Raman-microscopy images of domain or phase structure or composition may be obtained.The potential of Raman imaging of polymer blends has been demonstrated,[188] see Figure 34. Here, the filter, a narrow-bandpass, multilayer dielectric filter in the optical train, is set to observe the 1665 cm^{-1} νC=C— band, which is a characteristic of the nitrile rubber toughener dispersed in an epoxy matrix. The microprobe spectra recorded from the two regions suggested complete phase separation of the rubber polymer and the epoxy matrix.[188]

In a FTIR-microscopy study of a polymer-dispersed liquid crystal (PDLC) system,[189] functional group concentration (νC=O absorbance) colour-scale images of LC droplets dispersed in poly(n-butyl methacrylate) were recorded, in order to monitor the fluctuations that occurred over a period of time in the phase-separated system at room temperature. Chemical composition gradients were highlighted through the droplets as with time they became richer in LC due to molecular diffusion processes, while both optical and infrared images showed the droplet size growth with time via the process of coalescence. Colour scale contour plots were also generated from FTIR-microspectroscopy data in order to investigate the interphase region of epoxy resin-glass fibre reinforced composites,[190,191] in terms of interfacial chemistry and the effects of fibre-glass treatment[190] and under wet conditions.[191] Localised FTIR spectra have also been recorded of morphologically distinct parts of thin sections of plant tissue prepared by cryostatic sectioning on a microtome.[192] Chemical information on different regions was obtained for several food ingredients: cereals, oilseeds, spices, and flavour sources.

[187] Kalasinsky K.S., Magluilo J., and Schaefer T., *Forensic Sci. Int.*, 1993, **63**, 253.
[188] Garton A., Batchelder D.N., and Cheng C., *Appl. Spectrosc.*, 1993, **47**, 7, 922.
[189] Challa S.R., Wang S-Q., and Koenig J.L., *Appl. Spectrosc.*, 1995, **49**, 3, 267.
[190] Arvanitopoulos C.D. and Koenig J.L., *Appl. Spectrosc.*, 1996, **50**, 1, 1.
[191] Arvanitopoulos C.D. and Koenig J.L., *Appl. Spectrosc.*, 1996, **50**, 1, 11.
[192] Wetzel D.L. and Fulcher R.G., *Dev. Food Sci.*, 1990, **24**, 485.

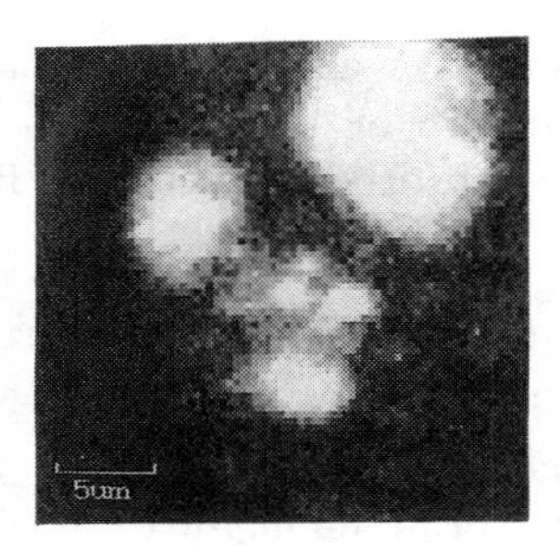

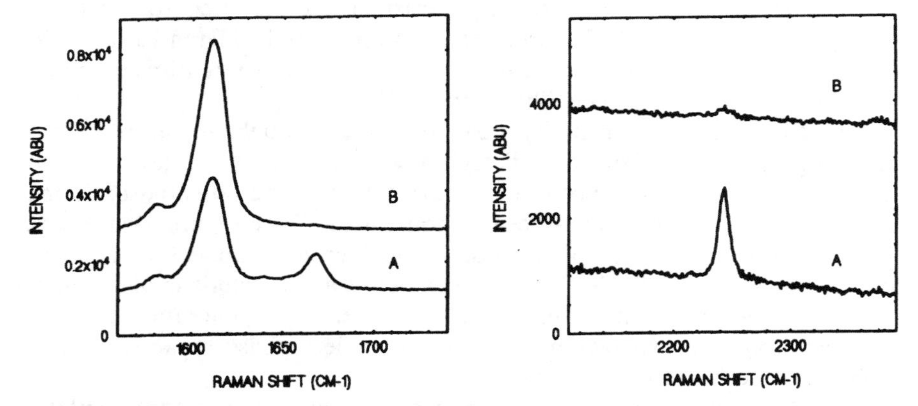

Figure 34 *Micro-Raman study of phase-separated rubber-toughened epoxy resin;* Top, *Raman image at* 1665 Δcm^{-1} *(minus* 1720 Δcm^{-1} *background);* Bottom left, *spectra of* $\nu C{=}C$ *region;* Bottom right, $\nu C{\equiv}N$, *region.* A, *corresponds to the light areas of the image;* B, *corresponds to the dark areas of the image* (Reproduced from ref. 188, Garton et al., *Appl. Spectrosc.*, 1993, **47**, 7, 922–927, by kind permission of the Society for Applied Spectroscopy; © 1993)

Chemical Composition Mapping

Mapping of chemical structure or composition variations through a product can be important; for example, it might be useful in monitoring the effects of environmental exposure on the product, it might be of value to probe the concentration profile of an additive in the product, or it might be useful in determining the influences of production or curing processes. The optimum vibrational spectroscopy technique for a particular investigation will, as we have already seen, be largely dictated by functional-group sensitivity. Again, there are many examples of this type of application to polymeric products. These include the effects of photo and thermal oxidation and degradation, additive distributions, and the consequences of the polymerisation process itself.

FTIR-microscopy has been used to probe the diffusion-in concentration profiles of a UV-stabiliser into 520 μm thickness polypropylene plaques.[193] An

[193] Hsu S.C., Lin-Vien D., and French R.N., *Appl. Spectrosc.*, 1992, **46**, 2, 225.

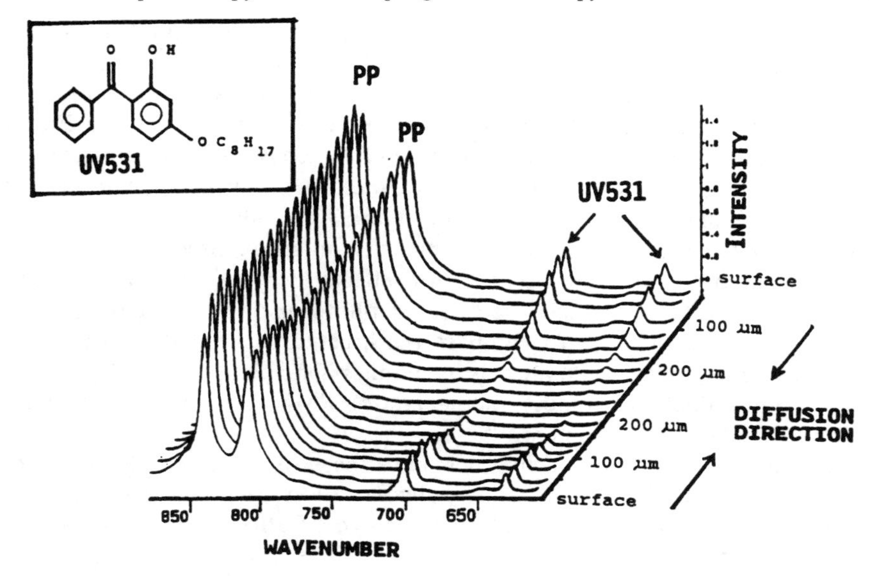

Figure 35 *FTIR-microscopy study of diffusion-in profile of UV stabiliser into a polymer film: A 3D plot of the infrared spectra of the stabiliser at different distances from the surfaces of the polypropylene plaque after a diffusion experiment at 60°C*
(Reproduced from ref. 193, Hsu et al., *Appl. Spectrosc.*, 1992, **46**, 2, 225–228, by kind permission of the Society for Applied Spectroscopy; © 1992)

automated sampling stage was used to collect spectra from consecutive adjacent 26 µm wide elements along the diffusion path; (the other aperture dimension was set to 360 µm), see Figure 35. These data were used to derive diffusion coefficients for the additive–polymer system; such knowledge is of value in assessing diffusion transport of additives to surfaces and subsequent loss by evaporation, or relating additive mobility to effectiveness within the polymer matrix.[193] In a similar type of study, FTIR-microscopy has been used to determine deterrent penetration into nitrocellulose-based propellant grains,[194] since this is thought to be related to ballistic performance. Diffusion profiles of the deterrents (di-n-butyl phthalate, dinitrotoluene, and methyl centralite) were built up from the absorbances of marker functional groups determined at 20–50 µm intervals from the edge to the centre of a microtomed section from a smokeless powder propellant grain. The sensitivity of the FTIR examination showed that in many instances the deterrent penetrated well beyond the visibly observed deterred skin layer, which was visibly highlighted by either brightfield or phase contrast illumination techniques.

[194] Varriano-Marston E., *J. Appl. Poly. Sci.*, 1987, **33**, 107.

The advantages of FTIR-microscopy have also been exploited in directly profiling the effects on polymer film samples of ageing in air.[195–197] Microtomed sections cut perpendicular to the oxidation gradient were examined in order to map any heterogeneous distribution of infrared absorbing species, see Figure 36. The spectra of Figure 37(a) were recorded from consecutive 15 μm layers of a sample of polypropylene that had been irradiated in artificial photo-ageing conditions, while the axonometric plot of Figure 37(b), which was derived from measurements of carbonyl absorption intensity on an oxidised polypropylene sample, is used to illustrate non-uniform effects from non-uniform processing.[196] The effects of artificial accelerated photo-oxidation and weathering have been studied in order to compare the kinetics associated with the two ageing processes for several materials including pigmented poly(vinyl chloride) (PVC), a 6 mm thick poly(methyl methacrylate) (PMMA) plaque, and carbon-black filled and cured elastomers.[198]

Both microspectroscopy techniques have been used in combination to monitor and map the effects of a polymerisation cycle on poly(ethylene terephthalate) (PET) in which a solid-state polymerisation step is employed to raise the intrinsic viscosity of the polymer.[133] The sensitivity of FTIR was utilised in measuring the hydroxyl and carboxyl end group concentrations, while the density was derived from a measure of the carbonyl stretching-vibration band-width in the Raman spectra. The investigations highlighted both chemical structure and physical property inhomogeneity through polymer chip (granule) samples.

The high sensitivity (resonance Raman) to conjugated polyenes has been well exploited in a Raman microline focus (MiFS) spectroscopic study of the thermal degradation of polyurethane (PU)-backed PVC sheet.[77,199] From a comparison of the data recorded across a section of the sheet, gross variations in the extent of degradation were observed across the PVC matrix, with the greatest level being seen close to the PU/PVC interface. This was attributed to catalytic dehydrochlorination by amine residues within the PU. Figure 38 shows the MiFS profiles for the —C=C— symmetric stretch (ν_2) band of a degraded (10 h at 120°C) PU-backed PVC sheet, as a function of distance across the PVC sheet. ~950 μm represents the PU contact side; ~150 μm represents the air contact side. The ν_2 band maximum near the PU interface at 1505 $\pm$ 1.5 Δcm^{-1} is higher than that at the air interface, which is at 1510 $\pm$ 1.5 Δcm^{-1}. This shift reflects the mechanistic differences associated with the degradation processes; the amine catalysed degradation produces a greater abundance of longer polyene sequences containing 14–16 —C=C— units sequences than the degraded surface at the air interface, where the most abundant species are polyenes of lengths 12–14 units.[199]

195 Gardette J-L., *Analusis Magazine* 1993, **21**, 5, M17.
196 Gardette J-L., *Spectroscopy Europe*, 1993, **5**, 2, 28.
197 Jouan X. and Gardette J-L., *Polymer Commun.*, 1987, **28**, 12, 329.
198 Lemaire J., Gardette J-L., and Lacoste J., *Makromol. Chem., Macromol. Symp.*, 1993, **70/71**, 419.
199 Bowden M., Bradley J.W., Dix L.R., Gardiner D.J., Dixon N.M., and Gerrard D.L., *Polymer*, 1994, **35**, 8, 1654.

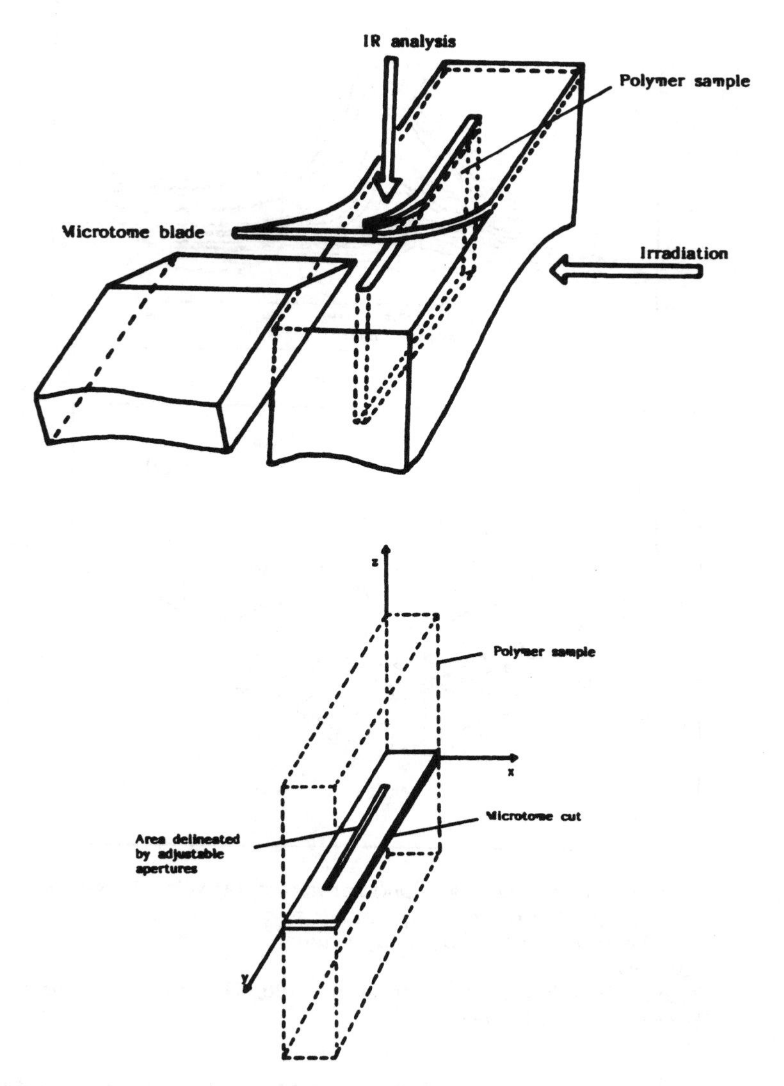

Figure 36 *FTIR-microscopy profiling of polymer ageing:* Top, *schematic of microtoming process;* Bottom, *schematic showing an aperture (*15 μm × ~600 μm*) delineated area on the microtomed section*
(Reprinted with permission from ref. 196, Gardette J-L., *Spectroscopy Europe*, **5**, 2, 28–32 (1993))

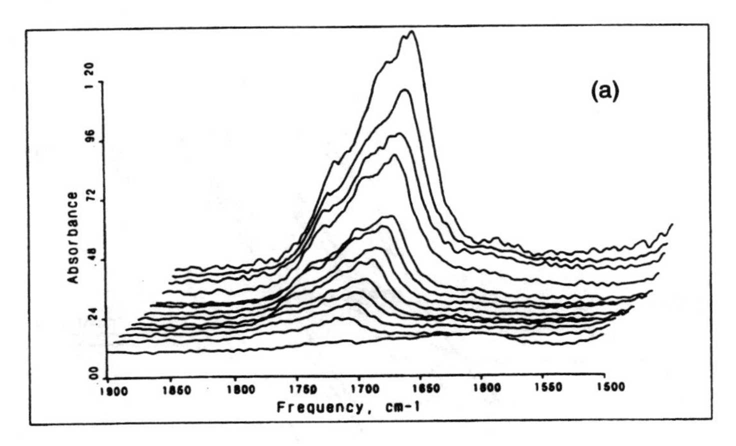

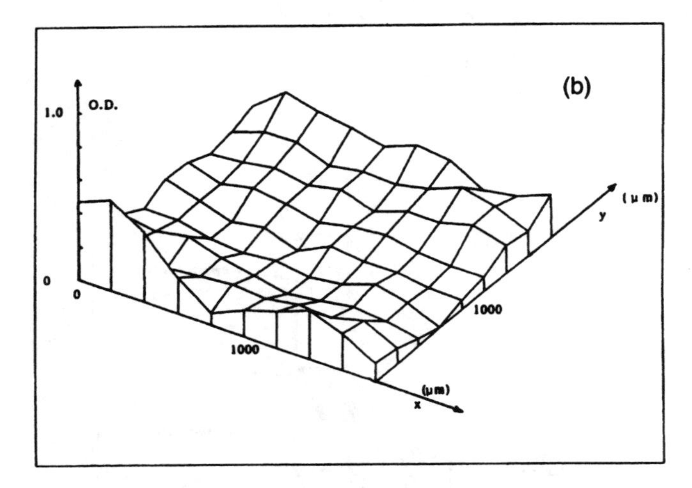

Figure 37 *FTIR-microscopy profiling of polymer ageing:* (a) *νC=O region of the first layers of a photo-oxidised polypropylene sample;* (b) *Axonometric plot of the νC=O optical density (OD) for a polypropylene sample oxidised in an oven at* 130°C *in air*
(Reprinted with permission from ref. 196, Gardette J-L., *Spectroscopy Europe*, **5**, 2, 28–32 (1993))

Water trees are a significant cause of electrical breakdown in power cables. A degradation process causes tree-like or bush-like hydrophilic structures to form on a microscopic scale that deteriorate the polyethylene insulation of underground cables.[200] FTIR-microscopy was found to be a key technique in elucidating the nature of the hydrophilic groups (*e.g.* carboxylate salts) within these structures in a pan-technique study of the composition, structure, and growth of water trees in polyethylene.[200] Raman-microscopy proved to be

[200] Ross R., *Kema Sci. Tech. Rep.*, 1990, **8**, 4, 209.

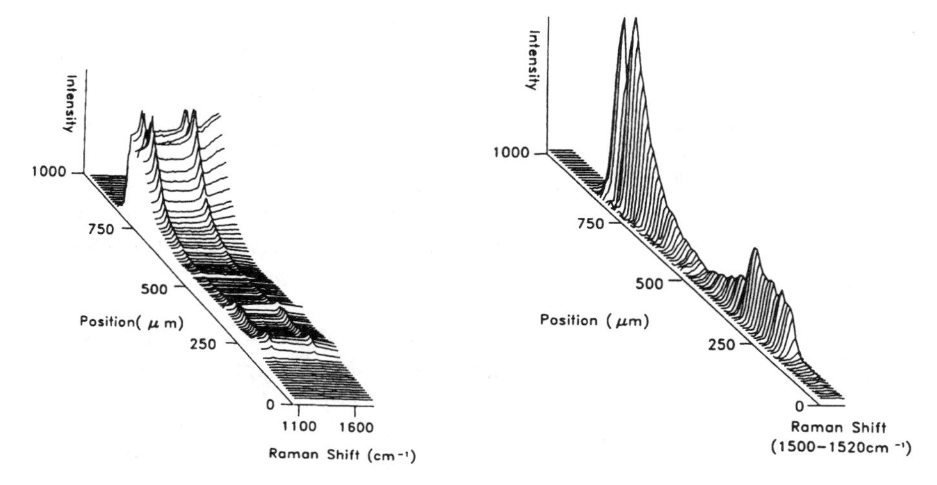

Figure 38 *Raman MiFS study of thermal degradation of PU-foam-backed PVC sheet:* Left, *the resonantly enhanced* ν_1 *C—C at* ~1130 Δcm^{-1} *and* ν_2 *C=C at* ~1510 Δcm^{-1} *are visible above the changing PVC fluorescence background;* Right, *profile of* ν_2 *band with background fluorescence subtracted* (Reprinted from ref. 199, *Polymer*, **35**, Bowden et al., *Thermal degradation of polyurethane-backed poly(vinyl chloride) studied by Raman microline focus spectrometry*, pp. 1654–1658, Copyright (1994), with kind permission from Elsevier Science Ltd., The Boulevard, Langford Lane, Kidlington OX5 1GB, UK)

unsuitable on two counts; sample luminescence was high, and the effect of water could not be studied since the laser irradiation expelled the water from the water tree. The sensitivity of infrared to polar functional groups was a clear benefit to the study.

Confocal Raman microspectroscopy has enabled high quality Raman spectra to be recorded rapidly from cells and chromosomes.[201,202] For example, single point spectra of the nucleus and cytoplasm of an intact human lymphocyte[201] and different interbands of a fixed polytene chromosome[202] have demonstrated the potential for the development of Raman imaging techniques to investigate cell biology. The confocal arrangement led to effective suppression of background Raman signals from substrates or buffers, while excitation using a 660 nm laser prevented radiation damage to the samples. High sensitivity was achieved using a high numerical aperture objective, and a microspectrometer utilising a notch filter and spectrograph in combination with a liquid nitrogen cooled CCD camera. Automated single point FTIR-microscopy measurements have been used to build up line profiles and axonometric and contour plot images showing the subcellular distributions of unsaturated acyl chains, lipid esters and proteins in atherosclerotic artery sections.[203] Individual spectra representative of a 20 μm × 20 μm area were collected over areas of about 1 or

[201] Puppels G.J., Otto C., and Greve J., *Trends Anal. Chem.*, 1991, **10**, 8, 249.
[202] Puppels G.J., Otto C., and Greve J., *Microbeam Anal.*, 1991, 85.
[203] Kodali D., Small D.M., Powell J., and Krishnan K., *Appl. Spectrosc.*, 1991, **45**, 8, 1310.

2 mm^2 from sections of about 5 μm thickness supported on BaF_2 windows. Other reported molecular mapping of variations in tissue biochemistry using scanning FTIR-microscopy have included *in situ* grey and white matter in frozen sections of brain tissue,[204] and artefacts in spleen tissue and phosphate concentration profile along a cryosection through the transition region between a cartilege and bone.[205] FTIR-microscopy has also proved of value in the study of biopolymer gels.[206,207] It has been used to determine the diffusion coefficient of bovine serum albumin in amylopectin gels,[206] and to determine amylopectin/gelatin polymer concentrations in the micro-phase-separated domains of a mixed gel.[207]

Near-infrared FTIR-microscopy using an InSb detector has been used to study the phase behaviour of an aqueous surfactant system.[208] Compositions were determined within a 30 μm^2 area in a sample approximately 25 μm thick in a special DIT (diffusive interfacial transport) cell. Quantitative binary phase-diagram information was obtained for the system octyldimethylphosphine oxide (C_8PO)/water.

FTIR molecular microspectroscopic mapping has also been reported for metamorphic rock phases by specular reflection spectra,[209] for lipid and protein across a corn kernel edge,[209] contour and axonometric displaying of the coating variation on a steel surface by reflectance mode microspectroscopy,[210] for latex coated paper[210] and photoresist coating thickness on a silicon wafer.[210]

Vibrational Spectroscopy – Microscopy Applications: Physical Characterisations

Mapping of the spatial distribution of bulk and/or surface characteristics of materials can be particularly useful in establishing the effects of process variables on their physical characteristics during product manufacture or fabrication. The morphological or physical effects which may be studied by vibrational spectroscopy–microscopy include molecular orientation, conformation and configuration, and strain in materials, as well as phase separation.

Morphology 'Fingerprinting'

Two simple illustrations of the potential of micro-Raman for this type of application to polymer analysis are shown in Figure 39. In the first, Figure

[204] Wetzel D.L. and LeVine S.M., *Proc. 8th Int. Conf. Four. Trans. Spectrosc.*, SPIE, 1991, **1575**, 435.
[205] Reffner J.A. and Wasacz F.M., *Proc. 8th Int. Conf. Four. Trans. Spectrosc.*, SPIE, 1991, **1575**, 437.
[206] Cameron R.E., Jalil M.A., and Donald A.M., *Macromol.*, 1994, **27**, 2708.
[207] Durrani C.M. and Donald A.M., *Macromol.*, 1994, **27**, 110.
[208] Marcott C., Munyon R.L., and Laughlin R.G., *Proc. 8th Int. Conf. Four. Trans. Spectrosc.*, SPIE, 1991, **1575**, 290.
[209] Reffner J.A., *Inst. Phys. Conf. Ser.* No. 98, Paper presented at EMAG-MICRO 89, London, UK, 13–15 September, 1989, ch. 13.
[210] Carl R.T. and Smith M.J., *Review of Progress in Nondestructive Evaluation*, Vol. 9, ed. Thompson D.O. and Chimenti D.E., Plenum Press, New York, USA, 1990, p. 1005.

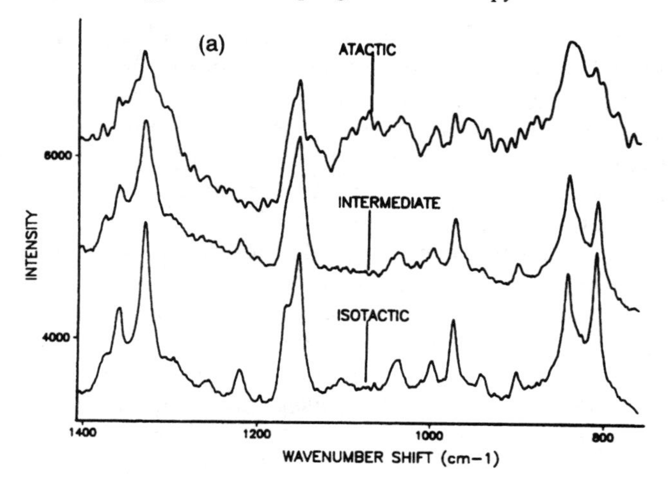

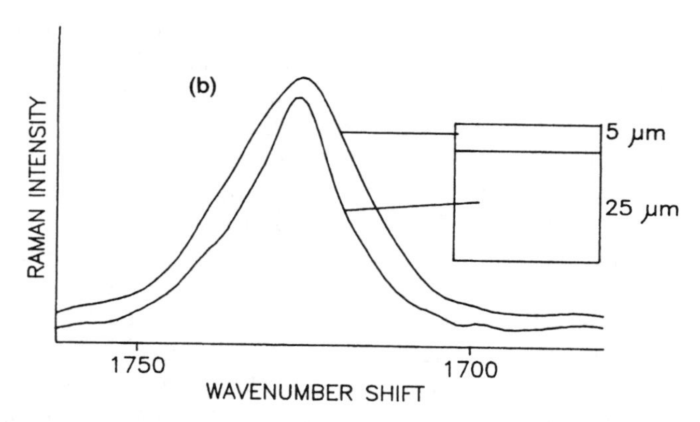

Figure 39 (a) *Raman-microscopy spectra of polypropylene particles obtained in situ on three different alumina-supported catalysts. The variation in polymer tacticity is evident in the three spectra;* (b) *Example of the use of Raman-microscopy to determine the degree of crystallinity of each layer of a bilayer polyester film, from measures of the band-width of the νC═O band; the* 5 μm *layer is essentially amorphous*

(Figure 39(a) reproduced from ref. 133, by kind permission of Huthig & Wepf Publishers, Zug, Switzerland)

39(a), micro-Raman was the favoured technique because of sampling constraints;[133,211] the requirement was for a rapid, no sample work-up, *in situ* analysis of catalyst stereospecificity effectiveness for the production of polypropylene. Polymer tacticity was readily assessed on the surface of particles of different alumina-supported catalyst systems. In the example of Figure 39(b),

[211] Chalmers J.M. and Everall N.J., in *Polymer Characterisation*, ed. Hunt B.J. and James M.I., Blackie Academic, Glasgow, UK, 1993, ch. 4.

the degree of crystallinity of each layer of a bilayer polyester (poly(ethylene terephthalate), PET) film is determined from a measure of the full-width at half-height (FWHH) of its carbonyl stretching mode band. The correlation or near correlation of the FWHH of the νC=O for PET with sample density, and by inference crystallinity, is well established,[125,212,213] although care needs to be exercised with, oriented samples,[125,213] particularly to ensure that the laser field is polarised parallel to the draw direction.[125]

In the application illustrated in Figure 40, both micro-vibrational spectroscopy techniques have been used to provide pictures of the morphology of the constituents in graphite-filled poly(tetrafluoro ethylene), PTFE, blocks.[133] FTIR ATR-microscopy more readily provided a measure of the molecular order of the polymer matrix, whereas Raman microprobe analysis provided a ready measure of the morphology of the filler. Absorption bands between 800 and 650 cm^{-1} associated with amorphous PTFE are easily observed in the sintered block sample; while Raman is well established as a tool for characterising carbon materials, see pp. 314–323.

The case study illustrated in Figure 41 is included to example the need for caution in studying physical property characteristics. The FTIR-microscopy spectrum of Figure 41(a) was obtained from a 'nib product defect' observed in a PEEK film. The 'nib' was isolated from the film and its thickness reduced to an appropriate level by compressing it in a diamond window compression cell. The spectrum of Figure 41(a) clearly showed that the 'nib', being essentially amorphous, had a much different morphology than that of the bulk film. However, the spectrum shown in Figure 41(b) was obtained from exactly the same sample in the cell, except that it was recorded the next day. Left overnight compressed in the cell the specimen had undergone some pressure-induced crystallisation, such that its spectrum now closely resembled that of the bulk film! The effects of sample preparation methods need to be very carefully considered when investigating the physical property characteristics of a sample.

Mapping

Microscopic mapping of the physical characteristics of polymers fabricated into structures is important for an understanding of their performance characteristics and predicting or monitoring their properties.[125,133]

Thermoplastic polymers, because they are light, flexible, tough, and corrosion resistant, compete with metals in many engineering applications.[214–216] For example, both polyethylene and polypropylene are widely used materials from

[212] Melveger A.J., *J. Polym. Sci.* 1972, **A2**, 10, 317.

[213] Everall N., Tayler P., Chalmers J.M., MacKerron D., Ferwerda R., and van der Maas J.H., *Polymer*, 1994, **35**, 15, 3184.

[214] Stevens S. M., *Infrared Technology and Applications*, SPIE, 1990, **1320**, 18.

[215] Wesley I., Magnaudeix B., Hodgson D., Hendra P., Hart P., Stevens S., and Jenkins N., *Plastics, Rubber and Composites Processing and Applications*, 1993, **20**, 211.

[216] Stevens S.M., in *Analytical Applications of Spectroscopy II* ed. Davies A.M.C. and Creaser C.S., Royal Society of Chemistry, Cambridge, UK, 1991, p. 79.

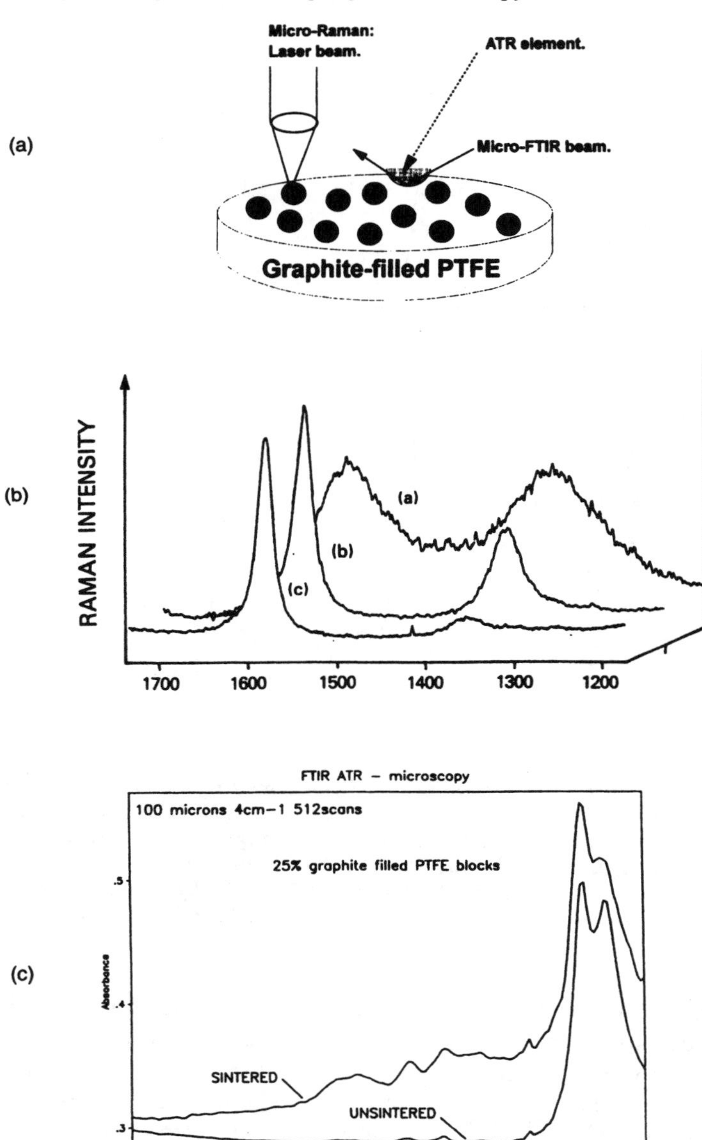

Figure 40 (a) *Schematic showing use of Raman-microscopy and FTIR ATR-microscopy to probe the morphologies of the filler and polymer respectively from the surface of a graphite-filled PTFE moulding;* (b) *Raman-microprobe spectra recorded from the surface of three PTFE blocks, each containing* 25 wt% *graphite filler but of differing levels of graphitisation;* (c) *FTIR-ATR spectra recorded from one such block before and after the sintering process*
(Figures 40(b) and 40(c) reproduced from ref. 133, by kind permission of Huthig & Wepf Publishers, Zug, Switzerland)

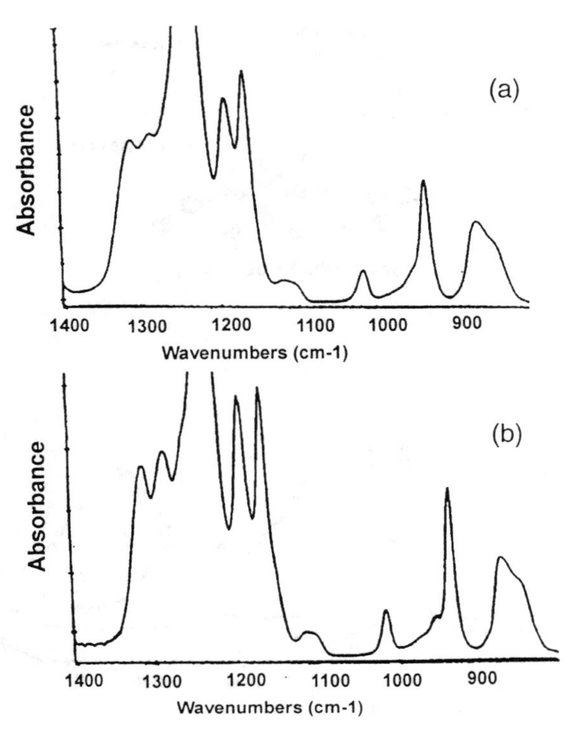

Figure 41 *FTIR-microscopy spectra recorded from a semi-crystalline PEEK polymer film defect, isolated from the film and compressed and held in a diamond window micro-compression cell:* (a) *spectrum as originally recorded from the compressed defect – showing it to be essentially amorphous in character;* (b) *spectrum as recorded after the defect had been left in the compression cell overnight – now exhibiting the effects of pressure-induced crystallisation*

which pipes are fabricated for the distribution of gas and water. Sections are fusion welded together. Investigating structural changes occurring through a weld zone provides insights into the thermal history of the weld assembly. FTIR-microscopy has been used to probe crystallinity changes through welded polyethylene samples.[214,216,217] The measurements were taken on sections microtomed perpendicular to the junction interface. The crystallinity in the weld region was shown to be less than that of the parent pipe, decreasing to a minimum at or very near to the interface. In a separate study, a triple monochromator Raman spectrometer has been used to observe changes in the wavenumber position of the longitudinal acoustic mode (LAM) vibration, which occurs close to the exciting line in the vicinity of 10 Δcm^{-1};[215] the peak maximum position of the 'accordion-like' LAM vibration of the extended chain

[217] Gueugnaut D., Wackernie P., and Forgerit J.P., *J. Appl. Polym. Sci.*, 1988, **35**, 1683.

correlates inversely with stem length, and hence crystalline lamellar thickness.[211,215] Although in this study[215] a microscope was not used, the conventional illumination still only examined a relatively small zone cylindrical volume of dimensions about 0.1 mm diameter and 0.5 mm length. Polarised FTIR-microscopy has also been used for the simultaneous quantitative measurement of molecular orientation and helical content through polypropylene hot plate welds[214,216,218] in a combined study with density, DSC, and XRD measurements. Under the conditions employed, the helical content was shown to be greater at the weld centre-line than in the parent material, the converse of the polyethylene situation.

Polarised micro-Raman has been employed in determining the micron-scale spatial distribution of molecular orientation in injection-moulded polypropylene.[219] The measurements revealed that the disorder of flow extended as far as about 2 mm distance into the (65 × 65 × 2 mm) sheet from the injection mould film gate. Figure 42 shows some results from a micro-Raman observation of the transition front in uniaxially stretched poly(vinylidene fluoride) (PVdF) at different strain rates.[220] The relative variation of the crystalline modification from form II(α) to form I(β) along the transition front was determined as R, where $R = 100.I_{840}/(I_{840} + I_{799})$ calculated from the intensities of the bands at 840 Δcm^{-1} and 799 Δcm^{-1} assigned to the nonplanar (TGTG′) form IIα conformation and to the planar zigzag (TTTT) Iβ conformation, respectively. However, for polarisation measurements of molecular orientation in the neck region of stretched polypropylene it was necessary to decrease the magnification of the objective lens from 100 × to 50 ×, since the 100 × lens was experimentally observed to scramble the polarisation of the laser beam, while the 50 × lens essentially maintained the polarisation.[220] Micro-Raman has also been employed to map this crystalline transformation in PVdF in the microindentation zone produced with increasing pressure during a microhardness test.[221] A similar study of the residual deformation produced by Vickers microindentations has also been reported for the polymer, poly(3,3-dimethyl oxetane).[222]

In other polymer studies, the potential of Raman-microscopy imaging has been demonstrated for investigating the phase composition of polymer blends,[188] while the orientation profiles in extrusion-moulded strands of a thermotropic liquid-crystalline polyester were obtained by polarised FTIR-microscopy.[223] In the latter study, the strands were embedded in an epoxy matrix before being microtomed in the extrusion direction to a thickness of 10 μm. Redundantly apertured transmission spectra were recorded using an aperture of 40–80 μm × 200 μm, and the dichroic ratio of a parallel skeletal

[218] Stevens S.M., *51st Ann. Tech. Conf. SPE*, 9–13 May 1993, New Orleans, USA, 1993, p. 378.
[219] Akashige E. and Usami T., *Anal. Sci. (Suppl. Proc. Int. Congr. Anal. Sci.)*, 1991, **7**, 2, 1655.
[220] Pastor J.M., Jawhari T., Merino J.C., and Fraile J., *Makromol. Chem., Macromol. Symp.*, 1993, **72**, 131.
[221] Jawhari T., Merino J.C., and Pastor J.M., *J. Mat. Sci.*, 1992, **27**, 2237.
[222] Jawhari T., Merino J.C., and Pastor J.M., *J. Mat. Sci.*, 1992, **27**, 2231.
[223] Kaito A., Kyotani M., and Nakayama K., *Macromol.*, 1991, **24**, 3244.

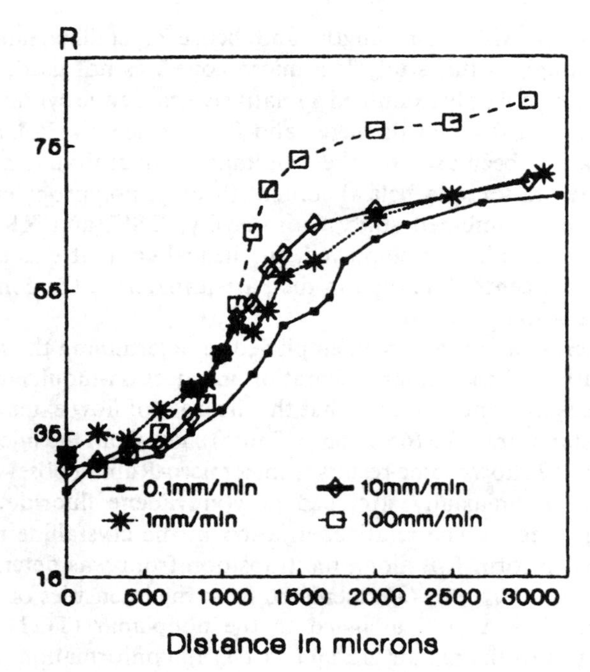

Figure 42 *Micro-Raman study of the transition front in uniaxially stretched poly(vinylidene fluoride): Plot of the crystalline degree R vs. the distance in microns along the transition front in the sample stretched at different strain rates. R was determined from,* $R = 100.I_{840}/(I_{840} + I_{799})$, *where* I_{840} *is the intensity of the Raman band at* 840 Δcm^{-1} *assigned to the non-planar (TGTG') conformation (form II(α)),* I_{799} *is the intensity of the Raman band at* 799 Δcm^{-1} *arising from the planar zigzag conformation (form I(β))*
(Reproduced from ref. 220, by kind permission of Huthig & Wepf Publishers, Zug, Switzerland)

vibration of an aromatic (naphthalene) ring measured to provide the orientation profiles in different diameter strands.

Polarised FTIR-microscopy affords the opportunity to map changes in the physical characteristics through the thickness of a sample, which, for example, might be aligned to changes in process variables during a film-line production run. It can also be used to investigate post-production problems associated with wind-up or relaxation, or contour differences in more complex structures such as carbonated soft drinks bottles.[133,224] However, for the technique to be of any value, any orientation or reorientation caused by microtoming must be very

[224] Chalmers J.M. and Everall N.J., largely unpublished work, but presented in parts as: (i) Chalmers J.M., Paper 021 presented at the 1994 Pittsburgh Conference, Chicago, USA, February 27–March 4, 1994; (ii) Chalmers J.M. and Everall N.J., Proc. European Seminar on Vibrational Spectroscopy in Materials Science, Louvain-la-Neuve, Belgium, 21–22 April, 1994; (iii) Chalmers J.M., Eaves J.G., Everall N.J. and Harvey T.G., Paper 303 at FACSS XX, Detroit, USA, October 17–22, 1993.

small compared to the inherent orientation of the sample. From a detailed study on PET it was concluded that the room temperature microtoming process used to prepare sections for analysis did induce a small amount of orientation into unoriented samples, and, more significantly, caused reorientation within one-way drawn samples when they were sectioned perpendicular to the draw direction; the technique was best suited to the analysis of biaxially oriented samples, which fortunately corresponds to those of most commercial interest.[133,224] For softer polyhydrocarbons such as polyethylene and polypropylene, however, this was not the case, since room temperature microtoming caused significant reorientation into the plane of sectioning regardless of the sample pre-history; these orientation/reorientation-induced effects may be minimised/eliminated by cryo-microtoming. The absorption bands used to map the molecular orientation in PET were the 1018 cm^{-1} (parallel character) band and the 875 cm^{-1} (perpendicular character) band. The technique has been exploited to enable three-dimensional mapping of the molecular orientation, molecular conformation and molecular configuration in PET samples.[133,224] The structures examined included thick film and bottles. Dichroic measurements were made on two transverse (*e.g.* through a film thickness or bottle wall) mutually perpendicular microtomed sections, the directions of which represented sample geometric axes. For a bottle these were the hoop and axes directions. Figure 43 illustrates some results recorded from an examination of a thick (~350 μm) curled film. The results with respect to position through the film thickness indicate significantly lower orientation near the 'concave' surface of the film. No significant differences were observed for either molecular conformation, determined from the 975 cm^{-1} (*trans* conformer) and 898 cm^{-1} (*gauche* conformer) bands, or molecular configuration, inferred from the bandwidth of the 975 cm^{-1} band, in unpolarised spectra. The underlying molecular reason for the film curl was clearly that the film was inhomogeneous with respect to molecular orientation through its thickness.

Electronic Devices

Raman microprobe analysis has proven a particularly sensitive technique for a wide range of studies on specialised electronic and optoelectronic devices.[225–228] Its advantages over infrared stem largely from a better spatial resolution and, since most of the materials are inorganic and many have vibrational modes which have large polarisability changes but small dipole moment changes, a ready access to low wavenumber data and higher intensity data.[131] The investigations include probing non-stoichiometry (see pp. 287–288) and disorder in local structure, crystalline morphology, orientation and damage, lattice strain effects, and local temperatures in devices under operational conditions.

[225] Nakashima S. and Hangyo M., *IEEE J. Quant. Electron.*, 1989, **25**, 5, 965.
[226] Nakashima S., in *Light Scattering in Semiconductor Structures and Superlattices*, ed. Lockwood D.J. and Young J.F., Plenum Press, New York, USA, 1991, p. 291.
[227] Tang W.C. and Rosen H.J., *Microbeam Anal.*, 1991, **26**, 101.
[228] Abstreiter G., *Appl. Surf. Sci.*, 1991, **50**, 73.

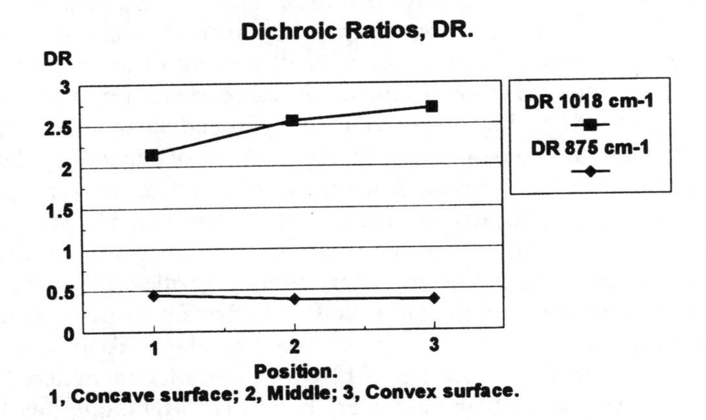

3
2
1
microtomed sections
~10 microns thickness
~350
microns

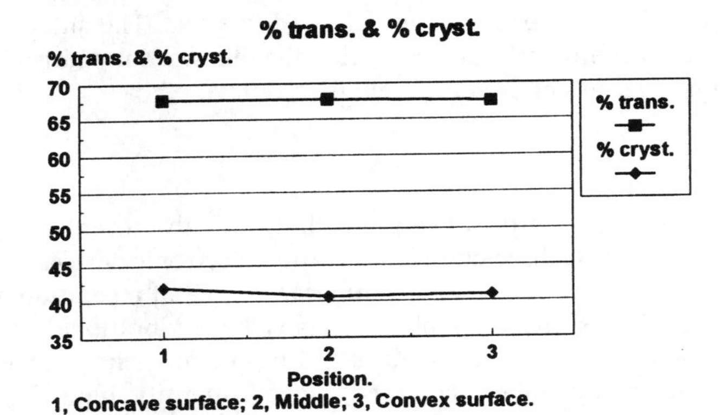

Figure 43 *FTIR-microscopy measurements of molecular orientation, molecular conformation and molecular configuration through the thickness of a 'curled' PET film.* Middle, *schematic of the film showing the regions from which the sections were microtomed (parallel to the 'circumferential' direction);* Top, *profile plots of the dichroic ratios of the* 1018 cm^{-1} *absorption band (parallel character) and* 875 cm^{-1} *(perpendicular character);* Bottom, *profile plots of the % trans conformer and % crystallinity*

Raman band energy (wavenumber), intensity, polarisation, and lineshape are all key parameters in diagnostic studies of electronic devices. Applications to semiconductors have been reviewed in ref. 225. Polarised micro-Raman has proven particularly useful in determining local crystalline orientations in silicon and other devices, because the technique is both non-destructive and offers high spatial resolution (~1 μm).[225,226,229,230] In a Raman microprobe study using polarised laser radiation with a focused spot size of 0.8 μm the crystallographic orientations were determined for single-crystalline silicon wafers with (100), (110) and (111) mirror polished surfaces.[230] The method has been employed to map from the 'seed-point' the crystallographic orientation of lateral epitaxy formed polycrystalline films on SiO_2 insulators overcoated onto silicon wafers;[226] the 'seed-point' is an opening in the oxide coating insulator layer through which the polycrystalline layer is directly attached to the silicon crystal. At this point the surface orientation of the recrystallised layer is the same as that of the Si substrate, but away from the seed-point the ⟨001⟩ axis was observed to rotate forwards about the ⟨010⟩ axis, when the Raman microprobe was scanned along the ⟨100⟩ direction of the silicon wafer, this being the direction the annealing laser was initially scanned to effect the recrystallised stripe in the surface layer.

Another key application of Raman microprobe analysis in this field is in the measurement of residual stress.[227,228,231–233] For example, micro-Raman has been used to obtain detailed information about stress distribution in local isolation structures.[231] The stress in the surface of the silicon substrate in the vicinity of a Si_3N_4 oxidation mask is profiled by measuring the position of the 520 Δcm^{-1} band. In strain-free crystalline silicon the three active vibration modes all have the same frequency; they consist of two transversal optical (TO) phonons along the (100) and (010) directions and one longitudinal optical (LO) phonon along the (001) direction. Under strain the symmetry of the crystal is lowered and the vibration mode degeneracy partly lifted and they shift to different frequencies.[231] The stress field has also been determined in machined surfaces of germanium crystals, by measuring the position and asymmetry of the longitudinal optical phonon spectra near 300 Δcm^{-1}.[232] In these studies the samples are probed with different wavelength lasers in order to obtain differential depth-related information. The 457.9 nm line of an argon line laser was preferred to the 488.0 nm line in the study of local isolation structure,[231] because of its short penetration depth in crystalline silicon, since the stress was expected to be largest nearest the surface. In the study on machined surfaces of germanium,[232] depth profiles of residual stress were generated from Raman data recorded using successively longer laser wavelengths. Both infrared and Raman have been used to evaluate local stress in plasma-enhanced chemical

[229] Tuschel D.D., *Microbeam Anal.*, 1991, **26**, 109.
[230] Mizoguchi K. and Nakashima S., *J. Appl. Phys.*, 1989, **65**, 7, 2583.
[231] De Wolf I., Vanheellemont A., Romano-Rodriguez A., Norstrom H., and Maes H.E., *J. Appl. Phys.*, 1992, **71**, 2, 898.
[232] Sparks R.G. and Paesler M.A., *J. Appl. Phys.*, 1992, **71**, 2, 891.
[233] Okada Y. and Nakajima S., *Appl. Phys. Lett.*, 1991, **59**, 9, 1066.

vapour deposition silicon nitride films on silicon wafers,[233] since it was found that the frequency of the Si—H stretching mode was sensitive to the film stress. For infrared it increased linearly with compressive stress within the films. The micro-Raman technique was used to probe local stress in Si_3N_4 films deposited over patterned structures, such as trenches of approximately 2 μm width and 2 μm depth. Figure 44[227] demonstrates the sensitivity of microprobe Raman to map strain and defect density across an $In_{0.25}Ga_{0.75}As$ film grown on patterned GaAs(100) substrate.[227] Substantial variations are clearly evident between the spectra, even though the three spectra are taken from regions separated by less than 2 μm. The GaAs-like LO optical phonon mode observed at $\pm 284.8\ \Delta cm^{-1}$ in the valley region is near that expected from unstrained films. At the ridge this mode is shifted to $\pm 289\ \Delta cm^{-1}$, indicating strain. Detection of other modes in the valley ($\sim \pm 258\ \Delta cm^{-1}$ and $\pm 237\ \Delta cm^{-1}$) and ridge ($\pm 268.7\ \Delta cm^{-1}$) regions are ascribed to the presence of structural disorder and defects.

The temperature-dependent shift of the Si optical phonon ($-0.021\ \Delta cm^{-1}$/K) has been used to determine local lattice temperatures in semiconductor devices under operational conditions;[227,228] the example in Figure 45 shows the ΔT *vs.* drain voltage plot determined from micro-Raman measurements on a 2.5 μm wide transistor operated at a gate voltage of 6V. Temperature profiles at the mirrors of high-power GaAs multi-quantum-well diode lasers have been mapped from micro-Raman spectroscopy;[228] local temperatures were extracted from the Stokes/anti-Stokes intensity ratio of the optical phonon. In another study,[227] micro-Raman was used to measure the temperature of the laser's emission facet. The time-evolution behaviour of an AlGaAs single-quantum-well laser diode facet temperature and laser output under conditions of a constant operating current were measured until catastrophic optical damage (COD) occurred. The measurements identified important features underlying the laser's lifetime to COD, such as the existence of a critical temperature where COD occurs.

Carbon, C and SiC, Fibres and Composites

Micro-Raman provides a unique probe for characterising carbons and mapping *in situ* the surface morphology of carbons in many composite materials, providing due attention is given to the experimental parameters, in particular laser power.

Graphite has a high degree of structural anisotropy. The crystal structure of hexagonal graphite is composed of stacked layers of strongly covalently bonded C atoms which occupy the lattice sites of a two-dimensional honeycomb network. Within the basal plane (two-dimensional hexagonal lattice) the in-plane nearest neighbour interatomic distance is 1.42 Å. The bonding between adjacent planes is much weaker. The interlayer distance has values from 3.354 Å (single crystal) to ~3.7 Å.[234] The first-order Raman spectrum of graphitised

[234] Tunistra F. and Koenig J.L., *J. Chem. Phys.*, 1970, **53**, 3, 1126.

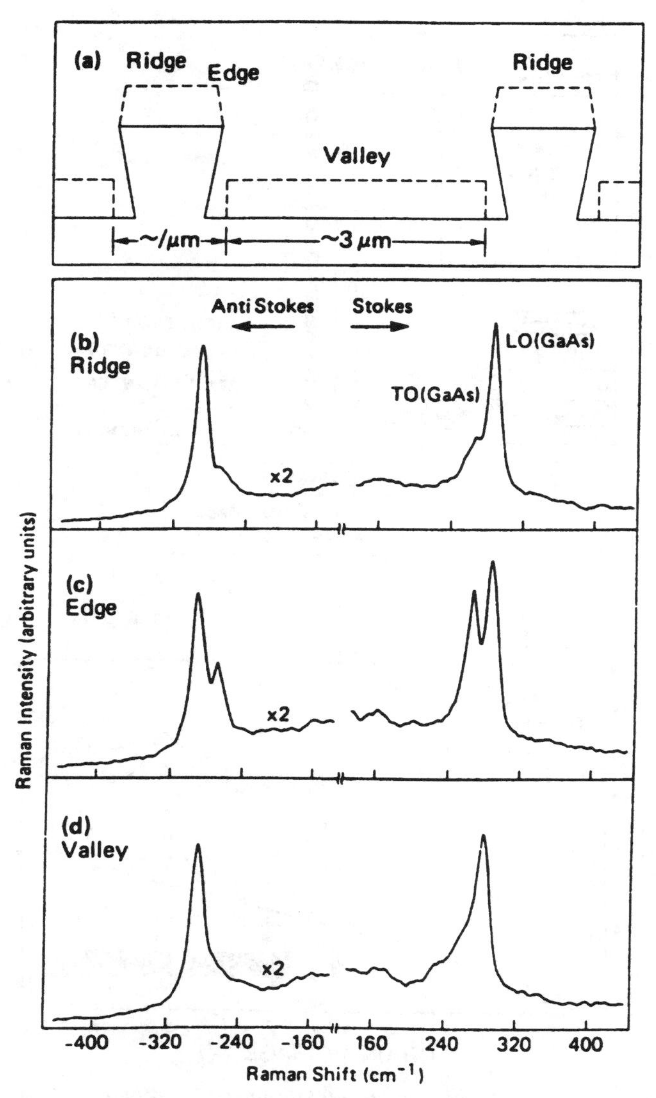

Figure 44 (a) *Schematic of a microstructure of $In_{0.25}Ga_{0.75}As$ film grown on patterned ridges on (100) GaAs substrate;* (b)–(d) *Micro-Raman spectra of $In_{0.25}Ga_{0.75}As$ on various regions of patterned (100) GaAs substrate;* (b) *from the centre of a ridge;* (c) *from the edge of a ridge;* (d) *from a valley between two ridges*
(Reproduced from ref. 227, by kind permission of San Francisco Press, Inc., CA, USA)

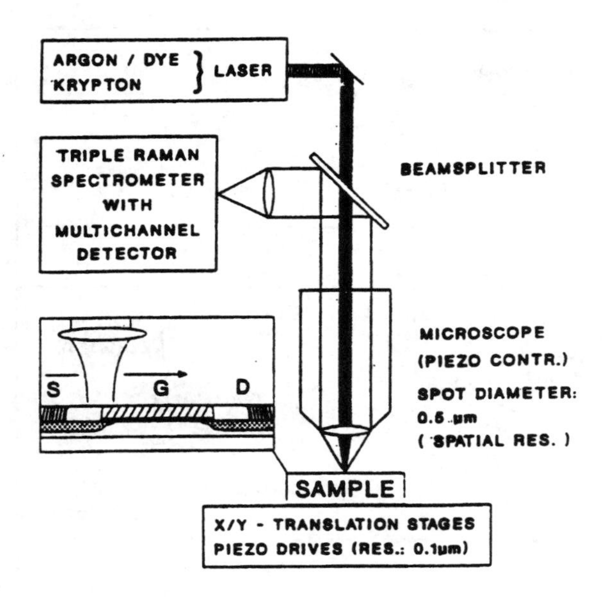

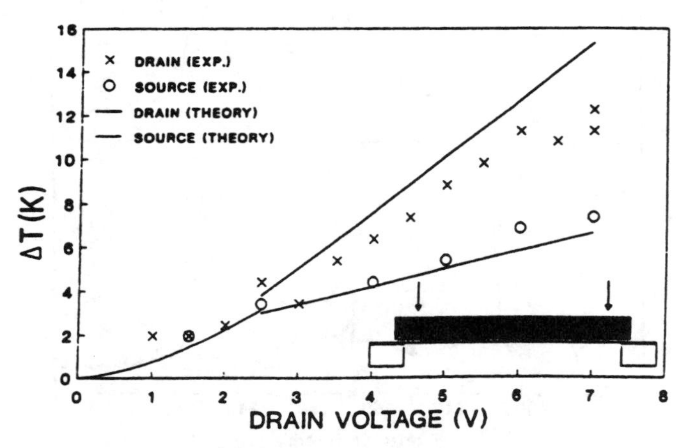

Figure 45 Top, *schematic of the experimental set-up used to profile the local temperature distribution in a semiconductor device (a short-channel Si-MOSFET);* bottom, *an example of the temperature profile vs. drain voltage measured for a MOSFET with a* 2.5 µm *wide gate electrode; the transistor was operated at a gate voltage of* 6V
(Reprinted from ref. 228, *Applied Surface Science*, **50**, G. Abstreiter, *Micro-Raman spectroscopy for characterisation of semiconductor devices,* pp. 73–78, ©1991 with kind permission of Elsevier Science – NL, Sara Burgerhartstraat 25, 1055 KV Amsterdam, The Netherlands)

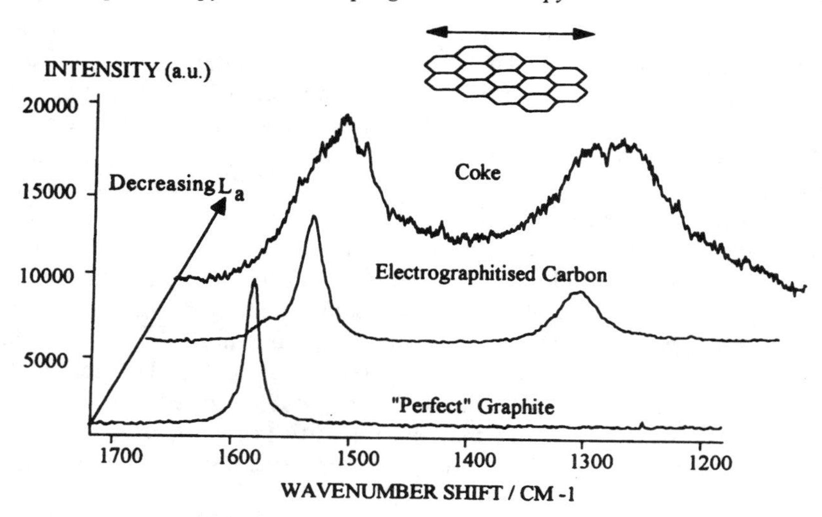

Figure 46 *Variation in the Raman spectra of graphitic powders as a function of lattice perfection, i.e. crystallite diameter L_a*
(Reproduced with kind permission of Plenum Publishing Corp., from Everall N.J., *Industrial Applications of Raman Spectroscopy*, pp 115–131, in *An Introduction to Laser Spectroscopy*, ed. Andrews D.L. and Demidov A., Plenum Press, New York (1995))

carbon is very sensitive to two-dimensional ordering.[234–236] It is characterised by a band at $\sim$1580 Δcm^{-1}, an in-plane lattice mode with E_{2g} symmetry (G band), which is present in all graphites, and disorder-induced bands which occur at $\sim$1350 Δcm^{-1} (D band) and $\sim$1630 Δcm^{-1}.[234–236] (There is also an E_{2g} lattice mode at $\sim$42 Δcm^{-1}.)[235] The $\sim$1350 Δcm^{-1} band becomes prominent in imperfect graphites, see Figure 46, where the size of the planes (L_a) decreases and there is relaxation of the symmetry rule.

A measure of 'structural disorder' may be inferred from the intensity ratio I_{1350}/I_{1580} . For instance, for a series of microcrystalline powder samples an inverse linear relationship with the X-ray determined L_a values has been reported.[234] Similar correlations have also been observed, but with different dependencies for differing carbon fibre types.[237,240] Other workers have established correlations between the intensity ratio and d_{002}, the X-ray determined distance between graphitic layers.[235] The width ($\tilde{\nu}$ cm^{-1}) and position ($\tilde{\nu}$ Δcm^{-1}) of the E_{2g} (1580 Δcm^{-1}) band have also been demonstrated to correlate with the

[235] Lespade P., Marchand A., Couzi M., and Cruege F., *Carbon*, 1984, **22**, 4/5, 375.
[236] Lespade P., Al-Jishi R., and Dresselhaus M.S., *Carbon*, 1982, **20**, 5, 427.
[237] Fitzer E., Rozploch F., and Kunkele F., *Carbon '88: Proc. Int. Conf. Carbon*, University of Newcastle, 18–23 Sept. 1988, ed. McEnaney B. and Mays T.J., IOP Publishing, Bristol, UK, 1988.

development of the two-dimensional graphitic assemblies, *i.e.* the first phase of the graphitisation process.[235] The ~1620 Δcm^{-1} 'shoulder' is another disorder-induced feature which appears to increase in intensity with decrease of crystallite size.[236] Many other factors, such as band half-widths and second- (*e.g.* ~2700 Δcm^{-1}) and third-order bands also reflect changes in graphitisation level, and must be taken into account in order to fully characterise two- and three-dimensional ordering in graphite.[235,238]

The intensity ratio $I_{1580}/I_{1350} \propto L_a$ is only valid for samples in which the graphite crystals are randomly oriented.[125] If an overall orientation exists in the sample, this will affect the measured band intensity ratio, and that from an edge plane will be different from that of a basal plane.[125,237,239,240] Some recent studies have indicated that the intensity of the ~1350 Δcm^{-1} band arises from edge effects rather than being *per se* a direct consequence of microcrystallite size,[239,241] and that the intensity ratio I_{1350}/I_{1580} depends on both the degree of graphitisation and the orientation of the graphitic planes.[125,240–242] This is illustrated in Figure 47, which compares the micro-Raman spectra recorded from two 7 μm diameter carbon fibres.[125,243] For the Rayon-based fibre, in which the crystallites are randomly oriented in all dimensions, identical spectra are recorded for the fibre surface and cross-section. However the equivalent spectra recorded from a partially graphitised poly(acrylonitrile) (PAN)-based fibre differed, since in PAN-based fibres the graphite planes are arranged such that their normals are perpendicular to the fibre long axis. Surface illumination predominantly samples the basal planes, while illumination of the fibre cross-section will predominantly sample the edges of graphitic planes; thus, the former appears to be of higher crystallinity, while the latter underestimates crystallite size.[125] Raman microscopy has been used to map the morphology in carbon–carbon composites, which consist of a graphitised carbon matrix reinforced by carbon fibres. Measurements of the influence of the fibre on matrix graphitisation has been studied for both completely[239] and partially[125,243] graphitised C–C composites.

Raman microscopy has also been used as a non-intrusive technique for measuring strains in high modulus carbon fibres and to map strains present in C-fibre reinforced composite materials. Both the first- and second-order Raman spectra of carbon fibres show strain dependencies. The position of the G band near 1580 Δcm^{-1} has been shown to correlate linearly with applied strain (%). Shifts of about −9.2 Δcm^{-1}/% have been reported for high modulus (HM)

[238] Cuesta A., Dhamelincourt P., Laureyns J., Martinez-Alonso A., and Tascon J.M.D., *Carbon*, 1994, **32**, 8, 1523.
[239] Katagiri G., Ishida H., and Ishitani A., *Carbon*, 1988, **26**, 4, 565.
[240] Fitzer E., Gantner E., Rozploch F., and Steinert D., *High Temp.-High Press.*, 1987, **19**, 5, 537.
[241] Wang Y., Alsemeyer D.C., and McCreery R.L., *Chem. Mater.*, 1990, **2**, 557.
[242] Fitzer E. and Rozploch F., *High Temp.-High Press.*, 1988, **20**, 449.
[243] Ragan S., Everall N., and Lumsdon J., *Extended Abstracts of "Carbone '90", Int. Carbon Conf.*, 10–12 July 1990, GFEC, Paris, France, 1990, p. 506.

(a) surface, (b) cross-section.

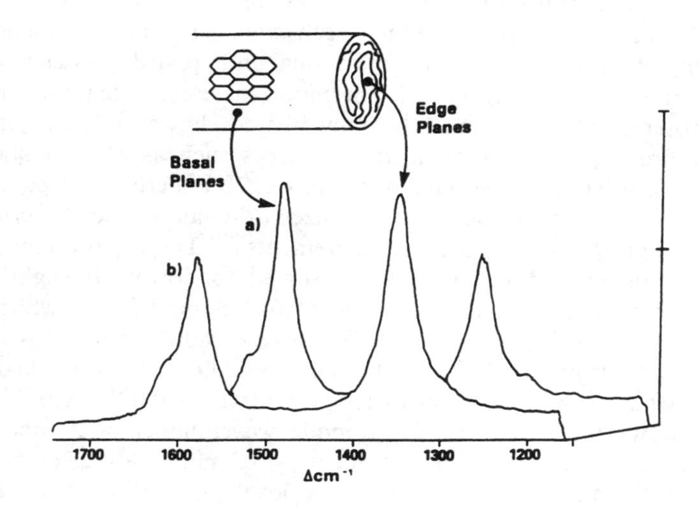

(a) surface, (b) cross-section.

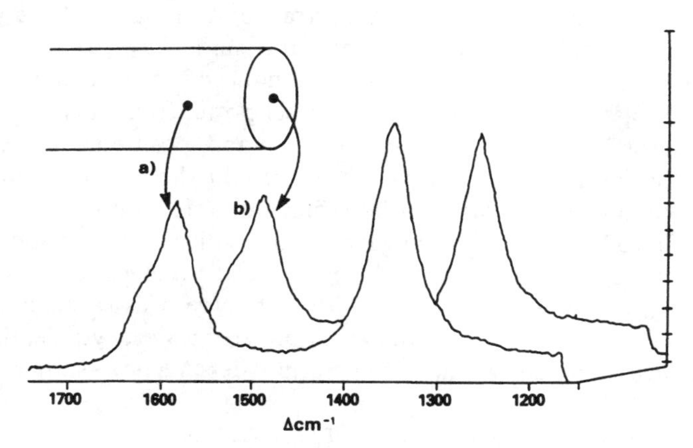

Figure 47 *Comparison of the micro-Raman spectra recorded from the surfaces* (a) *and cross-sections* (b) *of two* 7 μm *diameter carbon fibres.* Top, *a partially graphitised PAN-based fibre, in which the graphite crystallite basal planes are arranged with their normals perpendicular to the fibre long axis;* Bottom, *a Rayon-based fibre, in which the crystallites are randomly oriented in all directions*
(Reproduced from ref. 125, by kind permssion of IOS Press, Amsterdam)

fibres under tensile strain,[244–249] with similar (but blue-shifted) values being measured ($\sim$10 $\Delta cm^{-1}/\%$) for fibres under compressive strain,[248] but with shifts of only $\sim$5 $\Delta cm^{-1}/\%$ being observed for lower modulus fibres.[248,249] Second-order band shifts of >20 $\Delta cm^{-1}/\%$ have been measured for HM fibres.[244,245] This type of application has been extended from measurements on free-standing fibres to measurements of residual and applied stresses of carbon fibres present as a reinforcing agent in thermoplastic composite materials[245,246] and thermoset resins.[247] It must be remembered, however, that the band parameters are dependent upon intrinsic factors such as fibre crystallinity, graphite plane orientation, surface damage, *etc.*[248] Experimental parameters such as laser wavelength, power, and spot size at the sample are also critical to the fundamental precision of these measurements.[250] The E_{2g} (G band) mode band of disordered carbons has been observed to downshift slightly with increasing temperature (90–130 K → ~ -0.013 $\Delta cm^{-1}/K$),[251] while for a highly oriented pyrolitic graphite (HOPG) above $\sim$280 K a shift of ~ -0.028 $\Delta cm^{-1}/K$ has been reported.[252] The D band ($\sim$1350 Δcm^{-1}) for disordered carbons showed a smaller effect with temperature (~ -0.0017 $\Delta cm^{-1}/K$).[251] Figure 48 shows the variation of E_{2g} mode wavenumber as a function of temperature, recorded from a flake fragment ($\sim$0.5 mm × $\sim$0.25 cm^2) from a pressed disc of graphite.[250,253] To minimise localised heating effects a 10× microscope objective was used and the laser power kept at $\sim$75 mW. The temperature sensitivity determined was ~ -0.03 $\Delta cm^{-1}/K$. Clearly, laser power density needs to be a carefully controlled parameter if meaningful band shifts are to be measured in determining strain and similarly related properties. Laser-induced heating and consequent band shifts have been shown to be significant.[253,254] Figure 49 shows the laser-induced bandshift in a single ($\sim$40 μm diameter) graphite grain on changing laser power from 1mW to 6mW using a 50× microscope objective; the original band position was restored on returning the laser power to 1mW.[253] Figure 50 shows the G and D band positions for a single HMS4 carbon fibre as a function of incident laser power.[253] The 50× objective used gave a $\sim$5 μm diameter laser spot size at the sample; reducing the spot size to 2 μm changed the dependency from $\sim$0.7 $\Delta cm^{-1}/mW$ to $\sim$1.4 $\Delta cm^{-1}/mW$![253] Although there is also a small dependency of the position of the G band with excitation laser wavelength λ_0, the D band, which showed a much smaller temperature dependency, shows a large λ_0

[244] Galiotis C. and Batchelder D.N., *J. Mat. Sci. Lett.*, 1988, **7**, 545.
[245] Young R.J., Day R.J., Zakikhani M., and Robinson I.M., *Comp. Sci. Tech.*, 1989, **34**, 243.
[246] Young R.J. and Day R.J., *Br. Polym. J.*, 1989, **21**, 17.
[247] Young R.J., Huang Y-L., Gu X., and Day R.J., *Plast. Rubber Compos. Process Appl.*, 1995, **23**, 1, 11.
[248] Melanitis N. and Galiotis C., *J. Mat. Sci.*, 1990, **25**, 5081.
[249] Huang Y. and Young R.J., *Carbon*, 1995, **33**, 2, 97.
[250] Everall N.J., Lumsdon J., and Christopher D., *Proc. XII Int. Conf. Raman Spectrosc.*, 1990, 812.
[251] Fischbach D.B. and Couzi M., *Carbon*, 1986, **24**, 3, 365.
[252] Erbil A., Postman M., Dresselhaus G., and Dresselhaus M.S., *Abs. and Program, 15th Bienn. Conf. of Carbon*, Pennsylvania, USA, 1981, p. 48.
[253] Everall N.J., Lumsdon J., and Christopher D.J., *Carbon*, 1991, **29**, 2, 133.
[254] Everall N. and Lumsdon J., *J. Mater. Sci.*, 1991, **26**, 5269.

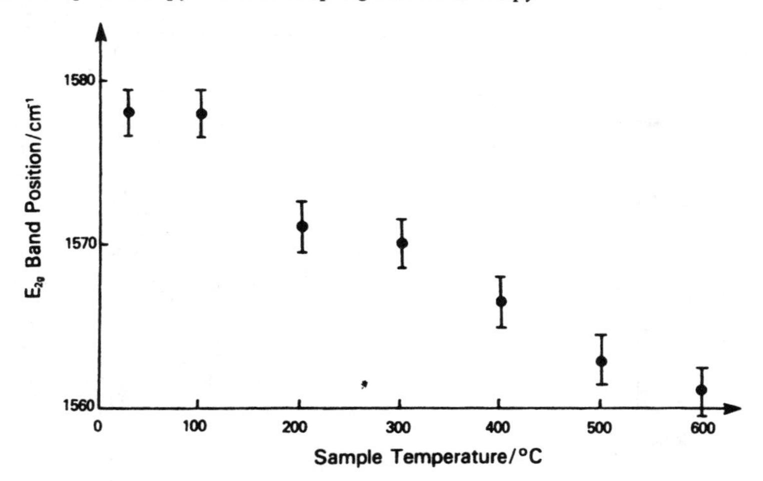

Figure 48 *Variation of the E_{2g} mode wavenumber position as a function of temperature recorded from a flake fragment (~0.5 mm × ~0.25 cm^2) from a pressed disc of Lonza KS-5–75 graphite*
(Reprinted from ref. 253, *Carbon*, **29**, Everall et al., *The effect of laser-induced heating upon the vibrational Raman spectra of graphites and carbon fibres*, pp. 133–137, Copyright (1991), with kind permission from Elsevier Science Ltd., The Boulevard, Langford Lane, Kidlington OX5 1GB, UK)

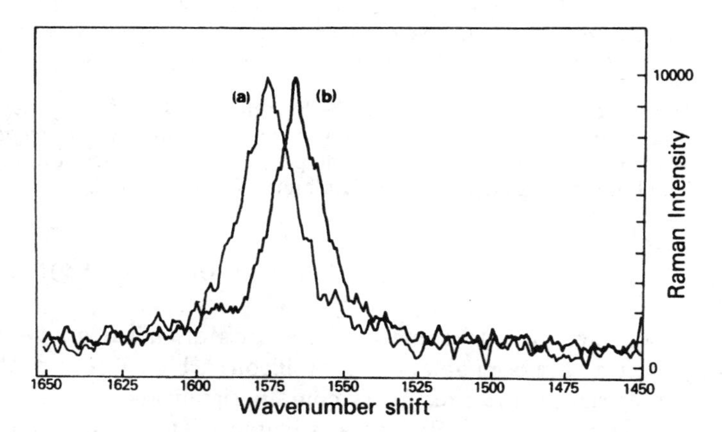

Figure 49 *Laser-induced bandshift in a single (~40 μm diameter) graphite grain:* (a) E_{2g} *mode,* 1 mW *laser beam power at the sample;* (b) E_{2g} *mode,* 6 mW *laser power at the sample*
(Reprinted from ref. 253, *Carbon*, **29**, Everall et al., *The effect of laser-induced heating upon the vibrational Raman spectra of graphites and carbon fibres*, pp. 133–137, Copyright (1991), with kind permission from Elsevier Science Ltd., The Boulevard, Langford Lane, Kidlington OX5 1GB, UK)

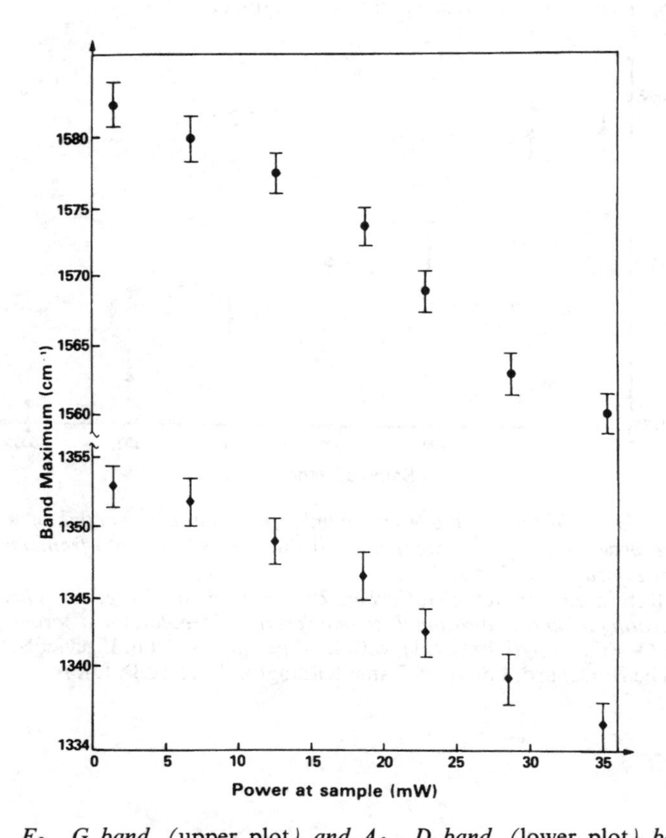

Figure 50 *E_{2g}, G band,* (upper plot) *and A_{1g}, D band,* (lower plot) *band maxima positions for a single HMS4 carbon fibre as a function of incident laser beam power, using a* 50 × *objective in the Raman-microscope to give a* ~50 μm *laser spot size at the sample*
(Reprinted from ref. 253, *Carbon*, **29**, Everall et al., *The effect of laser-induced heating upon the vibrational Raman spectra of graphites and carbon fibres*, pp. 133–137, Copyright (1991), with kind permission from Elsevier Science Ltd., The Boulevard, Langford Lane, Kidlington OX5 1GB, UK)

dependency, shifting by about $-70\ \Delta\text{cm}^{-1}$ on going from a λ_0 of 407 nm to 782 nm.[241]

For purposes analogous to some of the studies discussed above, Raman microscopy spectra have been obtained from silicon carbide fibres.[255,256] Silicon fibres from polycarbosilane precursors contain significant amounts of free carbon.[255] In one study micro-Raman was employed to examine the size of graphitic microcrystals as a function of distance into the fibre; transverse, oblique, and longitudinal sections for study were prepared from fibres embedded in an epoxy resin.[255] The same workers also measured the sensitivity

[255] Day R.J., Piddock V., Taylor R., Young R.J., and Zakikhani M., *J. Mater. Sci.*, 1989, **24**, 2898.
[256] Sasaki Y., Nishina Y., Sato M., and Okamura K., *J. Mater. Sci.*, 1987, **22**, 443.

of the G band (observed here at 1601 Δcm^{-1} in the unstrained fibres) to strain, determining a dependency of 6.6 $\Delta cm^{-1}/\%$ strain.[255] In another study,[256] integrated intensities of SiC, C, and SiO_2 components in the Raman spectra were used to ascertain their dependence on the heat treatment temperature of a polycarbosilane precursor from 700° to 1900°C. Raman microprobe analysis has also been employed to examine SiC deposited by chemical vapour deposition on carbon filament and tungsten wire substrates which have been incorporated into a titanium matrix.[257] The SiC/Ti composite materials were cut transversely, and polished cross-sections examined at different distances from the filament core. At the core/deposit boundary of the SiC filament formed on the tungsten-wire substrate, two strong narrow lines at 800 and 970 Δcm^{-1} were observed and assigned to the β form cubic SiC. Away from the core the SiC became more disordered as the fibre surface was approached. The fibre diameter was about 100 μm. Only disorganised carbon was observed close to the core/deposit boundary of the SiC formed on carbon filaments.[257]

Raman microscopy has more recently emerged as one of the principal tools for characterising thin diamond films.[258] These hard, chemically inert, and optically transparent films are produced by chemical vapour deposition (CVD) processes onto inorganic substrates such as silicon, germanium, and glass to produce novel electronic devices and optical coatings. The first-order Raman spectrum of natural diamond is characterised by a single sharp line at 1332 Δcm^{-1}. Raman spectroscopy can readily distinguish between graphitic and other sp^2 bonded amorphous carbons and sp^3 bonded crystalline diamond and diamond-like materials,[258,259] and has been used to highlight localised differences in polycrystalline CVD films,[259,260] to compare the sp^3:sp^2 bonding ratio in diamond-like amorphous carbon films,[261] or to profile, using MiFS, laser annealed spots in hydrogen-containing amorphous carbon films.[262]

Amorphous carbon films are widely used as protective overcoats ($\sim$15–40 nm thick) on magnetic storage discs. These have been the subject of Raman microprobe studies.[263,264] The investigations correlated the relative intensities and positions of the D and G bands with performance ratings determined from tribological testing.

Polymer Fibres

Raman spectroscopy offers distinct sampling advantages over infrared in the study of fibres where it is vital to maintain sample integrity and minimise optical artefacts.

[257] Couzi M., Cruege F., Martineau P., Mallet C., and Pailler R., *Microbeam Anal.*, 1983, 274.
[258] Knight D.S. and White W.B., *J. Mater. Res.*, 1989, **4**, 2, 385.
[259] Plano L.S. and Adar F., SPIE 822 *Raman and Luminescence in Technology*, 1987, 52.
[260] Shroder R.E., Nemanich R.J., and Glass J.T., SPIE 969 *Diamond Optics*, 1988, 79.
[261] Yoshikawa M., *Materials Sci. Forums*, 1989, **52/53**, 365.
[262] Bowden M., Gardiner D.J., and Southall J.M., *J. Appl. Phys.*, 1992, **71**, 1, 521.
[263] Lauer J.L., Blanchet T.A., and Ng Q., *STLE Tribology Transactions*, 1994, **37**, 3, 566.
[264] Marchon B., Heiman N., Khan M.R., Lautie A., Ager J.W., and Veirs D.K., *J. Appl. Phys.*, 1991, **69**, 8, 5748.

Raman spectra have been compared from two series of gel-spun ultra drawn ultrahigh molecular weight polyethylene (PE) fibres.[265] The series of different draw ratio fibres were manufactured by different processes. The molecular deformation properties of each series were compared by studying the strain dependency of the C—C stretching region, and contrasted with X-ray measurements. During straining, additional broad bands appeared to the low wavenumber side of both the asymmetric (~1062 Δcm^{-1}) and symmetric (~1130 Δcm^{-1}) bands. Both the narrow νC—C bands and the weaker broad strain-induced bands shifted to lower wavenumbers with increasing strain, the latter to a much greater extent. The researchers concluded that the shifts of bands increased linearly with increasing Young's modulus, and that there was evidence for the existence of two types of strained molecules. This study, although not definitive, provided new information important to modelling deformation processes.[266] Strain-induced shifts have been measured in the Raman spectra of untreated and surface-treated single gel-spun polyethylene fibres embedded in an epoxy resin as a method of analysing the micromechanics of composite deformation.[267]

In a polarised micro-Raman study of single filaments of spin-oriented and drawn fibres of PET, which had been previously characterised by X-ray diffraction, density and optical birefringence, various Raman bands and their intensities were correlated with conformation of the glycol linkage, orientation of the chains, and crystallinity.[268] (Reference 268 also reviews briefly previous Raman literature on the subject of PET orientation and crystallinity.) The low velocity wind-up samples which were subsequently drawn to differing extents, from 3.8:1 to 5.0:1 all had similar densities, similarly high birefringences, similar crystalline orientation factors and crystallinities (~30%); the spin-oriented series prepared with much higher and varying wind-up velocities, exhibited a gradual increase in birefringence and density, while only those wound at the highest velocities had measurable X-ray diffraction crystallinity. Individual fibres (~20 μm diameter) were mounted onto a rotating microscope stage and three independent Raman measurements recorded from a selected spot on each fibre sample. These represented the situations for which the input laser beam was polarised parallel to the fibre axis (defined as the Z direction) and the polarisation analyser set to record Raman polarisation parallel to this axis (labelled as ZZ), the analyser was then rotated and the spectrum equivalent to ZR recorded (R is one of the radial directions), and then the sample was rotated through 90° and the RR spectrum recorded. (No significant differences were observed between ZR and RZ intensities). The Raman bands between 950 and 1220 Δcm^{-1}, see Figure 51, were observed to correlate well with conformational changes and inter-chain effects rather than amorphous–crystalline differences.[268] For instance, the band at 1030 Δcm^{-1} assigned to glycol units in a

[265] Kip B.J., Van Ejik C.P., and Meier R.J., *J. Polym. Sci.: B: Polym. Phys.*, 1991, **29**, 99.
[266] Prevoresk D.C., *Chemtracts – Macromol. Chem.*, 1991, **2**, 282.
[267] Young R.J. and Gonzalez-Chi P.I., *J. Mater. Sci. Lett.*, 1994, **13**, 21, 1524.
[268] Adar F. and Noether H., *Polymer*, 1985, **26**, 1935.

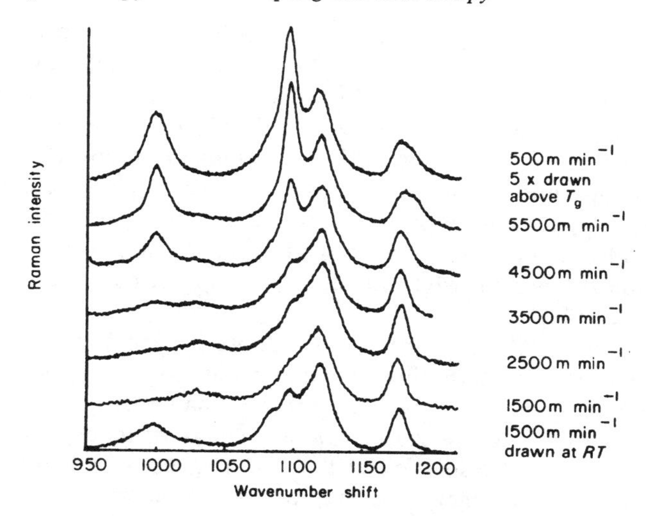

Figure 51 *Polarised micro-Raman study of spin-oriented and drawn PET filaments: ZZ component of five samples spin-oriented with take-up speeds* 1500, 2500, 3500, 4500 *and* 5500 m min^{-1} *and two drawn samples. The top spectrum was recorded from a fibre spin-oriented at* 500 m min^{-1} *over pins held at* 90°C *and* 150°C; *the bottom spectrum was recorded from a fibre spin-oriented at* 1500 m min^{-1} *and drawn* 2.5 × *at room temperature*
(Reprinted from ref. 268, *Polymer*, **35**, Adar F. and Noether H, *Raman microprobe spectra of spin-oriented and drawn filaments of poly(ethylene terephthalate)*, pp. 1935–1985, Copyright (1985), with kind permission from Elsevier Science Ltd., The Boulevard, Langford Lane, Kidlington OX5 1GB, UK)

gauche conformation is replaced by the *trans* band at 1000 Δcm^{-1}. Extended PET chains require the *trans* conformation. The 1030 Δcm^{-1} band appears in both the *ZZ* and *RR* spectra, but since the 1000 Δcm^{-1} band was only observed in the *ZZ* arrangement it implied simultaneous development of *trans* conformation and chain orientation, which was independent of crystallisation. The half-width data of the carbonyl band near 1730 Δcm^{-1} narrowed with density, following correlations established previously by other workers.[212] The $\nu C{=}O$ is a complex band containing contributions from several phases. In a follow-up study,[269] its components were separated using band-fitting software into a crystalline band with a maximum intensity at 1726 Δcm^{-1}, and two amorphous environment bands with maxima at 1735 Δcm^{-1} and 1721 Δcm^{-1}.

Crystal Modifications

Vibrational spectroscopy–microscopy techniques may be employed in investigating crystal types in organic and inorganic substances.

[269] Adar F., Armellino D., and Noether H., SPIE 1336 *Raman and Luminescence Spectroscopies in Technology II*, 1990, 182.

Characterising and distinguishing different polymorphic forms of a solid is particularly important in the pharmaceutical industry, since, although chemically identical, different polymorphic forms (crystalline arrangements) can have very different physical properties and affect a drug's performance.[270,271]

The complementary role of micro-FTIR thermomicroscopy alongside DSC, thermomicroscopy and thermophotometry has been well demonstrated in a study of the polymorphism of p-hexadecylaminobenzoic acid (HABA), a pharmaceutical intermediate. It can provide confirmation of retention of the primary chemical bonding, while yielding specific information on the distinct phases,[270,271] see Figure 52.

The grain structures and properties of nanophase inorganic materials can also vary significantly and are therefore important to characterise. A study of as-compacted nanophase TiO_2 has shown that Raman microprobe analysis can be useful in investigating the spatial inhomogeneity of its crystalline phases, anatase and rutile.[272] It also showed that the E_{2g} modes of both forms are sensitive to oxygen deficiency, and may be used to estimate deviation from stoichiometry brought about by sample annealing.

Lubrication/Wear Studies

Both micro-vibrational spectroscopy techniques have been successfully applied to the study of lubrication and wear processes. The micro-focused beams allow for the direct investigation of lubricants in model elastohydrodynamic (EHD) contacts in test chambers (FTIR;[273,274] Raman[275–277]), see Figure 53. In these dynamic *in lubro* lubricant response measurements, pressure profiles through an EHD contact are built up by measuring shifts in pressure sensitive bands in their vibrational spectra. Also, the build-up of antiwear additives may be observed, or component concentrations determined and, using polarised radiation, shear alignment in different regions of EHD contacts of grease components, such as soap thickeners or polymer viscosity-index improvers, can be investigated.[273,274] In the FTIR microreflectance studies, the infrared radiation is focused into the contact and a transflectance measurement made. Raman offers the advantage over infrared of higher spatial resolution and of using glass windows for the test chamber, as opposed to sapphire, diamond, or calcium fluoride; while infrared enables thinner films to be studied and is not limited to fluorescence-free basestocks.[273,274] Raman microscopy has also been conveniently applied to

[270] Reffner J.A. and Ferrillo R.G., *J. Thermal Anal.*, 1988, **34**, 19.
[271] Reffner J.A., in *Infrared Microspectrometry. Theory and Applications*, ed. Messerschmidt R.G. and Harthcock M.A., Marcel Dekker, New York, USA, 1988, ch. 13.
[272] Parker J.C. and Siegel R.W., *J. Mater. Res.*, 1990, **5**, 6, 1246.
[273] Cann P.M. and Spikes H.A., *STLE Tribology Transactions*, 1991, **34**, 2, 248.
[274] Cann P.M., Williamson B.P., Coy R.C., and Spikes H.A., *J. Phys. D: Appl. Phys.*, 1992, **25**, 1A, A124.
[275] Gardiner D.J., Baird E.M., and Craggs C., *Lubr. Sci.*, 1989, **1**, 4, 301.
[276] Gardiner D.J., Bowden M., and Graves P.R., *Phil. Trans. R. Soc. Lond. A*, 1986, **320**, 295.
[277] Gardiner D.J., Baird E., Gorvin A.C., Marshall W.E., and Dare-Edwards M.P., *Wear*, 1983, **91**, 111.

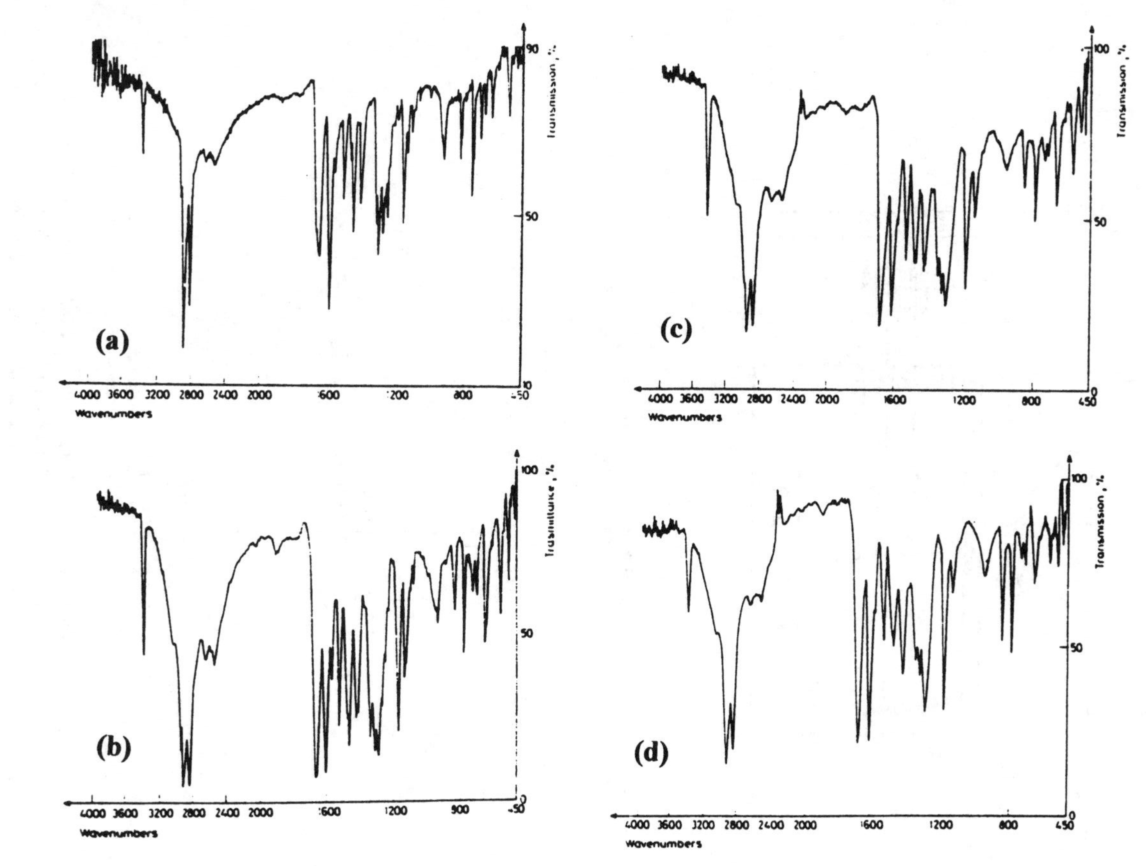

Figure 52 *FTIR-thermomicroscopy study of HABA polymorphism:* (a) *spectrum of Form I at room temperature;* (b) *spectrum of Form II at room temperature;* (c) *spectrum of Form III at* 90°C; (d) *spectrum of liquid crystal at* 112°C
(Reproduced from ref. 270, *Thermal Analysis of Polymorphism*, Reffner J.A. and Ferrillo R.G., *J. Thermal Anal.*, **34**, 19–36 (1988). Copyright (1988). Reprinted by permission of John Wiley & Sons, Ltd)

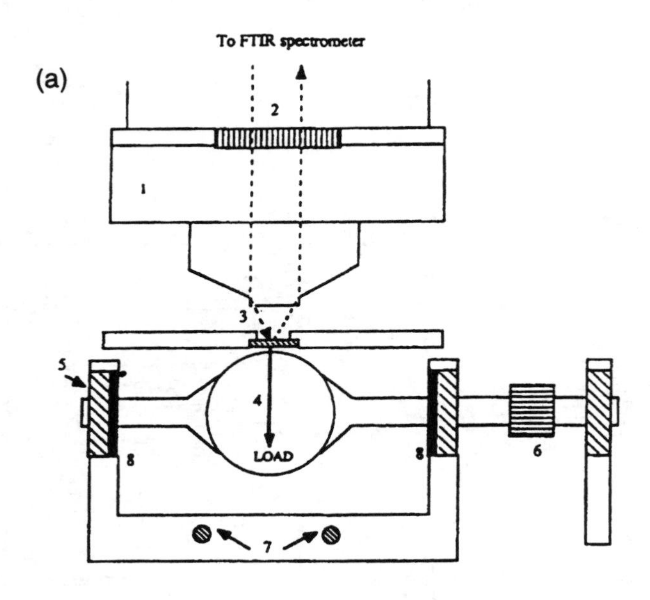

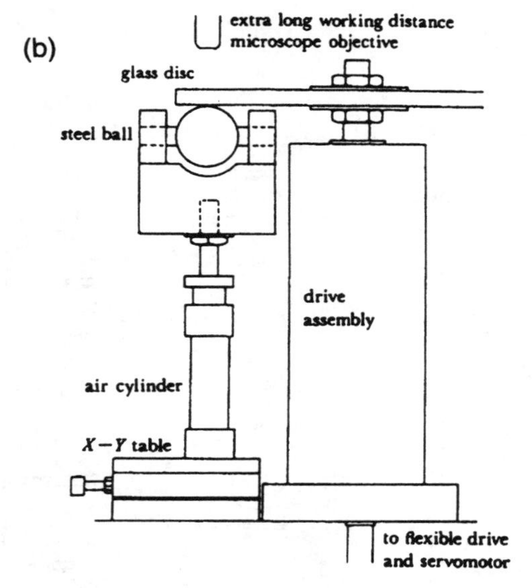

Figure 53 *Schematics of apparatus for studying elastohydrodynamic (EHD) contacts using* (a) *FTIR-microscopy and* (b) *Raman-microscopy. (In* (a): 1, *IR microscope;* 2, *polariser position;* 3, *diamond window;* 4, *steel ball;* 5, *bearings;* 6, *belt and pulley drive;* 7, *heating rods;* 8, *oil seals)*
((a) Reproduced from ref. 273, by kind permission of The Society of Tribologists and Lubrication Engineers
(b) Reproduced from ref. 276, by kind permission of The Royal Society of London)

simply measure fluid (polyphenylether) film thickness (0.1–10 μm) of an entrapped film between a diamond window and a 1 in diameter steel ball.[278]

Micro-infrared and micro-Raman have also been used in static measurements of post-wear test samples.[263,264,279–281] For example, grazing-angle incidence infrared microscopy has been used to record slider rail depletion of a perfluoro-polyether lube on thin-film magnetic discs, as a function of contact time in an oscillatory test.[279] Lubrication and wear mechanisms have been derived for a fluoropolymer coating by polarised FTIR-microscopy observations of the molecular structural changes and wear debris produced in an oscillatory friction test study.[280] This involved a rolling stainless steel ball rubbing against a thin film fluoropolymer coated stainless steel specimen; micro-FTIR was used to analyse both the film and the opposing ball surface after 100 and 2000 sliding cycles to investigate the tribological effects before extreme decomposition occurs. Extreme decomposition products, such as carbon and metal carbide are identified by micro-AES, micro-XPS and micro-Raman spectroscopy.[280,281] *In situ* micro-Raman analysis of wear debris generated on TiN and WC-Co during fretting vibration testing has been demonstrated. In the tests performed in air, species such as defective TiO_2 and amorphous or disordered WO_3 are observed in Raman spectra acquired at low laser powers (1 mW, 10 μm spot size), which were however *in situ* annealed to rutile TiO_2 and crystalline WO_3, respectively, with increased laser powers (5 μm and 10 μm).[282]

6 Closing Remarks

Although the treatment is still far from exhaustive, we have, in the second part of this chapter, given examples covering a diverse range of chemical and physical characterisations which have been undertaken by vibrational spectro-scopy–microscopy techniques. The importance of these techniques to industrial fingerprinting analyses, problem-solving, and research studies is clear. Many other more specialised approaches have been developed and demonstrated, such as systems for simultaneous recording of DSC and FTIR-microscopy data,[283] (see Chapter 7), to correlate thermal responses and spectral changes, and the use of FTIR-microscopy as a novel detection system in micro-flow injection analysis (FIA).[284]

The growth of applications continues to be extensive and invaluable to industrial problem-solving, and several new, key developments have come to the fore even while this chapter has been in preparation, many of them commercialised. While detailed analysis of the spatial distribution of chemical species (spectral maps) may now be built up readily using measurements made

[278] Hutchinson E.J., Shu D., LaPlant F., and Ben-Amotz D., *Appl. Spectrosc.*, 1995, **49**, 9, 1275.
[279] Mastrangelo C.J., Von Schell L., and Le Y., *Appl. Spectrosc.*, 1990, **44**, 8, 1415.
[280] Sugimoto I. and Miyake S., *J. Appl. Phys.*, 1990, **67**, 9, 4083.
[281] Sugimoto I. and Miyake S., *J. Appl. Phys.*, 1989, **65**, 2, 767.
[282] Mohrbacher H., Blanpain B., Celis J.-P., and Roos J.R., *J. Mater. Sci. Lett.*, 1995, **14**, 4, 279.
[283] Mirabella F.M., *Appl. Spectrosc.*, 1986, **40**, 3, 417.
[284] Kellner R. and Lendl B., *Anal. Methods Instrum.*, 1995, **2**, 1, 52.

with a motorised microscope stage,[119,190,191] real-time, mid-infrared imaging microscopy using focal-plane array detectors presents exciting opportunities for advancing the technique,[285,286] with a system having just become commercially available using a InSb focal-plane array detector.[287] 'Point-and-click' video-capture operations from interactive multimedia software display systems are now available, which, for example, simultaneously display different magnification visual images alongside a functional-group chemical map together with spectra related to cursor selected points on the image. Reference 288 is one example. National assets are being made available (or contemplated) so that the attributes of synchrotron FTIR-microscopy will be available from sources other than that at Brookhaven, USA,[113] *e.g.* the Daresbury ring in UK and the ring at Paris, France. Commercial Raman microscope systems fitted with solid state lasers operating in the red are coming to the fore as options, offering, for example, output powers of ~20 mW at 782 nm[289,290] or ~250 mW at 785 nm.[291] A high resolution LCTF imaging Raman microspectrometer has also been launched recently.[292]

A recent addition to the published literature is ref. 293, a book describing many developments and applications in Raman microscopy.

[285] Lewis E.N. and Levin I.W., *Appl. Spectrosc.*, 1995, **49**, 5, 672.

[286] Lewis E.N., Levin I.W., and Crocombe R.A., *Progress in Fourier transform spectroscopy: 10th Int. Conf. Four. Transf. Spectros., Mikrochim. Acta [Suppl.]*, 1997, **14**, 589.

[287] Bio-Rad Digilab 'Stingray' model 6000 FTIR spectrometer. Bio-Rad Digilab Division, Cambridge, USA.

[288] Perkin-Elmer *i*-series FTIR microscope, with IMAGE™ software, Perkin-Elmer Ltd., Beaconsfield, UK.

[289] Renishaw Transducer Systems Limited, Wotton-under-Edge, UK.

[290] Williams K.P.J., Wilcock I.C., Hayward I.P., and Whitley A., *Spectroscopy*, 1996, **11**, 3, 45.

[291] Kaiser Optical Systems, Inc., Ann Arbor, USA.

[292] 'Falcon'® imaging Raman microscope, ChemIcon Inc., Pittsburgh, USA.

[293] *Raman Microscopy. Developments and Applications*, ed. Turrell G. and Corset J., Academic Press, London, UK, 1996.

CHAPTER 7

Hyphenated Techniques

1 Introduction

Previous chapters have discussed the range and adaptability of vibrational spectroscopy essentially as stand-alone techniques for characterising samples and industrial problem-solving. Many materials and samples across all industries are not encountered as pure substances but as multi-component physical or chemical mixtures. These may be deliberate mixtures made for the effective end use of a product, *e.g.* pigments formulated as inks, pure drugs diluted in tablet or liquid form, or polymers with wide molecular weight ranges. Other such mixtures may be the unwanted results of side-reactions during a chemical process, waste materials or recycle streams, or the partitioned components of a product, such as the soluble extracts from polymers. Some may be easy to separate and examine, particularly utilising the advantages and attributes of IR and Raman microscopes. However, large numbers of samples will require separation by other techniques prior to examination by vibrational spectroscopy, or other molecular spectroscopy techniques, if an analysis is to be unambiguous and effective.

Early work involved using one technique to separate the fraction and then transferring the separated fractions to a spectrometer for examination. Many developments have taken place in recent years which combine separation techniques, such as chromatography, with spectroscopic detection and/or identification in a single unit. These combinations of techniques are commonly referred to as hyphenated or hybrid techniques. The latter description implies that a different type of instrument is created. Usually a hyphenated technique combines the best, (and perhaps worst!), principles of two or more techniques, whilst retaining the essential characteristics of both. Most often the effort goes into designing an optimal interface between the two hyphenated techniques, rather than a re-design of either the chromatograph or spectrometer, or their total integration. Examples discussed in this chapter include directly coupled, integrated and independent unit 'hyphenated' systems, with comments on their ease of use, the pitfalls for the unwary, and the efficacy of the combination. The link types (interface, hybridisation) between the highly selective (*e.g.* chromatography) and highly specific (spectroscopy) techniques may be categorised by essentially either a 'real-time' flow-through system or a delayed system, in which the separated analyte is trapped for subsequent interrogation.

2 Gas Chromatography–FTIR (GC–FTIR)

IR spectroscopy has been used for many years to identify compounds eluting from a GC column. In the earliest forms, fractions were condensed onto windows, or collected in cells, sometimes by complex apparatus. The total spectra were then subsequently recorded on grating instruments, each spectrum taking several minutes. This was a very time-consuming, labour-intensive process, and led to poor resolution of the chromatographic peaks. Early attempts with grating instruments used heated flow-through light-pipe type cells as accessories,[1,2] with which, 'on-the-fly', very low resolution, low signal-to-noise spectra could be recorded in a few seconds, or alternatively, an eluting peak could be trapped, and its spectrum then signal averaged to improve its S/N, whilst the rest of the chromatographic run by-passed the cell.[1,2] An early commercial GC–dispersive IR instrument was the Norcon RS-1 GC Analyser (Norcon Instruments, Inc.),[3] which could record mid-IR vapour phase spectra in 6 sec.[2] However, around this time, the advantages of FTIR were being rapidly brought to bear on both trap and particularly flow-through techniques.[4–7] Rapid scan times and higher sensitivities allowed for either smaller peak volumes or more scans per peak leading to enhanced higher spectral resolution, higher S/N spectra. Because capillary columns were initially thought to produce very sharp peaks, generally too quickly for FTIR to cope with, much of the earliest trapping and flow-through work involved peak volumes of a few micrograms, which typically elute from packed columns. Problems were therefore often encountered with column bleed interfering with the spectra, and difficulties were experienced with temperature programming above 200°C. However, developments in interior gold-coated light-pipe cell design[7–11] led to capillary column interfaces becoming commercially available around 1980, which, when incorporated into a FTIR spectrometer system equipped with a fast, sensitive, liquid-nitrogen cooled MCT detector, enabled

[1] Louw C.W. and Richards J.F., *Appl. Spectrosc.*, 1975, **29**, 1, 15.
[2] Katlafsky B. and Dietrich M.W., *Appl. Spectrosc.*, 1975, **29**, 1, 24.
[3] (a) Penzias G.J., *Anal. Chem.*, 1973, **45**, 6, 890; (b) Penzias G.J. and Boyle M.J., *Am. Lab.*, 1973 October, 53.
[4] Low M.J.D. and Freeman S.K., *Anal. Chem.*, 1967, **39**, 2, 195.
[5] Wall D.L. and Mantz A.W., *Appl. Spectrosc.*, 1977, **31**, 6, 552.
[6] Vidrine D.W., in *Advances in Applied Fourier Transform Spectroscopy*, ed. Mackenzie M.W., J. Wiley, Chichester, UK, 1988, ch. 2.
[7] Griffiths P.R. and de Haseth J.A., *Fourier Transform Infrared Spectrometry*, J. Wiley, New York, USA, 1986.
[8] Azzagara L.V., *Appl. Spectrosc.*, 1980, **34**, 2, 224.
[9] Yang P.W.J., Ethridge E.L., Lane J.L. and Griffiths P.R., *Appl. Spectrosc.*, 1984, **38**, 6, 813.
[10] Yang P.W.J. and Griffiths P.R., *Appl. Spectrosc.*, 1984, **38**, 6, 816.
[11] Herres W., *HRGC-FTIR: Capillary Gas Chromatography – Fourier Transform Infrared Spectroscopy. Theory and Applications,* Hüthig, Heidelberg, Germany, 1987.

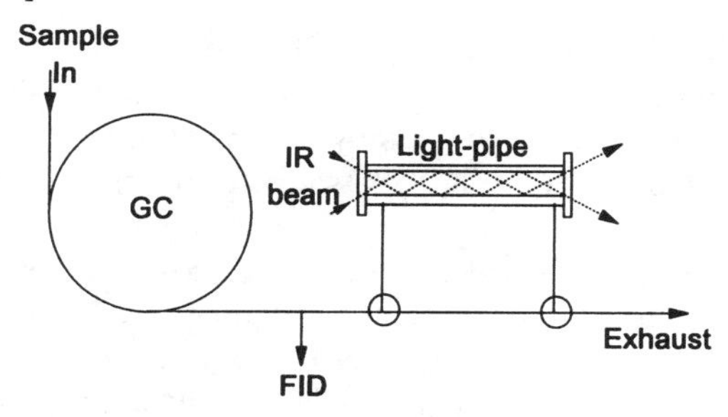

Figure 1 *Simplified schematic of flow-through GC-FTIR set-up*

the recording of a reasonable S/N spectrum, typically 8 scans at 8 cm^{-1}, from a single chromatographic peak eluting from a capillary column separation. Nowadays, in some systems, as many as 40 scans, at 8 cm^{-1}, may be recorded in 1 sec.

Coupled Systems

Early commercial GC–FTIR systems consisted of a gas chromatograph coupled by heated transfer lines to a light-pipe flow-through cell housed, often in a dedicated external sample compartment module, in a FTIR spectrometer fitted with its own MCT detector mounted on an external beam port of the spectrometer. Optimised interfaces of this type are still commonly employed today and available from most FTIR spectrometer manufacturers. The eluate passes from the end of the GC column, then, perhaps, through a splitter, from which 90% of the eluate flows into the IR cell, the remainder passing to a GC detector, such as a Flame Ionisation Detector (FID), see simplified schematic of Figure 1.

The separated components (chromatographic peaks) are swept by the carrier/ make-up gas, which maintains them as discrete entities, along the heated, transfer line into the infrared cell (usually a gold-coated light-pipe), which is located in the GC–FTIR optical module. The cell or light-pipe should ideally have a volume equal to the chromatographic peak at half-height.[12] Since the chromatographic peaks in a separation will vary considerably in width, a fixed 'working' cell path-length is used as a compromise, the volume of which 'optimally' should not exceed that of the half-width of the narrowest peak. Whilst many GC separations are achieved isothermally, temperature gradient programming of the chromatography is often desirable in order to make the chromatography peaks of as similar a width as feasible throughout the GC–FTIR run. The IR beam is diverted from its normal path in the sample compartment by a series of mirrors, to an external port. The beam passes through the light-pipe, then by a further series of mirrors on to a detector, which

[12] *Annual Book of ASTM Standards*, Vol. 03.06, ASTM, Philadelphia, USA, 1995, Designation E 1642–94.

is often a narrow band, high sensitivity MCT detector, dedicated to GC–FTIR experiments. The light-pipe and transfer line can be independently temperature controlled from a control unit, which may also be used to set trigger levels and through which peak hold (trapping) software commands are fed to the light-pipe system.

Infrared spectra may be recorded either continuously from an initial trigger or when a chromatographic peak is in the light-pipe. In the latter case, the collection of a spectrum is triggered either by a signal from the GC detector or by a change in IR absorption in the light-pipe. Using a GC trigger a delay time has to be calculated to allow for the time difference between the peak reaching the GC detector and entering the IR cell. Successive spectra can usually be displayed continuously during a run, as a continuously up-dated 'waterfall' or stacked-plot display of absorbance *vs.* wavenumber. However, their interpretation will usually be post-experiment, on data generated from stored interferograms or spectra, which also allows for optimal choice of background, *e.g.* one generated from scans recorded immediately prior to a peak detection. The response should be linear with concentration provided that the sample transmittance is greater than 70%, since the signal is directly related to transmittance; strongly absorbing samples are not expected to give a linear response. An alternative to the display of spectra, is the 'IR chromatogram(s)', which may also be generated either in real-time during the experiment or from the stored data. IR chromatograms are derived in various ways. Several functional group chromatograms may be generated and displayed simultaneously and rapidly, each being ostensibly a plot of absorbance intensity within a selected spectral window (region). Another common method of 'reconstructing' the chromatogram is the Gram–Schmidt vector orthogonalisation procedure, which operates directly on the recorded interferograms, distinguishing rapidly those with analyte information from 'baseline' interferograms.[6,7,11]

If the eluting peaks are not well resolved, peaks can be divided into time slices, and the averaged spectra stored separately for each slice. Spectroscopic separation can sometimes be achieved by subtracting the spectrum of a known component, or subtracting, say, data from the leading-edge of a peak from its tail. If a peak of interest is very small, such as that from a weak absorber or low concentration component, the peak can be held (trapped) in the cell for increased signal averaging to improve the signal-to-noise ratio of its spectrum. This is achieved by opening and closing by-pass valves. Whilst the peak is held in the cell, the rest of the eluate is diverted through a by-pass loop. In principle, holding the peak in the light-pipe allows for a very large number of scans to be averaged; however, some light-pipes only have a valve at one end, so the peak gradually diffuses out of the other end. This gradual reduction of concentration limits the number of useful scans that can be averaged. Higher sample loads are required for IR detection than for a FID, but very high sample loads are not necessary, providing good chromatography is achieved. Typical detection limits for an optimised system and a strongly infrared absorbing analyte will likely be in the low to tens of nanogram region, but probably at least twice as much for unambiguous fingerprinting.

Effect of GC Column Selection

Samples in industry submitted for GC–FTIR analysis will, in the majority of cases, have been examined previously by GC under optimised conditions. It may not always be possible to transfer these conditions directly onto the GC–FTIR system. Once a column with a phase of comparable polarity has been selected, the primary requirements are a high sample capacity, followed by good component resolution. For this reason, the type of column originally recommended was a thick film (5 μm), wide bore capillary. A typical column used in early capillary GC–FTIR work was a 25 m, 530 μm i.d., 5 μm thick 5% phenyl methyl silicone column. Although this column was excellent for pyrolysis work and low boiling species, samples with high boiling components (>200°C) were not eluted. This was due to the thicker film requiring temperatures of 50°C higher than would normally be employed with the corresponding 1 μm phase. However, with modern capillary GC–FTIR instruments column selection is not usually a problem, and columns with 0.32 mm i.d. are used regularly.[11,12] The column selected depends more on the retention of sample components on a particular phase than on a need for increased loading. In most cases when using a low area, high sensitivity MCT detector, high sample loads are not a prerequisite to providing peaks concentrated enough for analytically useful spectra to be collected.

Cold Spots

The most serious problems are encountered with higher boiling samples, and incorrectly heated transfer lines. The problem becomes particularly apparent at higher temperatures. The higher boiling components are detected by the FID but are missing in the IR-chromatogram, or appear much later and are broader. The temperature of the line should be increased to the usual maximum of about 250°C.[12] In many cases this is higher than the elution temperature of the sample. If the chromatogram does not improve, this points to a fault in the transfer line. A series of n-C_{10} to n-C_{24} hydrocarbons illustrate the effect very well, compare Figure 3 with Figures 2 and 4.

Incorrectly positioned thermocouples can lead to cold spots existing along the transfer line, which manifest themselves through broadened and 'missing' peaks. As can be seen from Figure 3, the 'poor' temperature-control of the transfer line caused loss of intensity, broadening, and 'disappearance' of peaks at higher carbon numbers. A line without cold spots should cause no transfer problems and work well at all temperatures.

Dedicated GC–FTIR Systems

Flow-through HP-IRD®

The system types discussed above are all combinations of GC instruments and FTIR spectrometers, in which the GC was essentially an adjunct to the spectrometer. To optimise the 'hyphenation' Hewlett Packard (HP) launched,

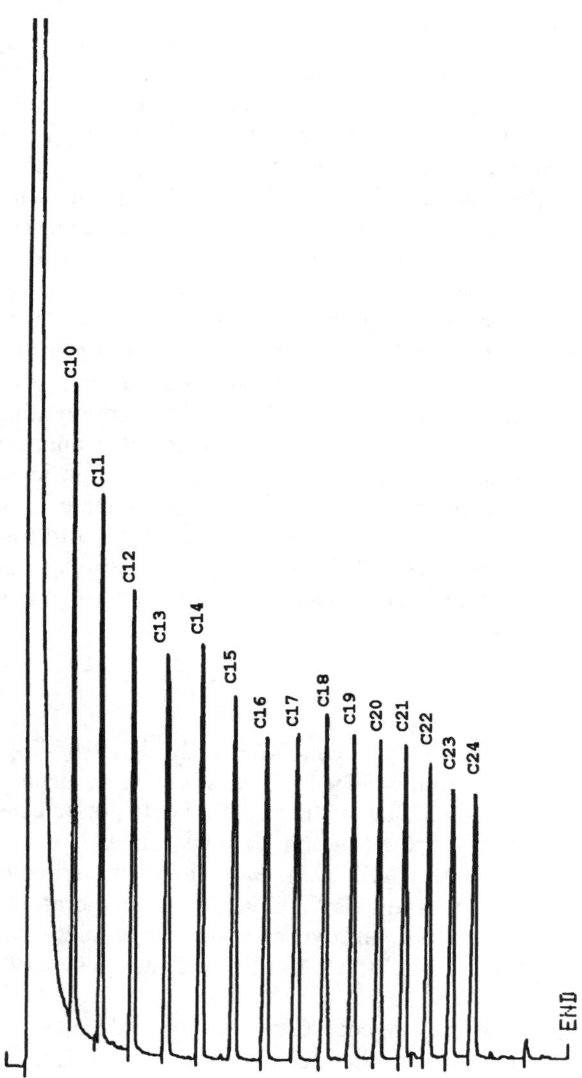

Figure 2 *FID chromatogram of n-C_{10} to n-C_{24} hydrocarbons*

in 1980s, the IRD®, which has a minimum length transfer line to a light-pipe. A FTIR spectrometer with optimised optics is built around the ~100 μl flow cell, see schematic of Figure 5. The peak passes without splitting through the cell, which puts the maximum sample in the cell. A work-station controls the total instrument, collecting and storing the spectra, which can also then be library searched. The effluent from the cell can be passed to a FID, although the cell volume may produce some peak broadening of the chromatogram. The eluate from the GC can be split prior to the IR cell and a small fraction passed through a Mass Spectrometer, for a combined GC–FTIR–MS examination of a single

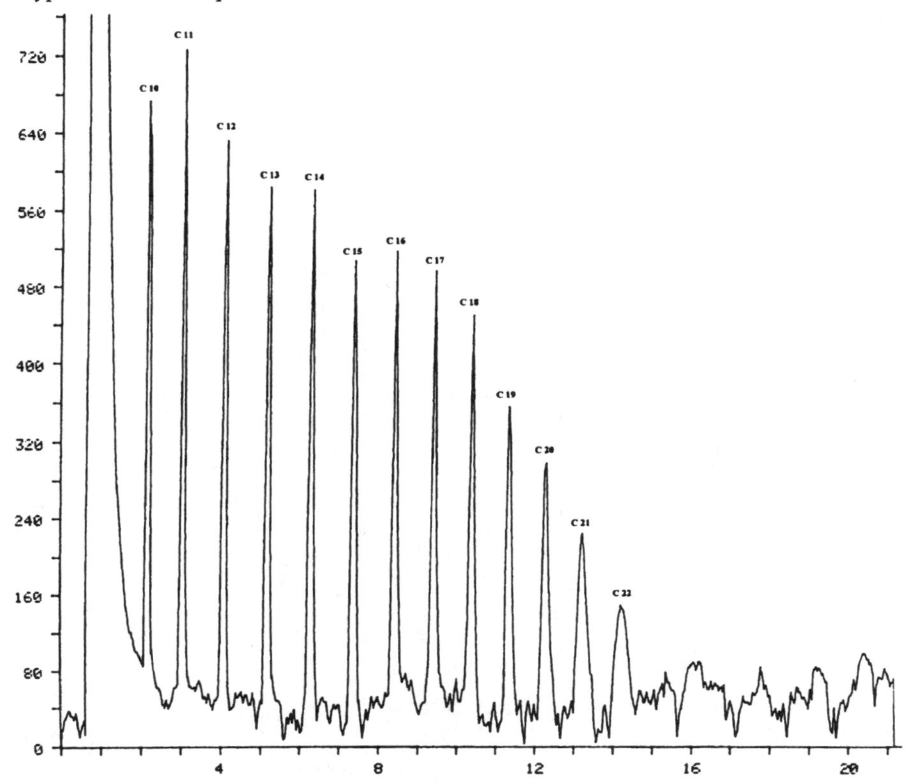

Figure 3 *IR chromatogram generated from n-C_{10} to n-C_{24} hydrocarbons separation, with cold spots present in the transfer line*

peak. The HP5890 GC may be used routinely with columns such as a 25 m, 0.32 mm i.d., 0.52 µm cross-linked 5% phenyl methyl silicone column. It is connected to the IRD using 0.32 mm deactivated silica capillary transfer lines. The transfer line temperature is normally set at 250°C. The gold-lined light-pipe is a glass tube 120 mm path-length, with an internal diameter of 1 mm. The GC eluate is passed through the heated light-pipe allowing 'on-the-fly' gas-phase FTIR chromatograms to be built up, together with spectra obtained from the components detected. The IRD can be used with 4, 8 and 16 cm^{-1} settings, but is routinely operated with 8 cm^{-1} optical resolution, giving 3.3 scans per sec.

The rights to market the IRD were recently acquired by Bio-Rad Inc., who have now relaunched the instrument.

Cryo-trapping Systems

The sensitivity of flow-through systems are limited by the length of time a peak is in the cell, and hence the number of infrared scans which are co-added and averaged. The IRD can be used routinely to identify 10 ng of sample in a GC

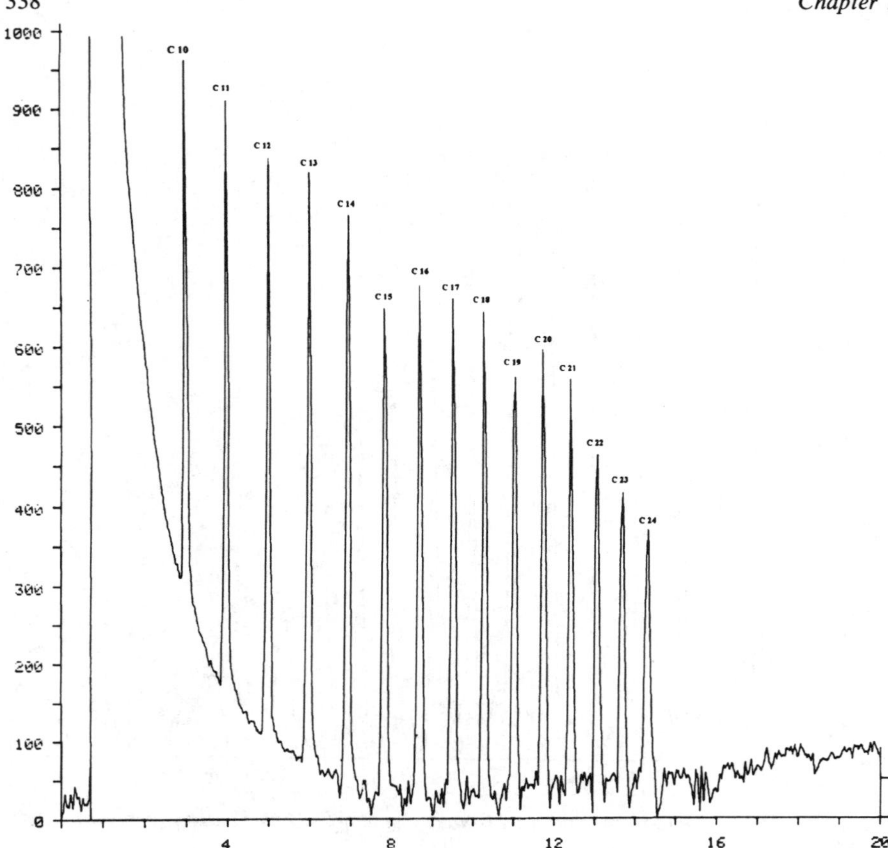

Figure 4 *IR chromatogram generated from n-C_{10} to n-C_{24} hydrocarbons separation with properly heated transfer line (250°C)*

peak, and has a detection sensitivity of about 1–5 ng. This inherent sensitivity limitation has been addressed by use of low-temperature sample storage interfaces between the GC and the FTIR spectrometer. These are of two main designs. The first deposits the GC eluate directly onto a moving low-temperature transparent window (typically ZnSe).[13,14] The second developed at Argonne National Laboratory,[15–18] utilises matrix isolation (MI) to trap the GC eluate in a glassy argon matrix on a moving reflective surface at cryogenic temperatures. In both cases, the GC eluate is stored at low temperature as a

[13] Haeffner A.M., Norton K.L., Griffiths P.R., Bourne S. and Curbelo R., *Anal. Chem.*, 1988, **60**, 2441.
[14] Bourne S., Haeffner A.M., Norton, K.L., and Griffiths P.R., *Anal. Chem.*, 1990, **62**, 2448.
[15] Reedy G.T., Bourne S., and Cunningham P.T., *Anal. Chem.*, 1979, **51**, 1535.
[16] Reedy G.T., Bourne S., and Cunningham P.T., *J. Chromatogr. Sci.*, 1979, **17**, 460.
[17] Reedy G.T., Ettinger D.G., Schneider J.F., and Bourne S., *Anal. Chem.*, 1985, **57**, 1602.
[18] Bourne S., Reedy G.T., Coffey P.J., and Mattson D., *Am. Lab.*, 1984, **16**, 90.

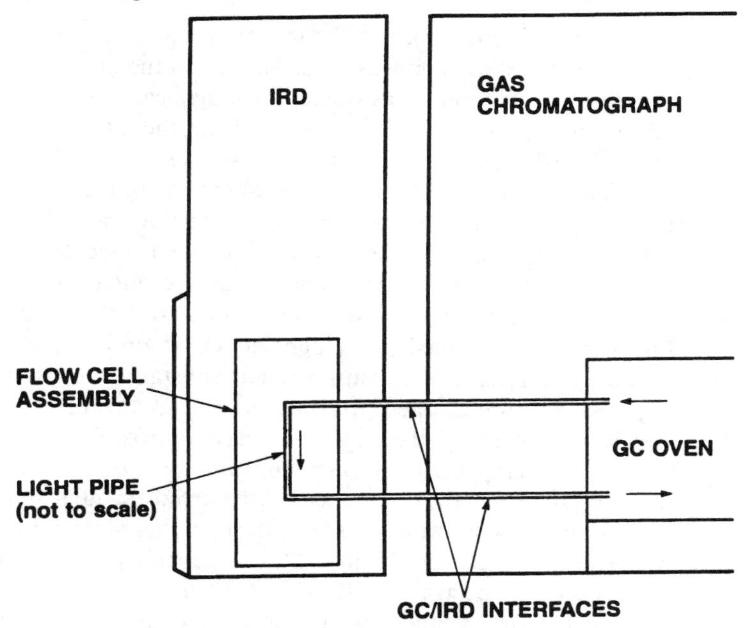

Figure 5 *Simplified schematic of Hewlett Packard 5965A IRD GC-FTIR system*

solid 'spot'. This allows more time for FTIR data acquisition and the use of more focused infrared beams, and, in the case of matrix isolation, produces much narrower spectral bands, resulting in increased sensitivity. Both systems approach the detection sensitivity obtained with a bench-top GC–MS instrument. Both these interface designs are now/or have been commercially available. The first, is incorporated into the Bio-Rad Digilab, Digilab Division, Tracer® (GC–FTIR), which was introduced in 1988, while the second formed the basis around which the Mattson Instruments Inc., Cryolect® (GC–MI-FTIR) was built. A close facsimile of the latter has recently been launched (1996) as the MITREK®, Reedy Scientific Instruments, Bourbonnais, IL, USA.

Bio-Rad Tracer®. The standard Tracer interface combines a Carlo Erba MEGA 5160 HRGC with a Bio-Rad Digilab FTS-40 FTIR spectrometer. The GC is equipped with both cold on-column and split/splitless injectors. Columns of 0.25 mm i.d., 0.25 μm film thickness, may be used routinely, enabling infrared identification of components at the 100 pg, or less, level, which is useful, for example, in supporting and complementing GC–MS studies on flavours and fragrances in the brewing industry,[19] or in isomer-specific identifications of the components from a complex mixture obtained from a steam distillation/

[19] Powell J. and Compton S., *Master Brewers Association of the Americas, Technical Quarterly*, 1991, **3**, 123.

extraction on a sub-bituminous coal;[20] in the latter example, a 25 m × 0.2 mm i.d. × 0.11 mm film thickness column was used. The end of the capillary column is connected to a deactivated fused silica transfer line by means of a glass press-fit connector (Supelco). The transfer line is guided from the GC oven into the vacuum chamber of the Tracer interface through a stainless steel tube. A fused silica deposition tip with an i.d. of 150 μm is fixed to the end of the transfer line, acting as a restricting element to reduce the operating flow rate to approximately 1 ml/min. Both the tip and the transfer line are heated to a default temperature of 250°C. The deposition tip is located approximately 30 μm above the surface of a moving zinc selenide window, which is cooled with liquid nitrogen to 77 K. The interface housing is a high vacuum chamber which is held at 10–5 Torr to prevent condensation of atmospheric substances.

Eluting GC components are trapped (frozen) onto the cold window as solid-phase deposits, with the width of the deposited trace approximately 100–150 μm. The window is continually moved during deposition by means of a stepper-motor drive. The deposited eluant components are passed through the FTIR beam a few seconds after deposition. In this way, the complete chromatogram is both immobilised and scanned 'on-the-fly'. This is generally carried out by averaging 4 scans, with a spectral resolution of 8 cm^{-1}, every 2 sec. All interferograms recorded are stored on the computer disk, which allows spectra, Gram–Schmidt, and functional group reconstructed chromatograms to be processed subsequently.

Post-run scanning of selected spots at specific retention times can also be performed by repositioning the window under the IR beam. This allows transmission spectra to be obtained both at higher resolution and with increased signal-to-noise ratios. However, since the IR beam is adjacent to the deposition tip, repeated positioning of a sample within the IR beam may result in sample degradation, due to contamination from components eluting from the transfer line.

Mattson Cryolect® Matrix-trapping. The Cryolect 4800 instrument combined GC–FID, MI-FTIR, and MS in a single instrument, with a HP5890 GC at the heart of the system. Similar injection procedures to those used on the Tracer are available. Following the GC separation, the column eluate is split, taking 20% to the flame ionisation detector (FID), 40% to the MS and 40% to the MI-FTIR interface. Post-column splitting ensures complete synchronisation between the FID, FTIR, and MS detectors. The three-way splitting arrangement is mounted permanently inside the GC oven and comprises a simple three-hole ferrule, with the flow and split ratios determined by the capillary lengths and diameters used. Also mounted inside the oven are two open cross divert systems, which allow flushing away of unwanted large components, *e.g.* solvents, which might contaminate the matrix-isolation interface or mass spectrometer source.

[20] Smyrl N.R., Hembree D.M., Davis W.E., Williams D.M., and Vance J.C., *Appl. Spectrosc.*, 1992, **46**, 2, 277.

Approximately 40% of the GC eluate is directed to the GC-MS interface, which can be supplied as standard with most bench-top mass spectrometers. The 40% of the GC eluate (for which helium is the carrier gas) for FTIR analysis is combined at an open cross with 15% argon in helium to give, typically, between 100:1 and 1000:1 argon to sample molecule ratios. The flows into the open cross are balanced to ensure that all the material eluting from the splitter is effectively transferred to the matrix-isolation interface. The eluate is then transferred to the matrix-isolation vacuum chamber via a heated transfer line (0.32 mm i.d. capillary) and heated deposition tip. The tip is held approximately 0.3 mm from the gold disk during deposition, and is retracted to a distance of approximately 3 cm when not depositing, to avoid contaminating the sample already laid down. The vacuum chamber pressure is maintained between 5×10^{-5} and 2×10^{-4} Torr with flow rates of approximately 1.8 ml/min from the transfer line from the open cross. The argon mixture is then sprayed onto the side of a gold-plated disk, which is typically held between 7 and 20 K. This is cooled by a closed cycle helium gas expander/refrigerator. As the GC run progresses, the cryogenic disk is slowly rotated using a stepper-motor drive system. The argon is laid down as a frozen spiral onto the gold-coated surface of the slowly rotating disk, trapping any eluted components. The system is set up to give, as closely as possible, a deposition 'spot size' equal to the IR beam diameter (approximately 350 μm). The deposition time is then directly related to the stepper-motor position and therefore provides the necessary time synchronisation between the GC–FID, GC–MS, and FTIR data. Some small variations in retention time are observed (normally less than 0.03 min) due to slightly different transfer line lengths to each detector.

Once the sample is deposited onto the cryogenic disk, this is rotated by 180° to position the matrix in the FTIR beam, and a rapid-scan, low resolution 'reconstruct' is performed. This is simply a rapid Gram–Schmidt absorption map of the deposited matrix, the FTIR 'chromatogram'. The reconstruct is then displayed and both peaks of interest and background positions are selected from the software with a cursor. High resolution, high sensitivity (transflectance) scans can then be obtained from the peak of interest. Alternatively, small peaks that do not give good responses in the reconstructed chromatogram can be located using the derived synchronous GC–MS or GC-FID retention times.

Comparison of Detector Systems and Their Spectra

As part of a comprehensive investigation[21] into the application of GC–FTIR in industrial and environmental analyses, representative sample sets have been analysed in parallel using commercial instruments. Both the Mattson Cryolect and the Bio-Rad Digilab Tracer utilise low temperature storage of the GC eluate to extend the time available for FTIR analysis, yielding greater sensitivity

[21] Jackson P., Dent G., Carter D., Schofield D.J., Chalmers J.M., Visser T., and Vrendenbregt M., *J. High Res. Chromatogr.*, 1993, **16**, 515.

than that possible by conventional 'light-pipe' GC–FTIR. In certain circumstances, both sample storage instruments give rise to spectra that exhibit features characteristic of the type of interface used. Chromatographic resolution was found not to be significantly degraded by use of either sample storage interface. Particular advantages were found in having parallel flame ionisation detection and mass spectrometry, allowing smaller components to be located and giving greater certainty of identification.

If particular care is not taken, both the Cryolect and Tracer instruments may be subject to interference from small amounts of background water. On the Tracer, this appears in the spectra as broad bands due to solid ice. On the Cryolect, water may appear as both ice and matrix-isolated molecules (including dimers, trimers, etc.), with sharp signals shifted above 3500 cm^{-1} in the FTIR spectrum, giving less interference and overlap with bands from other acidic functionalities. Water signals are minimised on both instruments by operating at high vacuum and by conditioning the system after routine maintenance. With both the Cryolect and the Tracer, one day is typically required to dry the vacuum chamber out after opening to atmosphere, but this is sample dependent; concentrated samples may be run less than one hour afterwards, whereas weak components may only be seen effectively after several days of dry operation.

The chromatographic resolution obtained with both interfaces depends on the GC peak widths, peak separation, cold surface movement, and spot size. With suitable variation of the disk rotation rate or window velocity, it is possible to retain the GC separation adequately without degrading chromatographic resolution. On the Cryolect, disk speed variation cannot currently be performed straightforwardly from the instrument software, so dual methodologies have been developed with separate PCs (computers), enabling disk speeds from 0.05 to 0.5 mm s^{-1} to be achieved. On the Tracer, the cold window scan rate can be controlled from the software, and in order to compensate for the increase of the chromatographic peak-width with longer retention times, a decreasing window velocity is usually programmed. In each case discussed here, the disk speed was optimised to maintain the chromatographic resolution, rather than optimised for sensitivity.

In order to test the effective gains in sensitivity of the two sample storage instruments over the optimised light-pipe system, a series of caffeine concentration standards was made up in a 3% ethanol in toluene solution. Solution strengths of 500, 100, 50, 10, 5, 1, 0.5, 0.05, and 0.01 mg l^{-1} were prepared and provided in duplicate to each instrument for analysis.[21]

Chromatograms and spectra were recorded on the HP IRD using both split and splitless injection. The split ratio was determined to be 38:1. In each case, 1 μl of sample was injected. With splitless operation, the instrument located signals from the first three samples, corresponding to 500, 100, and 50 ng of caffeine injected. With split injections, signals could only be obtained from the 500 ng sample, corresponding to 13 ng of caffeine on the column. This spectrum is shown in Figure 6(a). The peak absorbance value obtained from the carbonyl signal at *ca.* 1660 cm^{-1} from the 13 ng sample was 0.0042. Given the peak-to-

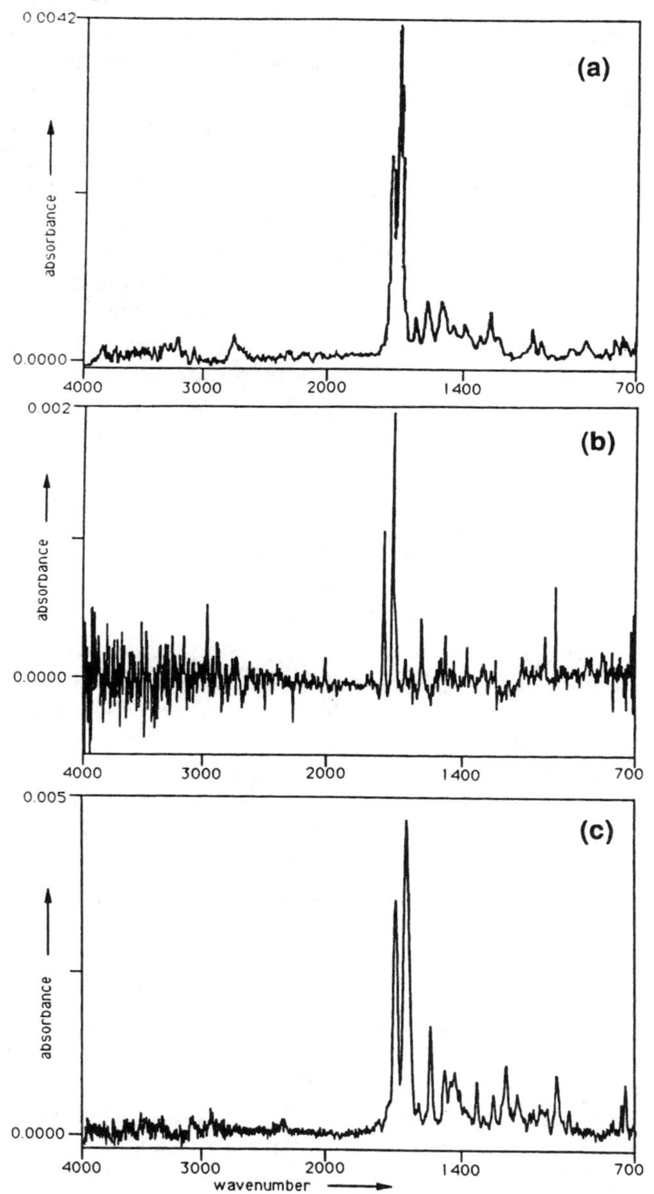

Figure 6 *FTIR spectra obtained from caffeine concentration standards:* (a) *HP-IRD (*4 scans, 8 cm^{-1} *resolution,* 13 ng *sample);* (b) *Cryolect (*2048 scans, 4 cm^{-1} *resolution,* 40 pg *sample);* (c) *Tracer (*512 scans, 4 cm^{-1} *resolution,* 35 pg *sample)*

(Reproduced from ref. 21, Jackson P., Dent G., Carter D., Schofield D.J., Chalmers J.M., Visser T., and Vrendenbregt M., *Investigation of High Sensitivity GC-FTIR as an Analytical Tool for Structural Identification, J. High Res. Chromatogr.*, 1993, **16**, 515, by kind permission of Hüthig Publishers, Heidelberg)

peak noise level of 0.0003 AU, this gives a signal-to-noise ratio of 14:1 from this sample and a detection limit for caffeine of approximately 10 ng.

Reconstructed chromatograms and FTIR spectra were recorded at 4 cm^{-1} resolution on the Cryolect following split injection, with the splitter ratio adjusted to 10:1. With 40% of the column eluate transferred to the FTIR interface, this gave an effective FTIR split ratio of 25:1. With 1 μl injected, the caffeine peak could be detected from the first five samples using a rapid low-sensitivity Gram–Schmidt chromatogram reconstruction, but the higher sensitivity available by analysing a single retention time meant that good FTIR spectra could actually be obtained from the first six samples. This corresponds to between 20 ng and 40 pg of caffeine collected on the cryogenic disk. For the lowest concentration sample, the disk was positioned at the appropriate synchronised retention time, since the signal-to-noise ratio of the reconstructed chromatogram was too low to allow unambiguous location of the caffeine peak position. Using 128 scans, the signal-to-noise ratio for the 1660 cm^{-1} carbonyl peak was 4.5:1, with a noise level of 0.0004 AU (measured at 2000 cm^{-1}). By acquiring 2048 scans, an improved signal-to-noise ratio of 18:1 was obtained, as shown in Figure 6(b), with a peak absorbance of 0.0018. At the other extreme, 20 ng of caffeine on the cryogenic disk gave rise to a peak absorbance of 0.5120, a signal-to-noise ratio of over 1200 from 128 scans.

Spectra and chromatograms were recorded using the Tracer following split and splitless injection, with samples corresponding to 500 ng to 50 pg injected. The split ratio, where used, was 25:1. Functional group chromatograms were also recorded, using the spectral information between 1600 and 1680 cm^{-1} to increase the chromatographic sensitivity. The spectrum in Figure 6(c) shows the Tracer FTIR spectrum from 35 pg of sample deposited. The peak absorbance is 0.0048, giving a signal-to-noise ratio of 32:1 with the noise level at 0.00015. The spectrum was acquired using 512 scans and 4 cm^{-1} resolution. The spectrum obtained from 18 ng exhibited a signal-to-noise ratio of 693:1 from 4 scans.

Both the Cryolect and the Tracer systems give recognisable caffeine spectra from tens of picograms of material, under normal operating conditions. With the Tracer, the smaller sample spot size (*ca.* 150 μm) results in a more concentrated sample, and the microscope optics give improved detection sensitivity compared to the Cryolect (350 μm spot, 0.5 × 0.5 mm area detector), but this is somewhat offset by the Cryolect's double-pass transflectance-type measurement and the sharper lines obtained using matrix-isolation.

These results illustrate the clear sensitivity advantage of the two sample storage instruments. Consequently, further samples were analysed using only the Cryolect and the Tracer, to test the chromatographic capabilities of the sample storage interfaces and to illustrate routine applications with typical samples.

Grob Test Mixture

A standard Grob test mixture was chosen to test the chromatographic capabilities of the respective GC-FTIR interface systems and to illustrate particular spectral characteristics on a variety of chemical types. The sample was made up from routine analytical reagents with approximately 55 mg l^{-1} of each component in dichloromethane as follows: (1) octan-2-one, (2) octan-1-ol, (3) 2,6-dimethylphenol, (4) 2,4-dimethylaniline, (5) naphthalene, (6) tridecane, and (7) tetradecane.

Two different experiments were performed on the Cryolect, one with 1 μl injected, the other with 3 μl injected, to illustrate the effects of varying matrix concentration. In both cases split injection was used, with a 10:1 split ratio resulting in 2.2 ng and 6.6 ng of each component trapped on the cryogenic disk. In addition to varying the sample amount, the matrix gas flow was doubled for the 1 μl sample to give an approximately 6 × difference in sample concentration within a thicker argon matrix. With the Tracer split injection was used, with a split ratio of 25:1 resulting in 2.2 ng deposited onto the cold window. Figure 7 shows the chromatograms obtained from the Grob test mix. Both the Cryolect and Tracer interfaces were capable of resolving components separated by GC–FID and GC–MS detection. Figure 7 shows the Tracer FTIR chromatogram as well as the GC–FID, GC–MS (total ion chromatogram, TIC) and GC–FTIR chromatograms recorded on the Cryolect. Some variations in peak intensities between the detection techniques are seen. This is, of course, due to the differing response factors between the respective detectors used. In addition, some differences can be seen between the Tracer and Cryolect FTIR chromatograms. In both cases, a Gram–Schmidt algorithm was used to calculate the total FTIR response. Slight variations in the algorithm are suspected, giving rise, for example, to a reduced response on the Cryolect from naphthalene (peak 5), which has significant bands only at low wavenumbers in the FTIR spectrum.

The spectra obtained from octan-2-one are shown in Figure 8. Features characteristic to the sample storage method are clearly exhibited. The most striking difference is between spectra recorded at different matrix/sample ratios on the Cryolect. At the lower concentration, Figure 8(a), the carbonyl signal is split, with components at 1715 and 1732 cm^{-1}. At the higher concentration, Figure 8(b), only a single, broader carbonyl signal is found. This type of splitting is common at lower concentrations, and is generally due to 'matrix effects'. These can include: effective freezing out of high-temperature, gas-phase structures; locking-out of particular conformations, formation of dimers, trimers, *etc.*, as well as direct matrix–sample interactions. At higher concentrations in the matrix, and with larger sample molecules, true matrix-isolation is less likely, and samples become more like condensed phase, 'amorphous' solids. The spectrum obtained from octan-2-one on the Tracer, Figure 8(c), is, as expected, characteristic of a solid-phase sample. There is, however, evidence for crystallisation, indicated by peak splitting (especially the CH_2-rocking doublet at 720 cm^{-1}) and fringing effects. In the spectra of naphthalene and the long-chain hydrocarbons, further spectral differences were attributed to the sample form; crystalline features, such as band splitting, found in the Tracer spectra; isolated molecules or amorphous-solid-like spectra from the Cryolect. In addition to exhibiting further evidence for crystallisation, spectra from octan-1-ol, 2,4-dimethylaniline and 2,6-dimethylphenol, obtained from the Tracer,

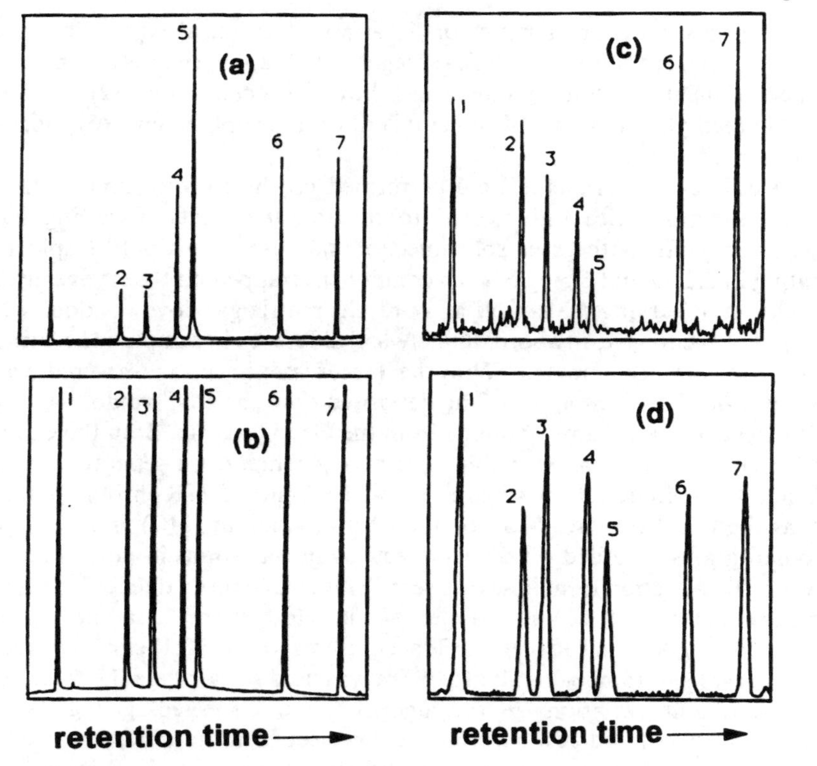

Figure 7 *Chromatograms obtained from the Grob test mixture:* (a) *Cryolect GC-FID;* (b) *Cryolect GC-MS TIC;* (c) *Cryolect GC-MI-FTIR;* (d) *Tracer GC-FTIR. Peaks:* 1, *octan-2-one;* 2, *octan-1-ol;* 3, *2,6-dimethylphenol;* 4, *2,4-dimethylaniline;* 5, *naphthalene;* 6, *tridecane;* 7, *tetradecane. (Note the differing relative sensitivities between the different detection techniques)*
(Reproduced from ref. 21, Jackson P., Dent G., Carter D., Schofield D.J., Chalmers J.M., Visser T., and Vrendenbregt M., *Investigation of High Sensitivity GC-FTIR as an Analytical Tool for Structural Identification, J. High Res. Chromatogr.*, 1993, **16**, 515, by kind permission of Hüthig Publishers, Heidelberg)

had significantly broadened νC—O, νC—N, νO—H and νN—H bands, reflecting the strong influence of hydrogen bonding on these vibrations. The effect of matrix-isolation on the Cryolect spectra is to dramatically narrow these signals, although they are very concentration dependent. Hydrogen-bonding effects are less prominent in the spectra of 2,6-dimethylphenol, due to the steric effect of the two methyl groups hindering interactions with the phenolic -OH.

Commercial Alcohols

A commercial sample of isoheptanol (a mixture of branched chain primary C_7 alcohols) was chosen to further test the chromatographic capabilities of the sample storage GC–FTIR interfaces. This sample also formed part of an earlier

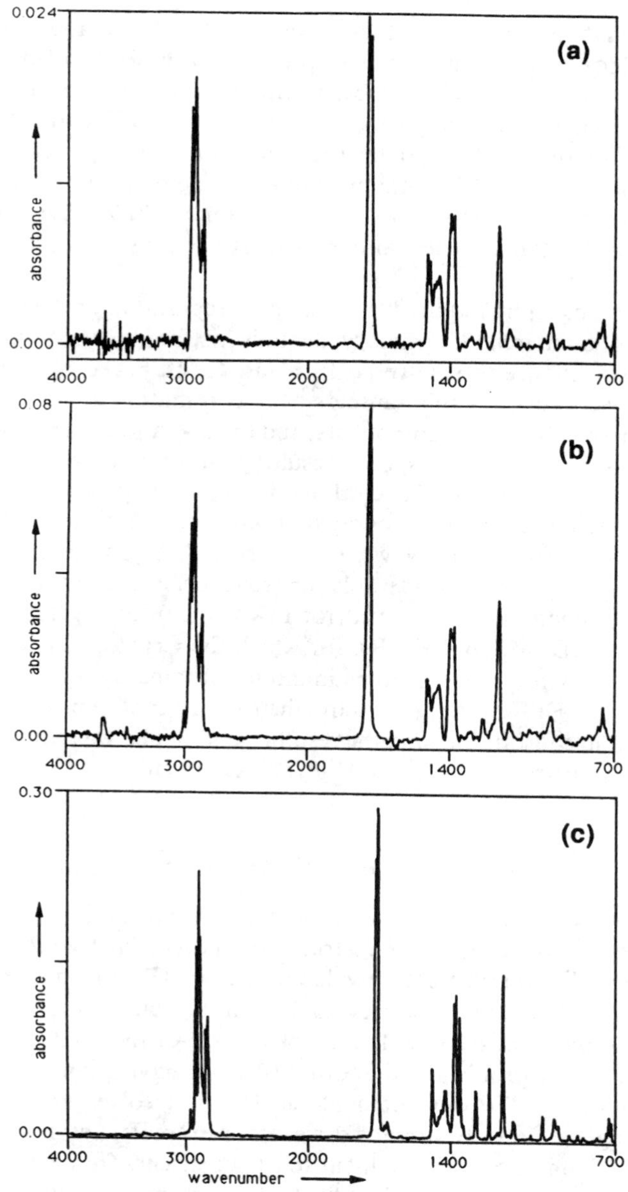

Figure 8 *FTIR spectra obtained from the Grob test mixture component octan-2-one:* (a) *Cryolect,* 32 scans, 4 cm^{-1} *resolution, low concentration* (2.2 ng) *in matrix;* (b) *Cryolect,* 32 scans, 4 cm^{-1} *resolution, high concentration* (6.6 ng) *in matrix;* (c) *Tracer,* 128 scans, 4 cm – 1 *resolution,* 2 ng *sample*
(Reproduced from ref. 21, Jackson P., Dent G., Carter D., Schofield D.J., Chalmers J.M., Visser T., and Vrendenbregt M., *Investigation of High Sensitivity GC-FTIR as an Analytical Tool for Structural Identification, J. High Res. Chromatogr.* 1993, **16**, 515, by kind permission of Hüthig Publishers, Heidelberg)

Cryolect study[22] by workers at the University of California, Riverside, where several specific isomeric structures were proposed. A more detailed analysis with optimised chromatography was subsequently performed by us, using Cryolect FTIR and MS data combined with ^{13}C NMR data. This identified the 8 major components in the mixture to be the following primary alcohols: (1) 2,4-dimethylpentanol, (2) 3,4-dimethylpentanol, (3) 2-ethylpentanol, (4) 2-methylhexanol, (5) 5-methylhexanol, (6) 3-methylhexanol, (7) 2-ethyl-3-methylbutanol, and (8) heptanol. The sample was studied as a 0.1% solution in dichloromethane.

The chromatography required to effect good separation of the 8 components found in this commercial C_7 primary alcohols mixture is demanding, requiring optimum performance to resolve peaks 3 and 4, and peaks 5, 6, and 7. In the earlier Cryolect work on this sample,[22] the chromatographic resolution was inadequate to resolve these components, and the operation of the cryogenic disk interface was at too low a speed, resulting in further degradation of the chromatographic resolution. Spectral manipulation methods, *e.g.* subtraction, were then required to obtain separated spectra from each individual component. Even so, peaks 3 and 4 were not resolved. Dynamic cryo-disk speed programming can very considerably improve peak separations.[23] Figure 9 shows the chromatograms obtained for this work from the Tracer and from the Cryolect (FID, MS and FTIR), for which the Cryolect was operated with 5 × the usual disk rotation rate to maintain the chromatographic resolution. In both cases, the FTIR interfaces were then capable of reproducing the GC chromatographic resolution, and the resolution obtained was equivalent to that obtained on a separate optimised GC–FID instrument.

Complementary GC–FTIR and GC–MS Examinations

At the time of the studies above, a prime motivator behind much of this work was the need to convince our industrial MS and chromatography colleagues that, although the complementary value of GC–FTIR information was not in doubt, it really had the sensitivity to become a routinely useful technique supporting long-established, well-equipped, state-of-the-art GC–MS laboratories. A secondary problem and another issue is, 'having got such good data, can you interpret it?'. Good data alone does not solve problems! Generic 'fingerprinting' is relatively straightforward; for example, ketones can easily be distinguished from ethers, as can di-substituted benzenes from one another, but full component, unambiguous identification of complex molecules can still be an issue, unless worked iteratively between the infrared and mass spectroscopists in combination with the problem owner. Libraries of spectra at cryogenic temperatures are scant, and unlikely to include many species of real

[22] Baumeister E.R., Zhang L., and Wilkins C.L., *J. Chromatogr. Sci.*, 1991, **29**, 331.
[23] Klawun C., Sasaki T.A., Wilkins C.L., Carter D., Dent G., Jackson P., and Chalmers J., *Appl. Spectrosc.*, 1993, **47**, 7, 957.

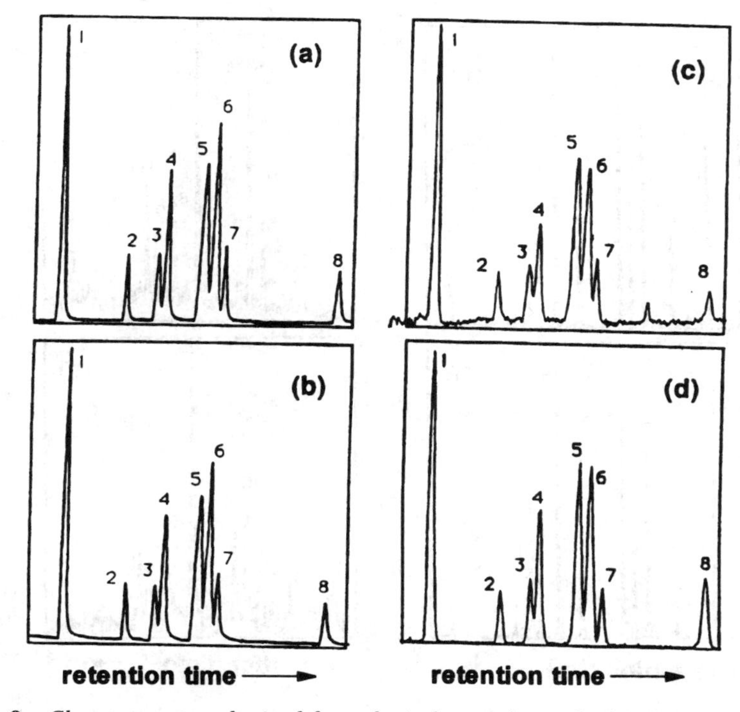

Figure 9 *Chromatograms obtained from the isoheptanol sample:* (a) *Cryolect GC-FID;* (b) *Cryolect GC-MS TIC;* (c) *Cryolect GC-MI-FTIR;* (d) *Tracer GC-FTIR. Peaks:* 1, *2,4-dimethylpentanol;* 2, *3,4-dimethyl pentanol;* 3, *2-ethylpentanol;* 4, *2-methylhexanol;* 5, *5-methylhexanol;* 6, *3-methylhexanol;* 7, *2-ethyl-3-methylbutanol;* 8, *heptanol*
(Reproduced from ref. 21, Jackson P., Dent G., Carter D., Schofield D.J., Chalmers J.M., Visser T., and Vrendenbregt M., *Investigation of High Sensitivity GC-FTIR as an Analytical Tool for Structural Identification, J. High Res. Chromatogr.*, 1993, **16**, 515, by kind permission of Hüthig Publishers, Heidelberg)

value to identifying the true 'unknowns' of industry. Also, the nuances of cryogenically recorded spectra are less familiar.

The identification of components in mixtures offers many challenges to analytical scientists, with increasing pressure, both from within industry and from government and legislative bodies, to identify smaller amounts with increased certainty. There is little doubt that combined separation science-mass spectrometry (MS) can meet many of these challenges. Nevertheless, even with the most powerful mass spectrometers available there is often a requirement for complementary evidence to confirm identification, differentiate closely related structures, or determine specific isomeric forms. Fourier transform infrared (FTIR) spectroscopy is particularly well suited to the identification of functional groups and in the fingerprinting of specific isomers, and is therefore the ideal complementary and confirmatory technique to MS analysis. For

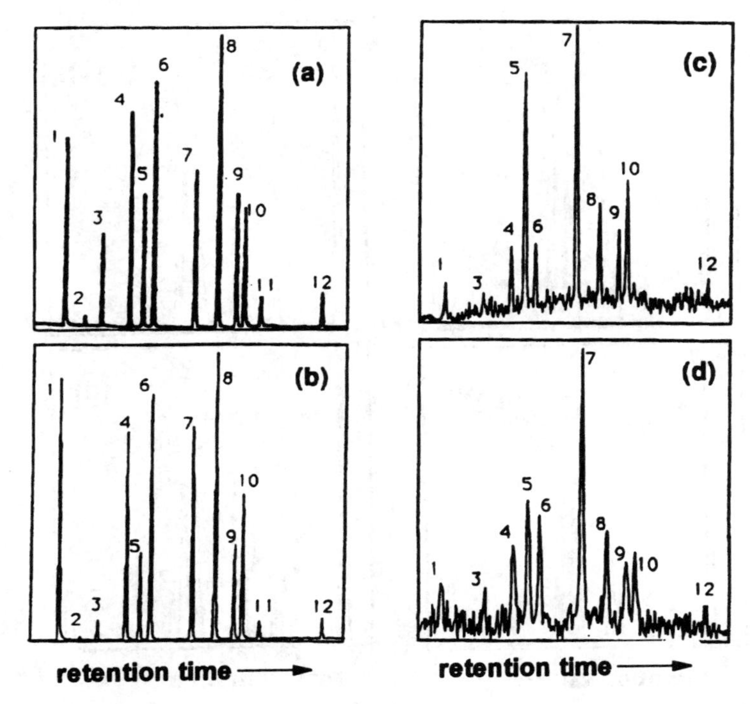

Figure 10 *Chromatograms obtained from the pesticides standard sample:* (a) *Cryolect GC-FID;* (b) *Cryolect GC-MS TIC;* (c) *Cryolect GC-MI-FTIR;* (d) *Tracer GC-FTIR. Peaks:* 1, *delta-HCH;* 2, *impurity;* 3, *heptachlor;* 4, *aldrin;* 5, *telodrin;* 6, *isodrin;* 7, *alpha-endosulfan;* 8, *dieldrin;* 9, *endrin;* 10, *beta-endosulfan;* 11, *impurity;* 12, *impurity*
(Reproduced from ref. 21, Jackson P., Dent G., Carter D., Schofield D.J., Chalmers J.M., Visser T., and Vrendenbregt M., *Investigation of High Sensitivity GC-FTIR as an Analytical Tool for Structural Identification, J. High Res. Chromatogr.*, 1993, **16**, 515, by kind permission of Hüthig Publishers, Heidelberg)

example, a nine-component 'drins standard' sample containing the following pesticide components in hexane was obtained and analysed:[21] *delta*-HCH, heptachlor, aldrin, telodrin, isodrin, *alpha*-endosulfan, dieldrin, endrin, and *beta*-endosulfan. This nine-component pesticides' sample was chosen to illustrate the utility of high sensitivity GC–FTIR in the exact identification of pesticide isomers. Furthermore, the sample also illustrates well the effectiveness of operation with multiple, parallel detectors, as on the Cryolect instrument. As shown in Figure 10, there are in fact twelve peaks detected in both the Cryolect GC–FID and GC–MS traces obtained from the sample. The GC–FTIR chromatograms obtained on both the Cryolect and the Tracer are also shown. As can be seen, the sensitivity obtained on the FTIR instruments using the fast-scan, Gram–Schmidt reconstruction approach does not allow the clear location of all twelve components found by the other detectors. On the Tracer, this limits the ability to acquire more sensitive signal-averaged spectra from specific components, since the operator does not know that there is a component there, and survey scanning of the entire chromatogram is usually prohibitively time consuming. On the Cryolect instrument, however, the presence of the minor components could be deduced from the parallel detectors, allowing high resolution, high sensitivity spectra to be recorded at the correct retention time. In this fashion, FTIR spectra could be obtained from every peak except peak 2, which was the smallest component and had a weak response. Figure 11 shows the FTIR spectra and the MS results from peaks 7 and 10, the *alpha*- and *beta*-endosulfan. As seen earlier, Cryolect matrix spectra exhibit sharper lines than in spectra recorded on the Tracer, with more splitting and multiplicity resolved. Once again, this increased resolution offsets the slightly better sensitivity of the Tracer, to yield spectra with comparable signal-to-noise ratios. The two endosulfan isomers give rise to essentially identical positive EI mass spectra, making unambiguous identification impossible by GC–MS alone, particularly in the situation where an unknown sample is analysed, and standard materials, methodologies, and calibrated retention times are not available. However, the GC–MS data is capable of identifying the correct molecular structure, and in this case the FTIR data can identify the specific isomer concerned by matching against library spectra.The fact that there are twelve components obtained from a nominally nine-component mixture is due to the presence of isomeric aldehyde impurities. These were clearly identified by the carbonyl signals present in the FTIR spectra, and the corresponding MS data allowed elucidation of their specific structures.

3 Liquid and Gel Permeation Chromatography–FTIR (HPLC–FTIR, GPC–FTIR)

Flow-through Techniques

GC–FTIR interfaces have been developed by many chromatography and FTIR instrument manufacturers. Similar interfaces between high performance liquid chromatography (HPLC) and FTIR spectrometers have not been so widely developed and popularised, although much research effort has been expended in this direction.[7] Initially, one might think that interfacing would be easier as the peaks from packed LC columns contain a greater volume of eluate, and flow rates are much lower. With flow-through GC–FTIR the sample is in a carrier gas which does not have any significant general IR absorption or specific bands to interfere with the sample spectrum. In normal phase LC–FTIR the eluants are organic solvents. At the path-lengths employed, typically 0.1 mm, organic solvents, with the exception of carbon tetrachloride (CCl_4), absorb strongly, and specific bands 'black-out' large regions of the spectrum. The recent move to microbore columns has reduced the peak volume, but this has not significantly diminished the major drawback. Many separations of ionic compounds are carried out by reverse phase HPLC. The mobile phase is largely aqueous with varying concentrations of methanol and acetonitrile. Inorganic phosphates are also used as buffers to adjust pH. As stated above, the cell path-lengths used in

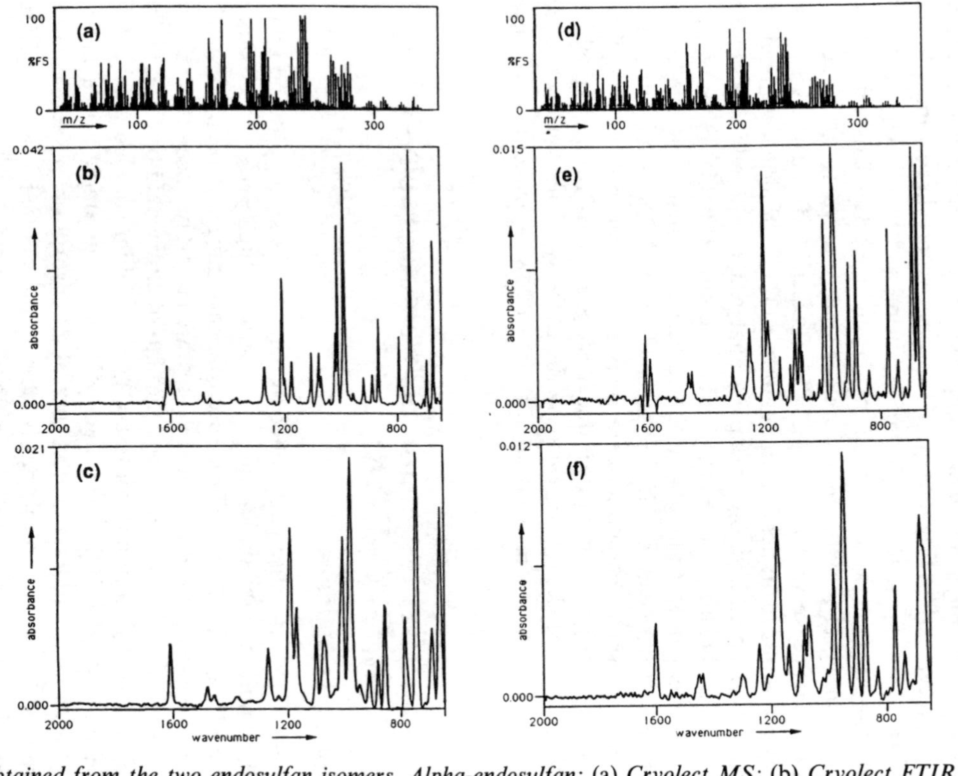

Figure 11 *Spectra obtained from the two endosulfan isomers. Alpha-endosulfan:* (a) *Cryolect MS;* (b) *Cryolect FTIR;* (c) *Tracer FTIR. Beta-endosulfan:* (d) *Cryolect MS;* (e) *Cryolect FTIR;* (f) *Tracer FTIR. All FTIR spectra,* 256 scans, 4 cm^{-1} *resolution*
(Reproduced from ref. 21, Jackson P., Dent G., Carter D., Schofield D.J., Chalmers J.M., Visser T., and Vrendenbregt M., *Investigation of High Sensitivity GC-FTIR as an Analytical Tool for Structural Identification, J. High Res. Chromatogr.*, 1993, **16**, 515, by kind permission of Hüthig Publishers, Heidelberg)

flow-through cells is often of the order of 0.1 mm; for aqueous-based separations, it would need to be reduced to less than 25 μm to be of any value. However, a balance has to be struck between strong solvent and buffer absorptions, weak sample absorptions, and the viscosity (resistance to flow) of the eluant. Successful fingerprinting of peaks has been achieved with Gel Permeation Chromatography (GPC), (or Size Exclusion Chromatography (SEC)), where molecular sizes are separated, although with these techniques common solvents are tetrahydrofuran (THF) or a chlorinated hydrocarbon. While the solvents have characteristic peak absorptions these can often be subtracted out, leaving wide enough windows in the mid-IR region to obtain useful spectra of the chromatography peaks. The stack plot of the spectra of two chromatography peaks shown in Figure 12 clearly shows both an aliphatic acid and an aliphatic ester.

If a successful system is established, a further problem can be encountered with background spectra. As FTIR is a single-beam technique, background spectra are recorded prior to an eluting peak and subtracted from the sample/solvent mixture. In GC–FTIR the carrier gas density, although maybe affecting the energy throughput, does not have specific bands. LC solvents not only have their own bands, but with gradient elution techniques mixed solvents are employed. The ratios of solvent can be very different throughout the chromatographic run. This leads to the solvent matching in the background and peak spectra being very difficult to achieve. With large peaks, the displacement of solvent immediately prior to the eluting peak can also cause significant refractive index changes.

Owing to the limited number of applications where flow-through cells are a viable option, recent development of commercial interfaces has been much more around sample deposition or solvent elimination techniques. Volatile organic solvents can be removed readily by heating or vaporising via a nebuliser. Reverse phase systems, however, are still problematic, particularly with inorganic buffers.

Deposition Techniques

Several attempts have been made to develop deposition techniques, particularly for automated on-line elimination of the solvent, though few have been commercialised. One of the earliest automated devices[24,25] deposited the solute onto powdered KCl prior to it being examined by diffuse reflectance (DRIFT) spectroscopy. The effluent from the column was sprayed into a short heated tube, increasing concentration by a factor of 10. The concentrate at the end of the tube was then allowed to drip onto a cup containing the powdered KCl. A microcomputer coupled to HPLC UV detector calculated the time delay of a 'peak' in the tube, and 'allowed' deposition of the peak; at all other times the

[24] Kuehl D. and Griffiths P.R., *J. Chromatogr. Sci.*, 1979, **17**, 471.
[25] Kuehl D. and Griffiths P.R., *Anal. Chem.*, 1980, **52**, 1394.

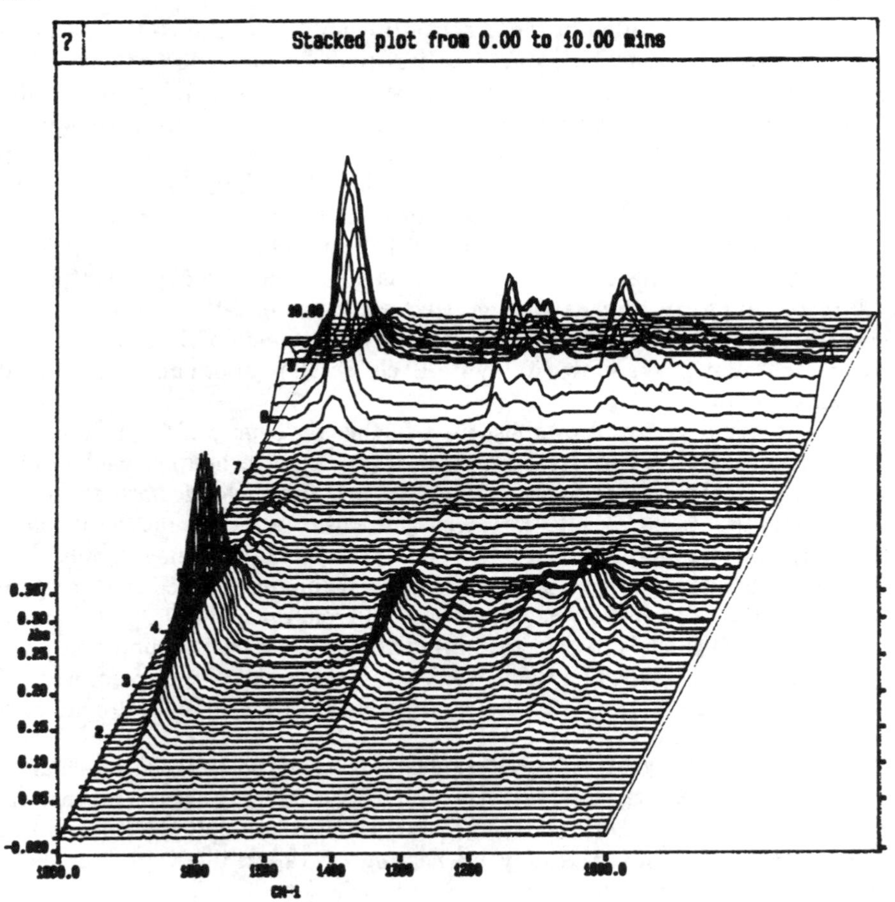

Figure 12 *Stack-plot of GPC-FTIR spectra from polymerisation of hydroxy stearic acid sample, showing aliphatic carboxylic acid and ester bands. Solvent, THF; flow-through transflectance cell,* 7 μm *spacer, effective path-length* 14 μm, *AgCl window*

concentrate was drawn off into an aspirator. The KCl was in cups on a carousel. Each cup had a vent to draw air or N_2 through to complete removal of any solvent. As each peak was detected by the UV detector a fresh cup was positioned under the concentrator tube. As each cup passed under the DRIFT accessory of the spectrometer the FTIR spectrum of the peak could be recorded. Sub-microgram detection limits were obtained for several compounds separated using organic solvents (hexane/propanol). A pesticide, Endrin, required 400 μg to be injected for a flow-through cell, whereas only 10 μg was required for DRIFT examination,[26] (*cf.* GC–MI–FTIR). Advantages of the deposition technique are similar to those for GC–FTIR, in that S/N enhancement can be

[26] Brown R.H., Knecht J., and Witek. H., *Proc. Soc. Photo-Opt. Instrum. Eng.*, 1981, **289**, 51.

achieved by re-examining any peak after a run, with an increased number of scans. The peaks are also preserved for additional NMR or MS examination, if required. The disadvantages come with aqueous solvents and reverse phase systems. Aqueous solvents are less volatile, and would dissolve the KCl. Inorganics (buffers) are also deposited with the peak, and the bands from these components can dominate the spectrum. The DRIFT method, although achieving relatively low detection limits, suffers from the disadvantage that discrete peaks have to be collected and the practical difficulties encountered in data collection.

Various methods have been devised which remove the aqueous solvent; these generally involve solvent extraction with phase separation,[27] or using thermospray nebulizers, similar to those employed in HPLC-MS interfaces.[28] The introduction of microbore columns and the development of nebulizer technology, coupled with using IR microscopes as detectors,[29–34] has resulted in developments with deposition methods (solvent elimination procedures) being concentrated on using windows rather than powders as infrared transparent supports for separated analytes. If the peaks are deposited onto IR transmitting windows, spectra can be recorded in transmission or by transflectance, (*cf.* GC–FTIR). The use of IR microscopes as detectors increases sensitivity and reduces the spot size needed to fill the beam.

A currently available commercial automated HPLC-FTIR system is the LC-TRANSFORM® from Lab Connections, Inc., Marlborough, MA, USA. The accessory comprises two modules: a sample collection module and a beam-condensing optics module, see Figure 13. In the sample collection module, effluent from a LC or GPC column is directed to a nozzle with gas flowing through it. This sprays the fluid onto a rotating disk. During this process the solvent evaporates, leaving a dry deposit on the Ge disk. This should create a track of solute peaks on the disk. Flow rate, temperature control, disk speed, and distance of the tip from the disk can all be varied to optimise the deposition. Ideally the peak separation and resolution should match the LC chromatogram. The disk is 1.5 mm thick Ge with an IR reflecting coating on the underside. At the end of a deposition run, the disk is removed from the sample collector and mounted in the optical module. This module is pre-positioned in the normal sample compartment of a FTIR spectrometer. The beam-condensing mirror arrangement directs the beam onto the disk. The beam passes through the sample and the germanium, is back reflected through the disk and sample, then directed onto the instrument detector, *i.e.* a transflectance measurement is made. In this way the beam passes twice through the sample, increasing path-

[27] Conroy C.M., Duff P.J., Griffiths P.J., and Azarraga L.V., *Anal. Chem.*, 1984, **56**, 2636.
[28] Blakley C.R. and Vestal M.L., *Anal Chem.*, 1983, **55**, 750.
[29] Jinno K. and Fujimoto C., *J. High Res. Chromatogr. Chromatogr. Commun.*, 1981, **4**, 532.
[30] Jinno K., Fujimoto C., and Ishii D., *J. Chromatogr.*, 1982, **239**, 625.
[31] Jinno K., Fujimoto C., and Hirata Y., *Appl. Spectrosc.*, 1982, **36**, 67.
[32] Jinno K., *Spectrosc. Lett.*, 1981, **14**, 659.
[33] Lange A.J., Griffiths P.R., and Fraser D.J.J., *Anal Chem.*, 1991, **63**, 782.
[34] Raynor M.W., Bartle K.D., and Cook B.W., *J. High. Res. Chrom.*, 1992, **15**, 361.

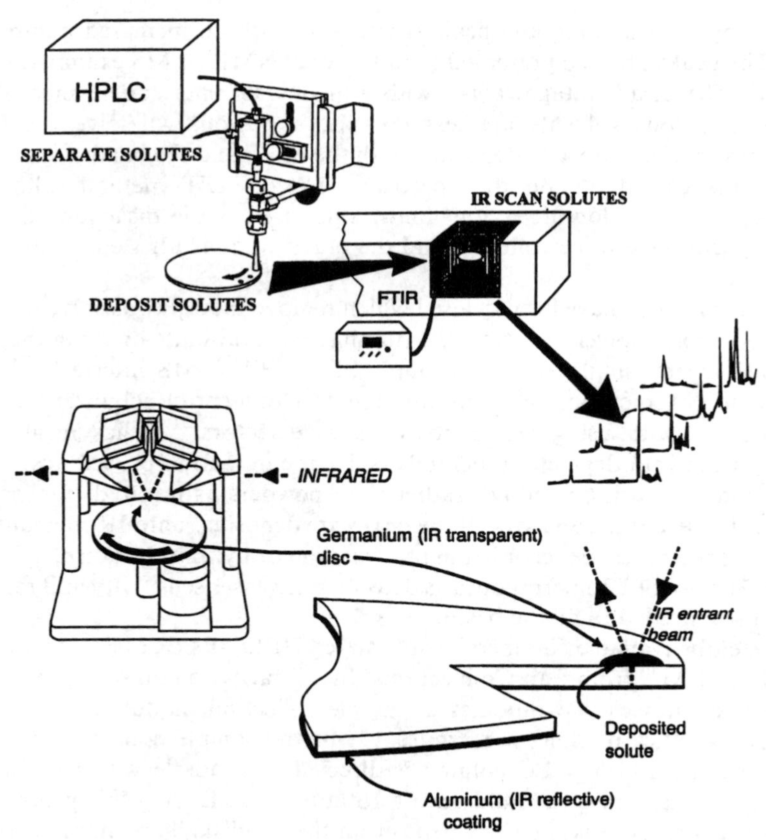

Figure 13 *Schematic diagrams of LC-Transform collection module and optics disc* (Reproduced by kind permission of Lab Connections, Inc., Marlborough, MA, USA)

length and hence sensitivity. An electronics control box rotates the disk at the same speed in the spectrometer as in the collection module, to maintain synchronisation. An IR chromatogram can be constructed which should be similar to the LC chromatogram, allowing for the different detector responses. The spectra of individual peaks can then be recorded from known spots on the disk. Spectrum averaging increases sensitivity to allow identification of quite small components in the chromatogram. A variation now available with the LC transform is the use of an ultrasonic nebuliser which enhances sensitivity and is capable of removing aqueous solvents. The detection limit is less than 100 ng. An example output from a GPC–FTIR separation is shown in Figure 14. The limit of sensitivity is the ability to match the sample size with the beam size. A spectrometer beam spot size of ~ 3 mm can be obtained by using a $3\times$ beam-condenser.

Significant increases in sensitivity should be possible if an IR microscope is

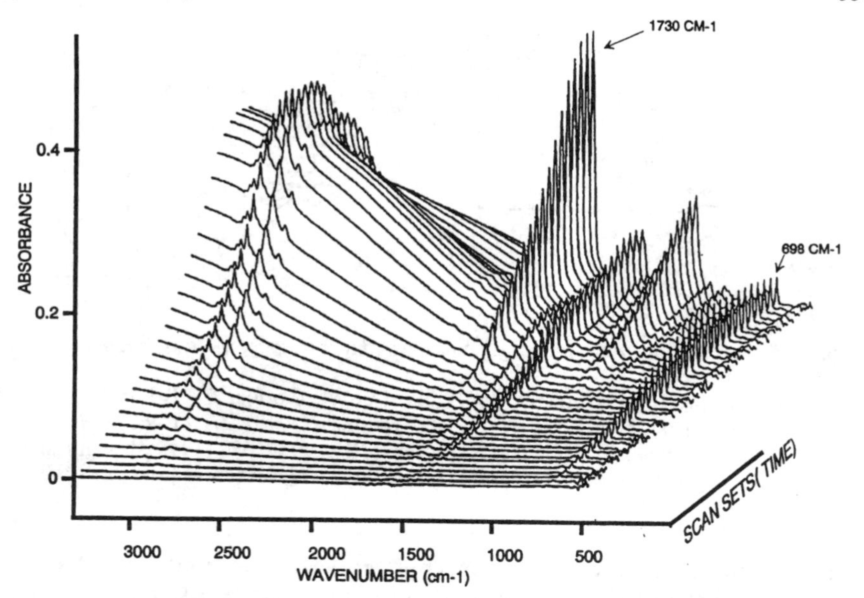

Figure 14 *Series of LC-Transform spectra taken from the sample collection disc, following GPC separation of a sample of polymethylmethacrylate/styrene, using THF as the mobile phase*
(Reproduced by kind permission of Lab Connections, Inc., Marlborough, MA, USA)

used as a detector. Work is presently being carried out in this area,[35–38] but no system has as yet been commercialised. Early investigations involved the adaption of the Mass Spectrometry electrospray technique. The eluant from the LC is subjected to a high positive potential relative to the surroundings. The electric field generated charges the surface of the emerging liquid, causing it to form charged droplets which are attracted to an earthed, infrared transparent plate, usually ZnSe. A sheath gas is used to evaporate the solvent from the droplets, with the solid component being deposited on the plate. The deposits on the plate are examined by FTIR microspectrometry. An IR chromatogram is generated via a mapping stage on the IR microscope. De Haseth and Robinson[39] used a modified MAGIC–LC–MS (Monodisperse Aerosol Generation Interface Combining LC and MS) to remove aqueous solvents prior to deposition. Developments in this area have involved the study of biological

[35] Somsen G.W., van de Nesse R.J., Gooijer C., Brinkman U.A.Th., Velthorst N.H., Visser T., Kootstra P.R., and de Jong A.P.J.M., *J. Chromatogr.*, 1991, **552**, 635.

[36] Somsen G.W., Gooijer C., Brinkman U.A.Th., Velthorst N.H., and Visser T., *SPIE*, 1994, **2089**, 536.

[37] Robertson R.M., de Haseth J.A., and Browner B., *Mikrochim. Acta*, 1988, **II**, 199.

[38] (a)Turula V.E. and de Haseth J.A., *Appl. Spectrosc.*, 1994, **48**, 10, 1255; (b) de Haseth J.A. and Turula V.E., *10th Int. Conf. Four. Trans. Spectrosc.*, Budapest, Hungary, 1995, Paper L13.

[39] (a) de Haseth J.A. and Robertson R.M., *Microchem. J.*, 1989, **40**, 1, 77; (b) de Haseth J.A. and Robertson R.M., *Microchem. J.*, 1989, **42**, 1365.

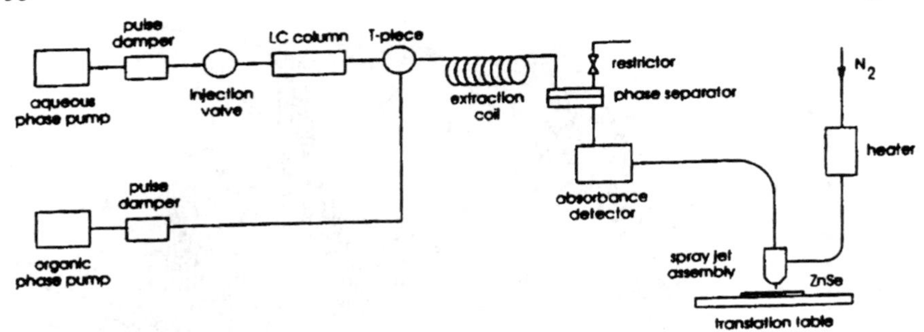

Figure 15 *Schematic diagram of chromatographic set-up of a post-column LC-FTIR interface*
(Reproduced from ref. 36, Somsen G.W., Gooijer C., Brinkman U.A.Th., Velthorst N.H., and Visser T., *Reversed Phase HPLC-FTIR by On-line Extraction and Solvent Elimination*, SPIE, **2089**, 536 (1994), by kind permission of the Society of Photo-Optical Instrumentation Engineers (SPIE))

samples, but again an instrument has not yet been commercialised. One of the drawbacks of the LC–FTIR deposition techniques for reverse phase (aqueous) systems is the presence of involatile, inorganic buffers. These are deposited with the peak of interest. Having very intense bands, these can frequently mask areas of interest in the analyte spectrum. Other researchers[35,36] have linked several techniques with a microbore column followed by an in-line extraction column coupled to a phase separator prior to a nebuliser; the liquid–liquid extraction of the aqueous effluent by a volatile organic solvent, circumvents many of these limitations. (Post-column solid phase extractions coupled with a flow-through cell has also been demonstrated for pharmaceutical analysis.)[40] The deposited peaks are again examined by microspectrometry, see schematic of Figure 15. Using this technique a 2.10^{-4} M solution of acenaphthenequinone, phenanthrenequinone, and the polar pesticides linuron and diuron, were separated. The mobile phase was acetonitrile–phosphate buffer, the extraction carried out in dichloromethane. By employing a UV detector post extraction, the chromatographic integrity is monitored. Initial experiments injected 500 ng per component. From the absorbance values recorded it is predicted that concentrations of 10^{-6} M could be successfully examined.

The LC-FTIR interfaces using deposition techniques can lead to increased sensitivity by increasing concentration due to the removal of solvent and buffers. Employing microspectrometry allows smaller samples to be examined. However the area which causes most problems is the physical nature of the sample after deposition. Detection techniques which employ reflection or transflection can suffer from spectrum artefacts caused by specular reflection from the top surface. Transmission techniques can suffer from variable thickness or crystallisation of the analyte 'spot', the latter leading to radiation scatter.

[40] DiNunzio J.E., *J. Chromatogr.*, 1992, **626**, 97.

With highly polished IR plates like Ge or ZnSe, some samples spread significantly or form rain-like droplets on deposition, the latter often leading to significant 'stray-light' effects, thereby making quantitative, and sometimes even semi-quantitative, comparisons dangerous, unless particular care and attention are taken with the mode of sample deposition.[41] GPC separation of high molecular weight polymers is usually quite successful, as the depositions are generally film-like unless the polymer is prone to crystallise readily. With lower molecular weight polymers and chemicals, a smooth, amorphous deposition can be difficult to achieve, certainly without practice. A skilled operator is required to optimise flow rates, temperatures, and deposition distances to achieve a spectroscopically acceptable deposit. Despite these difficulties, the techniques are currently proving particularly valuable in qualitative and semi-quantitative evaluations.

4 Supercritical Fluid Chromatography–FTIR (SFC–FTIR)

A resurgence of interest in the early 1980s in Supercritical Fluid Chromatography (SFC), seen as bridging the gap between GC and HPLC,[42,43] was welcomed by many IR spectroscopists as the answer to many of the problems encountered by HPLC–FTIR. The mobile phase is a fluid held above its critical temperature and pressure, and is therefore simple to volatilise. Early work was carried out on both wall-coated open tubular (WCOT) GC columns and packed HPLC columns. Chromatography advantages were expounded to be with high molecular weight, involatile, or thermally unstable materials, which could not be examined by GC or which demanded complex LC separations. Path-lengths of early flow-through cells varied between 1 and 10 mm (1 μl to 8 μl volume).[44–48] The mobile phase most commonly used was carbon dioxide, CO_2. At pressures above 73 atm (1073 psi) and 31.3°C, CO_2 is supercritical. Separations are carried out by pressure programming, similar to temperature programming in GC. Near its critical point CO_2 is virtually transparent to IR radiation. As the pressure is increased, significant bands do appear in the spectrum. These bands broaden with pressure, see Figure 16, reducing the effective window for observing absorbance bands associated with the chromatography peaks. With spectral subtraction techniques this was not seen as a significant problem, except, of course, for the region between 2500

[41] Liu M.X. and Dwyer J.L., *Appl. Spectrosc.*, 1996, **50**, 3, 349.
[42] Chester T.L., *J. Chromatogr. Sci.*, 1986, **24**, 226.
[43] Raynor M.W., Davies I.L., Bartle K.D., Williams A., Chalmers J.M., and Cook B.W., *European Spectrosc News*, 1987, **1**, 4, 18.
[44] Shafer K.H. and Griffiths P.R., *Anal Chem.*, 1983, **55**, 1939.
[45] Hughes M.E. and Fasching J.L., *J. Chromatogr. Sci.*, 1985, **24**, 535.
[46] Johnson C.C., Jordan J.W., Taylor L.T., and Vidrine D.W., *Chromatographia* 1985, **20**, 12, 717.
[47] Olesik S.V., French S.B., and Novotny M., *Chromatographia*, 1984, **18**, 9, 489.
[48] Bartle K.D., Raynor M.W., Clifford A.A., Davies I.L., Kithinji J.P., Shilstone G.F., Chalmers J.M., and Cook B.W., *J. Chromatogr. Sci.*, 1989, **27**, 283.

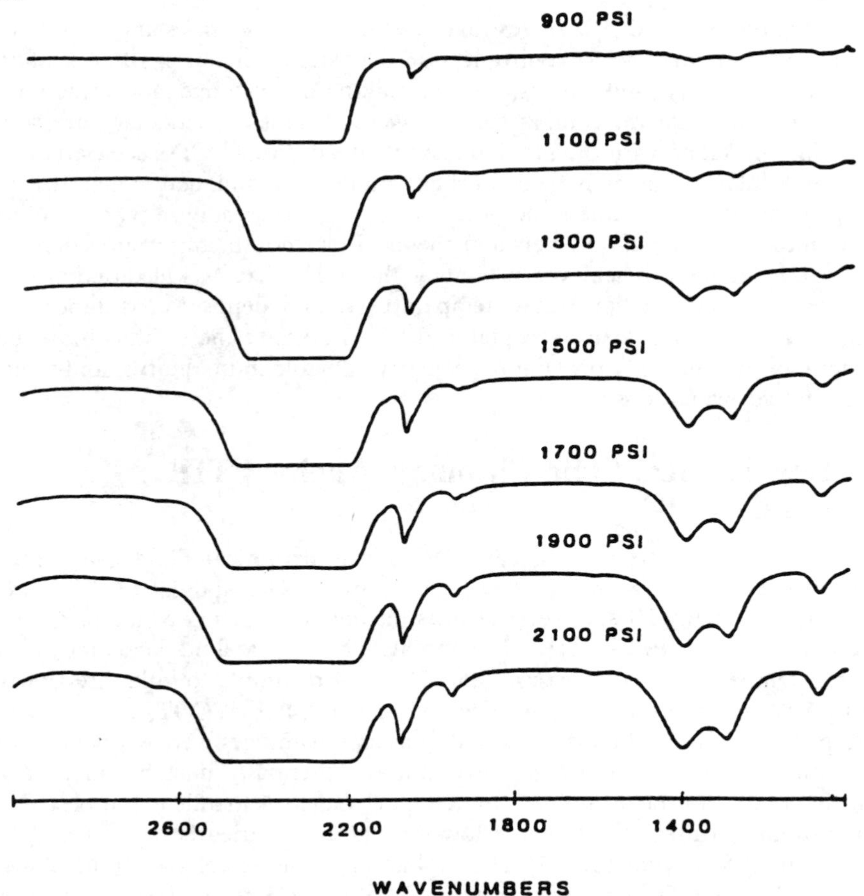

Figure 16 *Transmittance spectra of supercritical* CO_2, 1 mm *path-length cell at* 40°C *and pressures between* 900 *and* 2100 psi, *showing the intensification of bands with increasing pressure*
(Reproduced from ref. 44 by kind permission of the American Chemical Society)

and 2100 cm^{-1}. Spectra from injections of 1 μg per component were reported in 1983.[44] Although CO_2 was seen as near ideal by spectroscopists, separations were limited by its non-polar nature. Modifiers such as methanol could be added. However, the modifiers have strong absorption bands, which can obscure the spectrum and significantly limit the use of long path-length cells.[49] As a consequence, deposition techniques became more favoured and

[49] Jordan J.W. and Taylor J.W., *J. Chromatogr. Sci.*, 1986, **24**, 82.

practised. Once the pressure is released the mobile phase is extremely volatile, as are many modifiers. Both the DRIFT[50,51] and the FTIR-microspectroscopy transmission[52,53] (see Figure 17) mobile-phase elimination techniques have been successfully employed and are fairly easy to implement non-automated, off-line, with detection sensitivities in the 10–100 ng range. Xenon became a favoured mobile phase for a short period,[54,55] because it had no mid-IR bands and had very good solvent properties. However, the price of Xenon rose dramatically and the deposition technique (and system leaks) saw money literally evaporating away with each run!

In the mid to late 1980s, SFC was not seen as quantitatively reproducible by chromatographers and yielded few advantages over conventional separation techniques. Consequently it was not widely taken up in industrial laboratories. The development of SFC–FTIR has been very slow since then.[56] More recently supercritical fluids are increasing in popularity as extraction fluids, particularly for environmental concerns such as pesticides, or contaminants in soil, and in purification processes. The extracts can be complex mixtures, and supercritical fluid extraction (SFE) coupled with SFC–FTIR may see the latter re-emerge as a valuable identification tool.

5 Other Hyphenated/Combination Techniques

While the major developments in FTIR hyphenated techniques have been mostly in connection with GC and HPLC interfaces, other separation and characterisation techniques have not been ignored. Thin Layer Chromatography (TLC), Evolved Gas Analysis (EGA), Thermal Gravimetric Analysis (TGA), and Differential Scanning Calorimetry (DSC), coupled or combined with IR and Raman spectrometers, have all undergone significant evaluation, development and commercialisation. Although, as with some mobile-phase elimination approaches, for various applications, direct coupling does not always occur, good quality separation precedes the spectroscopic detection and/or identification.

Thin Layer Chromatography (TLC–FTIR, TLC–Raman)

TLC remains a widely used simple separation technique in many industrial laboratories. In one approach, one end of a paper strip is dipped into a solution

[50] Shafer K.H., Pentoney S.L., and Griffiths P.R., *J. High Res. Chromatogr. Chromatogr. Commun.*, 1984, **7**, 707.
[51] Shafer K.H., Pentoney S.L., and Griffiths P.R., *Anal. Chem.*, 1986, **58**, 58.
[52] Raynor M.W., Bartle K.D., Davies I.L., Williams A., Clifford A.A., Chalmers J.M., and Cook B.W., *Anal. Chem.*, 1988, **60**, 5, 427.
[53] Raynor M.W., Bartle K.D., Clifford A.A., Chalmers J.M., Katase T., Rouse C.A., Markides K.E., and Lee M.L., *J. Chromatogr.*, 1990, **505**, 179.
[54] French S.B. and Novotny M., *Anal. Chem.*, 1986, **58**, 164.
[55] Raynor M.W., Shilstone G.F., Bartle K.D., Clifford A.A., Cleary M., and Cook B.W., *J. High Resolut. Chromatogr.*, 1989, **12**, 5, 300.
[56] Smith R.D., Wright B.W., and Yonker C.R., *Anal. Chem.*, 1988, **60**, 23, 1323A.

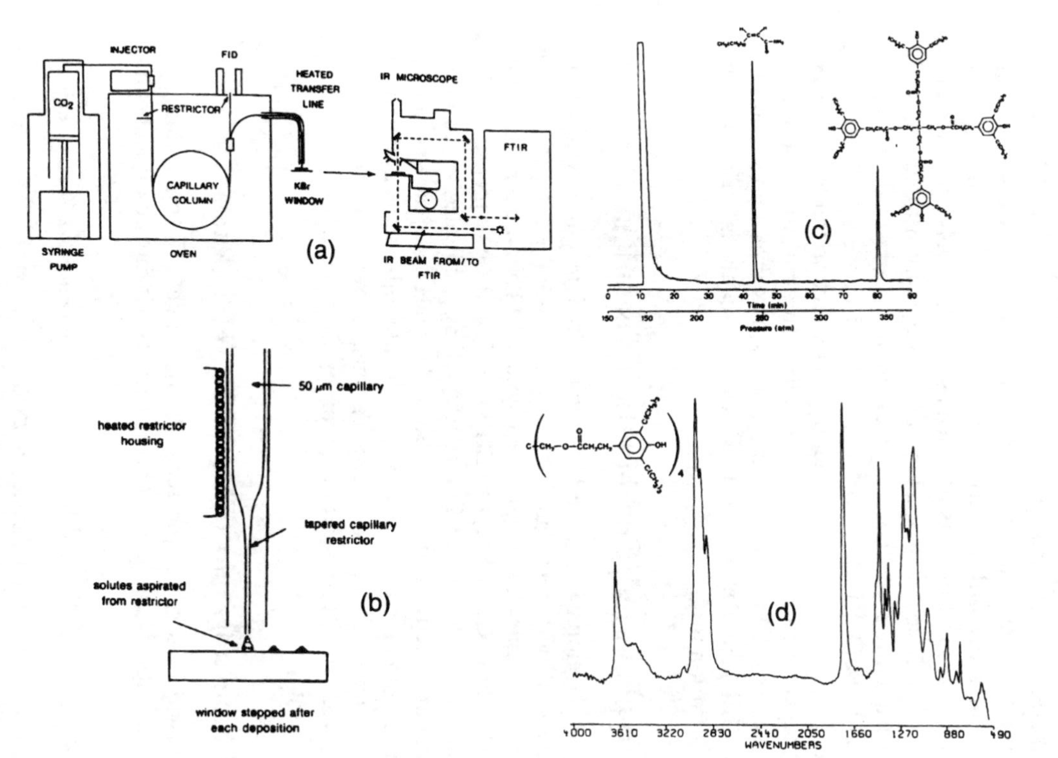

Figure 17 *SFC-FTIR microscopy:* (a) *schematic of the instrumental components;* (b) *schematic of basic design of a solvent elimination interface;* (c) *capillary SFC-FID of polymer additives extracted from a polypropylene sample. The peaks are due to a slip agent, erucamide, and an antioxidant, Irganox 1010 (Ciba Geigy). The mobile phase was CO_2 at* 140°C, *pressure programmed from* 150–350 ats *at* 3 ats min^{-1}, *after an initial* 12 min *isobaric period;* (d) *FTIR-microscopy spectrum of the Irganox 1010 additive spot,* 1000 scans *at* 4 cm^{-1} *resolution* (Reproduced from Raynor M.L., Davies I.L., Bartle K.D., Williams A., Chalmers J.M., and Cook B.W., *European Chromatography News*, 1987, **4**, 1, 18, by kind permission of John Wiley & Sons Ltd., Chichester, UK)

of a mixture. The solvent runs up the paper by capillary action evaporating from the top. Molecules in the mixture move at differing rates, separating and forming discrete bands on the paper. More sophisticated approaches might use silica, alumina, cellulose, or polyamide stationary phases supported on glass or aluminium plates, but the principle is the same. *In situ* identification of the TLC bands or spots has always been a challenge for vibrational spectroscopy. Conventional TLC substrates have intense infrared absorption bands, and in some cases strong Raman bands, and fluorescence, whether from the adsorbent, analyte or indicator, compounds the problem for some Raman studies. Notwithstanding, a variety of approaches have been evaluated, with varying degrees of success.

An ingenious early method was based around the Wick-Stick^{TM}, see p. 247, in which the portion of TLC substrate onto which a separated analyte was adsorbed was isolated (cut out) and placed into the base of the vial, see Figure 3 of Chapter 6, for subsequent solvent elution of the analyte to the tip of the KBr wedge, although if methanol was used, silica exhibited a slight solubility which could be enough to dominate any resulting infrared spectrum. A similar approach was developed around the use of pellets, made from compacted, dried KCl powder, for subsequent examination by DRIFT spectroscopy. High S/N spectra were obtained from additives, at the 50–100 μg level, which had previously been extracted from polypropylene granules,[57] although, one additive was missed initially, since it had a high vapour pressure at room temperature and rapidly volatilised from the non-adsorbent/non-bonding KCl substrate! *In situ detection* by DRIFTS and PAS of spots has been demonstrated by a number of workers on both commercial and purpose-designed TLC plates, but all have suffered to varying extents from the intense absorption features of the substrate or their consequent specular reflectance components. None has proven routinely, widely applicable to the *identification* of 'unknowns', particularly since most industrial chemists tend, for efficiency, to want to examine their own separations by well-established methods, without needing special modifications to suit the spectroscopist. A TLC–FTIR interface has been made available, integral with a dedicated MCT detector, for attaching to an external port of a spectrometer by Bruker Spectrospin GmbH. It is based on a DRIFT unit in which Fresnel (front-surface) reflection is minimised at maximum signal throughput.[58] The instrument will accommodate TLC plates 10 cm × 10 cm size on a computer-controlled translation stage. However, the spectra that are recorded are essentially those of adsorbed species, and they may therefore differ from their normal condensed phase counterparts to such an extent that specialised libraries are required for spectral-matching purposes. Zirconium oxide is a material which has recently been investigated as being both a suitable stationary phase for TLC separations and providing a suitable support for

[57] Chalmers J.M., Mackenzie M.W., Sharp J.L., and Ibbett R.N., *Anal. Chem.*, 1987, **59**, 415.
[58] Glauninger G., Kovar K.A., and Hoffman V., *Fresenius J. Anal. Chem.*, 1990, **338**, 6, 710.

enabling *in situ* DRIFT[59] and diffuse reflectance FTIR-microscopy[60] identifications of separated analyte spots.

While TLC–Raman is not new,[61–63] FT–Raman clearly offers some potential for circumventing fluorescence problems. Limited success has been reported for fingerprinting polymer additives[64] and pharmaceuticals[65] at the tens of nanograms level on commercial TLC plates. The limitations were primarily to do with fluorescence, which was plate dependent, and sensitivity, which was related to spot size (spread of the analyte). Significant enhancement of signals has been demonstrated by application of FT–SERS (subsequent spraying of the TLC plate with a silver colloidal solution), to the problem,[65] with good quality spectra reported for sub-microgram amounts of materials; while, sub-femtogram analyte levels have been reported as being detected from TLC plates with a FT–Raman–SERS microspectrometer system.[66]

Evolved Gas Analysis (EGA–FTIR)

The industrial need to study evolved gases directly has led to the development of several specialised accessories and procedures by which the gases may be monitored in real time in a FTIR spectrometer. Changes taking place can be followed by analysing the vapours or evolved gases (EGA). Early applications included characterising the combustion and pyrolysis products of materials related to the tobacco industry.[67–69] Although GC–FTIR software routines can be used to generate evolved gas detection profiles, they may not be specific, since multiple evolved components might exist in the EGA cell at any one time; consequently, special software routines were developed to generate specific gas evolution profiles against increasing temperature.[67] Automobile exhaust gases have also been subjected to direct FTIR analysis.[70] Other examples include thermal breakdown of materials and head-space analysis of paint systems[71] and polymer pyrolysis.[7] In one type of accessory, the sample stage comprises a test tube into which sample volumes of up to 5 ml may be placed, which can then be

59 Danielson N.D., Katon J.E., Bouffard S.P., and Zhu Z., *Anal. Chem.*, 1992, **64**, 2183.
60 Bouffard S.P., Katon J.E., Sommer A.J., and Danielson N.D., *Anal. Chem.*, 1994, **66**, 13, 1937.
61 Grasselli J.G., Hazle M.A., and Wolfram L.E., in *Molecular Spectroscopy*, ed. West A., Heyden, London, UK, 1977, ch. 14.
62 Adams D. and Gardener J., *J. Chem. Soc. Perkin Trans. II*, 1978, **15**, 2278.
63 (a) Huvenne J.P., Vergoten G., Charlier J., Moschetto Y., and Fleury G., *C.R. Hebd. Seances Acad. Sci., Ser. C*, 1978, **286**, 24, 633; (b) Huvenne J.P., Vergoten G., and Fleury G., *Proc. 6th Int. Conf. Raman Spectrosc.*, **2**, ed. Schmid E.D., Krishnan R.S., and Kiefer W., Heyden, London, UK, 1978, 518.
64 Everall N.J., Chalmers J.M., and Newton I.D., *Appl. Spectrosc.*, 1992, **46**, 4, 597.
65 Rau A., *J. Raman Spectrosc.*, 1993, **24**, 4, 251.
66 Caudin J.P., Beljebbar A., Sockalingum G.D., Angiboust J.F., and Mainfait M., *Spectrochim. Acta A*, 1995, **51**, 1977.
67 Lephardt J.O. and Fenner R.A., *Appl. Spectrosc.*, 1980, **34**, 2, 174.
68 Lephardt J.O. and Fenner R.A., *Appl. Spectrosc.*, 1980, **35**, 1, 95.
69 Lephardt J.O., *Appl. Spectrosc. Rev.*, 1982/1983, **18**, 2, 265.
70 Herget W.F. and Lowry S.R., *Proc. SPIE-Int. Soc. Opt. Eng.*, 1991, **1433**, 275.
71 (a) Rossiter V., Milward R., and Simpson M., *International Labmate*, 1989, **XIV**, IV, 27; (b) AABSPEC EGA/500, *Applications note*, Aabspec Instrumentation Ltd., Dublin, Ireland.

heated/temperature programmed up to 700°C, in different carrier gas environments if required. This chamber is connected to a low volume light-pipe, which is separately temperature controlled and mounted in the FTIR spectrometer.[71] In another system, fast thermolysis ($<400°C\ s^{-1}$; FTIR spectral collection rates of 10 scans/sec, 2 spectra per file at 4 cm^{-1} resolution, limit the utility of using higher rates), typically of 1–2 mg of a material, is achieved with filament heating. This system has been used to study the combustion of RDX explosive.[72]

Thermogravimetric Analysis (TGA–FTIR)

TGA is concerned with measuring the change in mass of a sample as it is heated. As such, TGA is a quantitative rather than a qualitative technique, determining weight loss profiles against temperature or time, in controlled atmospheres. TGA is a very well established technique in many industries for studying material weight loss. For example, it is commonly used in the polymer industry, perhaps to evaluate residual monomers, or product decomposition or breakdown.[73–76] In the pharmaceutical, fine chemicals and colour industries, TGA may be used to study purity, solvent retention,[77] stability, and shelf-life properties. Weight loss might be due to solvent, loose (absorbed) water, or hydrate (bound water). TGA can reveal compositional changes, but when a weight loss occurs TGA tells us how much but not unambiguously what gas/gases or effluents are being evolved. For many systems, such as coal and combustion products from inorganics or from materials entrapped in a charcoal filter,[78] decomposition or release processes can be complex. By coupling a TGA to FTIR (or MS), the evolved gases can be analysed to yield much more compositional information,[73–78] see Figure 18.

TGA–FTIR systems are available through many of the major manufacturers of FTIR spectrometers. In a manner somewhat analogous to GC–FTIR, the TGA balance is connected by a heated tube to a heated flow-through gas cell. However, rather than using a light-pipe, cells have lengths typically of 6–10 cm, and may be (low number) multi-passing. They are designed to be inert and minimise condensation of materials onto windows. They may typically be maintained at temperatures between 200°C and 300°C, although the TGA furnaces may be heated to temperatures as high as 1500°C, or higher. The spectra recorded are vapour phase spectra, often of small molecules resulting from thermal degradation, *e.g.* CO, CO_2, H_2O, HCl, benzene. Many polymers

[72] Brill T.B., *Anal. Chem.*, 1989, **61**, 15, 897A.
[73] Schild H.G., *J. Polym. Sci. Part A: Polym. Chem.*, 1993, **31**, 2403.
[74] Wieboldt R.C., Adams G.E., Lowry S.R., and Rosenthal R.J., *Am. Lab.*, 1988, **20**, 1, 70.
[75] Kinoshita R., Teramoto Y., and Yoshida H., *Thermochim. Acta*, 1993, **222**, 45.
[76] Kinoshita R., Teramoto Y. ,and Yoshida H., *J. Thermal Anal.*, 1993, **40**, 605.
[77] Johnson D.J. and Compton D.A.C., *Spectroscopy*, 1988, **3**, 6, 47, and, *Spectroscopy Int.*, 1989, **1**, 1, 45.
[78] Kempfert K.D., Nicolet FT-IR *Application Note AN-8820*, Nicolet Instruments Inc., Madison, USA.

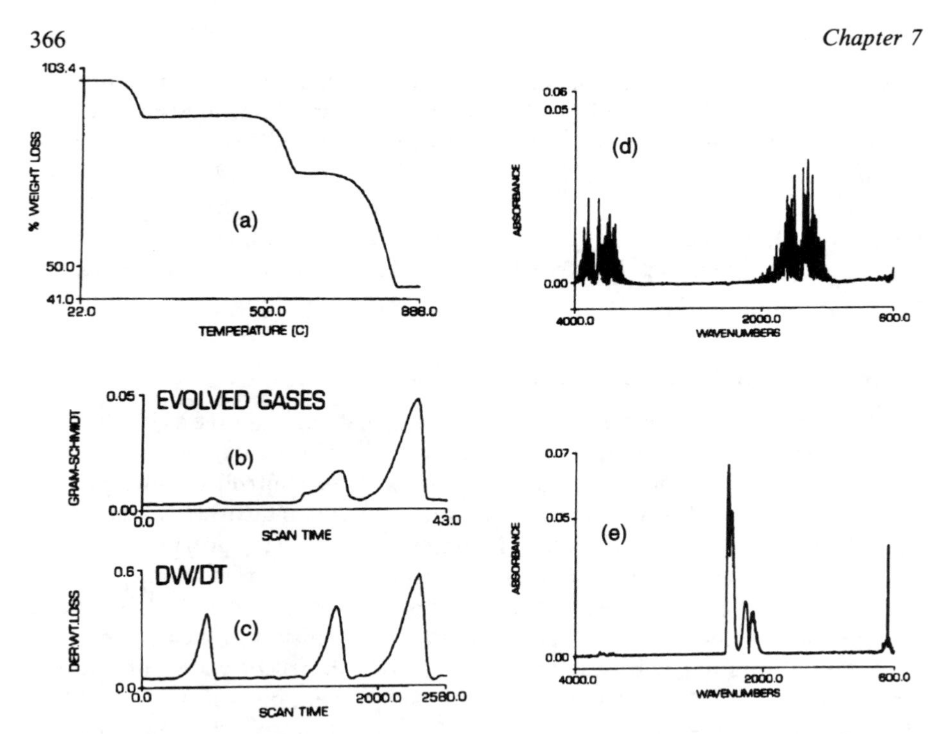

Figure 18 *TGA-FTIR of calcium oxalate:* (a) *profile of weight vs. temperature;* (b) *evolved gases profile, calculated using a Gram–Schmidt orthogonalisation method;* (c) *first-derivative weight loss vs. scan time;* (d) *water vapour evolved during first weight loss;* (e) *CO and CO2 evolved during second weight loss* (Reproduced from Compton D.A.C., *FTS®/IR Notes* No. 47, May 1987, by kind permission of Bio-Rad, Digilab Division, Cambridge, MA, USA)

may break down (de-polymerise) to the original monomers from which they were synthesised. Although the TGA balances can be typically sensitive to 400 ng changes, the volumes of the gas cells are relatively large, and decomposition occurs gradually rather than in discrete steps. This can lead to mixtures of vapours being present in the cell at the same time, and careful temperature-programming and software manipulation may be required for optimised analysis. Also, if one is expecting to monitor low concentrations of evolved CO_2 and H_2O, particular attention must be paid in the FTIR measurement to selection of an appropriate single-beam background. The spectrum in Figure 19 illustrates the effects of both a poor evolved gases separation and a poor choice of background, the latter leading to a significant imbalance in species. The carbonyl band at 1700 cm^{-1} is accompanied by a free νOH band at 3600 cm^{-1}. Some hydrogen-bonded —NH remains near 3300 cm^{-1}. The fine structure around 2800 cm^{-1} shows the presence of HCl vapour. The negative-going peaks in the 3800–3600 cm^{-1} region and the CO_2 band at 2370 cm^{-1} are caused by selecting an inappropriate background spectrum. Nevertheless, the %T scale shown covers only 1.22%, demonstrating the good S/N and sensitivity of the technique.

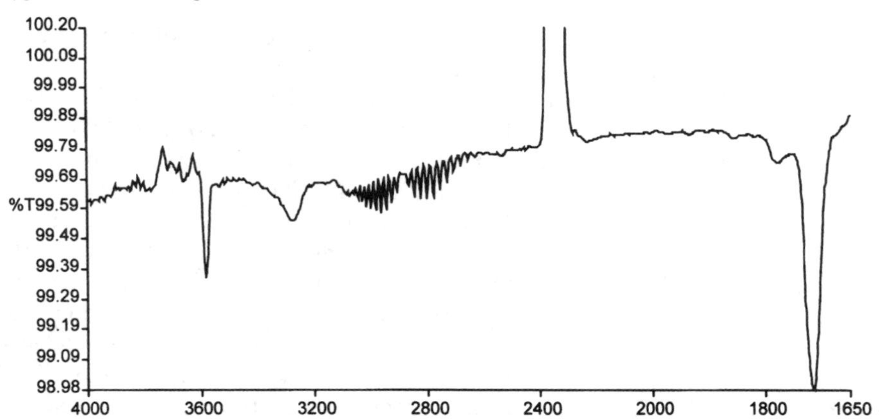

Figure 19 *Spectrum from a TGA-FTIR investigation of a dye sample, showing poor separation of the evolved gases together with some consequences of an imbalance between the sample and the background single-beam spectra*

Differential Scanning Calorimetry (DSC–FTIR, DSC–FT-Raman)

DSC is used to measure heat flow changes in materials as a function of temperature. As the temperature is raised, phase transformations occur which may be endothermic or exothermic. The technique may be considered as complementary to TGA, since many of the changes that are observed do not coincide with weight loss processes, *e.g.* glass transition temperature (T_g), crystallisation, and melting. The industrial uses for DSC, like TGA, are extensive, particularly for polymers, polymorphism, degradation, sublimation, and melting point characterisations. Experienced thermal analysts can correctly infer many changes taking place from the breadth, position and intensity of the heat flow peaks. However, not all transitions are simple to interpret, and as mentioned in Section 6 of Chapter 6 and (as with TGA–FTIR) complementary information about the heat-induced transitions taking place in a sample should be available through the coupling of DSC with FTIR[79–81] or Raman[82] spectroscopy.

One possible method involves examining a sample by DSC, then using a FTIR-microscope system to characterise the sample's state. Repeated DSC scans could be stopped after each transition and a spectrum recorded. However, this would only be appropriate for irreversible transitions. Any material which sublimed on a DSC pan lid could also be characterised by FTIR-microscopy. However, proper insight into the thermal property responses only comes with

[79] Mirabella F.M., *Appl. Spectrosc.*, 1986, **40**, 3, 417.
[80] Mirabella, F.M., in *Infrared Microspectroscopy: Theory and Applications*, ed. Messerschmidt R.G and Harthcock M.A., Marcel Dekker, New York, USA, 1988, ch. 5.
[81] Johnson D.J., Compton D.A.C., and Canale P.L., *Thermochim. Acta*, 1992, **195**, 5.
[82] Harju M.E., Valkonen J., and Jayasooriya U., *Spectrochim. Acta*, 1991, **47A**, 9/10, 1395.

coincident observations. Simultaneous DSC and FTIR observations have been achieved by examining samples in specially prepared crucibles.

A coupled DSC–FTIR which comprised a Mettler FP84 thermal analysis DSC microscopy cell, a Digilab IRMA microscope and a Nicolet 6000 FTIR spectrometer has been employed to record simultaneous DSC traces and FTIR spectra from polymers.[79,80] The cells used had a 2.5 mm hole in their base for the infrared beam to pass through. They were constructed with either NaCl[79] or KBr[80] windows for the DSC–FTIR observations, or made specifically with sapphire for optical microscopy-DSC characterisations.[80] The studies demonstrated include spectra collected isothermally during the melting endotherm of polypropylene,[79] and during the melting/recrystallisation of a low density polyethylene and the degradation in air of a poly(ethylene vinyl alcohol).[80] In another system, FTIR reflectance microspectroscopy procedures were used *in situ* to monitor thermal transitions in PET.[81] DSC has also been coupled with FT-Raman for *in situ* investigations of phase transitions.[82] Simultaneous thermal analysis and vibrational spectroscopic data were recorded relating to crystalline phase transitions of ammonium nitrate.

6 Closing Remarks

In this chapter we have discussed in some detail several of the major hyphenated techniques, while describing others only briefly. For some such as LC–FTIR and particularly GC–FTIR, few industrial application references have been listed, since these are widespread and too numerous to include, and examples are easy to find in the published literature. Developments continue, and today's research project is tomorrow's commercial interface or instrument. Just round the corner perhaps is Flow Injection Analysis (FIA–FTIR)[83,84] or Capillary Electrophoresis (CE–Raman).[85,86] 'Double hyphenations' of TGA–FTIR–MS[87] and DSC/TGA–FTIR[88,89] have been demonstrated, while preliminary results for a GC detector based on SERS have been reported.[90]

Some emphasis has been placed in the chapter on the commonly used areas where both chromatographer and spectroscopist can come to grief if each does not understand the other's requirements and constraints. The chapter has tried to instil the principle that, in the modern world, industrial problem-solving is almost always carried out more fully, effectively, and successfully if the complementary natures of techniques are appreciated and employed to the full. The true hyphenated specialist is perhaps rare compared to the single

[83] Kellner R. and Lendl B., *Anal. Meth. Instrum.*, 1995, **2**, 1, 52.
[84] Rosenberg E. and Kellner R., *Vib. Spectrosc.*, 1993, **5**, 33.
[85] Liu K.-L.K., Davis K.L., and Morris M.D., *Anal. Chem.* 1994, **66**, 21, 3744.
[86] Walker P.A., Kowalchyk W.K., and Morris M.D., *Anal. Chem.*, 1995, **67**, 23, 4255.
[87] Cook B., Cleary M., Bosley N., and Groves I., *Lab. Equip. Digest*, October 1989, 31.
[88] Compton D.A.C. and Loeb D., *Proc. 20th N. Am. Therm. Anal. Soc. Conf.*, Minneapolis, USA, ed. Keating M.Y., 1991, p. 315.
[89] Akinade K.A., Canpbell R.M., and Compton D.A.C., *J. Mat. Sci.*, 1994, **29**, 3082.
[90] Carron K.T. and Kennedy B.J., *Anal. Chem.*, 1995, **67**, 18, 3353.

technique expert (and worse still, devotee!), but the latter will gain far more information from, and discover more about their own technique by comparing and contrasting the information available from complementary sources and, when necessary, carrying out collaborative, iterative examinations and arguments.

Subject Index